IET CONTROL, ROBOTICS AND SENSORS SERIES 111

The Inverted Pendulum in Control Theory and Robotics

Other volumes in this series:

Volume 8 **A History of Control Engineering, 1800–1930** S. Bennett
Volume 18 **Applied Control Theory, 2nd Edition** J.R. Leigh
Volume 20 **Design of Modern Control Systems** D.J. Bell, P.A. Cook and N. Munro (Editors)
Volume 28 **Robots and Automated Manufacture** J. Billingsley (Editor)
Volume 33 **Temperature Measurement and Control** J.R. Leigh
Volume 34 **Singular Perturbation Methodology in Control Systems** D.S. Naidu
Volume 35 **Implementation of Self-tuning Controllers** K. Warwick (Editor)
Volume 37 **Industrial Digital Control Systems, 2nd Edition** K. Warwick and D. Rees (Editors)
Volume 39 **Continuous Time Controller Design** R. Balasubramanian
Volume 40 **Deterministic Control of Uncertain Systems** A.S.I. Zinober (Editor)
Volume 41 **Computer Control of Real-Time Processes** S. Bennett and G.S. Virk (Editors)
Volume 42 **Digital Signal Processing: Principles, devices and applications** N.B. Jones and J.D.McK. Watson (Editors)
Volume 44 **Knowledge-based Systems for Industrial Control** J. McGhee, M.J. Grimble and A. Mowforth (Editors)
Volume 47 **A History of Control Engineering, 1930–1956** S. Bennett
Volume 49 **Polynomial Methods in Optimal Control and Filtering** K.J. Hunt (Editor)
Volume 50 **Programming Industrial Control Systems Using IEC 1131-3** R.W. Lewis
Volume 51 **Advanced Robotics and Intelligent Machines** J.O. Gray and D.G. Caldwell (Editors)
Volume 52 **Adaptive Prediction and Predictive Control** P.P. Kanjilal
Volume 53 **Neural Network Applications in Control** G.W. Irwin, K. Warwick and K.J. Hunt (Editors)
Volume 54 **Control Engineering Solutions: A practical approach** P. Albertos, R. Strietzel and N. Mort (Editors)
Volume 55 **Genetic Algorithms in Engineering Systems** A.M.S. Zalzala and P.J. Fleming (Editors)
Volume 56 **Symbolic Methods in Control System Analysis and Design** N. Munro (Editor)
Volume 57 **Flight Control Systems** R.W. Pratt (Editor)
Volume 58 **Power-plant Control and Instrumentation: The control of boilers and HRSG systems** D. Lindsley
Volume 59 **Modelling Control Systems Using IEC 61499** R. Lewis
Volume 60 **People in Control: Human factors in control room design** J. Noyes and M. Bransby (Editors)
Volume 61 **Nonlinear Predictive Control: Theory and practice** B. Kouvaritakis and M. Cannon (Editors)
Volume 62 **Active Sound and Vibration Control** M.O. Tokhi and S.M. Veres
Volume 63 **Stepping Motors, 4th Edition** P.P. Acarnley
Volume 64 **Control Theory, 2nd Edition** J.R. Leigh
Volume 65 **Modelling and Parameter Estimation of Dynamic Systems** J.R. Raol, G. Girija and J. Singh
Volume 66 **Variable Structure Systems: From principles to implementation** A. Sabanovic, L. Fridman and S. Spurgeon (Editors)
Volume 67 **Motion Vision: Design of compact motion sensing solution for autonomous systems** J. Kolodko and L. Vlacic
Volume 68 **Flexible Robot Manipulators: Modelling, simulation and control** M.O. Tokhi and A.K.M. Azad (Editors)
Volume 69 **Advances in Unmanned Marine Vehicles** G. Roberts and R. Sutton (Editors)
Volume 70 **Intelligent Control Systems Using Computational Intelligence Techniques** A. Ruano (Editor)
Volume 71 **Advances in Cognitive Systems** S. Nefti and J. Gray (Editors)
Volume 72 **Control Theory: A guided tour, 3rd Edition** J.R. Leigh
Volume 73 **Adaptive Sampling with Mobile WSN** K. Sreenath, M.F. Mysorewala, D.O. Popa and F.L. Lewis
Volume 74 **Eigenstructure Control Algorithms: Applications to aircraft/rotorcraft handling qualities design** S. Srinathkumar
Volume 75 **Advanced Control for Constrained Processes and Systems** F. Garelli, R.J. Mantz and H. De Battista
Volume 76 **Developments in Control Theory towards Glocal Control** L. Qiu, J. Chen, T. Iwasaki and H. Fujioka (Editors)
Volume 77 **Further Advances in Unmanned Marine Vehicles** G.N. Roberts and R. Sutton (Editors)
Volume 78 **Frequency-Domain Control Design for High-Performance Systems** J. O'Brien
Volume 80 **Control-oriented Modelling and Identification: Theory and practice** M. Lovera (Editor)
Volume 81 **Optimal Adaptive Control and Differential Games by Reinforcement Learning Principles** D. Vrabie, K. Vamvoudakis and F. Lewis
Volume 83 **Robust and Adaptive Model Predictive Control of Nonlinear Systems** M. Guay, V. Adetola and D. DeHaan
Volume 84 **Nonlinear and Adaptive Control Systems** Z. Ding
Volume 86 **Modeling and Control of Flexible Robot Manipulators, 2nd edition** M.O. Tokhi and A.K.M. Azad
Volume 88 **Distributed Control and Filtering for Industrial Systems** M. Mahmoud
Volume 89 **Control-based Operating System Design** A. Leva *et al.*
Volume 90 **Application of Dimensional Analysis in Systems Modelling and Control Design** P. Balaguer
Volume 91 **An Introduction to Fractional Control** D. Valério and J. Costa
Volume 92 **Handbook of Vehicle Suspension Control Systems** H. Liu, H. Gao and P. Li
Volume 93 **Design and Development of Multi-Lane Smart Electromechanical Actuators** F.Y. Annaz
Volume 94 **Analysis and Design of Reset Control Systems** Y. Guo, L. Xie and Y. Wang
Volume 95 **Modelling Control Systems Using IEC 61499, 2nd Edition** R. Lewis and A. Zoitl
Volume 96 **Cyber-Physical System Design with Sensor Networking Technologies** S. Zeadally and N. Jabeur (Editors)
Volume 99 **Practical Robotics and Mechatronics: Marine, space and medical applications** I. Yamamoto
Volume 100 **Organic Sensors: Materials and applications** E Garcia-Breijo and P Cosseddu (Editors)
Volume 102 **Recent Trends in Sliding Mode Control** L. Fridman JP. Barbot and F. Plestan (Editors)
Volume 104 **Control of Mechatronic Systems** L. Guvenc, B.A Guvenc, B. Demirel, M.T Emirler
Volume 105 **Mechatronic Hands: Prosthetic and robotic design** P.H. Chappell
Volume 107 **Solved Problems in Dynamical Systems and Control** D. Valério, J.T. Machado, A.M. Lopes and A.M. Galhano

The Inverted Pendulum in Control Theory and Robotics

From theory to new innovations

Edited by
Olfa Boubaker and Rafael Iriarte

The Institution of Engineering and Technology

Published by The Institution of Engineering and Technology, London, United Kingdom

The Institution of Engineering and Technology is registered as a Charity in England & Wales (no. 211014) and Scotland (no. SC038698).

First published 2017

The Institution of Engineering and Technology
Michael Faraday House
Six Hills Way, Stevenage
Herts, SG1 2AY, United Kingdom

www.theiet.org

British Library Cataloguing in Publication Data
A catalogue record for this product is available from the British Library

ISBN 978-1-78561-320-3 (hardback)
ISBN 978-1-78561-321-0 (PDF)

Typeset in India by MPS Limited

Contents

Foreword

For at least 50 years, the inverted pendulum has been the most popular benchmark, among others, for teaching and research in control theory. Achieving its stability problem has become then a common engineering challenge. The fundamental focus of this book is then to enhance the wealth of this robotic system and provide an overall picture of historical, current trends and challenging problems in control theory and possible applications in mobile robotics, based on its simple structure and its rich nonlinear model.

The book is a timely and comprehensive reference guide for graduate students, researchers, engineers and practitioners in the areas of control theory and robotics. Readers will find it to be a clear and understandable textbook for control system courses at colleges and universities. It is written for senior electrical, mechanical, aerospace or mechatronics students.

Written by eminent scientists in the field of control theory and robotics from 12 countries (China, France, Italy, Japan, Luxembourg, Mexico, Oman, Saudi Arabia, Tunisia, Turkey, United Kingdom and USA), this book covers the latest advanced control approaches based on the stabilization principle as well as the trajectory generation methods of the inverted pendulum. Approaches include sliding-mode control, PID control and intelligent control techniques, among others. Observer design problems are also considered. The efficiency of the controllers-based observers approaches are proved for overcoming the complications introduced by the presence of uncertain dynamics, external disturbance, faults, non-holonomic kinematic constraints as well as under-actuated configurations.

Not only does this book provide the readers with control fundamentals and algorithms but it also discusses key applications using different experimental platforms including the cart-inverted pendulum, the inertia-wheel pendulum, the mobile-wheeled inverted pendulum, etc.

In total, 14 chapters written by active researchers in the field are compiled in this book to provide an overall picture of the most challenging problems to be solved in control theory from design to experiments. The book begins with a survey chapter where the available varieties of the inverted pendulum system are depicted, highlighting the richness of the dynamics of such system and then providing an overall picture of different control design approaches and trendy robotic problems related to its simple structure. The book covers then the topic in four parts: Part I, organized in four chapters, presents applications of robust state estimation and control with applications to pendulum-cart systems; Part II, structured in four chapters, depicted different control approaches designed for under-actuated mechanical systems with applications to the inertia-wheel pendulum; Part III is organized in three chapters where nonlinear controllers are designed for mobile inverted pendulum systems; finally, Part IV, structured in three chapters, proposes robust controllers-based observers via Takagi–Sugeno or linear approaches.

The following researchers are particularly acknowledged for their considerable efforts:

- We are grateful to Professor Leonid Fridman from Facultad de Ingeniería, UNAM, Mexico, for the fruitful collaboration. It would not have been possible to compile the book without his precious help, particularly the contributions proposed in Chapters 2, 3, 6 and 7 and experimental results presented in Chapters 2 and 3.
- We thank the Emeritus Professors Toshio Fukuda from Nagoya University, Japan, Mohamed Darouach from Research Centre for Automatic Control of Nancy, France, and MagdiSadek Mahmoud from King Fahd University of Petroleum and Minerals, Saudi Arabia, for their valuable supports and contributions at Chapters 9, 12 and 14.
- We are grateful to Professor Guillermo Fernandez-Anaya from Universidad Iberoamericana, Mexico, and the Emeritus Professor José Ragot from Centre de Recherche en Automatique de Nancy, France, for their helpful and professional efforts to provide precious comments and reviews.
- The authors of Chapter 2, Diego Gutierrez and Héctor Ríos, the authors of Chapter 3, Jesús Mendoza and Ismael Castillo, and the author of Chapter 5, Ahmed Chemori, are particularly thanked for all the efforts provided to present experimental data.
- All our thanks go to the Professor Giuseppe Carbone from University of Cassino and South Latium, Italy, for his valuable support and for choosing our book to present the new prototype of the balancing inverted pendulum of LARM laboratory of University of Cassino at Chapter 10.

Finally, the editors would like to express their gratefulness to all authors of the book for their valuable contributions and all reviewers for their helpful and professional efforts to provide precious comments and feedback.

Olfa Boubaker and Rafael Iriarte

Chapter 1

The inverted pendulum: history and survey of open and current problems in control theory and robotics

Olfa Boubaker[1]

Abstract

The inverted pendulum is a classical problem in dynamics and control theory widely used as a fundamental system for testing emerging control algorithms. In spite of its very simple structure appropriate for developing real-time implementation, the inverted-pendulum model is considered the richest one among common robotic benchmarks. Not only it can describe many engineering problems but also it can explain several biological examples. Various mathematical models and experimental designs for the inverted pendulum exist offering an attractive tool for education and research. The objective of this survey is to present an overview of the available varieties of such system, highlighting the richness of its dynamics and then providing an overall picture of different control design approaches and trendy robotic problems related to its simple structure. In total, 300 references in the open literature, dating back to Galileo's first experiments written in 1602, are compiled to provide an overall picture of historical, current and challenging developments based on the stabilization principle of the inverted pendulum.

1.1 Introduction

A pendulum, *pendere* in Latin, is a string, cord or massless rod of length l that allows a suspended mass m, also called *Bob*, to swing back and forth, as shown by Figure 1.1, in a stable manner with limited amplitude. Simple pendulum system, like crane payload and grandfather pendulum clock, see Figure 1.2, have its center of mass down unlike inverted-pendulum system which is an inherently unstable system, has its center of mass higher than its pivot as shown by Figure 1.3 and must be actively balanced in order to remain upright. A torque T must be applied at its pivot point in order to oscillate the pivot point vertically and creating a remaining torque on the pendulum.

[1]National Institute of Applied Sciences and Technology (INSAT), Tunisia

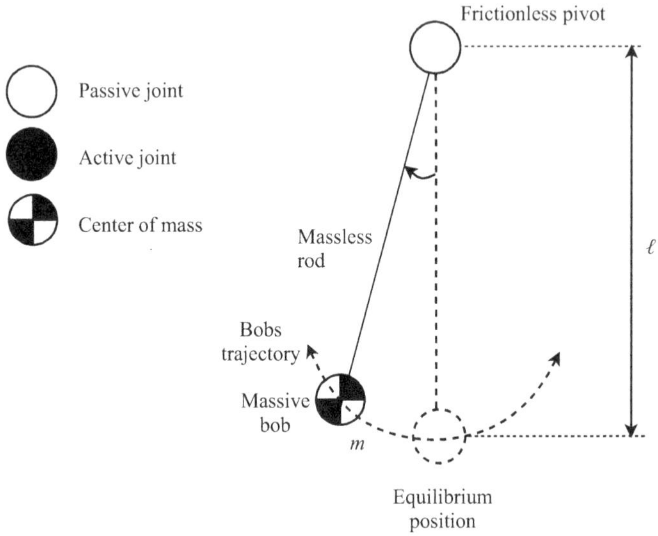

Figure 1.1 The pendulum system

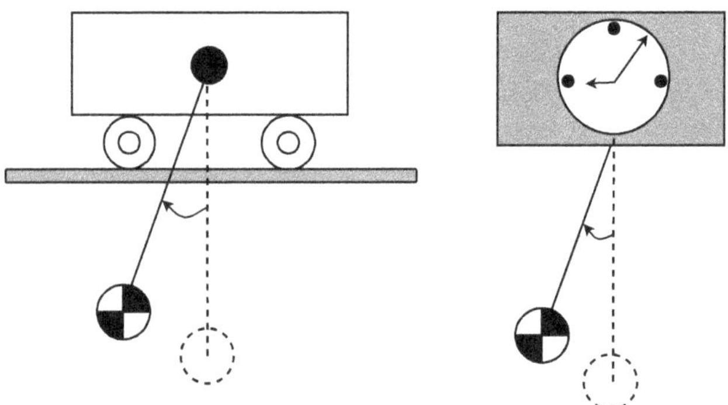

Figure 1.2 Simple pendulum systems

Since Galileo's first experiments written in his first letter in 1602 [1], the pendulum system was introduced in all almost physical courses to illustrate basic concepts of periodic motion and conservation of energy. The valuable hypothesis of Galileo of the isochronisms assuming that the period of the pendulum is approximately independent of the amplitude of the swing was the most fundamental one. Galileo has also found that the period is independent of the suspended mass, and proportional to the square root of the length of the pendulum.

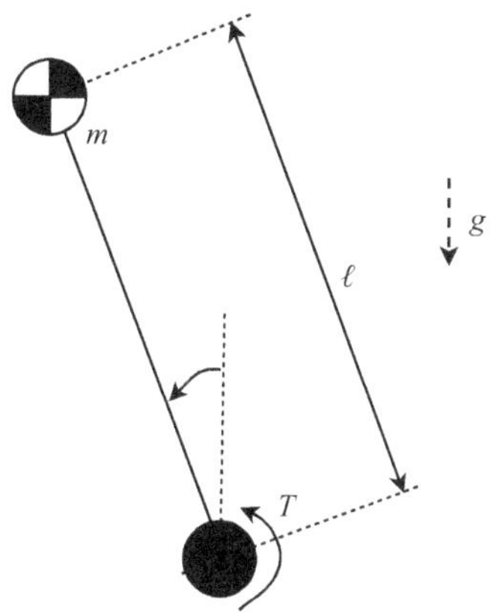

Figure 1.3 Inverted-pendulum system

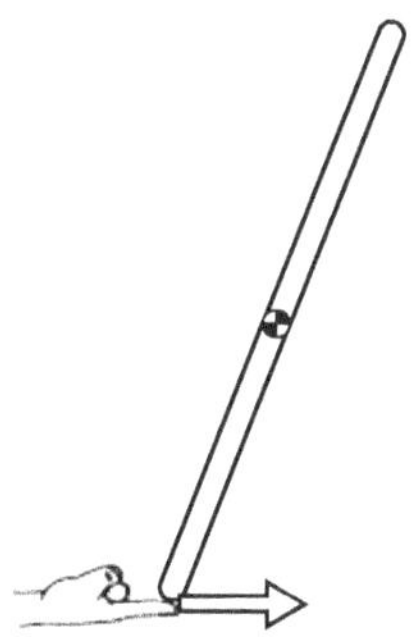

Figure 1.4 The stick-balancing problem

The pendulum has played a significant role not only in the development of science and education but also in culture and politics [2,3]. Besides, its initial utilization as a clock by Christian Huygens in 1656 as a first measure of time, the simple pendulum was studied by the leading scientists of the seventeenth century, like Newton and Hooke for establishing a lot of physical laws. In that framework, we can cite the conservation laws, the collision laws, the value of the acceleration due to gravity g, the finding of the variation in g from equatorial to polar regions and consequently discovering the oblate shape of the earth and mainly the decisive evidence for Newton's synthesis of terrestrial and celestial mechanics [3].

A simple demonstration of stabilizing the center of mass of an inverted pendulum higher than its pivot is balancing an upturned broomstick on the end of one's finger by applying a horizontal force F as shown by Figure 1.4. This concept can be also displayed by a simple device that allows observing the movement of a liquid placed in a basin at the top of the structure.

For the inverted pendulum, it has been necessary to wait until the beginning of the twentieth century to have the first findings around its dynamical system. As claimed in [4], the first results were initiated by Stephenson, who demonstrated in 1908 that the inverted pendulum could be stabilized by applying rapid, vertical, harmonic oscillations to its base [5], and who established in 1909, the stability conditions for the double and the triple inverted pendulums [6]. In 1932, Lowenstern developed the general equations of motion for the inverted pendulums [7]. It is just in the 1960s that the dynamics of inverted pendulums were well understood. In their paper *History of Inverted Pendulum* [8], the authors present several of the early treatments from the literature between 1960 and 1970 and state that Roberge demonstrated a first solution to the single inverted-pendulum system at M.I.T. in his aptly named bachelor's thesis, *The Mechanical Seal* [9], supervised by Leonard Gould. They also claim that Donald, Higdon and Cannon described, in 1963, systems with multiple independent inverted pendula at Stanford University [10] and that Schaefer and Cannon discussed, in 1966, jointed and flexible inverted-pendulum systems [11]. Several interesting papers are then followed [12–14]. It is just with the takeoff of the control theory and robotics at the end of the 1960s that the inverted pendulum becomes the most popular system, among others, for teaching and researches. Solutions of high interest are then proposed to solve various control problems. At the first stage, the stabilization problem around the upright position [15–19] and the swing-up problem [20–23] are solved independently. At the last stage, complete solution is then proposed [24–29]. Some solutions are then following to solve the tracking problem [30,31] and the observer-based controller problem with separation principle [32,33]. The parameter identification problem is also considered in some papers [34–36]. Few papers dealt then with the delay-control problem [37,38], the bifurcation analysis [39–41] and possible chaos dynamics of the inverted pendulum [42,43].

It is important to note that the simple structure of the inverted-pendulum model allowing carrying out experimental validations is one of the key motivations of using this benchmark in education and in research. Furthermore, as shown by Figure 1.5, many engineering systems are based on the stabilization principle of the inverted pendulum. For the liftoff of rockets and flight of missiles, for example, the first achievement is done by Norris at California Institute of Technology, in 1956, in his Engineer's thesis [44]. Standing and walking of bipeds using the stabilization principle of the inverted pendulum was initiated by the pioneers Hemami, Vukobratovic, Raibert and Kajita in their works presented in [45–50], whereas self-balancing problem of wheeled inverted pendulums was started in [51–53].

Since more than seven decades, the inverted pendulum is considered the most popular benchmark in control theory and robotics. For example, Scopus bibliographic database shows more than 6,000 of references written around the dynamical model of this important system. The analysis of the inverted-pendulum system and the design of different control laws have proposed by several textbooks [54–66], monographs [67] and survey papers [68–72].

The objective of this first chapter is to present the history and survey of open and current problems in control theory and robotics written around the inverted-pendulum system. The chapter is organized as follows: in Section 1.2, an overview

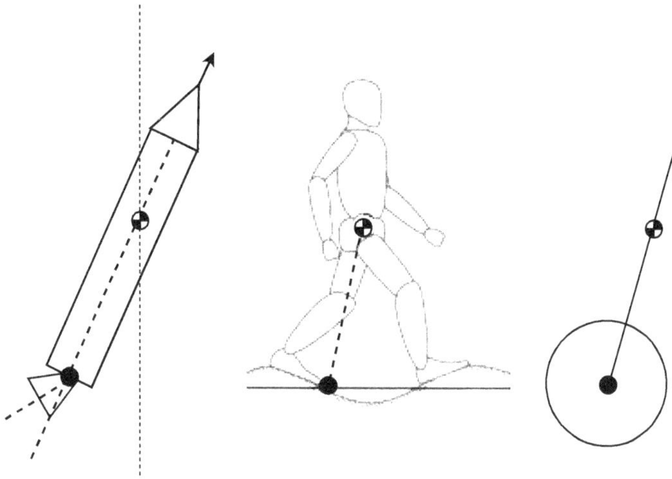

Figure 1.5 Engineering systems based on the stabilization principle of the inverted pendulum

of inverted-pendulum benchmarks will be presented. Experimental platforms will be then extended in Section 1.3. In Section 1.4, advances and challenging achievements in control theory will be deeply expanded. Finally, an overview of major accomplishments in the robotics field based in the stabilization principle of the inverted pendulum will be advanced in Section 1.5.

1.2 Inverted-pendulum benchmarks: an overview

Several benchmark systems of high interest exist in the literature of control theory and are frequently used as experimental models to validate the efficiency of control algorithms. In the field of robotics, some examples of high interest exist. Their related mathematical models are generally derived using two traditional formulations: the Lagrange–Euler formulation and the recursive Newton–Euler formulation [73–80]. The Lagrange–Euler approach uses the Lagrangian function which is the difference between the kinetic energy and the potential energy of the robotic system which is composed on different compounds, while Newton–Euler approach describes the combined translational and rotational dynamics of the different rigid bodies with respect to the each link's center of mass. In the literature, the Lagrange–Euler formulation is generally preferred since the controllers are conducted to obtain the time evolution of the generalized coordinates of the robotic system. Using Lagrange approach or Newton–Euler approach, the analytic developments must lead to a dynamical robotic system.

The reader can refer to Chapters 2, 10 and 14 of this book to find mathematical developments based on Lagrange–Euler method expanded for establishing examples of pendulum dynamical models.

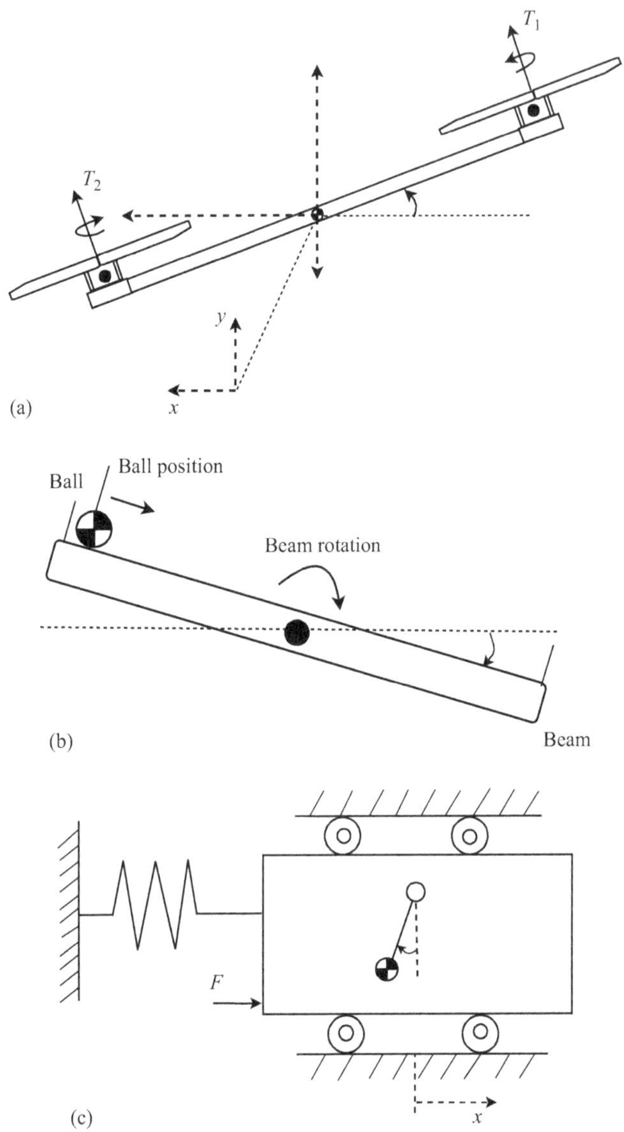

Figure 1.6 (a) VTO Aircraft benchmark, (b) ball and beam benchmark, (c) TORA benchmark, and (d) the bicycle benchmark

In the field of robotics, the most popular benchmarks are the vertical take-off and landing (VTOL) aircraft [81], the translational oscillator with a rotational actuator (TORA) [82], the ball and beam system [83] and the bicycle [84], respectively, shown by Figure 1.6. However, in spite of its simple structure, the inverted pendulum is

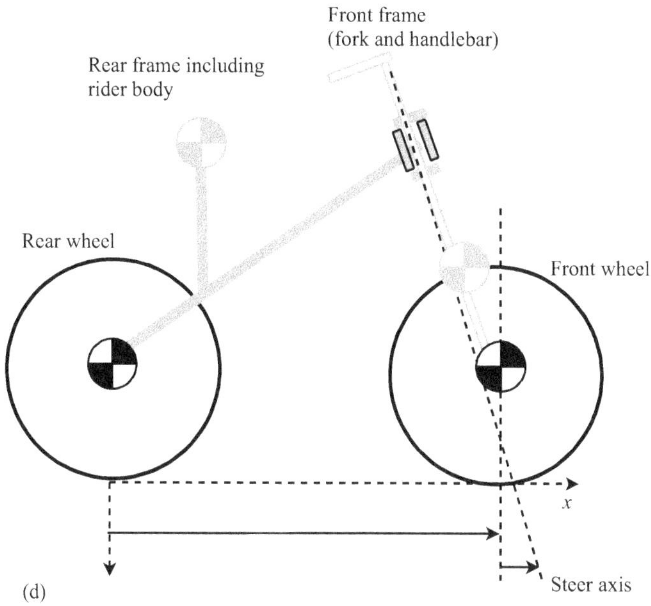

Figure 1.6 (Continued)

Table 1.1 Common varieties of the inverted-pendulum benchmark

Benchmark		Synoptic diagram	Mathematical model
Inverted pendulum on a pivot	Pendupot	Figure 1.7	[62,65,85,86]
	Acrobot		[62,65,85,87,88]
	Inertia wheel pendulum	Figure 1.8	[62,65,67,85]
	Spherical pendulum	Figure 1.9	[89,90]
Inverted pendulum on a cart	Inverted pendulum on a cart	Figure 1.10	[62,65,85]
	Double-inverted pendulum	Figure 1.11	[19,91]
	Twin-inverted pendulum	Figure 1.12	[92]
	Triple pendulum	Figure 1.13	[91,93]
	Quadruple pendulum	Figure 1.14	[91,94]
	Furuta or rotational pendulum	Figure 1.15	[62,65,85,91,95,96]
Wheeled inverted pendulum		Figure 1.16	[97,98]

considered, among the last examples, the most fundamental one. Various models for the inverted pendulum exist in the literature. They can be classified as follows:

- **Inverted pendulums on a pivot:** n rigid bodies (serial or parallel) are joined together and fixed on a pivot where the system inputs are the m voltages of the motors at each joint and the outputs are the n pendulum angles. In most cases, under-actuated inverted pendulums are considered such as $n > m$.

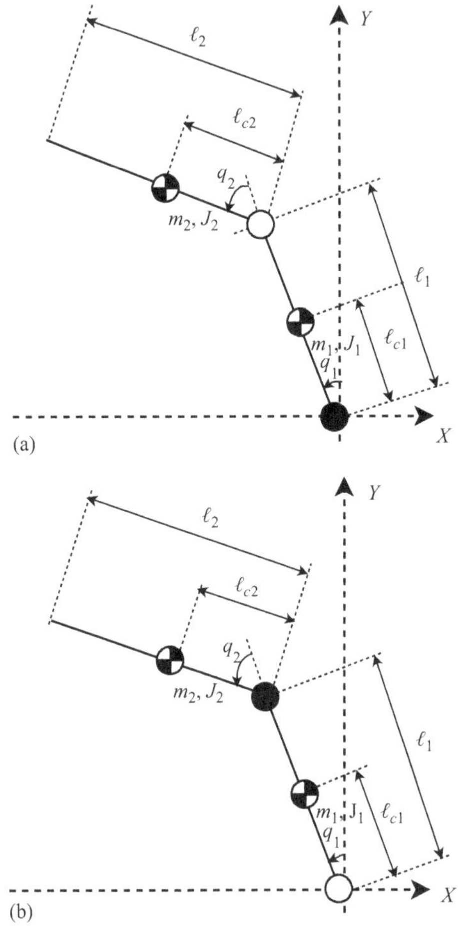

Figure 1.7 (a) The pendubot and (b) the acrobot

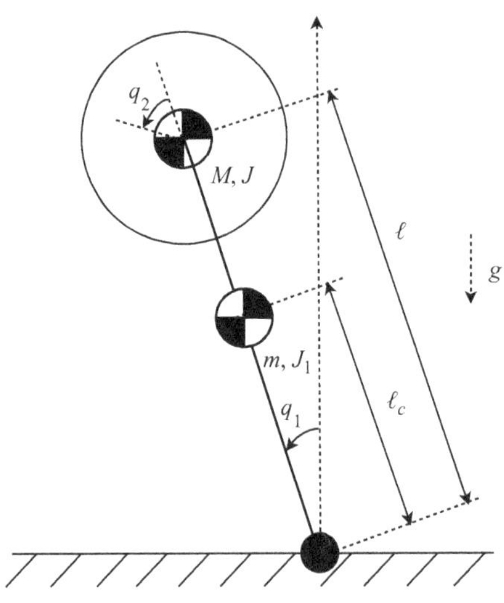

Figure 1.8 The inertia wheel pendulum

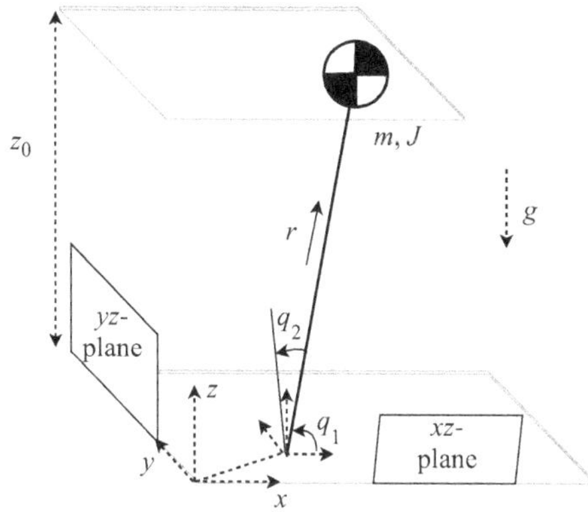

Figure 1.9 The spherical pendulum

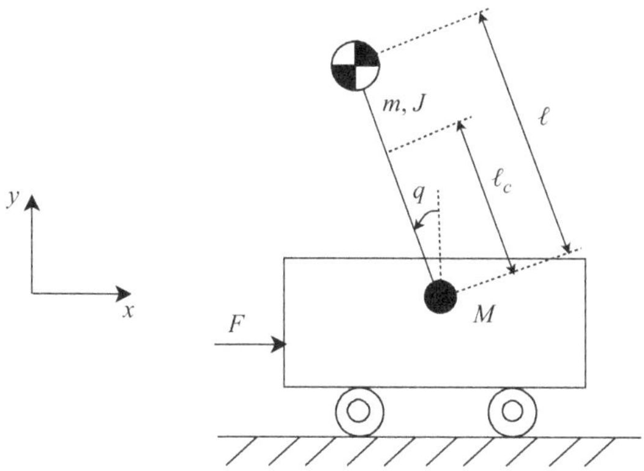

Figure 1.10 The cart-inverted pendulum

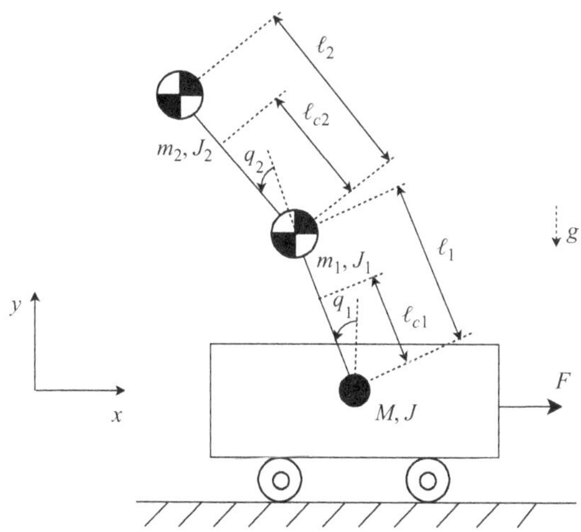

Figure 1.11 The double-inverted pendulum on a cart

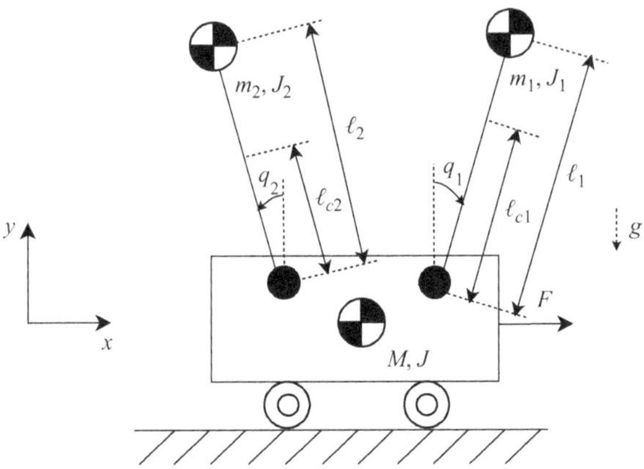

Figure 1.12 The twin-inverted pendulum on a cart

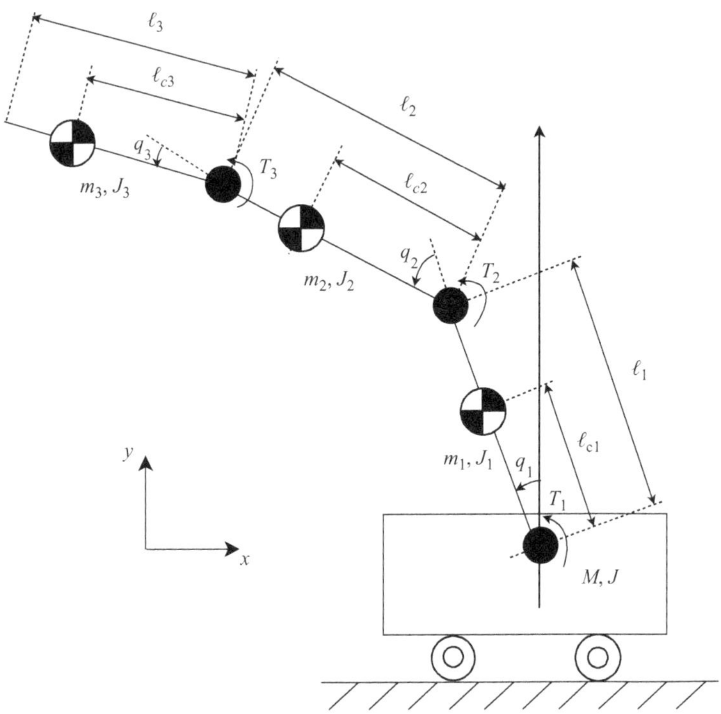

Figure 1.13 Triple-inverted pendulum

- **Inverted pendulums on a cart:** a multilink system of n rigid bodies is attached to an inherently stable moving base (the cart). Hence, an under-actuated mechanical system is obtained such as the system input is the force acting upon the cart and where the $(n + 1)$ outputs are the cart position at the n pendulum angles.

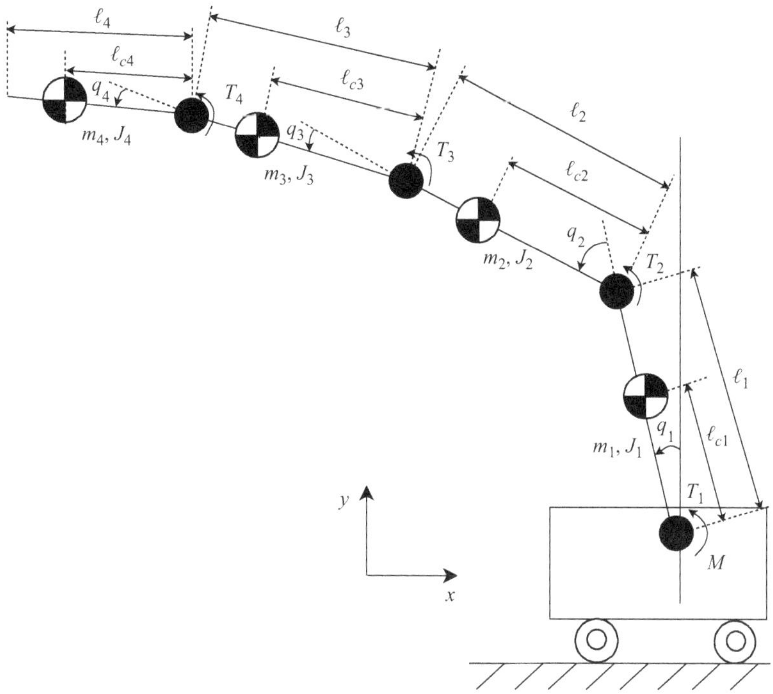

Figure 1.14 Quadruple-inverted pendulum

- **Wheeled inverted pendulum:** a single-link rigid body is self-balanced by an unstable moving base mounted on one or two wheels. The inputs and the outputs vary depending from one design to another but an under-actuated mechanical system is always considered. The system can act on 2D or on 3D space.

For better understanding, the common varieties of the inverted-pendulum benchmark are summarized in Table 1.1 where a synoptic diagram and the related references to related mathematical models are given for each variety.

1.2.1 Notation

For Figures 1.7–1.16, M and J design the mass and the moment of inertia of the stable or unstable base, respectively. m_i, l_i, l_{ci}, J_i design, for $i = 1, \ldots, n$, the mass, the length, the center of mass and the moment of inertia of the pendulum i, respectively. When only one pendulum is considered, the index i is omitted. The acceleration due to gravity is designed by g.

It is obvious that the most familiar models in the literature are the inverted pendulum on a cart for which Chapters 2–4 of this book are devoted and the inertial wheel

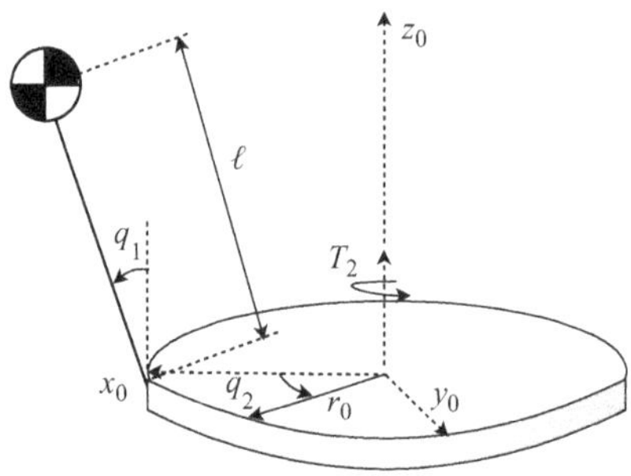

Figure 1.15 The Furuta pendulum (rotational pendulum)

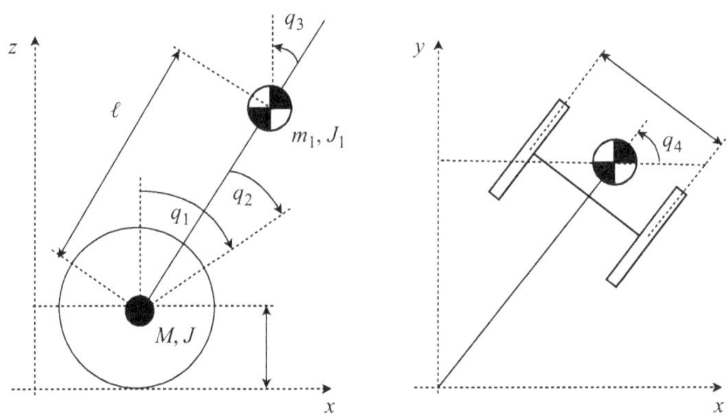

Figure 1.16 The wheeled inverted pendulum

pendulum for which Chapters 5–7 are dedicated. New structures for the inverted pendulums are also recently proposed like the fly inverted pendulum [99], the inverted bicycle [100] and the spherical inverted pendulum on a five-bar mechanism [101].

1.3 Experimental laboratories

The simple structure of the inverted-pendulum model allowing carrying out experimental validations in education and in research was one of the key motivations of using this benchmark. Numerous real and virtual platforms exist.

1.3.1 Commercial devices

Several companies around the world are dedicated to manufacture professional products. In this framework, we can cite the LEGO products [102–106], the Quanser products [107] fully compatible with MATLAB®/Simulink™ and LabVIEW™ software including translational and rotary motion control inverted-pendulum systems [101–110], the Googol Technology products [111] including transitional, circular and self-balancing wheeled inverted-pendulum platforms [112–114] and supporting MATLAB environment, the INTECO inverted pendulum on a cart [115], the Instruments Digital dual-inverted pendulum/crane system fully compatible with MATLAB software [116], the Model 505 of inverted pendulum of Educational Control Products [117] and the Rex Controls Furuta pendulum [118].

The reader can find more details about the mechanical system and the physical parameters of the inverted pendulum on a cart of INTECO available at National Autonomous University of Mexico at Chapters 2 and 3 of this book and about the inverted pendulum of Googol Technology in Chapter 4.

1.3.2 Universities platforms

A number of universities and research laboratories around the world have designed and built their own real inverted platforms to validate the efficiency of several emerging control techniques and verify their implementation. In this framework, we can cite the inertia flywheel pendulum mechatronic kit of Huafan University (Taiwan) [99], the balancing inverted pendulum on a parallel manipulator of LARM laboratory of University of Cassino (Italy) [119], the Furuta pendulum of Massachusetts Institute of Technology (USA) [120], the flying inverted pendulum of the Institute for Dynamic Systems and Control (Switzerland) [121], the inverted pendulum of Department of Aeronautical and Mechanical Engineering of Western Michigan University (USA) [122], the single/double inverted pendulum of the University of Adelaide, School of Mechanical Engineering (Australia) [123], the rotary inverted pendulum of the Rensselaer Polytechnic Institute of Troy (USA) [124] and so on.

Furthermore, full details about the inertia wheel pendulum of the LIRMM Laboratory of Informatics, Robotics and Microelectronics (France) can be found in Chapter 5, whereas the balancing inverted pendulum of LARM laboratory of University of Cassino (Italy) is fully described in Chapter 10.

1.3.3 Virtual models

To avoid unreliable and expensive real platforms, real experiments are often substituted by virtual reality enabled by tools. One may find good examples of virtual control laboratories of pendulum-car benchmark using for example SimMechanics in MATLAB/Simulink environment [125] or solid works using MATLAB–Simulink, ControlDesk, and a DS1104 board from dSPACE software [126]. Open source inverted-pendulum virtual models are also available [127] making this an ideal student project for an undergraduate controls curriculum.

Nowadays, many students communicate with the educational material through internet-based technologies. Time savings, sharing resources of expensive equipment and individual access to experimentation are the main factors that motivate remote laboratories over the world. Remote pioneered projects based on the inverted pendulum can be found in [128,129].

1.4 Advances in control theory

1.4.1 Why the inverted pendulum is the most popular benchmark?

Although the structure of the pendulum systems already presented seems to be quite simple, many standard techniques in control theory are ineffective when tested on this benchmark. The inverted-pendulum system is a typical nonlinear dynamic system including a stable equilibrium point when the pendulum is in a pending position and an unstable equilibrium point when the pendulum is in an upright position [63]. When the system is moved up from the pending position to the upright position, the model is nonlinear with the pendulum angle. The control problem of the system is much more difficult than it appears to be for many reasons, e.g., geometric properties of the system are lost when the pendulum moves through horizontal positions [130]. Indeed, the output-zeroing manifold does not contain any equilibrium points, the relative degree of the system is not constant and the controllability distribution of the system does not have a constant rank. The system is therefore not linearizable [131]. The control problem to be solved is generally composed of two main specific tasks: the first one is the swing-up control problem and the second one is the stabilization problem around the unstable equilibrium point [132]. Finding a continuous feedback that can render the vertical upward position of the pendulum globally asymptotically stable has been considered a major problem for a long time.

1.4.2 Emerging control problems

The inverted pendulum was also used to illustrate major emerging problems in control theory which are mainly:

- **Swing-up:** The control objective is to upswinging the pendulum from its stable position to its unstable position. It was first proposed by Furuta in 1976 [134]. Many papers are proposed in this framework but one of the most interesting was written by Astrom and Furuta in which properties of the simple strategies for swinging up based on energy control are presented, and the idea to give the nonlinear system the energy of the upright position is clearly explained [23].
- **Stabilization problem around the upright position:** This problem is a very popular example in control education, generally introduced to undergraduate-level students from a linearization of the nonlinear model using a first-order Taylor approach at the unstable equilibrium point. As a result, a linear system is obtained, and a feedback gain is computed using a linear control strategy such as LQR or pole placement [19,54–57,67,68]. The importance of such an approach in education is that many typical problems in control theory can be highlighted such as undesirable dynamics exhibited by unstable and nonminimum phase systems and difficulties of controlling single input/multioutput systems. In research, the stabilization problem is generally solved using a nonlinear controller [62,63]. One of the important objectives is to stabilize the system around its unstable equilibrium point and having at the same time a larger basin of attraction for initial conditions. But the most important problem is to guarantee the global stabilization

Table 1.2 Main control problems

System	Swing-up problem	Stabilization problem	Tracking problem	Observer-based control problem
Pendubot	[135,136]	[137,138]	[139]	[140]
Acrobot	[87,141,142]	[137,143]	[144,145,148,149]	[146,147,150,151]
Inertia wheel pendulum	[144,148]	[145,149]	[146–147,150–152]	[153,154]
Inverted pendulum on a cart	[155–158]	[159–172]	[173–175]	[176–178]
Twin-inverted pendulum	–	[179]	–	–
Double-inverted pendulum	[180]	[181–187]	[188]	[189]
Triple-inverted pendulum	[190]	[191–195]	–	[196–198]
Furuta pendulum	[199]	[200–203]	[204–207]	[208,209]
Spherical pendulum	[210]	[211]	[212,213]	–

of the system dynamics. One possibility to achieve global stabilization of the pendulum is by means of hybrid controllers [59,60]. Typically, two different control laws are employed, one takes care of the local stabilization of the pendulum in the upright position and the other swings the pendulum up. Another way is to stabilize the system by means of continuous feedback. In this way, for almost all initial conditions, the upright position is attractive (though not stable).

- **The tracking problem:** The objective is to effectively follow a desired reference input for outputs, the position and velocity variables in our case. The importance for such unstable system is taking into account the stochastic disturbances is evident.
- **Observer-based control problem:** It is well known that output feedback control or observer design is a very important problem in control systems, since in many practical nonlinear control systems, state variables are often unavailable. Observer-based controllers have attracted some interest in nonlinear systems. For inverted-pendulum systems, such problem is widely studied when velocity variables are often considered unavailable states. The main further is the development of the separation property; that is, the controller and the observer can be independently designed.

Table 1.2 gives an overview of different solutions proposed in the literature for solving main control problems presented below. On the other hand, inverted-pendulum benchmarks are used to illustrate almost all emerging ideas in control theory techniques. Table 1.3 depicted few high cited scientific articles for each control approach.

It is also important to note that one of the complex problems to achieve for inverted pendulum was the global asymptotic stabilization of the origin by a single and smooth continuous feedback performing both the upswing and the stabilization control. A complete solution to this problem can be found in [168,171].

On other side, accurate mathematical models of physical systems are essential for understanding the behavior of actual systems under different operating conditions

Table 1.3 Nonlinear control techniques illustrated by the inverted-pendulum benchmark

Control technique	Mathematical model	Reference
Bang–bang control	Furuta pendulum	[20]
Fuzzy logic control	Inverted pendulum on a cart	[214]
Neural network control		[215]
PID control		[216]
Optimal control	Triple-inverted pendulum	[217]
Energy-based control	Acrobot	[142]
Adaptive control	Wheeled inverted pendulum	[218]
Sliding mode control		[219]
	Furuta pendulum	[220]
Predictive control	Twin-inverted pendulum	[223]
Feedback linearization	Inertia wheel pendulum	[145,149]
Passivity-based control		
Back-stepping control	Cart inverted pendulum	[221]
Hybrid control		[222]
Feedforward control	Double-inverted pendulum on a cart	[224]

and for designing control systems. In mechanical systems, especially the inverted pendulum, difficulty of the exact determination of system parameters adversely affects the accuracy of the models that's why an experimental validation procedure of the control laws is often accompanied by a system parametric identification/estimation. A literature search showed that very few papers are devoted for such framework [34–36,225–228].

1.4.3 Challenging control problems

In this section, challenging and still open control problems will be presented. They include but are not limited to the following fields:

- **Robustness-based sliding mode techniques:** Besides stability, robustness is another important requirement to be considered in the study of uncertain nonlinear inverted-pendulum systems. Since uncertainty is source of instability, especially for tracking problems, common solutions are no more effective. Control design methods are then referred to confront system uncertainties and/or stochastic perturbations. Several promising directions can be chosen in such situations. In our sense, sliding mode observers-based controllers, very appropriate for robotic tracking problems with unavailable velocities measurements, will continue to have an enlarging impact in the following years in such framework. Sliding mode observers were initially proposed in [229] for systems with a relative degree equal to one and Lipschitz uncertainties. Then, they are extended for many other specific situations with high interest in [230–235]. In this book, solutions for solving problems related to robustness-based sliding mode techniques are proposed in Chapters 2–4, 6, 7, 9 and 13.

- **Robustness-based T–S fuzzy models:** T–S fuzzy model-based observers/controllers are widely popular models providing simple and effective solutions to the control of plants that are complex, uncertain and ill-defined such as inverted-pendulum systems. Several research works using the inverted-pendulum systems are still recently organized in such directions [236–241]. In this book, solutions to the robustness-based T–S fuzzy models are found in Chapters 12 and 13.
- **Robust fault-tolerant control:** Nowadays, fault diagnosis and fault-tolerant control are considered major problems in the modern control theory [242]. Fault can occur in any part of the system and do degrade the feedback system performances, which leads to instability and even can result into dangerous situations. The inverted pendulum as inherent unstable nonlinear system with possible parameter uncertainties and possible faults in sensors and actuators may be considered a typical example to validate new approaches for robust fault-tolerant control [243–245]. In this book, robust fault-tolerant control problem is solved in Chapter 13.
- **Fractional-order control:** Fractional calculus is an old theory for solving differential equations. However, for only a few years, emerging and effective methods in differentiation and integration of noninteger-order equations make fractional-order systems and also fractional-order controllers more and more attractive for control engineering applications. Fractional-order controllers are generally applied to enhance the system control performance. Some initiatives have been launched in this direction deriving some fractional-order inverted-pendulum models and/or using fractional-order-controller-based inverted-pendulum systems [246–249]. A solution to fractional-order proportional–integral–derivative (PID) controller is proposed in Chapter 4.
- **Event-triggered control:** In recent time, event-triggered control has received much attention since it has been demonstrated to reduce the amount of computations while still maintain a desired closed-loop behavior. It also allows the reduction of communications in control systems in which the information is transmitted through a shared network. Besides, the event-based control strategies have proved to be a good alternative to deal with delays and packet losses in communication networks. In addition, the event-based strategy offers several advantages such as protecting the actuators from frequent wear and tear and also reducing the waste of electric energy due to the switching voltage. In such framework, several recent papers have used the inverted pendulum as a typical example to prove such a challenging controller [250–253].
- **Time-delayed control:** It is well known that some important classes of systems, like oscillators and double integrators cannot be stabilized by a static output feedback, may be stabilized by inserting an artificial time delay in the feedback. Furthermore, static output-feedback controllers have advantages over observer-based controllers, since the design of observer-based controller is a complicated procedure mainly when uncertainties are considered. Several research papers had attempted to solve such a problem using time-delayed feedback [254]. Furthermore, in the presence of time delays, the well-known control techniques stabilizing the inverted-pendulum benchmarks may result in inadequate behavior and may

even cause instability. The existence of time delay creates infinite dimensional dynamics and complicates the theoretical analysis. The dynamics of the inverted pendulum with delayed feedback control is handled recently by some researches [255–258].

- **Bifurcation and chaos analysis:** Specific bifurcation conditions and chaotic generators have been verified for the inverted-pendulum systems using several phenomena. Such phenomena include impacts generally designed by piecewise linear functions in the related mathematical models [259–261] or delays in control laws [262–266].
- **Generation and stabilization of self-oscillations:** Friction is a very complex phenomenon that occurs in pendulum systems. When friction effects are not modeled, feedback controllers may fail in real-time experiments. In most research works, the pendulum system was studied assuming that the only kind of friction is viscous friction, whereas a more accurate model of the friction should also include static and Coulomb (sliding) friction. Indeed, it was proven in [267] that it is always possible to obtain stable oscillations for the inverted-pendulum models associated to a limit cycle when such phenomena occur. Such dynamics may also bear through a supercritical Poincaré–Andronov–Hopf bifurcation induced by time delays in control variables [264]. Orbital stabilization of pendulum systems can also induce a periodic reference orbit in finite time [268–271]. Generation and stabilization of self-oscillations in presence of external perturbations and uncertainties are proposed in Chapters 6 and 7, respectively.

1.5 Major accomplishments in robotics

In robotics, inverted-pendulum systems are mainly used in three directions: (1) testing emerging control laws of the complex class of under-actuated mechanical systems, (2) designing and controlling wheeled self-balancing robots and (3) modeling and controlling walking pattern generation for bipedal and humanoid robots.

1.5.1 Modeling and controlling under-actuated mechanical systems

All inverted-pendulum models presented in this survey paper belong to the class of underactuated mechanical systems. A mechanical system is said to be under-actuated when it have fewer independent inputs than generalized coordinates to be controlled and have then unactuated degrees of freedom [85,261,264]. Because the number of inputs is less than the number of degrees of freedom, the controllability of the mechanical under-actuated system is not always satisfied, geometric properties of the system are lost and the system is therefore not linearizable. For such systems, the unactuated generalized coordinates may be controlled by the actuated coordinates through the dynamic coupling. This coupling is often inherently nonlinear, resulting in generally nonintegrable dynamic constraints and therefore second-order nonholonomic [272–275]. The study of such dynamical systems is much more difficult than it appears, and pendulum models are widely used to test different emerging control approaches proposed to control this class of systems so important in many applications [276–281]. In this book, Chapters 5–9 are devoted to solve control problems for such class of systems.

1.5.2 Self-balanced mobile-wheeled robots

Mobile robots are more and more omnipresent today. They are used in a diversity of applications including transport and entertainment. Especially, wheeled robots have a simpler mechanical structure than legged robots and are known to be energy efficient. Wheeled inverted pendulum devices are self-balancing robots that are statically unstable variety. They have generally two coaxial wheels mounted on either side of an intermediate body, with a center of mass above the wheel axes. Their particular configuration makes them highly maneuverable to hang around and quite suitable for indoor environments. Furthermore, they have the ability to maintain stability on inclines, by leaning into the slope and can accelerate quickly without tipping over. Nowadays, several well-known devices belong to the class of wheeled inverted pendulums. The most famous ones are the Segway [282,283], the JOE [284], the UW-car [285], the PMP [286], the B2 [287], the I-Pentar [288], a self-balancing humanoid robot, the Balbot [289] with a single-wheel moving robot in any direction and the ball-driven balancing robot [290], a mobile robot mounted on a moving ball. Examples of mobile platforms are found in Chapters 9, 10 and 14. Such devices must actively stabilize themselves to prevent overbalancing. Although they are more difficult to control than statically stable wheeled robots, they are still much easier to control than legged robots. Control law design for tracking trajectories in presence of any disturbances and uncertainties remain until now a challenging problem for such devices [291].

1.5.3 Modeling and controlling walking pattern generation

Unlike wheeled robots, legged robots are able to step over obstacles. However, they are more complex to model and stabilize during a walking cycle due to the greater number of degrees of freedom and the impact contact force with the ground, see bipedal robot model given in Chapter 11.

Based on the inverted-pendulum principle stabilization analogy with the human gait pattern, a number of inverted-pendulum models are used as walking pattern generation for humanoid robots. The most well-known models are the 3D linear inverted-pendulum models (LIPM) [292], the Kajita cart-table model [293], the variable impedance LIPM [294], the spring-loaded inverted pendulum [295] and the angular momentum pendulum model [296]. The most well-known humanoid robots using LIPM principles are the famous Japanese Honda female Android HRP-4C [297] and WABIAN-2R Humanoid robot [298,299]. Such models are also extensively used, in medicine, to find solutions for walking pattern generation impairments of elderly and stroked patients [300].

1.6 Conclusion

Control methods and technologies arising from the model of the inverted pendulums attribute to the applications to both biological and mechanical control problems such stabilizing walking pattern generation and self-balancing wheeled transporters. By means of the rich properties of the inverted pendulum, this system will continue to play an important role in detecting abilities of innovative control methods dealing with nonlinearities, delays, bifurcations, chaos, minimum phase dynamics, oscillations and limit cycles, uncertainties and so on.

References

[1] Galilei G. (Favaro A., ed.) *The works of Galileo Galilei (National Edition)*. Florence: Barbera; 1890–1909 (reprinted 1929–1939 and 1964–1966).

[2] Matthews M.R. *Time for science education: How teaching the history and philosophy of pendulum motion can contribute to science literacy*. New York: Kluwer Academic Publishers; 2000.

[3] Matthews M.R., Stinner A., Gauld C.F. (eds.) *The pendulum: Scientific, historical, philosophical and educational perspectives*. Netherlands: Springer; 2005.

[4] Castro A. *Modeling and dynamic analysis of a two-wheeled inverted pendulum*. Master Science thesis, Georgia Institute of Technology Publication, Atlanta, USA; 2012.

[5] Stephenson A. 'A new type of dynamic stability'. *Memories and Proceeding of the Manchester Literary and Philosophical Society*. 1908; **52**: 1–10. Manchester.

[6] Stephenson A. 'On induced stability'. *Philosophical Magazine*. 1909; **17**(1): 765–768.

[7] Lowenstern E.R. 'Stabilizing effect of imposed oscillations on a dynamical system'. *Philosophical Magazine*. 1932; **13**: 458–486.

[8] Lundberg K.H., Barton T.W. 'History of inverted pendulum systems'. *IFAC Proceedings Volumes*. 2010; **42**(24): 131–135.

[9] Roberge J.K. *The mechanical seal*. Bachelor's thesis, Massachusetts Institute of Technology, Cambridge; 1960.

[10] Donald T.H., Robert H.C. 'On the control of unstable multiple-output mechanical systems'. *ASME Publication 63-WA-148*; American Society of Mechanical Engineers, New York, 1963.

[11] Schaefer J.F., Cannon R.H. (eds.) 'On the control of unstable mechanical systems'. *Proceedings of the third Automatic and Remote Control*; London, June 1966. pp. 12–24.

[12] Phelps F.M., Hunter Jr J.H. 'An analytical solution of the inverted pendulum'. *American Journal of Physics*. 1965; **33**(4): 285–295.

[13] Bagdanoff J.L., Citron S.J. 'Experiments with an inverted pendulum subject to random parametric excitation'. *The Journal of the Acoustical Society of America*. 1965; **38**: 447–452.

[14] Blitzer L. 'Inverted pendulum'. *American Journal of Physics*. 1965; **33**(12): 1076–1078.

[15] Young L.R. 'Bang-bang aspects of manual control in high-order systems'. *IEEE Transactions on Automatic Control*. 1965; **10**(3): 336–341.

[16] Furuta K., Okutani T., Sone H. 'Computer control of a double inverted pendulum'. *Computer and Electrical Engineering*. 1978; **5**: 67–84.

[17] Furuta K., Ochiai T., Ono N. 'Attitude control of a triple inverted pendulum'. *International Journal of Control*. 1984; **39**(6): 1351–1365.

[18] Gurumoorthy R., Sanders S.R. (eds.) 'Controlling non-minimum phase non-linear systems—The inverted pendulum on a cart example'. *American Control Conference*; San Francisco, California, June 1993. pp. 680–685.

[19] Bogdanov A. 'Optimal control of a double inverted pendulum on a cart'. *Tech. Rep. CSE-04-006*; Oregon Health and Science University, OGI School of Science and Engineering, Beaverton, OR, 2004.

[20] Furuta K., Yamakita M., Kobayashi S. (eds.) 'Swing up control of inverted pendulum'. *International Conference Industrial Electronics, control and Instrumentation (IECON '91)*; Kobe, Japan, October 1991. pp. 2193–2198.

[21] Spong M.W. 'The swing up control problem for the acrobot'. *IEEE Control Systems Magazine*. 1995; **15**(1): 49–55.

[22] Yoshida K. (ed.) 'Swing-up control of an inverted pendulum by energy-based methods'. *Proceedings of the American Control Conference*; San Diego, CA, June 1999. pp. 4045–4047.

[23] Astrom K.J., Furuta K. 'Swinging up a pendulum by energy control'. *Automatica*. 2000; **36**(2): 287–295.

[24] Mori S., Nishihara H., Furuta K. 'Control of unstable mechanical system control of pendulum'. *International Journal of Control*. 1976; **23**(5): 673–692.

[25] Wei Q., Dayawansa W.P., Levine W.S. 'Nonlinear controller for an inverted pendulum having restricted travel'. *Automatica*. 1995; **31**(6): 841–850.

[26] Spong M.W., Corke P., Lozano R. 'Nonlinear control of the reaction wheel pendulum'. *Automatica*. 2001; **37**(11): 1845–1851.

[27] Aracil J., Gordillo F. (eds.) 'The inverted pendulum: A benchmark in nonlinear control'. *Proceedings of the Sixth Biannual World Automation Congress-Intelligent Automation and Control Trends, Principles, and Applications*; Seville, June 2004. pp. 468–482.

[28] Muskinja N., Tovornik B. 'Swinging up and stabilization of a real inverted pendulum'. *IEEE Transactions on Industrial Electronics*. 2006; **53**(2): 631–639.

[29] Srinivasan B., Huguenin P., Bonvin D. 'Global stabilization of an inverted pendulum-Control strategy and experimental verification'. *Automatica*. 2009; **45**(1): 265–269.

[30] Berkemeier M.D., Fearing R.S. 'Tracking fast inverted trajectories of the underactuated Acrobot'. *IEEE Transactions on Robotics and Automation*. 1999; **15**(4): 740–750.

[31] Wai R.J., Chang L.J. 'Adaptive stabilizing and tracking control for a nonlinear inverted-pendulum system via sliding-mode technique'. *IEEE Transactions on Industrial Electronics*. 2006; **53**(2): 674–692.

[32] Ma X.J., Sun Z.Q., He Y.Y. 'Analysis and design of fuzzy controller and fuzzy observer'. *IEEE Transactions on Fuzzy Systems*. 1998; **6**(1): 41–51.

[33] Tong S., Li H.X., Wang W. 'Observer-based adaptive fuzzy control for SISO nonlinear systems'. *Fuzzy Sets and Systems*. 2004; **148**(3): 355–376.

[34] Kobayashi M., Shinohara S., Tamura K. 'Swing-up and stabilizing control of an inverted pendulum with adaptive identifier'. *Transactions of the Japan Society of Mechanical Engineers Series C*. 1992; **58**(556): 3606–3613.

[35] Brock S. 'Identification of the parameters in inverted pendulum model'. *7th International Workshop on Advanced Motion Control (AMC)*. 2002: 316–321.

[36] White W.N., Fales R.C. (eds.) 'Control of a double inverted pendulum with hydraulic actuation: A case study'. *Proceedings of the American Control Conference*; San Diego, CA, June 1999. pp. 495–499.

[37] Atay F.M. 'Balancing the inverted pendulum using position feedback'. *Applied Mathematics Letters*. 1999; **12**(5): 51–56.

[38] Landry M., Campbell S.A., Morris K., Aguilar C.O. 'Dynamics of an inverted pendulum with delayed feedback control'. *SIAM Journal on Applied Dynamical Systems*. 2005; **4**(2): 333–351.

[39] McLaughlin J.B. 'Period-doubling bifurcations and chaotic motion for a parametrically forced pendulum'. *Journal of Statistical Physics*. 1980; **24**(2): 375–388.

[40] Sieber J., Krauskopf B. 'Bifurcation analysis of an inverted pendulum with delayed feedback control near a triple-zero eigenvalue singularity'. *Nonlinearity*. 2004; **17**(1): 85–103.

[41] Sieber J., Krauskopf B. 'Complex balancing motions of an inverted pendulum subject to delayed feedback control'. *Physica D: Nonlinear Phenomena*. 2004; **197**(3–4): 332–345.

[42] Levien R.B., Tan S.M. 'Double pendulum: An experiment in chaos'. *American Journal of Physics*. 1993; **61**: 1038–1044.

[43] Kim S.Y., Hu B. 'Bifurcations and transitions to chaos in an inverted pendulum'. *Physical Review E—Statistical Physics, Plasmas, Fluids, and Related Interdisciplinary Topics*. 1998; **58**(3 Suppl. A): 3028–3035.

[44] Norris J.C. *An analysis of a compound pendulum rocket suspension*. Thesis of California Institute of Technology, California; 1951.

[45] Hemami H., Weimer F.C., Koozekanani S.H. 'Some aspects of the inverted pendulum problem for modeling of locomotion systems'. *IEEE Transactions on Automatic Control*. 1973; **18**(6): 658–661.

[46] Hemami H., Golliday Jr C.L. 'The inverted pendulum and biped stability'. *Mathematical Biosciences*. 1977; **34**(1–2): 95–110.

[47] Vukobratovic M., Stepaneko Y. 'On the stability of anthropomorphic systems'. *Mathematical Biosciences*. 1972; **14**(1): 1–38.

[48] Raibert M.H. *Legged robots that balance*. Cambridge, Mass: MIT Press; 1986.

[49] Shuuji K., Kazuo T. (eds.) 'Study of dynamic biped locomotion on rugged terrain-derivation and application of the linear inverted pendulum mode.' *IEEE International Conference on Robotics and Automation*; Sacramento, CA, USA, April 1991. pp. 1405–1411.

[50] Shuuji K., Kazuo T. (eds.) 'Kazuo Experimental study of biped dynamic walking in the linear inverted pendulum mode'. *IEEE International Conference on Robotics and Automation*; Nagoya, Aichi, Japan, May 1995. pp. 2885–2891.

[51] Yamafuji K., Kawamura T. 'Postural control of a monoaxial bicycle'. *Journal of the Robotics Society of Japan*. 1989; **7**(4): 74–79 (in Japanese).

[52] Matsumoto O., Kajita S., Tani K. (eds.) 'Attitude estimation of the wheeled inverted pendulum using adaptive observer'. *Ninth Academic Conference of the Robotics Society of Japan*; 1991. pp. 909–910 (in Japanese).

[53] Koyanagi E., Iida S., Kimoto K., Yuta S. (eds.) 'A wheeled inverse pendulum type self-contained mobile robot and its two-dimensional trajectory control'. *Proceedings of the Second International Symposium on Measurement and Control in Robotics*; Tsukuba Science City, Japan, November 1992. Society of Instrument and Control Engineers of Japan; 1992. pp. 891–98.

[54] Cannon R.H. *Dynamics of physical systems*. New York: McGraw-Hill; 1967. pp. 703–710.

[55] Dorf R.C. *Modern control systems*. Reading: Addison-Wesley; 1967. pp. 276–279.

[56] Ogata K. *Modern control engineering*. Englewood Cliffs: Prentice-Hall; 1970. pp. 277–279.

[57] Kwakernaak H., Sivan R. *Linear optimal control systems*. New York: Wiley; 1972.

[58] Siebert W.McC. *Circuits, signals, and systems*. Cambridge: MIT Press; 1986. pp. 177–182.

[59] Morse A.S. (ed.) *Control using logic-based switching*. London: Springer; 1997.

[60] Yamamoto Y., Hara S. *Learning, control & hybrid systems*. Berlin: Springer; 1999.

[61] Siebert W.McC. *Circuits, signals, and systems*. Cambridge: MIT Press; 1998.

[62] Fantoni I., Lozano R. 'Non linear control for underactuated mechanical systems'. *Communications and Control Engineering Series*; Springer-Verlag, London, 2000.

[63] Khalil H.K. *Nonlinear systems*. New Jersey: Prentice-Hall; 2002.

[64] Li Z., Yang C., Fan L. *Advanced control of wheeled inverted pendulum systems*. London, UK: Springer; 2013.

[65] Xin X., Liu Y. *Control design and analysis for underactuated robotic systems*. London: Springer Science & Business Media; 2014.

[66] Aguilar L.T., Boiko I., Fridman L., Iriarte R. *Self-oscillations in dynamic systems: A new methodology via two-relay controllers*. Switzerland: Springer; 2015.
[67] Block D., Astrom K., Spong M. *The reaction wheel pendulum*. Princeton, NJ: Morgan & Claypool Publication; 2007.
[68] Horacek P. 'Laboratory experiments for control theory courses: A survey'. *Annual Reviews in Control*. 2000; **24**: 151–162.
[69] Hahn H.H., Spong M.W. (eds.) 'Remote laboratories for control education'. *Proceedings of the IEEE Conference on Decision and Control*; Sydney, NSW Australia, 12–15 December 2000. pp. 895–900.
[70] Acosta J.A. 'Furuta's pendulum: A conservative nonlinear model for theory and practise'. *Mathematical Problems in Engineering*. 2010; 2010: 29 pp.
[71] Boubaker O. (eds.) 'The inverted pendulum: A fundamental benchmark in control theory and robotics'. *International Conference on Education and e-Learning Innovations (ICEELI)*; Sousse, Tunisia, July 2012. pp. 1–6.
[72] Boubaker O. 'The inverted pendulum benchmark in nonlinear control theory: A survey'. *International Journal of Advanced Robotic Systems*. 2013; **10**: 1–9.
[73] Craig J.J. *Introduction to robotics*. Addison Wesley: Pearson Education, Inc, Upper Saddle River; 1986.
[74] Spong M., Vidyasagar M. *Robot dynamics and control*. John Wiley and Sons, New Jersey; 1989.
[75] Shabana A.A. *Computational dynamics (Second Edition)*. New York: Wiley; 2001.
[76] Sciavicco L., Siciliano B. *Modelling and control of robot manipulators*. Berlin: Springer-Verlag; 1996.
[77] Khalil W., Dombre E. *Modeling, identification and control of robots*. New York, NY: Taylor & Francis; 2002.
[78] Hahn H. *Rigid body dynamics of mechanisms*. Berlin: Springer; 2002. pp. 143.
[79] Featherstone R. *Rigid body dynamics algorithms*. Berlin: Springer; 2008.
[80] B. Siciliano, O. Khatib. *Springer handbook of robotics*. Berlin: Springer; 2008.
[81] Martin P., Devasia S., Paden B. 'A different look at output tracking: Control of a VTOL aircraft'. *Automatica*. 1996; **32**: 101–107.
[82] Jankovic M., Fontaine D., Kokotovic P.V. 'TORA example: Cascade and passivity-based control designs'. *IEEE Transactions on Control Systems Technology*. 1996; **4**: 292–297.
[83] Andreev F., Auckly D., Gosavi S., Kapitanski L., Kelkar A., White W. 'Matching, linear systems, and the ball and beam'. *Automatica*. 2002; **38**: 2147–2152.
[84] Aström K.J., Klein R.E., Lennartsson A. 'Bicycle dynamics and control'. *IEEE Control Systems Magazine*. 2005; **25**: 26–47.
[85] Olfati-Saber R. *Nonlinear control of underactuated mechanical systems with application to robotics and aerospace vehicles*. Doctoral dissertation, Massachusetts Institute of Technology; 2000.

[86] Xin X., Tanaka S., She J., Yamasaki T. 'New analytical results of energy-based swing-up control for the Pendubot'. *International Journal of Non-Linear Mechanics*. 2013; **52**: 110–118.

[87] Spong M.W. 'The swing up control problem for the acrobot'. *IEEE Control Systems*. 1995; **15**(1): 49–55.

[88] Brown S.C., Passino K.M. 'Intelligent control for an acrobot'. *Journal of Intelligent and Robotic Systems: Theory and Applications*. 1997; **18**(3): 209–248.

[89] Liu G., Nesic D., Mareels I. 'Modelling and stabilisation of a spherical inverted pendulum'. *IFAC Proceedings Volumes*. 2005; **38**(1): 1130–1135.

[90] Liu G., Nesic D., Mareels I. 'Modelling and stabilisation of a spherical inverted pendulum'. *IFAC Proceedings Volumes*. 2005; **38**(1): 1130–1135.

[91] Jadlovsky S., Sarnovsky J. 'Modelling of classical and rotary inverted pendulum systems–a generalized approach'. *Journal of Electrical Engineering*. 2013; **64**(1): 12–19.

[92] Alamir M., Murilo, A. 'Swing-up and stabilization of a twin-pendulum under state and control constraints by a fast NMPC scheme'. *Automatica*. 2008; **44**(5): 1319–1324.

[93] Eltohamy K.G., Kuo C.Y. 'Nonlinear optimal control of a triple link inverted pendulum with single control input'. *International Journal of Control*. 1998; **69**(2): 239–256.

[94] Li H., Miao Z., Wang J. 'Variable universe adaptive fuzzy control on the quadruple inverted pendulum'. *Science in China, Series E: Technological Sciences*. 2002; **45**(2): 213–224.

[95] Yavin Y. 'Control of a rotary inverted pendulum'. *Applied Mathematics Letters*. 1999; **12**(2): 131–134.

[96] Gordillo F., Acosta J.A., Aracil J. 'A new swing-up law for the Furuta pendulum'. *International Journal of Control*. 2003; **76**(8): 836–844.

[97] Pathak K., Franch J., Agrawal S.K. 'Velocity and position control of a wheeled inverted pendulum by partial feedback linearization'. *IEEE Transactions on Robotics*. 2005; **21**(3): 505–513.

[98] Chan R.P.M., Stol K.A., Halkyard C.R. 'Review of modelling and control of two-wheeled robots'. *Annual Reviews in Control*. 2013; **37**(1): 89–103.

[99] Lin J., Chen S.Y., Gau W.H. 'Design and implementation of a novel inertia flywheel pendulum mechatronic kit'. *JVC/Journal of Vibration and Control*. 2015; **21**(16): 3417–3430.

[100] Feng Y., Du R., Xu Y. 'Steering angle balance control method for rider-less bicycle based on ADAMS'. *Smart Innovation, Systems and Technologies*. 2017; **53**: 15–31.

[101] Soto I., Campa R. 'Modelling and control of a spherical inverted pendulum on a five-bar mechanism'. *International Journal of Advanced Robotic Systems*. 2015; **12**(7): 95.

[102] Gawthrop P.J., McGookin E. (eds.) 'Using LEGO in control education'. *Seventh IFAC Symposium on Advances in Control Education*; Madrid, Spain, June 2006. pp. 31–38.

[103] Le T., Oh P. (eds.) 'NXT mobile inverted pendulum'. *Proceedings of the ASME Design Engineering Technical Conference and Computers and Information in Engineering Conference*; Brooklyn, New York, USA, August 2008. pp. 515–524.

[104] Bobtsov A.A., Pyrkin A.A., Kolyubin S.A. (eds.) 'Adaptive stabilization of a reaction wheel pendulum on moving LEGO platform'. *IEEE International Conference on Control Applications (CCA) & Intelligent Control (ISIC)*; St. Petersburg, Russia, July 2009. pp. 1218–1223.

[105] Kapitonov A.A., Bobtsov A.A., Kapitanyuk Y.A., *et al.* 'Course of lab activities on control theory based on the LEGO NXT'. *IFAC Proceedings Volumes*. 2014; **47**(3): 9063–9068.

[106] Canale M., Casale-Brunet S. 'A multidisciplinary approach for model predictive control education: A LEGO Mindstorms NXT-based framework'. *International Journal of Control, Automation and Systems*. 2014; **12**(5): 1030–1039.

[107] Quanser. http://www.quanser.com [Accessed March 2017].

[108] Al-Hiddabi S.A. 'Implementation of stable inversion control on a nonlinear non-minimum phase system: Cart-pendulum experiment'. *Proceedings of the SICE Annual Conference*; Okayama, Japan, August 2005. pp. 729–733.

[109] Al-Jodah A., Zargarzadeh H., Abbas M.K. (eds.) 'Experimental verification and comparison of different stabilizing controllers for a rotary inverted pendulum'. *IEEE International Conference on Control System, Computing and Engineering (ICCSCE 2013)*; Penang, Malaysia, November 2013. pp. 417–423.

[110] Siva-Kumar M., Dasu B., Ramesh G. 'Design of LQR based stabilizer for rotary inverted pendulum system'. *International Journal of Control Theory and Applications*. 2016; **9**(29): 21–29.

[111] Googol Technology Limits. http://www.googoltech.com/web/eng/product_list_inner.jsp?module=Educational%20Products&content=Content [Accessed March 2017].

[112] Fu C., Liu Y., Tan W., Zhang J. (eds.) 'μ Synthesis with regional pole assignment for a double inverted pendulum'. *Proceedings of the 29th Chinese Control Conference (CCC'10)*; Beijing, China, July 2010. pp. 1895–1900.

[113] Wu J., Zhang W., Wang S. 'A two-wheeled self-balancing robot with the fuzzy PD control method'. *Mathematical Problems in Engineering*. 2012; 13pp. doi:10.1155/2012/469491.

[114] Googol Technology. *Inverted pendulum experimental manual suitable for GLIP series (Second Edition)*. 2006.

[115] INTECO. User manual. http://ee.sharif.edu/~linearcontrol/Files/Lab/Pend_man.pdf [Accessed March 2017].

[116] Feedback. Digital pendulum. http://www.feedback-instruments.com/products/education/control_instrumentation/digital_pendulum [Accessed March 2017].

[117] Educational Control Products. http://www.ecpsystems.com/docs/ECP_InvPend_Model_505.pdf. [Accessed March 2017].

[118] Rex Controls. https://www.rexcontrols.com/furuta-pendulum [Accessed March 2017].

[119] Cafolla D., Carbone G., Ceccarelli M. 'Balancing of a 3-DOFs parallel manipulator'. in Zhang D., Wei B. (eds.) *Dynamic Balancing of Mechanisms and Synthesizing of Parallel Robots*. Switzerland: Springer; 2015. pp. 173–191.

[120] Houck A.C., Katzschmann R.K., Ramos J.L.A.S. *Furuta pendulum*. Department of Mechanical Engineering, Massachusetts Institute of Technology; 2013.

[121] Hehn M., D'Andrea R. 'A flying inverted pendulum'. *IEEE International Conference on Robotics and Automation (ICRA)*; Shanghai, China, May 2011. pp. 763–770.

[122] Hovingh A., Roon M. *Design and control of an inverted pendulum* [online]. 2007. Available from http://homepages.wmich.edu/~kamman/IP_FINAL_PROJECT_REPORT.pdf.

[123] Driver J., Thorpe D. *Design, build and control of a single/double rotational inverted pendulum*. Australia: The University of Adelaide, School of Mechanical Engineering; 2004.

[124] Awtar S., King N., Allen T., *et al.* 'Inverted pendulum systems: Rotary and arm-driven-a mechatronic system design case study'. *Mechatronics*. 2002; **12**(2): 357–370.

[125] Sham J.S., Solihin M.I., Heltha F., Muzaiyanah H. 'Modelling and simulation of an inverted pendulum system: Comparison between experiment and CAD physical model'. *ARPN Journal of Engineering and Applied Sciences*. 2015; **10**(20): 9752–9757.

[126] Antonio-Cruz M., Silva-Ortigoza R., Sandoval-Gutiérrez J., *et al.* 'Modeling, simulation, and construction of a Furuta pendulum test-bed'. *International Conference on Electronics, Communications and Computers (CONIELECOMP)*; Cholula, Puebla Mexico, February 2015. pp. 72–79.

[127] Turner M., Cooley T.R. (eds.) 'A low cost and flexible open source inverted pendulum for feedback control laboratory courses'. *122nd ASEE Annual Conference and Exposition: Making Value for Society*; Seattle, Washington, June 2015.

[128] Hahn H.H., Spong M.W. (eds.) 'Remote laboratories for control education'. *Proceedings of the IEEE Conference on Decision and Control*; Darling Harbour, Australia, December 2000. pp. 895–900.

[129] Ling K., Lai Y., Chew K. (eds.) 'An online Internet laboratory for control experiments'. *IFAC/IEEE Symposium on Advances in Control Education*; Queensland, Australia, December 2000. pp. 173–176.

[130] Zhao J., Spong M.W. 'Hybrid control for global stabilization of the cart-pendulum system'. *Automatica*. 2001; **37**(12): 1941–1951.

[131] Chung C.C., Hauser J. 'Nonlinear control of a swinging pendulum'. *Automatica*. 1995; **31**(6): 851–862.

[132] Muskinja N., Tovornik B. 'Swinging up and stabilization of a real inverted pendulum'. *IEEE Transaction on Industrial Electronics*. 2006; **53**(2): 631–639.

[133] Kheir N.A., Astrom K.J., Auslander D., *et al.* 'Control systems engineering education'. *Automatica*. 1996; **32**(2): 147–166.
[134] Mori S., Nishihara H., Furuta K. 'Control of unstable mechanical systems: Control of pendulum'. *International Journal of Control*. 1976; **23**(5): 673–69.
[135] Qian D., Yi J., Zhao D. (eds.) 'Hierarchical sliding mode control to swing up a pendubot'. *Proceedings of the American Control Conference*; New York, July 2007. pp. 5254–5259.
[136] Xin X., Tanaka S., She J., Yamasaki T. 'New analytical results of energy-based swing-up control for the Pendubot'. *International Journal of Non-Linear Mechanics*. 2013; **52**: 110–118.
[137] Lai X.Z., She J.H., Yang S.X., Wu M. 'Comprehensive unified control strategy for underactuated two-link manipulators'. *IEEE Transactions on Systems, Man, and Cybernetics, Part B (Cybernetics)*. 2009; **39**(2): 389–398.
[138] Zhang M., Tarn T.J. 'Hybrid control of the Pendubot'. *IEEE/ASME Transactions on Mechatronics*. 2002; **7**(1): 79–86.
[139] Begovich O., Sanchez E.N., Maldonado M. 'Takagi–Sugeno fuzzy scheme for real-time trajectory tracking of an underactuated robot'. *IEEE Transactions on Control Systems Technology*. 2002; **10**(1): 14–20.
[140] Eom M., Chwa D. 'Robust swing-up and balancing control using a nonlinear disturbance observer for the pendubot system with dynamic friction'. *IEEE Transactions on Robotics*. 2015; **31**(2): 331–343.
[141] Mahindrakar A.D., Banavar R.N. 'A swing-up of the acrobot based on a simple pendulum strategy'. *International Journal of Control*. 2005; **78**(6): 424–429.
[142] Xin X., Kaneda M. 'Analysis of the energy-based swing-up control of the Acrobot'. *International Journal of Robust and Nonlinear Control*. 2007; **17**(16): 1503–1524.
[143] Horibe T., Sakamoto N. (eds.) 'Swing up and stabilization of the Acrobot via nonlinear optimal control based on stable manifold method'. *IFAC Symposium on Nonlinear Control Systems*; California, USA, August 2016. pp. 374–379.
[144] Berkemeier M.D., Fearing R.S. 'Tracking fast inverted trajectories of the underactuated Acrobot'. *IEEE Transactions on Robotics and Automation*. 1999; **15**(4): 740–750.
[145] Zhang A., She J., Lai X., Wu M. 'Motion planning and tracking control for an acrobot based on a rewinding approach'. *Automatica*. 2013; **49**(1): 278–284.
[146] Anderle M., Celikovsky S. (eds.) 'High gain observer for embedded Acrobot'. *IFAC Proceedings Volumes*; Cape Town, South Africa, August 2014. pp. 2818–2823.
[147] Jeong S., Lee S., Hong Y.D., Chwa D. 'Adaptive robust swing-up and balancing control of acrobot using a fuzzy disturbance observer'. *Journal of Institute of Control, Robotics and Systems*. 2016; **22**(5): 346–352.
[148] Srinivas K.N., Behera L. 'Swing-up control strategies for a reaction wheel pendulum'. *International Journal of Systems Science*. 2008; **39**(12): 1165–1177.
[149] Spong M.W., Corke P., Lozano R. 'Nonlinear control of the reaction wheel pendulum'. *Automatica*. 2001; **37**(11): 1845–1851.

[150] Estrada A., Aguilar L.T., Iriarte R., Fridman L. 'Two relay controller for real time trajectory generation and its application to inverted orbital stabilization of inertia wheel pendulum via quasi-continuous HOSM'. *Asian Journal of Control*. 2012; **14**(1): 58–66.

[151] Iriarte R., Aguilar L.T., Fridman L. 'Second order sliding mode tracking controller for inertia wheel pendulum'. *Journal of the Franklin Institute*. 2013; **350**(1): 92–106.

[152] Moreno-Valenzuela J., Aguilar-Avelar C., Puga-Guzman S., Santibanez V. 'Two adaptive control strategies for trajectory tracking of the inertia wheel pendulum: Neural networks vis à vis model regressor'. *Intelligent Automation and Soft Computing*. 2017; **23**(1): 63–73.

[153] Mohammadzadeh A., Hashemzadeh F. 'A new robust observer-based adaptive type-2 fuzzy control for a class of nonlinear systems'. *Applied Soft Computing Journal*. 2015; **37**: 204–216.

[154] Khalid N., Memon A.Y. 'Output feedback control of a class of under-actuated nonlinear systems using extended high gain observer'. *Arabian Journal for Science and Engineering*. 2016; **41**(9): 3531–3542.

[155] Bradshaw A., Shao J. 'Swing-up control of inverted pendulum systems'. *Robotica*. 1996; **14**(4): 397–405.

[156] Mason P., Broucke M., Piccoli B. 'Time optimal swing-up of the planar pendulum'. *IEEE Transactions on Automatic Control*. 2008; **53**(8): 1876–1886.

[157] Yang J.H., Shim S.Y., Seo J.H., Lee Y.S. 'Swing-up control for an inverted pendulum with restricted cart rail length'. *International Journal of Control, Automation and Systems*. 2009; **7**(4): 674–680.

[158] Dındış G., Karamancıoglu A. 'A self-learning robust swing-up algorithm'. *Transactions of the Institute of Measurement and Control*. 2016; **38**(4): 395–401.

[159] Shiriaev A., Ludvigsen H., Egeland O., Pogromsky A. (eds.) 'On global properties of passivity based control of the inverted pendulum'. *IEEE Conference on Decision and Control*; Arizona, USA, December 1999. pp. 2513–2518.

[160] Lozano R., Fantoni I., Block D.J. 'Stabilization of the inverted pendulum around its homoclinic orbit'. *Systems & Control Letters*. 2000; **40**(3): 197–204.

[161] Angeli D. 'Almost global stabilization of the inverted pendulum via continuous state feedback'. *Automatica*. 2001; **37**(7): 1103–1108.

[162] Shiriaev A.S., Egeland O., Ludvigsen H., Fradkov A.L. 'VSS-version of energy-based control for swinging up a pendulum'. *Systems & Control Letters*. 2001; **44**(1): 45–56.

[163] Chatterjee D., Patra A., Joglekar H.K. 'Swing-up and stabilization of a cart–pendulum system under restricted cart track length'. *Systems & Control Letters*. 2002; **47**(4): 355–364.

[164] Bugeja M. 'Non-linear swing-up and stabilizing control of an inverted pendulum system'. *The IEEE Region 8 EUROCON 2003. Computer as a Tool*; Ljubljana, Slovenia, September 2003. pp. 437–441.

[165] Ibañez C.A., Frias O.G., Castañón M.S. 'Lyapunov-based controller for the inverted pendulum cart system'. *Nonlinear Dynamics*. 2005; **40**(4): 367–374.

[166] Muskinja N., Tovornik B. 'Swinging up and stabilization of a real inverted pendulum'. *IEEE Transaction on Industrial Electronics*. 2006; **53**(2): 631–639.

[167] Alamir M. 'A low dimensional contractive NMPC scheme for nonlinear systems stabilization: Theoretical framework and numerical investigation on relatively fast systems'. *Lecture Notes in Control and Information Sciences*. 2007; **358**(1): 523–535.

[168] Gordillo F., Aracil J. 'A new controller for the inverted pendulum on a cart'. *International Journal of Robust and Nonlinear Control*. 2008; **18**(17): 1607–1621.

[169] Siuka A., Schoberl M. 'Applications of energy based control methods for the inverted pendulum on a cart'. *Robotics and Autonomous Systems*. 2009; **57**: 1012–1017.

[170] Park M.S., Chwa, D. 'Swing-up and stabilization control of inverted-pendulum systems via coupled sliding-mode control method'. *IEEE Transactions on Industrial Electronics*. 2009; **56**(9): 3541–3555.

[171] Srinivasan B., Huguenin P., Bonvin D. 'Global stabilization of an inverted pendulum–Control strategy and experimental verification'. *Automatica*. 2009; **45**(1): 265–269.

[172] Adhikary N., Mahanta C. 'Integral backstepping sliding mode control for underactuated systems: Swing-up and stabilization of the Cart–Pendulum System'. *ISA Transactions*. 2013; **52**(6): 870–880.

[173] Psillakis H.E., Alexandridis A.T. 'Inverted pendulum angle tracking control subject to uncertainties and stochastic perturbations'. *IFAC Proceedings Volumes*. 2006; **39**(15): 456–461.

[174] Yu H., Liu Y., Yang T. 'Closed-loop tracking control of a pendulum-driven cart-pole underactuated system'. *Proceedings of the Institution of Mechanical Engineers, Part I: Journal of Systems and Control Engineering*. 2008; **222**(2): 109–125.

[175] Llama M.A., De La Torre W., Jurado F., Garcia-Hernandez R. 'Robust Takagi–Sugeno fuzzy dynamic regulator for trajectory tracking of a pendulum-cart system'. *Mathematical Problems in Engineering* [online]. 2015. Available from http://dx.doi.org/10.1155/2015/247682.

[176] Wu Q.H., Jiang L., Wen J.Y. 'Decentralized adaptive control of interconnected non-linear systems using high gain observer'. *International Journal of Control*. 2004; **77**(8): 703–712.

[177] Lee J., Mukherjee R., Khalil H.K. 'Output feedback stabilization of inverted pendulum on a cart in the presence of uncertainties'. *Automatica*. 2015; **54**: 146–157.

[178] Xie W.B., Wang Y.L., Zhang J., Huang L. 'Membership function dependent observer-based controller design for A T-S fuzzy system'. *Asian Journal of Control*. 2017 (Article in Press).

[179] Alamir M., Murilo A. 'Swing-up and stabilization of a twin-pendulum under state and control constraints by a fast NMPC scheme'. *Automatica*. 2008; **44**(5): 1319–1324.

[180] Yamakita M., Nonaka K., Furuta K. (eds.) 'Swing up control of a double pendulum'. *American Control Conference*; San Francisco, California, June 1993. pp. 2229–2232.

[181] Guckenheimer J. 'A robust hybrid stabilization strategy for equilibria'. *IEEE Transactions on Automatic Control*. 1995; **40**(2): 321–326.

[182] Yi J., Yubazaki N., Hirota K. 'Stabilization control of series-type double inverted pendulum systems using the SIRMs dynamically connected fuzzy inference model'. *Artificial Intelligence in Engineering*. 2001; **15**(3): 297–308.

[183] Henmi T., Deng M., Inoue A. (eds.) 'Swing-up control of a serial double inverted pendulum'. *Proceedings of the IEEE American Control Conference*; Boston, MA, USA, June 2004. pp. 3992–3997.

[184] Inoue A., Deng M.C., Henmi T., Ueki N., Hirashima Y. 'Swing-up controller design for cart-type double inverted pendulum'. *Control Theory and Applications*. 2004; **21**(5): 709–716.

[185] Graichen K., Treuer M., Zeitz M. 'Swing-up of the double pendulum on a cart by feedforward and feedback control with experimental validation'. *Automatica*. 2007; **43**(1): 63–71.

[186] Tao C.W., Taur J., Chang J.H., Su S.F. 'Adaptive fuzzy switched swing-up and sliding control for the double-pendulum-and-cart system'. *IEEE Transactions on Systems, Man, and Cybernetics, Part B (Cybernetics)*. 2010; **40**(1): 241–252.

[187] Formal'skii A.M. 'Global stabilization of a double inverted pendulum with control at the hinge between the links'. *Mechanics of Solids*. 2008; **43**(5): 687–697.

[188] Yang W., Wu J., Yang S., Tao Y. 'Adaptive NN tracking control of double inverted pendulums with input saturation'. in Guo C., Hou ZG., Zeng Z. (eds.) *Advances in Neural Networks. Lecture Notes in Computer Science*. Berlin, Heidelberg: Springer; 2013 (vol. 7952). pp. 147–154.

[189] Hassanzadeh I., Nejadfard A., Zadi M. 'A multivariable adaptive control approach for stabilization of a cart-type double inverted pendulum'. *Mathematical Problems in Engineering*. 2011 (2011): Article ID 970786, 14 pp.

[190] Zhang Y.L., Cheng H.F., Li H.X. 'The swing-up and stabilization of the triple inverted pendulum'. *Control Theory and Applications*. 2011; **28**(1): 37–45.

[191] Medrano-Cerda G.A. 'Robust stabilization of a triple inverted pendulum-cart'. *International Journal of Control*. 1997; **68**(4): 849–865.

[192] Sehgal S., Tiwari S. (eds.) 'LQR control for stabilizing triple link inverted pendulum system'. *Second International Conference on Power, Control and Embedded Systems (ICPCES 2012)*; Motilal Nehru National Institute of Technology Allahabad, India, December 2012. doi:10.1109/ICPCES.2012.6508052.

[193] Gluck T., Eder A., Kugi A. 'Swing-up control of a triple pendulum on a cart with experimental validation'. *Automatica*. 2013; **49**(3): 801–808.
[194] Zhang X.L., Fan H.M., Zang J.Y., Zhao L., Hao S. 'Nonlinear control of triple inverted pendulum based on GA–PIDNN". *Nonlinear Dynamics*. 2014; **79**(2): 1185–1194.
[195] Zhang X.L., Fan H.M., Zang J.Y., Zhao L., Hao S. 'The stabilization and 3D visual simulation of the triple inverted pendulum based on CGA-PIDNN'. *International Journal of Control, Automation and Systems*. 2015; **13**(4): 1010–1019.
[196] Tsachouridis V.A., Medrano-Cerda G.A. 'Discrete-time H∞ control of a triple inverted pendulum with single control input'. *IEE Proceedings: Control Theory and Applications*. 1999; **146**(6): 567–576.
[197] Medrano-Cerda G.A. 'Robust stabilization of a triple inverted pendulum-cart'. *International Journal of Control*. 1997; **68**(4): 849–865.
[198] Molazadeh V.R., Banazadeh A., Shafieenejad I. (eds.) 'Design of the LQR controller and observer with fuzzy logic GA and GA-PSO algorithm for triple an inverted pendulum and cart system'. *International Conference on Advanced Mechatronic Systems (ICAMechS)*; Kumamoto, Japan, August 2014. pp. 295–300.
[199] Gordillo F., Acosta J.A., Aracil J. 'A new swing-up law for the Furuta pendulum'. *International Journal of Control*. 2003; **76**(8): 836–844.
[200] Ibanez C.A., Azuela J.H.S. 'Stabilization of the Furuta pendulum based on a Lyapunov function'. *Nonlinear Dynamics*. 2007; **49**(1): 1–8.
[201] Ibanez, C.A., Ramirez, H.S. 'A linear differential flatness approach to controlling the Furuta pendulum'. *IMA Journal of Mathematical Control and Information*. 2007; **24**(1): 31–45.
[202] Acosta J.A., Aracil J., Gordillo F. 'Nonlinear control strategies for the Furuta pendulum'. *Control and Intelligent Systems*. 2001; **29**(3): 101–107.
[203] Noguchi K., Izutsu M., Kamamichi N., Shiotsuki T., Ishikawa J., Furuta K. (eds.) 'Swinging up and stabilization control of double Furuta pendulums by safe manual control'. *IEEE/RSJ International Conference on Intelligent Robots and Systems (IROS 2009)*; St. Louis, MO, USA, October 2009. pp. 4232–4237.
[204] Ramirez-Neria M., Sira-Ramirez H., Garrido-Moctezuma R., Luviano-Juarez A. 'Linear active disturbance rejection control of underactuated systems: The case of the Furuta pendulum'. *ISA Transactions*. 2014; **53**(4): 920–928.
[205] Aguilar-Avelar C., Moreno-Valenzuela J. 'A composite controller for trajectory tracking applied to the Furuta pendulum'. *ISA Transactions*. 2015; **57**: 286–294.
[206] Moreno-Valenzuela J., Aguilar-Avelar C., Puga-Guzman S.A., Santibanez V. 'Adaptive neural network control for the trajectory tracking of the Furuta pendulum'. *IEEE Transactions on Cybernetics*. 2016; **46**(12): 3439–3452.
[207] Aguilar-Avelar C., Moreno-Valenzuela J. 'New feedback linearization-based control for arm trajectory tracking of the Furuta pendulum'. *IEEE/ASME Transactions on Mechatronics*. 2016; **21**(2): 638–648.

[208] Lenain R., Robertsson A., Johansson R., Shiriaev A., Berducat M. (eds.) 'A velocity observer based on friction adaptation'. *IEEE International Conference on Robotics and Automation*; Pasadena, California, May 2008. pp. 3365–3370.

[209] Stamnes O.N., Aamo O.M., Kaasa G.O. 'A constructive speed observer design for general Euler–Lagrange systems'. *Automatica*. 2011; **47**(10): 2233–2238.

[210] Shiriaev A.S., Ludvigsen H., Egeland O. 'Swinging up the spherical pendulum via stabilization of its first integrals'. *Automatica*. 2004; **40**(1): 73–85.

[211] Shiriaev A.S., Freidovich L.B., Gusev S.V. 'Transverse linearization for controlled mechanical systems with several passive degrees of freedom'. *IEEE Transactions on Automatic Control*. 2010; **55**(4): 893–906.

[212] Consolini L., Tosques M. 'On the exact tracking of the spherical inverted pendulum via an homotopy method'. *Systems and Control Letters*. 2009; **58**(1): 1–6.

[213] Ping Z. 'Tracking problems of a spherical inverted pendulum via neural network enhanced design'. *Neurocomputing*. 2013; **106**: 137–147.

[214] Ma X.J., Sun Z.Q., He Y.Y. 'Analysis and design of fuzzy controller and fuzzy observer'. *IEEE Transactions on Fuzzy Systems*, 1998; **6**(1): 41–51.

[215] Anderson C.W. 'Learning to control an inverted pendulum using neural networks'. *IEEE Control Systems Magazine*. 1989; **9**(3): 31–37.

[216] Chang, W.D., Hsieh, J.G., Hwang, R.C. 'A self-tuning PID control for a class of nonlinear systems based on the Lyapunov approach'. *Journal of Process Control*. 2002; **12**(2): 233–242.

[217] Eltohamy K.G., Kuo C.Y. 'Nonlinear optimal control of a triple link inverted pendulum with single control input'. *International Journal of Control*. 1998; **69**(2): 239–256.

[218] Yang C., Li Z., Li J. 'Trajectory planning and optimized adaptive control for a class of wheeled inverted pendulum vehicle models'. *IEEE Transactions on Cybernetics*. 2013; **43**(1): 24–36.

[219] Huang J., Guan Z.H., Matsuno T., Fukuda T., Sekiyama K. 'Sliding-mode velocity control of mobile-wheeled inverted-pendulum systems'. *IEEE Transactions on Robotics*. 2010; **26**(4): 750–758.

[220] Ferreira A., Bejarano F.J., Fridman L.M. 'Robust control with exact uncertainties compensation: With or without chattering'. *IEEE Transactions on Control Systems Technology*. 2011; **19**(5): 969–975.

[221] Ebrahim A., Murphy G.V. (eds.) 'Adaptive backstepping controller design of an inverted pendulum'. *Proceedings of the Annual Southeastern Symposium on System Theory*; Tuskegee, AL, USA, March 2005. pp. 172–174.

[222] Zhao J., Spong M.W. 'Hybrid control for global stabilization of the cart-pendulum system'. *Automatica*. 2001; **37**(12): 1941–1951.

[223] Alamir M., Murilo A. 'Swing-up and stabilization of a twin-pendulum under state and control constraints by a fast NMPC scheme'. *Automatica*. 2008; **44**(5): 1319–1324.

[224] Graichen K., Treuer M., Zeitz M. 'Swing-up of the double pendulum on a cart by feedforward and feedback control with experimental validation'. *Automatica*. 2007; **43**(1): 63–71.

[225] Garcia-Alarcon O., Puga-Guzman S., Moreno-Valenzuela J. 'On parameter identification of the Furuta pendulum'. *Procedia Engineering*. 2012; **35**: 77–84.

[226] Bicakci S., Akdas D., Deniz Karaoglan A. 'Optimizing Karnopp friction model parameters of a pendulum using RSM'. *European Journal of Control*. 2014; **20**(4): 180–187.

[227] Wang B., Hou Z.X., Gao X.Z., Shan S.Q. 'The application of inertial measurement unit in inertia parameter identification'. *Advanced Materials Research*. 2011; **201–203**: 974–981.

[228] Ronquillo-Lomeli G., Rios-Moreno, G.J., Gomez-Espinosa A., Morales-Hernandez L.A., Trejo-Perea M. 'Nonlinear identification of inverted pendulum system using Volterra polynomials'. *Mechanics Based Design of Structures and Machines*. 2016; **44**(1–2): 5–15.

[229] Davila J., Fridman L., Levant A. 'Second-order sliding-mode observer for mechanical systems'. *IEEE Transactions on Automatic Control*. 2005; **50**(11): 1785–1789.

[230] Moreno J.A., Osorio M. (eds.) 'A Lyapunov approach to second-order sliding mode controllers and observers'. *Proceedings of the IEEE Conference on Decision and Control*; Cancun, Mexico, December 2008. pp. 2856–2861.

[231] Moreno J.A., Osorio M. 'Strict Lyapunov functions for the super-twisting algorithm'. *IEEE Transactions on Automatic Control*. 2012; **57**(4): 1035–1040.

[232] Salgado I., Kamal S., Bandyopadhyay B., Chairez I., Fridman L. 'Control of discrete time systems based on recurrent super-twisting-like algorithm'. *ISA Transactions*. 2016; **64**: 47–55.

[233] Guzman E., Moreno J.A. 'Super-twisting observer for second-order systems with time-varying coefficient'. *IET Control Theory and Applications*. 2015; **9**(4): 553–562.

[234] Chalanga A., Kamal S., Fridman L.M., Bandyopadhyay B., Moreno J.A. 'Implementation of super-twisting control: Super-twisting and higher order sliding-mode observer-based approaches'. *IEEE Transactions on Industrial Electronics*. 2016; **63**(6): 3677–3685.

[235] Mobayen, S., Tchier, F., Ragoub, L. 'Design of an adaptive tracker for n-link rigid robotic manipulators based on super-twisting global nonlinear sliding mode control'. *International Journal of Systems Science*. 2017; 48(9): 1990–2002 (Article in Press): 1–13.

[236] Mansouri B., Manamanni N., Guelton K., Kruszewski A., Guerra T.M. 'Output feedback LMI tracking control conditions with H∞ criterion for uncertain and disturbed T–S models'. *Information Sciences*. 2009; **179**(4): 446–457.

[237] Muhammad M., Buyamin S., Ahmad M.N., Nawawi S.W. 'Takagi–Sugeno fuzzy modeling of a two-wheeled inverted pendulum robot'. *Journal of Intelligent and Fuzzy Systems*. 2013; **25**(3): 535–546.

[238] Xu J.X., Guo Z.Q., Lee T.H. 'Design and implementation of a Takagi–Sugeno-type fuzzy logic controller on a two-wheeled mobile robot'. *IEEE Transactions on Industrial Electronics*. 2013; **60**(12): 5717–5728.

[239] Chiu C.S. 'A dynamic decoupling approach to robust T–S fuzzy model-based control'. *IEEE Transactions on Fuzzy Systems*. 2014; **22**(5): 1088–1100.

[240] Tsai P.W., Chen C.W. 'A novel criterion for nonlinear time-delay systems using LMI fuzzy Lyapunov method'. *Applied Soft Computing Journal*. 2014; **25**: 461–472.

[241] Llama M.A., De La Torre W., Jurado F., Garcia-Hernandez R. 'Robust Takagi–Sugeno fuzzy dynamic regulator for trajectory tracking of a pendulum-cart system'. *Mathematical Problems in Engineering* [online]. 2015. Available from http://dx.doi.org/10.1155/2015/247682.

[242] Blanke M., Kinnaert M., Lunze J., Staroswiecki M. *Diagnosis and fault-tolerant control (Third Edition)*. Berlin Heidelberg: Springer; 2016.

[243] Schulte H., Poschke F. (eds.) 'H∞ criteria for robust actuator fault reconstruction for nonlinear systems in Takagi–Sugeno's form using sliding modes'. *23rd Mediterranean Conference on Control and Automation (MED 2015)*; Torremolinos, Spain, June 2015. pp. 224–230.

[244] Chen Z., Ding S.X., Luo H., Zhang K. 'An alternative data-driven fault detection scheme for dynamic processes with deterministic disturbances'. *Journal of the Franklin Institute*. 2017; **354**(1): 556–570.

[245] Faraji-Niri M., Jahed-Motlagh M.R., Barkhordari-Yazdi M. 'Stabilization of active fault-tolerant control systems by uncertain non homogeneous Markovian jump models'. *Complexity*. 2016; **21**: 318–329.

[246] Bettayeb M., Boussalem C., Mansouri R., Al-Saggaf U.M. 'Stabilization of an inverted pendulum-cart system by fractional PI-state feedback'. *ISA Transactions*. 2014; **53**(2): 508–516.

[247] Zhong G., Deng H., Li J. 'Chattering-free variable structure controller design via fractional calculus approach and its application'. *Nonlinear Dynamics*. 2015; **81**(1–2): 679–694.

[248] Bouafoura M.K., Braiek N.B. 'A state feedback control design for generalized fractional systems through orthogonal functions: Application to a fractional inverted pendulum'. *Asian Journal of Control*. 2013; **15**(3): 773–782.

[249] Mishra S.K., Chandra D. 'Stabilization and tracking control of inverted pendulum using fractional order PID controllers'. *Journal of Engineering (United States)* [online]. 2014. Available from http://dx.doi.org/10.1155/2014/752918.

[250] Durand S., Castellanos J.F.G., Marchand N., Sanchez W.F.G. 'Event-based control of the inverted pendulum: Swing up and stabilization'. *Journal of Control Engineering and Applied Informatics*. 2013; **15**(3): 96–104.

[251] Aranda-Escolastico E., Guinaldo M., Gordillo F., Dormido S. 'A novel approach to periodic event-triggered control: Design and application to the inverted pendulum'. *ISA Transactions*. 2016; **65**: 327–338.

[252] Mahmoud M.S., Memon A.M. 'A periodic triggering mechanisms for networked control systems'. *Information Sciences*. 2015; **296**: 282–306.

[253] Etienne L., Khaled Y., Gennaro S.D., Barbot J.P. 'Asynchronous event-triggered observation and control of linear systems via impulsive observers'. *Journal of the Franklin Institute*. 2017; **354**(1): 372–391.

[254] Fridman E., Shaikhet L. 'Delay-induced stability of vector second-order systems via simple Lyapunov functionals'. *Automatica*. 2016; **74**: 288–296.

[255] Xu Q., Stepan G., Wang Z. 'Balancing a wheeled inverted pendulum with a single accelerometer in the presence of time delay'. *Journal of Vibration and Control*. 2017; **23**(4): 604–614.

[256] Ghommam J., Mnif F. 'Predictor-based control for an inverted pendulum subject to networked time delay'. *ISA Transactions*. 2017; **67**: 306–316.

[257] Li Z., Bai Y., Huang C., Yan H. 'Further results on stabilization for interval time-delay systems via new integral inequality approach'. *ISA Transactions*. 2017; 68: 170–180.

[258] Zhou Y., Wang Z. 'A robust optimal trajectory tracking control for systems with an input delay'. *Journal of the Franklin Institute*. 2016; **353**(12): 2627–2649.

[259] Shaw S.W., Rand R.H. 'The transition to chaos in a simple mechanical system'. *International Journal of Non-Linear Mechanics*. 1989; **24**(1): 41–56.

[260] Shaw J., Shaw S.W. 'Onset of chaos in a two-degree-of-freedom impacting system'. *Journal of Applied Mechanics*. 1989; **56**(1): 168–174.

[261] Du Z., Li Y., Shen J., Zhang W. 'Impact oscillators with homoclinic orbit tangent to the wall'. *Physica D: Nonlinear Phenomena*. 2013; **245**(1): 19–33.

[262] Sieber J., Krauskopf B. 'Bifurcation analysis of an inverted pendulum with delayed feedback control near a triple-zero eigenvalue singularity'. *Nonlinearity*. 2004; **17**(1): 85–103.

[263] Landry M., Campbell S.A., Morris K., Aguilar C.O. 'Dynamics of an inverted pendulum with delayed feedback control'. *SIAM Journal on Applied Dynamical Systems*. 2005; **4**(2): 333–351.

[264] Yang R., Peng Y., Song Y. 'Stability and Hopf bifurcation in an inverted pendulum with delayed feedback control'. *Nonlinear Dynamics*. 2013; **73**(1–2): 737–749.

[265] Wang Z., Hu H. 'Stability and bifurcation of delayed dynamic systems: From theory to application'. *Advances in Mechanics*. 2013; **43**(1): 3–20.

[266] Boussaada I., Morerescu I.C., Niculescu S.I. 'Inverted pendulum stabilization: Characterization of codimension-three triple zero bifurcation via multiple delayed proportional gains'. *Systems and Control Letters*. 2015; **82**: 1–9.

[267] Campbell S.A., Crawford S., Morris K. 'Friction and the inverted pendulum stabilization problem'. *Journal of Dynamic Systems, Measurement and Control*. 2008; **130**(5): 7 pp.

[268] Santiesteban R., Floquet T., Orlov Y., Riachy S., Richard J.P. 'Second-order sliding mode control of underactuated mechanical systems II: Orbital stabilization of an inverted pendulum with application to swing up/balancing control'. *International Journal of Robust and Nonlinear Control*. 2008; **18**(4–5): 544–556.

[269] Aguilar L., Boiko I., Fridman L., Freidovich L. 'Generating oscillations in inertia wheel pendulum via two relay controller'. *International Journal of Robust and Nonlinear Control.* 2012; **22**(3): 318–330.

[270] Aguilar L., Boiko I., Fridman L., Iriarte R. 'Generating self-excited oscillations for underactuated mechanical systems via two relay controller'. *International Journal of Control.* 2009; **82**(9): 1678–1691.

[271] Aguilar L., Boiko I., Fridman L., Iriarte R. 'Generating self-excited oscillations via two-relay controller'. *IEEE Transactions on Automatic Control.* 2009; **54**(2): 416–420.

[272] Oriolo G., Nakamura Y. (eds.) 'Control of mechanical systems with second-order nonholonomic constraints: Underactuated manipulators'. *Proceedings of the IEEE Conference on Decision and Control*; Brighton, UK, December 1991. pp. 2398–2403.

[273] Spong M. (ed.) 'Energy based control of a class of underactuated mechanical system'. *13th IFAC World Congress*; San Francisco, USA, June 1996. pp. 431–436.

[274] Spong M. 'Underactuated mechanical systems'. in Siciliano B., Valavanis K.P. (eds.) *Control Problems in Robotics and Automation.* London: Springer-Verlag; 1998. pp.135–150.

[275] Reyhanoglu M., Van Der Schaft A., McClamroch N.H., Kolmanovsky I. 'Dynamics and control of a class of underactuated mechanical systems'. *IEEE Transactions on Automatic Control.* 1999; **44**(9): 1663–1671.

[276] Ortega R., Spong M.W., Gomez-Estern F., Blankenstein G. 'Stabilization of a class of underactuated mechanical systems via interconnection and damping assignment'. *IEEE Transactions on Automatic Control.* 2002; **47**(8): 1218–1233.

[277] Riachy S., Orlov Y., Floquet T., Santiesteban R., Richard J.P. 'Second-order sliding mode control of underactuated mechanical systems I: Local stabilization with application to an inverted pendulum'. *International Journal of Robust and Nonlinear Control.* 2008; **18**(4–5): 529–543.

[278] Liu Y., Yu H. 'A survey of underactuated mechanical systems'. *IET Control Theory & Applications.* 2013; **7**(7): 921–935.

[279] Andary S., Chemori A., Krut S. 'Control of the underactuated inertia wheel inverted pendulum for stable limit cycle generation'. *Advanced Robotics.* 2009; **23**(15): 1999–2014.

[280] Choukchou-Braham A., Cherki B., Djemai M., Busawon K. *Analysis and control of underwatcuated mechanical systems.* London, UK: Springer; 2014.

[281] He G., Chen R., Zhang Y. 'Globally stabilizing a class of underactuated mechanical systems on the basis of finite-time stabilizing observer'. *Journal of Intelligent and Robotic Systems: Theory and Applications.* June 2017; **86**(3–4): 353–366.

[282] *Segway Human Transporter.* 2004. Available from http://www.segway.com [Accessed 29 March 2017].

[283] Nguyen H.G., Morrell J., Mullens K., *et al.* (eds.) 'Segway robotic mobility platform'. *Proceedings of SPIE—The International Society for Optical Engineering*; USA, February 2004. pp. 207–220.

[284] Grasser F., D'Arrigo A., Colombi S., Rufer A.C. 'JOE: A mobile, inverted pendulum'. *IEEE Transactions on Industrial Electronics*. 2002; **49**(1): 107–114.

[285] Huang J., Ding F., Fukuda T., Matsuno T. 'Modeling and velocity control for a novel narrow vehicle based on mobile wheeled inverted pendulum'. *IEEE Transactions on Control Systems Technology*. 2013; **21**(5): 1607–1617.

[286] Sasaki M., Yanagihara N., Matsumoto O., Komoriya K. (eds.) 'Forward and backward motion control of personal riding-type wheeled mobile platform'. *IEEE International Conference on Robotics and Automation*; New Orleans, LA, USA, April 2004. pp. 3331–3336.

[287] Vermeiren L., Dequidt A., Guerra T.M., Rago-Tirmant H., Parent M. 'Modeling, control and experimental verification on a two-wheeled vehicle with free inclination: An urban transportation system'. *Control Engineering Practice*. 2011; **19**(7): 744–756.

[288] Jeong S., Takahashi T. 'Wheeled inverted pendulum type assistant robot: Design concept and mobile control'. *Intelligent Service Robotics*. 2008; **1**(4): 313–320.

[289] Lauwers T., Kantor G., Hollis R. (eds.) 'A dynamically stable single-wheeled mobile robot with inverse mouse-ball drive'. *IEEE International Conference on Robotics and Automation*; Orlando, FL, USA, May 2006. pp. 2889–2895.

[290] Kumagai M., Ochiai T. (eds.) 'Development of a robot balancing on a ball'. *International Conference on Control, Automation and Systems (ICCAS 2008)*; Seoul, Korea, October. 2008. pp. 433–438.

[291] Villacres J., Viscaino M., Herrera M., Camacho O. 'Real-time implementation of different controllers for a two-wheeled inverted pendulum'. *International Journal of Circuits, Systems and Signal Processing*. 2016; **10**: 281–290.

[292] Kajita S., Kanehiro F., Kaneko K., Yokoi K., Hirukawa H. (eds.) 'The 3D linear inverted pendulum model: A simple modeling for a biped walking pattern generator'. *IEEE/RSJ International Conference on Intelligent Robots and Systems (IROS)*; Maui, USA, October 2001. pp. 239–246.

[293] Kajita S., Kanehiro F., Kaneko K., *et al.* (eds.) 'Biped walking pattern generation by using preview control of zero-moment point'. *IEEE International Conference on Robotics and Automation (ICRA)*; Taipei, China, September 2003. pp. 1620–1626.

[294] Sugihara T., Nakamura Y. (eds.) 'Variable impedance inverted pendulum model control for a seamless contact phase transition on humanoid robot'. *IEEE International Conference on Humanoid Robots (Humanoids2003)*, Karlsruhe, Germany, September 2003.

[295] Ghigliazza R.M., Altendorfer R., Holmes P., Koditschek D. 'A simply stabilized running model'. *SIAM Journal on Applied Dynamical Systems*. 2004; **2**(2): 187–218.

[296] Siravuru A., Viswanathan S.P., Sreenath K., Sanyal A.K. 'The reaction mass biped: Geometric mechanics and control'. *Journal of Intelligent and Robotic Systems: Theory and Applications*. 2017. doi:10.1007/s10846-017-0508-7.
[297] Kajita S., Morisawa M., Miura K., *et al.* (eds.) 'Biped walking stabilization based on linear inverted pendulum tracking.' *Proceedings of the IEEE/RSJ 2010 International Conference on Intelligent Robots and Systems*; Taipei, Taiwan, October 2010. pp. 4489–449.
[298] Shiguematsu Y.M., Kryczka P., Hashimoto K., Lim H.O., Takanishi A. 'Heel-contact toe-off walking pattern generator based on the linear inverted pendulum'. *International Journal of Humanoid Robotics*. 2016; **13**(1): 25 pp.
[299] Luo J.W., Fu Y.L., Wang S.G. '3D stable biped walking control and implementation on real robot'. Advanced Robotics. 2017; 1–16.
[300] Huang H.W., Ju M.S., Lin C.C.K. 'Flexor and extensor muscle tone evaluated using the quantitative pendulum test in stroke and Parkinsonian patients'. *Journal of Clinical Neuroscience*. 2016; **27**: 48–52.

Part I

Robust state estimation and control: application to pendulum-cart systems

Chapter 2

State estimation and parameter identification via sliding-mode techniques: pendulum–cart system

Diego Gutiérrez[1] and Héctor Ríos[2]

In this chapter, the problems of state estimation and parameter identification for the pendulum–cart system are addressed. Different high-order sliding-modes techniques are applied for such a mechanical system. The mathematical model is studied and a couple of high-order sliding-modes observers are proposed to estimate the state, in spite of disturbances; exactly and in a finite or fixed time, respectively. Then, using the exact state estimation, two parameter identification algorithms based on sliding-modes techniques are introduced to identify the unknown parameters of the system, i.e. the friction coefficients. Some experiments and comparisons are presented to illustrate the effectiveness of the presented algorithms. Finally, some concluding remarks are given at the end of the chapter.

2.1 Introduction

The state-estimation problem is one of the most important problems in control theory. There exist many kinds of observers in the literature like adaptive observers, dissipative observers, high-gain observers, sampled-data observers, interval observers, etc. Some of these are the sliding-modes-based observers, which have been successfully developed in the last decades (see, e.g. [1–3]) presenting finite-time convergence and insensitivity, more than just robustness, to some classes of disturbances.

The first results used first-order sliding-modes (FOSM) output injections (see, e.g. [4,5]), and to guarantee asymptotic convergence of the estimation error, they assumed that the system was minimum phase and relative degree one with respect to the unknown input. Later on, an output linear injection term was added to the FOSM output injection improving the numerical issues (see, e.g. [6,7]), providing, based on the same assumptions, asymptotic convergence of the estimation error. Nevertheless,

[1]Departamento de Control y Robótica, División de Ingeniería Eléctrica, Facultad de Ingeniería, Universidad Nacional Autónoma de México, Mexico.
[2]CONACYT – TECNM/Instituto Tecnológico de La Laguna, División de Estudios de Posgrado e Investigación, México.

these assumptions limited the applicability of the proposed observers since they are not satisfied even for mechanical systems.

Afterwards, high-order sliding-modes (HOSM) observers were introduced based on strong observability and detectability conditions [8]. Such observers relaxed the condition related to the relative degree and ensure exact and finite-time state estimation (see, e.g. [9–13]). Continuous finite-time observers have been proposed based on homogeneity in [14–17]; and based on the implicit Lyapunov function theorem, a constructive homogeneous observer is proposed in [18]. These continuous observers provide exact and finite-time state estimation in absence of unknown inputs and uncertainties. Recently, the concept of fixed-time convergence, or uniform convergence with respect to the initial conditions, has been introduced by [19]. This property implies the existence of a bound for the convergence time, and such a bound is independent of the initial estimation error. There exist a couple of works related to this property and given in the differentiation framework: the given by [20] for the design of a first-order uniform robust exact differentiator, and that one in [21] for arbitrary order.

Regarding the parameter identification problem, it has been extensively studied during the last decades. One of the more important reasons is the need for accurate and efficient control for systems. The challenge of providing better models of physical phenomena leads to that the parameter identification problem becomes fundamental in industrial applications. System identification techniques are also used in signal processing applications (such as communications [22], geophysical engineering [23] and mechanical engineering [24]), in non-technical fields such as biology [25], environmental sciences and econometrics to improve the knowledge on the identified object, prediction and control.

The identification theory basically deals with the problem of the efficient extraction of signal and system dynamics properties based on available data measurements. In the literature there exist many methods to identify parameters, and the most popular ones belong to the group of least squares (LS) methods; for instance, the non-recursive method of LS, recursive methods of LS, method of weighted LS, exponential forgetting with constant forgetting factor, exponential forgetting with variable forgetting factor, etc. There exist also many modifications of the LS methods; for example, the method of generalized LS, method of extended LS, method of bias correction, instrumental variables method; and some others like Bayes method, maximum likelihood methods, extended Kalman filter, modulating functions methods, subspaces methods, etc. (see, e.g. [26–29]).

In the context of finite-time convergence [30], a recursive finite-time convergent algorithm has been presented in [31]. Such an algorithm is a non-linear recursive version of the LS algorithm, where the non-linear injection terms, designed based on the generalized super-twisting algorithm (STA) [32], lead to finite-time convergence. In this line of research, in [33,34], the filtered equivalent output injection of the STA together with a standard LS recursive algorithm have been used for parameter identification of mechanical systems. A finite-time and non-recursive LS algorithm is presented by [35] to identify constant parameters. Such an algorithm is based on adaptive control and it requires to solve matrix-valued ordinary differential equations and to check the invertibility of a matrix (persistence of excitation condition) online.

The observation and identification problems have been also studied in different types of pendulums. In the context of state estimation, in [36], Kalman filters are introduced to estimate the state of the Virgo inverted pendulum. To estimate the state of the Furuta pendulum and the wheeled inverted pendulum, some linear observers have been presented by [37,38], respectively. Some extended high-gain observers have been implemented in the inertia wheeled pendulum and in the inverted pendulum–cart by [39,40], respectively. For a mobile wheeled inverted pendulum, in [41,42], non-linear observers have been proposed to estimate the state or disturbances. On the other hand, in order to identify the pendulum parameters, in [43], some Kalman filters are used to identify the dynamic characteristic of the structures equipped with a pendulum tuned mass damper. For the Furuta pendulum, in [44], a standard LS algorithm is implemented. Based on a Volterra polynomials, in [45], a non-linear identification is proposed for the inverted pendulum systems. For the single and the double pendulum, some parameter identification algorithm based on harmonic balance-based algorithm are proposed by [46,47], respectively.

In this chapter, the problems of state estimation and parameter identification for the pendulum–cart system are addressed. A couple of HOSM observers are proposed to estimate the state, in spite of disturbances; exactly and in a finite or fixed time, respectively. Based on the exact state estimation, two parameter identification algorithms based on sliding-modes techniques are introduced to identify the unknown parameters of the system, i.e. the friction coefficients. The corresponding convergence proofs are given based on Lyapunov function approach. Some experiments and comparisons illustrate the effectiveness of the presented algorithms.

The outline of this work is as follows. The problem statement is given in Section 2.2 and some preliminary results are presented in Section 2.3. The corresponding analysis, design, and robustness properties of the observers are presented in Section 2.4 and the identification algorithms with the corresponding conditions to ensure the convergence of the parameters are given in Section 2.5, both of them present experimental results. Finally, some concluding remarks are discussed in Section 2.6.

Notation: Let $\mathbb{R}$ denote the set of real numbers. Let $\|q\|$ denote the Euclidean norm of a vector $q \in \mathbb{R}^n$; $\overline{1,n}$ a sequence integers $1, \ldots, n$; and I an identity matrix. The induced norm for a matrix $Q \in \mathbb{R}^{m\times n}$ is given as $\|Q\| := \sqrt{\lambda_{\max}(Q^TQ)}$, where $\lambda_{\max}$ (respectively, $\lambda_{\min}$) is the maximum (respectively, the minimum) eigenvalue. For a Lebesgue measurable function $u : \mathbb{R}_{\geq 0} \to \mathbb{R}^m$ define the norm $\|u\|_{(t_0,t_1)} :=$ ess $\sup_{t\in(t_0,t_1)} \|u(t)\|$, then $\|u\|_\infty := \|u\|_{(0,+\infty)}$ and the set of functions u with the property $\|u\|_\infty < +\infty$ is denoted as $\mathscr{L}_\infty$. Define the function $\lceil\cdot\rfloor^\gamma := |\cdot|^\gamma \mathrm{sign}(\cdot)$, for any $\gamma \in \mathbb{R}_{\geq 0}$.

2.2 Problem statement

Consider the pendulum–cart system depicted in Figure 2.1. This figure describes a pendulum rotating in a vertical plane around an axis located on a cart. The cart can move along a horizontal rail, lying in the plane of rotation. The system state is a vector $x = (x_1, x_2, x_3, x_4)^T$, where x_1 is the cart position, x_2 is the angle between

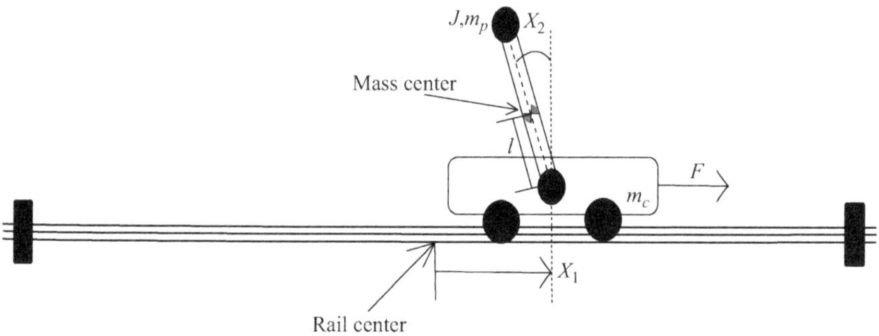

Figure 2.1 Pendulum–cart system

the upward direction and the pendulum, measured counterclockwise ($x_2 = 0$ for the upright position of the pendulum), x_3 is the cart velocity, and x_4 is the pendulum angular velocity. A control force F, parallel to the rail, is applied to the cart. It is produced by a DC flat motor controlled by a pulse-width-modulation (PWM) voltage signal u, and $F = p_1 u$. The system control u takes values in the interval $[-0.5, 0.5]$. The total mass of the pendulum and cart is denoted by m. The distance from the axis of rotation of the pendulum to the centre of mass of the system is l. The moment of inertia of the pendulum with respect to its axis on the cart is represented by J. The cart friction is compound of two forces: the static and the viscous friction proportional to the cart velocity, $f_c x_3$ and $f_s\mathrm{sign}(x_3)$, respectively. There is also a friction torque in the angular motion of the pendulum, proportional to the angular velocity, $f_p x_4$ [48].

The system dynamics is described by the following equations:

$$\dot{x}_1 = x_3, \tag{2.1}$$

$$\dot{x}_2 = x_4, \tag{2.2}$$

$$\dot{x}_3 = f_1(x, u) + \xi_1(t, x, u), \tag{2.3}$$

$$\dot{x}_4 = f_2(x, u) + \xi_2(t, x, u), \tag{2.4}$$

$$y = (y_1, y_2)^T = (x_1, x_2)^T, \tag{2.5}$$

where the measurable output is given by $y \in \mathbb{R}^2$, the nominal part of the system dynamics is represented by the functions $f_1 : \mathbb{R}^4 \times \mathbb{R} \to \mathbb{R}$ and $f_2 : \mathbb{R}^4 \times \mathbb{R} \to \mathbb{R}$, i.e.

$$f_1(x, u) := \frac{aw_1(x, u) + l\cos(x_2)w_2(x)}{J + m_i l\sin^2(x_2)}, \tag{2.6}$$

$$f_2(x, u) := \frac{l\cos(x_2)w_1(x, u) + w_2(x)}{J + m_i l\sin^2(x_2)}, \tag{2.7}$$

$$w_1(x, u) := p_1 u - m_i x_4^2 \sin(x_2) - f_c x_3 - f_s\mathrm{sign}(x_3), \tag{2.8}$$

$$w_2(x) := m_i g\sin(x_2) - f_p x_4, \tag{2.9}$$

$$m_i := l(m_c + m_p),\ a := l^2 + \frac{J}{m_c + m_p}, \tag{2.10}$$

Table 2.1 Parameters of the pendulum–cart system

Name	Description	Units
m_c	Cart mass	0.5723 (kg)
m_p	Pendulum mass	0.12 (kg)
l	Distance from the rotation axis to the mass center	0.01955717 (m)
f_c	Dynamic friction coefficient	0.5 (N s/m)
f_s	Dry friction coefficient	1.1975875 (N)
f_p	Rotational friction coefficient	0.6650×10^{-4} (N ms/rad)
p_1	Control-force-to-PWM-signal ratio	9.4 (N)
J	Moment of inertia of the pendulum	0.0038583 (kgm^2)

while the uncertainties and perturbations are concentrated in the terms ξ_1 and ξ_2. The solutions of the system (2.1)–(2.4) are understood in the Filippov's sense [49]. It is assumed that the uncertainties $\xi_1(t,x,u)$ and $\xi_2(t,x,u)$ are Lebesgue-measurable and uniformly bounded in any region of the state space $x \in \mathbb{R}^4$, i.e. $\|\xi_1(t,x,u)\|_\infty \leq \xi_1^+$ and $\|\xi_2(t,x,u)\|_\infty \leq \xi_2^+$, with known positive constants ξ_1^+ and ξ_2^+.

The aims of this chapter are to estimate the corresponding velocities, i.e. the states x_3 and x_4; and identify the friction parameters[1], i.e. f_c, f_p, and f_s; by means of only the measurable output y and the input u despite the bounded uncertainties ξ_1 and ξ_2.

The parameters of system (2.1)–(2.4) are given in Table 2.1.

2.3 Preliminaries

Let us consider a time-varying differential equation [50]:

$$\frac{dx(t)}{dt} = f(t, x(t)), \qquad t \geq t_0, \qquad t_0 \in \mathbb{R}, \tag{2.11}$$

where $x(t) \in \mathbb{R}^n$ is the state vector; $f : \mathbb{R}_{\geq 0} \times \mathbb{R}^n \to \mathbb{R}^n$ is a continuous function with respect to x and measurable with respect to t, and $f(t,0) = 0$. It is assumed that solution of the system (2.11) for an initial condition $x_0 \in \mathbb{R}^n$ at time instant $t_0 \in \mathbb{R}$ is denoted as $x(t, t_0, x_0)$ and it is defined on some finite time interval $[t_0, t_0 + T)$ where $0 \leq T < \infty$ [the notation $x(t)$ will be used to reference $x(t, t_0, x_0)$ if the origin of x_0 and t_0 is clear from the context].

Let Ω be an open subset of $\mathbb{R}^n$, such that $0 \in \Omega$.

Definition 2.1. *[19],[50],[51]. At the steady state $x = 0$ the system (2.11) is said to be:*

- *Uniformly stable (US) if for any $\varepsilon > 0$ there is $\delta(\varepsilon)$ such that for any $x_0 \in \Omega$, if $|x_0| \leq \delta(\varepsilon)$ then $|x(t, t_0, x_0)| \leq \varepsilon$ for all $t \geq t_0$, for any $t_0 \in \mathbb{R}$.*

[1] It is well known that most of the time the value of the friction coefficients is obtained by means of experimental evaluation and thus it is difficult to provide an estimation.

- *Uniformly asymptotically stable (UAS) if it is US and for any* $\kappa > 0$ *and* $\varepsilon > 0$ *there exists* $T(\kappa, \varepsilon) \geq 0$ *such that for any* $x_0 \in \Omega$, *if* $|x_0| \leq \kappa$ *then* $|x(t, t_0, x_0)| \leq \varepsilon$ *for all* $t \geq t_0 + T(\kappa, \varepsilon)$, *for any* $t_0 \in \mathbb{R}$.
- *Uniformly finite-time stable (UFTS) if it is US and finite-time converging from* Ω, *i.e. for any* $x_0 \in \Omega$ *there exists* $0 \leq T^{x_0} < +\infty$ *such that* $x(t, t_0, x_0) = 0$ *for all* $t \geq t_0 + T^{x_0}$, *for any* $t_0 \in \mathbb{R}$. *The function* $T_0(x_0) = \inf\{T^{x_0} \geq 0 : x(t, t_0, x_0) = 0, \forall t \geq t_0 + T^{x_0}\}$ *is called the settling-time of the system (2.11).*
- *Uniformly fixed-time stable (UFxTS) if it is UFTS and the settling-time function* $T_0(x_0)$ *is bounded, i.e.* $\exists T_{max} > 0 : T_0(x_0) \leq T_{max}$, *for all* $x_0 \in \Omega$ *and for any* $t_0 \in \mathbb{R}$.

If $\Omega = \mathbb{R}^n$, *then* $x = 0$ *is said to be globally US (GUS), UAS (GUAS), UFTS (GUFTS), or UFxTS (GUFxTS), respectively.*

2.4 High-order sliding-mode observers

In this section, a general HOSM observer is presented to estimate the corresponding velocities. Then, two particular cases are presented in order to provide finite and fixed-time convergence of the estimated state to the real one.

The general HOSM observer has the following form:

$$\dot{\hat{x}}_1 = \hat{x}_3 + k_1\phi_1(e_{y_1}), \tag{2.12}$$

$$\dot{\hat{x}}_2 = \hat{x}_4 + k_2\phi_2(e_{y_2}), \tag{2.13}$$

$$\dot{\hat{x}}_3 = f_{n_1}(\hat{x}, y, u) + k_3\phi_3(e_{y_1}), \tag{2.14}$$

$$\dot{\hat{x}}_4 = f_{n_2}(\hat{x}, y, u) + k_4\phi_4(e_{y_2}), \tag{2.15}$$

where $e_{y_1} := y_1 - \hat{x}_1$ and $e_{y_2} := y_2 - \hat{x}_2$ are the output errors, the known nominal functions $f_{n_1} : \mathbb{R}^4 \times \mathbb{R}^4 \times \mathbb{R} \to \mathbb{R}$ and $f_{n_2} : \mathbb{R}^4 \times \mathbb{R}^4 \times \mathbb{R} \to \mathbb{R}$ take the following structure

$$f_{n_1}(\hat{x}, y, u) := \frac{a w_{n_1}(\hat{x}, y, u) + l\cos(y_2) w_{n_2}(y)}{J + m_i l \sin^2(y_2)}, \tag{2.16}$$

$$f_{n_2}(\hat{x}, y, u) := \frac{l\cos(y_2) w_{n_1}(\hat{x}, y, u) + w_{n_2}(y)}{J + m_i l \sin^2(y_2)}, \tag{2.17}$$

$$w_{n_1}(\hat{x}, y, u) := p_1 u - m_i \hat{x}_4^2 \sin(y_2), \tag{2.18}$$

$$w_{n_2}(y) := m_i g \sin(y_2), \tag{2.19}$$

the non-linear output injections

$$\phi_1(e_{y_1}) := \mu_{11}\lceil e_{y_1}\rfloor^{\frac{1}{2}} + \mu_{12}\lceil e_{y_1}\rfloor^{1} + \mu_{13}\lceil e_{y_1}\rfloor^{\frac{3}{2}}, \tag{2.20}$$

$$\phi_2(e_{y_2}) := \mu_{21}\lceil e_{y_2}\rfloor^{\frac{1}{2}} + \mu_{22}\lceil e_{y_2}\rfloor^{1} + \mu_{23}\lceil e_{y_2}\rfloor^{\frac{3}{2}}, \tag{2.21}$$

$$\phi_3(e_{y_1}) := \mu_{31} \lceil e_{y_1} \rfloor^0 + \mu_{32} \lceil e_{y_1} \rfloor^1 + \mu_{33} \lceil e_{y_1} \rfloor^2, \tag{2.22}$$

$$\phi_4(e_{y_2}) := \mu_{41} \lceil e_{y_2} \rfloor^0 + \mu_{42} \lceil e_{y_2} \rfloor^1 + \mu_{43} \lceil e_{y_2} \rfloor^2, \tag{2.23}$$

and some positive constants $k_i, \mu_{ij} \geq 0$ for $i = \overline{1,4}$ and $j = \overline{1,3}$. Note that the properties of the general HOSM observer may change according to the selection of the parameters μ_{ij}, i.e.

- For $\mu_{i1} = \mu_{i3} = 0$, $i = \overline{1,4}$, the observer (2.12)–(2.15) recovers a classic linear observer.
- For $\mu_{i2} = \mu_{i3} = 0$, $i = \overline{1,4}$, the observer (2.12)–(2.15) recovers a well-known non-linear observer, i.e. the super-twisting observer [11].
- For $\mu_{i2} = 0$, $i = 1, 2$, the observer (2.12)–(2.15) recovers another non-linear observer, i.e. the fixed-time observer [20].

Let us define the state estimation error $e = (e_1, e_2, e_3, e_4)^T \in \mathbb{R}^4$ where $e_i := x_i - \hat{x}_i$ for $i = \overline{1,4}$. Thus, the error dynamics is given as follows:

$$\dot{e}_1 = e_3 - k_1\phi_1(e_{y_1}), \tag{2.24}$$

$$\dot{e}_2 = e_4 - k_2\phi_2(e_{y_2}), \tag{2.25}$$

$$\dot{e}_3 = F_1(x, \hat{x}, y, u) + \xi_1(t, x, u) - k_3\phi_3(e_{y_1}), \tag{2.26}$$

$$\dot{e}_4 = F_2(x, \hat{x}, y, u) + \xi_2(t, x, u) - k_4\phi_4(e_{y_2}). \tag{2.27}$$

where $F_1(x, \hat{x}, y, u) := f_1(x, u) - f_{n_1}(\hat{x}, y, u)$ and $F_2(x, \hat{x}, y, u) := f_2(x, u) - f_{n_2}(\hat{x}, y, u)$. Due to the boundedness assumption over the uncertain terms ξ_1 and ξ_2, and the stability properties of the pendulum–cart system[2], it is possible to find an upper bound such that

$$\left\| F_1(x, \hat{x}, y, u) + \xi_1(t, x, u) \right\|_\infty \leq f_1^+, \tag{2.28}$$

$$\left\| F_2(x, \hat{x}, y, u) + \xi_2(t, x, u) \right\|_\infty \leq f_2^+, \tag{2.29}$$

with known positive constants f_i^+ for $i = 1, 2$.

In the following, the error dynamics is analysed for two different non-linear observers, i.e. the super-twisting and the fixed-time observer, respectively.

2.4.1 Super-twisting observer

Consider the general HOSM observer (2.12)–(2.15) for the case in which $\mu_{i2} = \mu_{i3} = 0$, and $\mu_{i1} = 1$ for $i = \overline{1,4}$, i.e. the super-twisting observer [11]:

$$\dot{\hat{x}}_1 = \hat{x}_3 + k_1 \lceil e_{y_1} \rfloor^{\frac{1}{2}}, \tag{2.30}$$

$$\dot{\hat{x}}_2 = \hat{x}_4 + k_2 \lceil e_{y_2} \rfloor^{\frac{1}{2}}, \tag{2.31}$$

[2] Note that for any initial condition $x_0 \in \mathbb{R}^4$ the trajectories of the pendulum–cart system (2.1)–(2.4) are US for any bounded ξ_1 and ξ_2.

$$\dot{\hat{x}}_3 = f_{n_1}(\hat{x}, y, u) + k_3 \lceil e_{y_1} \rfloor^0, \tag{2.32}$$

$$\dot{\hat{x}}_4 = f_{n_2}(\hat{x}, y, u) + k_4 \lceil e_{y_2} \rfloor^0. \tag{2.33}$$

The following theorem describes the convergence properties of the super-twisting observer.

Theorem 2.1. *[32] Let observer* (2.30)–(2.33) *be applied to pendulum–cart system* (2.1)–(2.4). *If the observer parameters are designed as follows:*

$$k_1 > 0,\ k_2 > 0,\ k_3 > 3f_1^+ + 2\frac{f_1^+}{k_1^2},\ k_4 > 3f_2^+ + 2\frac{f_2^+}{k_2^2}, \tag{2.34}$$

then the error dynamics is UFTS.

Proof. The error dynamics between pendulum–cart system (2.1)–(2.4) and the super-twisting observer (2.30)–(2.33) is given as follows:

$$\dot{e}_1 = e_3 - k_1 \lceil e_1 \rfloor^{\frac{1}{2}}, \tag{2.35}$$

$$\dot{e}_2 = e_4 - k_2 \lceil e_2 \rfloor^{\frac{1}{2}}, \tag{2.36}$$

$$\dot{e}_3 = F_1(x, \hat{x}, y, u) + \xi_1(t, x, u) - k_3 \lceil e_1 \rfloor^0, \tag{2.37}$$

$$\dot{e}_4 = F_2(x, \hat{x}, y, u) + \xi_2(t, x, u) - k_4 \lceil e_2 \rfloor^0. \tag{2.38}$$

Based on [32], it is possible to show that the function

$$V = V_1 + V_2 = \zeta_1^T P_1 \zeta_1 + \zeta_2^T P_2 \zeta_2, \tag{2.39}$$

with

$$\zeta_1^T = \left[\lceil e_1 \rfloor^{\frac{1}{2}}, e_3 \right],\ \zeta_2^T = \left[\lceil e_2 \rfloor^{\frac{1}{2}}, e_4 \right], \tag{2.40}$$

$$P_1 = \frac{1}{2} \begin{bmatrix} 4k_3 + k_1^2 & -k_1 \\ -k_1 & 2 \end{bmatrix},\ P_2 = \frac{1}{2} \begin{bmatrix} 4k_4 + k_2^2 & -k_2 \\ -k_2 & 2 \end{bmatrix}, \tag{2.41}$$

is a Lyapunov function for the error dynamics (2.35)–(2.38), and its derivative along the trajectories of such an error dynamics is given by

$$\dot{V} = -\frac{1}{|e_1|^{\frac{1}{2}}} \zeta_1^T Q_1 \zeta_1 - \frac{1}{|e_2|^{\frac{1}{2}}} \zeta_2^T Q_2 \zeta_2 + (F_1 + \xi_1) q_1^T \zeta_1 + (F_2 + \xi_2) q_2^T \zeta_2, \tag{2.42}$$

where

$$Q_1 = \frac{k_1}{2} \begin{bmatrix} 2k_3 + k_1^2 & -k_1 \\ -k_1 & 1 \end{bmatrix},\ Q_2 = \frac{k_3}{2} \begin{bmatrix} 2k_4 + k_2^2 & -k_2 \\ -k_2 & 1 \end{bmatrix}, \tag{2.43}$$

$$q_1^T = [-k_1, 2],\ q_2^T = [-k_2, 2]. \tag{2.44}$$

Recalling that $|F_1 + \xi_1| \le f_1^+$ and $|F_2 + \xi_2| \le f_2^+$, it may be shown that $\dot{V}$ is upper bounded as follows

$$\dot{V} \le -\frac{1}{|e_1|^{\frac{1}{2}}}\zeta_1^T \bar{Q}_1 \zeta_1 - \frac{1}{|e_2|^{\frac{1}{2}}}\zeta_2^T \bar{Q}_2 \zeta_2, \tag{2.45}$$

where

$$\bar{Q}_1 = \frac{k_1}{2}\begin{bmatrix} 2k_3 + k_1^2 - 2f_1^+ & -k_1 - \frac{2f_1^+}{k_1} \\ -k_1 - \frac{2f_1^+}{k_1} & 1 \end{bmatrix}, \tag{2.46}$$

$$\bar{Q}_2 = \frac{k_3}{2}\begin{bmatrix} 2k_4 + k_2^2 - 2f_1^+ & -k_2 - \frac{2f_2^+}{k_2} \\ -k_2 - \frac{2f_2^+}{k_2} & 1 \end{bmatrix}. \tag{2.47}$$

Then, $\dot{V}$ is negative definite if $\bar{Q}_1 > 0$ and $\bar{Q}_2 > 0$. Hence, it is easy to show that this is true if the inequalities (2.34) hold. On the other hand, it follows that

$$|e_1|^{\frac{1}{2}} \le \|\zeta_1\| \le \left(\frac{V_1}{\lambda_{\min}(P_1)}\right)^{\frac{1}{2}}, \ |e_2|^{\frac{1}{2}} \le \|\zeta_2\| \le \left(\frac{V_2}{\lambda_{\min}(P_2)}\right)^{\frac{1}{2}}, \tag{2.48}$$

where $\lambda_{\min}(P_i)$ is the minimum eigenvalue of the matrix P_i. Therefore, $\dot{V}$ may be upper bounded as follows

$$\dot{V} \le -\gamma_1 V_1^{\frac{1}{2}} - \gamma_2 V_2^{\frac{1}{2}} \le -\gamma V^{\frac{1}{2}}, \tag{2.49}$$

with

$$\gamma_1 := \frac{\lambda_{\min}^{\frac{1}{2}}(P_1)\lambda_{\min}(\bar{Q}_1)}{\lambda_{\max}(P_1)}, \ \gamma_2 := \frac{\lambda_{\min}^{\frac{1}{2}}(P_2)\lambda_{\min}(\bar{Q}_2)}{\lambda_{\max}(P_2)}, \ \gamma := \min(\gamma_1, \gamma_2), \tag{2.50}$$

and $\lambda_{\max}(P_i)$ the maximum eigenvalue of the matrix P_i. Thus, since the solution of the differential equation $\dot{v} \le -\gamma v^{\frac{1}{2}}$, $v(0) \ge 0$, is given by $v(t) = (v^{\frac{1}{2}}(0) - \frac{\gamma}{2}t)^2$ for all $t \ge 0$, by the comparison principle [50] one obtains that $V(t)$, and therefore the error dynamics (2.35)–(2.38), converges to zero at most in a finite time given by $T = 2V^{\frac{1}{2}}(e(0))/\gamma$. □

2.4.2 Fixed-time observer

Let us consider the general HOSM observer (2.12)–(2.15) with $\mu_{i1} = 1$ for $i = 1, 2$; $\mu_{i1} = 0.5$ for $i = 3, 4$; $\mu_{i2} = 0$ for $i = 1, 2$; $\mu_{i2} = 2\alpha$ for $i = 3, 4$; $\mu_{i3} = \alpha$ for $i = 1, 2$; $\mu_{i3} = 1.5\alpha^2$ for $i = 3, 4$; and $\alpha > 0$; i.e. the fixed-time observer [20]:

$$\dot{\hat{x}}_1 = \hat{x}_3 + k_1\left(\lceil e_{y_1} \rfloor^{\frac{1}{2}} + \alpha \lceil e_{y_1} \rfloor^{\frac{3}{2}}\right), \tag{2.51}$$

$$\dot{\hat{x}}_2 = \hat{x}_4 + k_2\left(\lceil e_{y_2} \rfloor^{\frac{1}{2}} + \alpha \lceil e_{y_2} \rfloor^{\frac{3}{2}}\right), \tag{2.52}$$

$$\dot{\hat{x}}_3 = f_{n_1}(\hat{x}, y, u) + k_3\left(\frac{1}{2}\left\lceil e_{y_1}\right\rfloor^0 + 2\alpha\left\lceil e_{y_1}\right\rfloor^1 + \frac{3}{2}\alpha^2\left\lceil e_{y_1}\right\rfloor^2\right), \tag{2.53}$$

$$\dot{\hat{x}}_4 = f_{n_2}(\hat{x}, y, u) + k_4\left(\frac{1}{2}\left\lceil e_{y_2}\right\rfloor^0 + 2\alpha\left\lceil e_{y_2}\right\rfloor^1 + \frac{3}{2}\alpha^2\left\lceil e_{y_2}\right\rfloor^2\right). \tag{2.54}$$

The following theorem describes the convergence properties of the fixed-time observer.

Theorem 2.2. *[20] Let observer* (2.51)–(2.54) *be applied to pendulum–cart system* (2.1)–(2.4). *If the observer parameters are in the set:*

$$\left\{0 < k_1 \leq 2\sqrt{f_1^+},\ k_3 > \frac{k_1^2}{4} + \frac{(2f_1^+)^2}{k_1^2}\right\} \cup \left\{k_1 > 2\sqrt{f_1^+},\ k_3 > 2f_1^+\right\}, \tag{2.55}$$

$$\left\{0 < k_2 \leq 2\sqrt{f_2^+},\ k_4 > \frac{k_2^2}{4} + \frac{(2f_2^+)^2}{k_2^2}\right\} \cup \left\{k_2 > 2\sqrt{f_2^+},\ k_4 > 2f_2^+\right\}, \tag{2.56}$$

then the error dynamics is UFxTS.

Proof. The error dynamics between pendulum–cart system (2.1)–(2.4) and the fixed-time observer (2.51)–(2.54) is given as follows:

$$\dot{e}_1 = e_3 - k_1\bar{\phi}_1(e_1), \tag{2.57}$$

$$\dot{e}_2 = e_4 - k_2\bar{\phi}_2(e_2), \tag{2.58}$$

$$\dot{e}_3 = F_1(x, \hat{x}, y, u) + \xi_1(t, x, u) - k_3\bar{\phi}_3(e_1), \tag{2.59}$$

$$\dot{e}_4 = F_2(x, \hat{x}, y, u) + \xi_2(t, x, u) - k_4\bar{\phi}_4(e_2). \tag{2.60}$$

where

$$\bar{\phi}_1(e_1) := \left\lceil e_{y_1}\right\rfloor^{\frac{1}{2}} + \alpha\left\lceil e_{y_1}\right\rfloor^{\frac{3}{2}}, \tag{2.61}$$

$$\bar{\phi}_2(e_2) := \left\lceil e_{y_2}\right\rfloor^{\frac{1}{2}} + \alpha\left\lceil e_{y_2}\right\rfloor^{\frac{3}{2}}, \tag{2.62}$$

$$\bar{\phi}_3(e_1) := \frac{1}{2}\left\lceil e_{y_1}\right\rfloor^0 + 2\alpha\left\lceil e_{y_1}\right\rfloor^1 + \frac{3}{2}\alpha^2\left\lceil e_{y_1}\right\rfloor^2, \tag{2.63}$$

$$\bar{\phi}_4(e_2) := \frac{1}{2}\left\lceil e_{y_2}\right\rfloor^0 + 2\alpha\left\lceil e_{y_2}\right\rfloor^1 + \frac{3}{2}\alpha^2\left\lceil e_{y_2}\right\rfloor^2. \tag{2.64}$$

Based on [20], one can show that the function

$$V = V_1 + V_2 = \zeta_1^T P_1 \zeta_1 + \zeta_2^T P_2 \zeta_2, \tag{2.65}$$

with

$$\zeta_1^T = \left(\bar{\phi}_1(e_1), e_3\right),\ \zeta_2^T = \left(\bar{\phi}_2(e_2), e_4\right), \tag{2.66}$$

and P_1, P_2 symmetric and positive definite matrices; is a Lyapunov function for the error dynamics (2.57)–(2.60). Some suitable matrices $P_i = P_i^T > 0$ are the solutions of the following linear matrix inequalities [20]

$$\begin{bmatrix} A_1^T P_1 + P_1 A_1 + \varepsilon I + (2f_1^+)^2 C^T C & P_1 B \\ B^T P_1 & -1 \end{bmatrix} \leq 0, \tag{2.67}$$

$$\begin{bmatrix} A_2^T P_2 + P_2 A_2 + \varepsilon I + (2f_2^+)^2 C^T C & P_2 B \\ B^T P_2 & -1 \end{bmatrix} \leq 0, \tag{2.68}$$

where

$$A_1 := \begin{bmatrix} -k_1 & 1 \\ -k_3 & 0 \end{bmatrix}, \; A_2 := \begin{bmatrix} -k_2 & 1 \\ -k_4 & 0 \end{bmatrix}, \; B := [0, 1]^T, \; C := [1, 0], \tag{2.69}$$

with k_i for $i = \overline{1, 4}$ defined by (2.55)–(2.56), and some $\varepsilon > 0$. Thus, it is possible to show that

$$\dot{V} = -\gamma_{11} V_1^{\frac{1}{2}} - \gamma_{12} |e_1|^{\frac{1}{2}} V_1 - \gamma_{21} V_2^{\frac{1}{2}} - \gamma_{22} |e_2|^{\frac{1}{2}} V_2, \tag{2.70}$$

where

$$\gamma_{11} := \frac{\varepsilon}{2\lambda_{\max}^{\frac{1}{2}}(P_1)}, \; \gamma_{12} := \frac{3\varepsilon}{2\lambda_{\max}(P_1)}, \tag{2.71}$$

$$\gamma_{21} := \frac{\varepsilon}{2\lambda_{\max}^{\frac{1}{2}}(P_2)}, \; \gamma_{22} := \frac{3\varepsilon}{2\lambda_{\max}(P_2)}. \tag{2.72}$$

Therefore, this implies that the trajectories of the error dynamics (2.57)–(2.60) converge to zero at most in a finite time given by $T_1 = 2V^{\frac{1}{2}}(e(0))/\rho$, with $\rho := \min(\gamma_{11}, \gamma_{21})$. Hence, the uniform finite-time stability of the error dynamics is concluded. To provide the fixed-time convergence, it is needed to prove the uniform convergence with respect to the initial conditions.

Let us consider the following function

$$\bar{V}_1 = \frac{\delta}{2}\alpha^2 k_3 |e_1|^3 - e_1 \lceil e_3 \rfloor^{\frac{4}{3}} + \frac{\delta}{2}|e_3|^2 \tag{2.73}$$

$$\bar{V}_2 = \frac{\delta}{2}\alpha^2 k_4 |e_2|^3 - e_2 \lceil e_4 \rfloor^{\frac{4}{3}} + \frac{\delta}{2}|e_4|^2, \tag{2.74}$$

where $\delta > 0$ sufficiently large. Thus, taking $\bar{V} = \bar{V}_1 + \bar{V}_2$, its derivative along the trajectories of the error dynamics is given by

$$\dot{\bar{V}} = -\frac{1}{2}\left[\left(\frac{\bar{V}_1}{2c_1}\right)^{\frac{7}{2}} + \left(\frac{\bar{V}_2}{2c_2}\right)^{\frac{7}{2}}\right], \qquad \forall \bar{V}_1 > \varepsilon, \quad \forall \bar{V}_2 > \varepsilon, \tag{2.75}$$

for a sufficient large ε, and some positive constants c_1 and c_2. Hence, any trajectory of the error dynamics (2.57)–(2.60) converges to the compact set $\Gamma_\varepsilon := \{e \in \mathbb{R}^4 | \bar{V}_1 \leq \varepsilon, \; \bar{V}_2 \leq \varepsilon\}$ containing the origin, robustly and uniformly with respect to the initial error $e(0) \in \mathbb{R}^4$, and the convergence time is upper bounded by a time T_2 independent of the initial error. The estimation of such a time is given in [20].

Figure 2.2 Real pendulum–cart system

Therefore, combining the two previous properties, i.e. finite-time and uniform convergence, one can conclude that the error dynamics (2.57)–(2.60) is UFxTS. □

Remark 2.1. *The detailed convergence proofs for the super-twisting and fixed-time observers can be seen in [32] and [20], respectively. In this chapter, the proofs provided for Theorems 2.1 and 2.2 represent just brief sketches. Moreover, the given results are particularly provided for the pendulum–cart system.*

2.4.3 Experimental results

In order to illustrate the workability of the previous HOSM observers, some experiments have been done over the real pendulum–cart system by INTECO and depicted in Figure 2.2.

The control input u is designed in two steps:

1. **Swing-Up**. Based on an algorithm for generation of oscillations, the control input u is able to take the pendulum from its stable equilibrium point, i.e. $x = [0, \pi, 0, 0]^T$, to certain region close to its unstable equilibrium point, i.e. $x = [0, 0, 0, 0]^T$.
2. **Tracking Control**. Once a region around the unstable equilibrium point has been reached, a PID controller is applied to track the reference $r = 0.3\sin(0.05t)$ for the cart position and to maintain the pendulum in the upper position.

The super-twisting and the fixed-time observers, described by (2.30)–(2.33) and (2.51)–(2.54), respectively, are implemented in MATLAB® Simulink®. The parameters for both observers were chosen according to the theorems described for each algorithm and tuned in the experiments to obtain the minimum possible errors; these parameters are given in Table 2.2.

The experiments have been done on MATLAB Simulink platform over a RTDAC4/ PCIc data-acquisition board connected to the educational pendulum–cart system. All the experiments have been implemented with the Euler's integration method and with a sampling time equal to 1×10^{-2} s.

Table 2.2 Parameters of the observers

Parameter	Super-twisting obs.	Fixed-time obs.
k_1	6.7	6.7
k_2	6.7	6.7
k_3	22	22
k_4	22	22
f_1^+	1	1
f_2^+	1	1
α	–	1.5

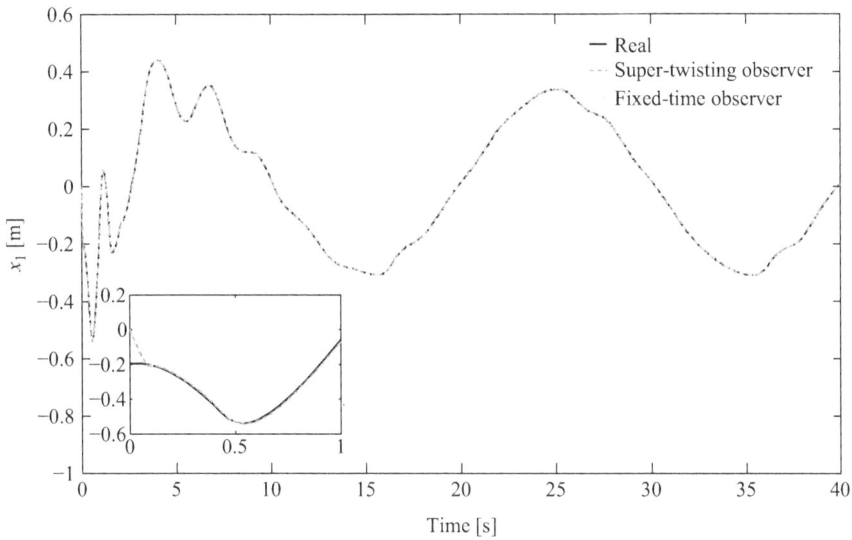

Figure 2.3 Cart position estimation

The results are presented in Figures 2.3–2.6. It is worth saying that the estimated positions, i.e. $\hat{x}_1$ and $\hat{x}_2$, are compared with the positions provided by a couple of position encoders given in the pendulum–cart system, while the estimated velocities, i.e. $\hat{x}_3$ and $\hat{x}_4$, are compared with the velocities given by a numerical differentiator included in the pendulum–cart toolbox. Then, it is easy to see that both observers have similar performance and they provide very good estimation results.

In order to illustrate the performance of both observers the following index is implemented:

$$I = \left(\frac{1}{\Delta T} \int_{t-\Delta T}^{t} |e(\tau)|^2 d\tau \right)^{1/2}, \tag{2.76}$$

where e is the state estimation vector and ΔT is a time window width; in the experiment, ΔT was selected as 2. This performance index calculates the Root Mean Square (RMS) value of the input e.

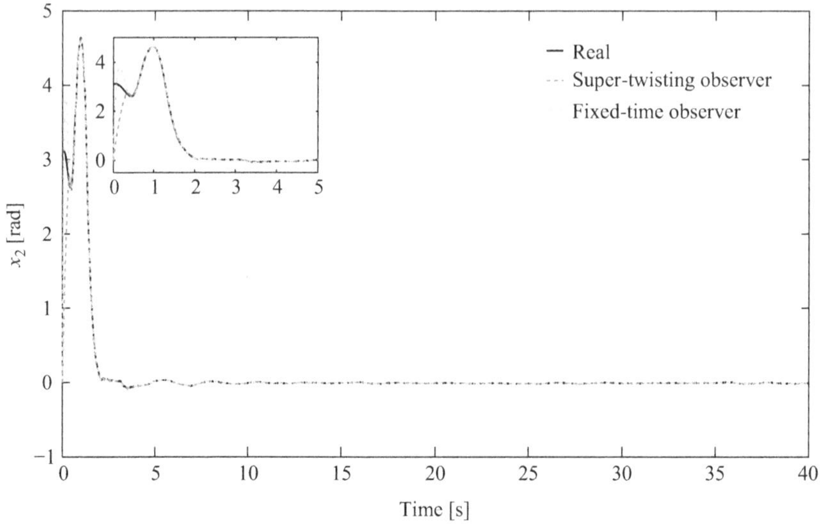

Figure 2.4 Pendulum position estimation

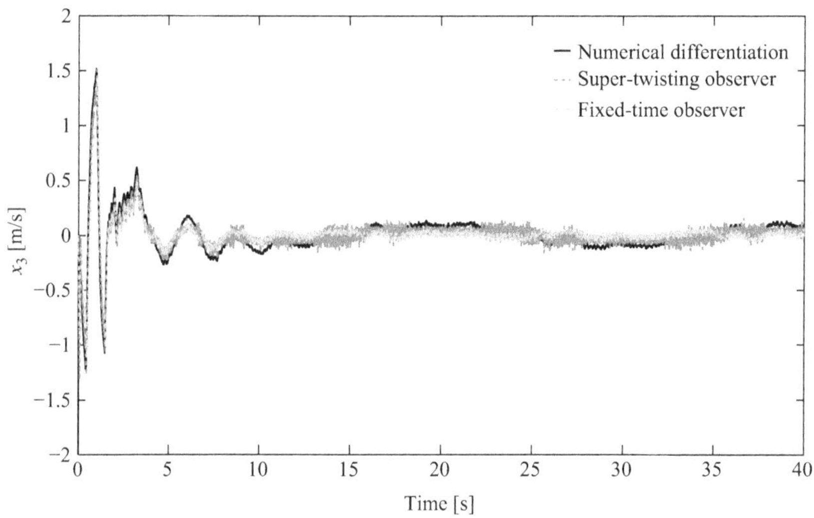

Figure 2.5 Cart velocity estimation

The behaviour of the performance index during the experiment is shown in Figure 2.7. One can see that the fixed-time observer presents a bigger overshoot than the super-twisting observer and this last one has a lower performance index after the transient.

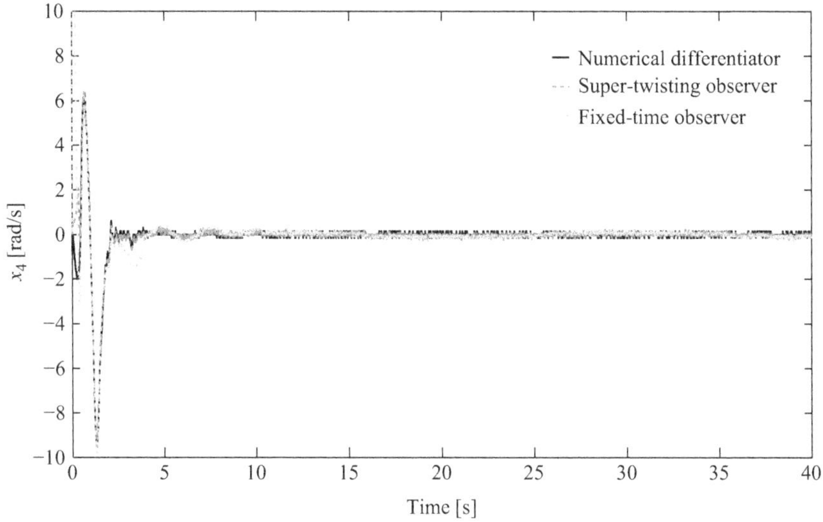

Figure 2.6 Pendulum velocity estimation

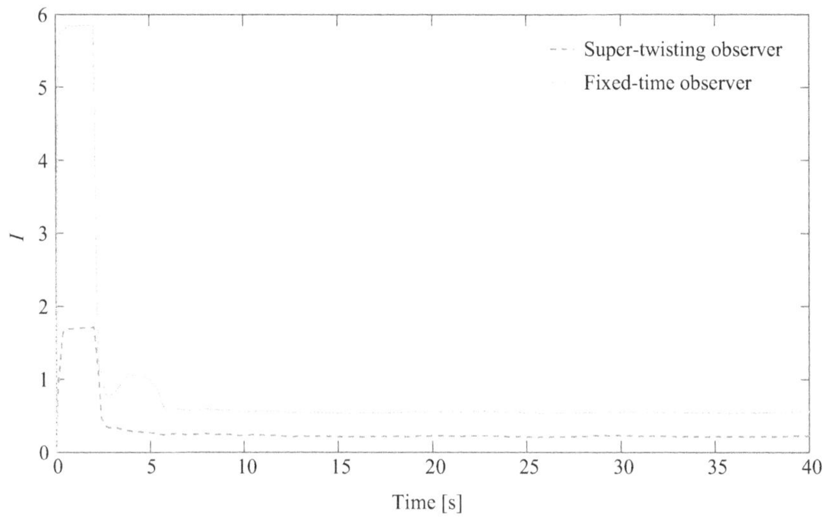

Figure 2.7 Performance index of state estimation

The results show that both observers present good results in order to estimate the pendulum–cart states, but super-twisting observer presents a better performance with the chosen index and the tuning of the fixed-time observer parameters can be a little more difficult because it presents more than the super-twisting.

2.5 Parameter identification algorithms

A common issue in the control design problem is the lack of knowledge in the parameters of the system, for instance, the friction coefficients in mechanical systems; and even a complete ignorance of them. In order to solve such a problem parameter identification algorithms are required.

Let us assume that $\xi_1 = \xi_2 = 0$, then the pendulum–cart dynamics can be rewritten in its regressor form as follows [29]:

$$\dot{x}_1 = x_3, \tag{2.77}$$

$$\dot{x}_2 = x_4, \tag{2.78}$$

$$\dot{x}_3 = f_{n_1}(x,u) + \Gamma_1(t)\theta, \tag{2.79}$$

$$\dot{x}_4 = f_{n_2}(x,u) + \Gamma_2(t)\theta, \tag{2.80}$$

where

$$f_{n_1}(x,u) := \frac{aw_{n_1}(x,u) + l\cos(x_2)w_{n_2}(x)}{J + m_i l\sin^2(x_2)},$$

$$f_{n_2}(x,u) := \frac{l\cos(y_2)w_{n_1}(x,u) + w_{n_2}(x)}{J + m_i l\sin^2(x_2)}, \tag{2.81}$$

$$w_{n_1}(x,u) := p_1 u - m_i x_4^2 \sin(x_2),\ w_{n_2}(x) \coloneqq m_i g\sin(x_2), \tag{2.82}$$

$$\Gamma_1(t) := \frac{-1}{J + m_i l\sin^2(x_2)}(ax_3, a\text{sign}(x_3), lx_4\cos(x_2)), \tag{2.83}$$

$$\Gamma_2(t) := \frac{-1}{J + m_i l\sin^2(x_2)}(l\cos(x_2)x_3, l\cos(x_2)\text{sign}(x_3), x_4), \tag{2.84}$$

$$\theta^T := (f_c, f_s, f_p), \tag{2.85}$$

where $\theta \in \mathbb{R}^3$ represents the unknown parameter vector, and $\Gamma_1^T : \mathbb{R} \to \mathbb{R}^3, \Gamma_2^T : \mathbb{R} \to \mathbb{R}^3$ are the regressor vectors. From (2.77) to (2.80), it is easy to see that it is enough to consider only dynamics (2.79) in order to identify the corresponding parameter vector θ.

Thus, it is assumed that Γ_1 is bounded, i.e. $0 < \alpha_1 = \|\Gamma_1\| < \infty$, and that it is persistently exciting (PE), i.e. there exist constants $\alpha_2 > 0$ and $\tau_0 > 0$ such that

$$\int_t^{t+\tau_0} \Gamma_1^T(\tau)\Gamma_1(\tau)d\tau \geq \alpha_2 I, \quad \forall t \geq 0. \tag{2.86}$$

In the following, two different non-linear identification algorithms are presented in order to identify the parameter vector θ, i.e. a homogeneous algorithm and a finite-time algorithm, respectively.

2.5.1 Homogeneous algorithm

The homogeneous parameter identification algorithm takes the following form [52]:

$$\dot{\hat{x}}_3 = -\bar{k}_1 \lceil \bar{e}_3 \rfloor^{\gamma} + \Gamma_1(t)\hat{\theta} + f_{n_1}(\hat{x}, y, u), \tag{2.87}$$

$$\dot{\hat{\theta}} = -\bar{k}_2 \lceil \bar{e}_3 \rfloor^{2\gamma-1} {\Gamma_1}^T(t), \tag{2.88}$$

where $\bar{k}_1$ and $\bar{k}_2$ are positive gains, $\bar{e}_3 := \bar{x}_3 - \hat{x}_3$ is a state estimation error where $\hat{x}_3$ may be given by the super-twisting or the fixed-time observer; and the parameter $\gamma \in (0.5, 1)$. Note that for the case $\gamma = 1$ the algorithm is reduced to the well-known linear adaptive observer [53].

Define the identification error as $\tilde{\theta} := \hat{\theta} - \theta$. Thus, based on homogeneity for time varying-systems (for more details, see [52,54]), the following theorem describes the convergence properties of the Homogeneous algorithm.

Theorem 2.3. *[52] Let the Homogeneous algorithm* (2.87)–(2.88) *be applied to the dynamics* (2.79)*, and the regressor vector* Γ_1 *be PE. Then for any* $\bar{k}_1, \bar{k}_2 > 0$ *and* $\gamma \in (0.5, 1)$ *the identification error* $\tilde{\theta}$ *is GUAS.*

Proof. Define $\tilde{e}_3 := \bar{x}_3 - x_3$. Then, the error dynamics is given by

$$\dot{\tilde{e}}_3 = -\bar{k}_1 \lceil \bar{e}_3 \rfloor^{\gamma} + \Gamma_1(t)\tilde{\theta} + f_{n_1}(\hat{x}, y, u) - f_{n_1}(x, u) \tag{2.89}$$

$$\dot{\tilde{\theta}} = -\bar{k}_2 \lceil \bar{e}_3 \rfloor^{2\gamma-1} {\Gamma_1}^T(t). \tag{2.90}$$

Then, according to the state estimation properties of the super-twisting and the fixed-time observers described in the previous sections, one can ensure that after a finite time $\hat{x}_3 \equiv x_3$, and thus, this implies that after a finite time $\tilde{e}_3 \equiv \bar{e}_3$ and also $f_{n_1}(\hat{x}, y, u) \equiv f_{n_1}(x, u)$. Therefore, the error dynamics can be rewritten as follows:

$$\dot{\tilde{e}}_3 = -\bar{k}_1 \lceil \tilde{e}_3 \rfloor^{\gamma} + \Gamma_1(t)\tilde{\theta} \tag{2.91}$$

$$\dot{\tilde{\theta}} = -\bar{k}_2 \lceil \tilde{e}_3 \rfloor^{2\gamma-1} {\Gamma_1}^T(t). \tag{2.92}$$

One can show that the system (2.91)–(2.92) is r-homogeneous, in the sense given by [52], with degree $d = \gamma - 1$ for the vector of weight $(r_1, r_2, r_3, r_4) = (1, \gamma, \gamma, \gamma)$. Note that $d < 0$ for all $\gamma \in (0.5, 1)$. Then, based on homogeneity theory one a show that the system (2.91)–(2.92) is GUAS.

First, note that the system (2.91)–(2.92) can be rewritten as follows:

$$\dot{\xi}(t) = A(t)\zeta(t) + f(t, \tilde{e}_3), \tag{2.93}$$

where $\zeta(t) = (\tilde{e}_3, \tilde{\theta}^T)^T \in \mathbb{R}^4$, $A(t)$ is a bounded and continuous matrix, for almost all $t \geq 0$, that is given by

$$A(t) = \begin{bmatrix} -\bar{k}_1 & \Gamma_1(t) \\ -\bar{k}_2 {\Gamma_1}^T(t) & 0 \end{bmatrix}, \tag{2.94}$$

and the non-linear term $f : \mathbb{R}_{\geq 0} \times \mathbb{R} \to \mathbb{R}^4$ is given as follows:

$$f(t, \tilde{e}_3) = \begin{bmatrix} -\bar{k}_1 \left(\lceil \tilde{e}_3 \rfloor^{\gamma} - \tilde{e}_3 \right) \\ -\bar{k}_2 \left(\lceil \tilde{e}_3 \rfloor^{2\gamma - 1} - \tilde{e}_3 \right) \Gamma_1{}^T(t) \end{bmatrix}. \tag{2.95}$$

Define

$$W_1(\tilde{e}_3, \tilde{\theta}) = \frac{\varsigma_1}{p} V_1{}^p(\tilde{e}_3, \tilde{\theta}) + \frac{\varsigma_2}{q} V_1{}^q(\tilde{e}_3, \tilde{\theta}), \tag{2.96}$$

$$W_2(t, \zeta) = \zeta^T P(t) \zeta, \tag{2.97}$$

where

$$V_1(\tilde{e}_3, \tilde{\theta}) = \frac{1}{2\gamma} |\tilde{e}_3|^{2\gamma} + \frac{1}{2\bar{k}_2} \tilde{\theta}^T \tilde{\theta}, \tag{2.98}$$

$$\varsigma_1 = 2\bar{k}_1 \mu c_2 (2\gamma)^{p-1}, \ \varsigma_2 = \frac{2\mu c_2 \left(\bar{k}_1^2 + \bar{k}_2^2 \alpha_1^2 \right) (2\gamma)^{q-1}}{\bar{k}_1}, \tag{2.99}$$

$$p = \frac{\gamma + 1}{2\gamma}, \ q = \frac{3 - \gamma}{2\gamma}, \tag{2.100}$$

and a positive constant μ; while $P(t)$ is a continuously differentiable, symmetric, bounded, and positive definite matrix, i.e. $0 < c_1 I \leq P(t) \leq c_2 I$, for all $t \geq 0$, which satisfies the following matrix differential equation:

$$\dot{P}(t) = -P(t)A(t) - A^T(t)P(t) - Q(t), \quad \forall t \geq 0, \tag{2.101}$$

where $Q(t) \geq c_3 I > 0$, for all $t \geq 0$, is a continuous, symmetric, and positive definite matrix; and c_1, c_2 and c_3 are positive constants. The existence of $P(t)$ and $Q(t)$ follows from the persistence of excitation condition of $\Gamma_1(t)$.

Let us introduce the following Lyapunov function candidate:

$$W(t, \tilde{e}_3, \tilde{\theta}) = W_1(\tilde{e}_3, \tilde{\theta}) + W_2(t, \zeta(t)). \tag{2.102}$$

It is possible to show that its derivative along the trajectories of the system (2.91)–(2.92), for all $|\tilde{e}_3| > 1$ satisfies

$$\dot{W}(t, \tilde{e}_3, \tilde{\theta}) \leq -\eta \, \|\zeta\|, \tag{2.103}$$

where $\eta := c_3 - (c_2/\mu)$. On the other hand, for all $|\tilde{e}_3| \leq 1$, $\dot{W}$ satisfies

$$\dot{W}(t, \tilde{e}_3, \tilde{\theta}) \leq -\eta \, \|\zeta\| + 2\mu c_2 \bar{k}_2^2 \alpha_1^2. \tag{2.104}$$

Therefore, uniform stability is concluded for the error dynamics (2.91)–(2.92), and then by homogeneity, one may conclude that the error dynamics (2.91)–(2.92) is GUAS [52]. □

2.5.2 Finite-time algorithm

The finite-time parameter identification algorithm takes the following form [31]:

$$\dot{\hat{x}}_3 = -\bar{k}_1 \left(\lceil \bar{e}_3 \rfloor^{\frac{1}{2}} + \lceil \bar{e}_3 \rfloor^{1} \right) + \Gamma_1(t)\hat{\theta} + f_{n_1}(\hat{x}, y, u), \tag{2.105}$$

$$\dot{\hat{\theta}} = -\bar{k}_2 \left(\frac{1}{2} \lceil \bar{e}_3 \rfloor^{0} + \frac{3}{2} \lceil \bar{e}_3 \rfloor^{\frac{1}{2}} + \lceil \bar{e}_3 \rfloor^{1} \right) \Gamma_1{}^T(t), \tag{2.106}$$

where $\bar{k}_1$ and $\bar{k}_2$ are positive gains, $\bar{e}_3 = \bar{x}_3 - \hat{x}_3$ is a state estimation error with $\hat{x}_3$ given by the super-twisting or the fixed-time observer.

The following theorem describes the convergence properties of the finite-time algorithm.

Theorem 2.4. *[31] Let the finite-time algorithm* (2.105)–(2.106) *be applied to the dynamics* (2.79), *and the regressor vector* Γ_1 *be PE. Then for any* $\bar{k}_1, \bar{k}_2 > 0$ *the identification error* $\tilde{\theta}$ *is UFTS.*

Proof. The error dynamics is given by

$$\dot{\tilde{e}}_3 = -\bar{k}_1 \tilde{\phi}_1(\bar{e}_3) + \Gamma_1(t)\tilde{\theta} + f_{n_1}(\hat{x}, y, u) - f_{n_1}(x, u) \tag{2.107}$$

$$\dot{\tilde{\theta}} = -\bar{k}_2 \tilde{\phi}_2(\bar{e}_3)\Gamma_1{}^T(t), \tag{2.108}$$

where

$$\tilde{\phi}_1(\bar{e}_3) := \lceil \bar{e}_3 \rfloor^{\frac{1}{2}} + \lceil \bar{e}_3 \rfloor^{1}, \tag{2.109}$$

$$\tilde{\phi}_2(\bar{e}_3) := \frac{1}{2} \lceil \bar{e}_3 \rfloor^{0} + \frac{3}{2} \lceil \bar{e}_3 \rfloor^{\frac{1}{2}} + \lceil \bar{e}_3 \rfloor^{1}. \tag{2.110}$$

Using the same arguments that in the previous proof, for the homogeneous algorithm, one can ensure that after a finite time $\hat{x}_3 \equiv x_3$, and thus, after a finite time $\tilde{e}_3 \equiv \bar{e}_3$ and $f_{n_1}(\hat{x}, y, u) \equiv f_{n_1}(x, u)$. Hence, the error dynamics can be rewritten as follows:

$$\dot{\tilde{e}}_3 = -\bar{k}_1 \tilde{\phi}_1(\tilde{e}_3) + \Gamma_1(t)\tilde{\theta} \tag{2.111}$$

$$\dot{\tilde{\theta}} = -\bar{k}_2 \tilde{\phi}_2(\tilde{e}_3)\Gamma_1{}^T(t). \tag{2.112}$$

Let us consider the following Lyapunov function candidate:

$$V(t, \tilde{e}_3, \tilde{\theta}) = \zeta^T P(t) \zeta, \tag{2.113}$$

where $\zeta^T := (\tilde{\phi}_1, \tilde{\theta}^T)$, and with $P(t)$ symmetric, bounded, and positive definite matrix, i.e. $0 < c_1 I \leq P(t) \leq c_2 I$, for all $t \geq 0$, satisfying

$$\dot{P}(t) = -\tilde{\phi}_1'(\tilde{e}_3) \left[P(t)A(t) + A^T(t)P(t) + Q(t) \right], \quad \forall t \geq 0, \tag{2.114}$$

$$A(t) = \begin{bmatrix} -\bar{k}_1 & \Gamma_1(t) \\ -\bar{k}_2 \Gamma_1{}^T(t) & 0 \end{bmatrix}, \tag{2.115}$$

where $Q(t) \geq c_3 I > 0$, for all $t \geq 0$, is a symmetric and positive definite matrix. The function V is continuous and continuously differentiable everywhere in $\mathbb{R}^4$, except on the set $\mathscr{S} = \left\{\left(\tilde{e}_3, \tilde{\theta}\right) \in \mathbb{R}^4 \mid \tilde{e}_3 = 0\right\}$, where it is not Lipschitz continuous. The derivative of V is not well defined for the partial derivative $\tilde{\phi}_1'(\tilde{e}_3)$, but if $\tilde{\theta}$ stays at zero just in isolated, i.e. the measure of the set $\mathscr{S}$ is zero, an analysis can be done. When $\tilde{e}_3 = 0$ in isolated points

$$\tilde{\phi}_2(\tilde{e}_3) = \tilde{\phi}_1'(\tilde{e}_3)\,\tilde{\phi}_1(\tilde{e}_3)\,, \tag{2.116}$$

$$\dot{\zeta} = \begin{bmatrix} \tilde{\phi}_1'(\tilde{e}_3)\left(-k_1\tilde{\phi}_1(\tilde{e}_3) + \Gamma_1(t)\,\tilde{\theta}\right) \\ -k_2\tilde{\phi}_2(\tilde{e}_3)\,\Gamma_1^T(t) \end{bmatrix} = \tilde{\phi}_1'(\tilde{e}_3)\,A(t)\,\zeta. \tag{2.117}$$

Note that this derivative is well defined. The time derivative of V along the trajectories of the system (2.111)–(2.112) is given by

$$\dot{V}\left(t, \tilde{e}_3, \tilde{\theta}\right) = \zeta^T \left[\dot{P}(t) + \tilde{\phi}_1'(\tilde{e}_3)P(t)A(t) + \phi_1'(\tilde{e}_3)A^T(t)P(t)\right]\zeta, \tag{2.118}$$

at the points where V is differentiable, i.e. for all points except those on the set $\{\tilde{e}_3 = 0\}$. This implies that

$$\dot{V}\left(t, \tilde{e}_3, \tilde{\theta}\right) = -\tilde{\phi}_1'(\tilde{e}_3)\,\zeta^T Q(t)\,\zeta = -\left(\frac{1}{2\,|\tilde{e}_3|^{\frac{1}{2}}} + 1\right)\zeta^T Q(t)\,\zeta. \tag{2.119}$$

Note that $|\tilde{e}_3|^{\frac{1}{2}} \leq \|\zeta\|_2$, with $\|\zeta\|_2^2 = |\tilde{e}_3| + 2\,|\tilde{e}_3|^{\frac{3}{2}} + \tilde{e}_3^2 + \|\tilde{\theta}\|_2^2$ as the Euclidean norm of ζ. Thus, $\dot{V}$ may be upper bounded as follows

$$\dot{V}\left(t, \tilde{e}_3, \tilde{\theta}\right) \leq -\left(\frac{1}{2}c_3\|\zeta\|_2 + c_3\|\zeta\|_2^2\right), \tag{2.120}$$

$$\leq -\frac{c_3}{2c_2^{\frac{1}{2}}}V^{\frac{1}{2}}\left(t, \tilde{e}_3, \tilde{\theta}\right) - \frac{c_3}{c_2}V\left(t, \tilde{e}_3, \tilde{\theta}\right). \tag{2.121}$$

Note that the previous expression is the derivative of V when the trajectories of system (2.111)–(2.112), i.e. $\varphi(t, t_0, \tilde{e}_3(t_0), \tilde{\theta}(t_0))$, are outside of the set $\mathscr{S}$, and it indicates that it is negative definite in $\mathscr{S}^c$, the complement of $\mathscr{S}$. To stay on the set $\mathscr{S}$ for an interval of time, it is necessary that $\tilde{e}_3(t) = 0$ and $\Gamma_1(t)\,\tilde{\theta}(t) = 0$ during that interval. From the PE condition, it follows that the condition $\Gamma_1(t)\,\tilde{\theta}(t) = 0$ cannot be satisfied during any interval $t \in [t, t + T_0]$, such that the trajectories of system (2.111)–(2.112) cannot stay in the set $\mathscr{S}$ for a long time, in fact if the system has sufficiently excitation, $\tilde{e}_3(t)$ crosses zero and does not stay there. Following the reasoning presented in [32], V is decreasing. It follows from Zubov's theorem [55] that the origin is asymptotically stable. Note that V is absolutely continuous, and it is therefore the integral of its derivative [56], then

$$
\begin{aligned}
&V\left(t,\varphi(t,t_0,\tilde{e}_3(t_0),\tilde{\theta}(t_0))\right) - V\left(t_0,\tilde{e}_3(t_0),\tilde{\theta}(t_0)\right) \\
&= \int_{t_0}^{t} \dot{V}\left(\tau,\varphi(\tau,t_0,\tilde{e}_3(t_0),\tilde{\theta}(t_0))\right) d\tau \le -\frac{c_3}{2c_2^{\frac{1}{2}}} \int_{t_0}^{t} V^{\frac{1}{2}}\left(\tau,\varphi(\tau,t_0,\tilde{e}_3(t_0),\tilde{\theta}(t_0))\right) d\tau.
\end{aligned}
\tag{2.122}
$$

The Bihari's inequality [55] implies that

$$
V\left(t,\varphi(t,t_0,\tilde{e}_3(t_0),\tilde{\theta}(t_0))\right) \le \left(V^{\frac{1}{2}}\left(t_0,\tilde{e}_3(t_0),\tilde{\theta}(t_0)\right) - \frac{c_3}{4c_2^{\frac{1}{2}}}(t-t_0)\right)^2, \tag{2.123}
$$

such that V, and also $\varphi(t,t_0,\tilde{e}_3(t_0),\tilde{\theta}(t_0))$, converge to zero in a finite time smaller than

$$
T\left(t_0,\tilde{e}_3(t_0),\tilde{\theta}(t_0)\right) = \frac{4c_2^{\frac{1}{2}}}{c_3} V^{\frac{1}{2}}\left(t_0,\tilde{e}_3(t_0),\tilde{\theta}(t_0)\right). \tag{2.124}
$$

This concludes the proof. □

Remark 2.2. *The detailed convergence proofs for the Homogeneous and finite-time algorithms can be seen in [52] and [31], respectively. In this chapter, the given proofs for Theorems 2.3 and 2.4 are only brief sketches; and the provided results are particularly given for the pendulum–cart system.*

2.5.3 Experimental results

In order to depict the workability of the previous identification algorithms, some experiments have been done over the same real pendulum–cart system by INTECO. All the experiments have been implemented with the Euler's integration method and with a sampling time equal to 1×10^{-2} s.

The control input u is designed as $u = 0.245\sin(3t)$ in order to excite the system from the initial position $x = [0,\pi,0,0]^T$. Then, the Homogeneous and the finite-time algorithms are implemented in MATLAB Simulink. The parameters for both algorithms were chosen as the observer parameters and are given in Table 2.3.

Table 2.3 Parameters of the identification algorithm

Parameter	Homogeneous alg.	Finite-time alg.
$\bar{k}_1$	20	2.3
$\bar{k}_2$	20	2.3
γ	0.75	–

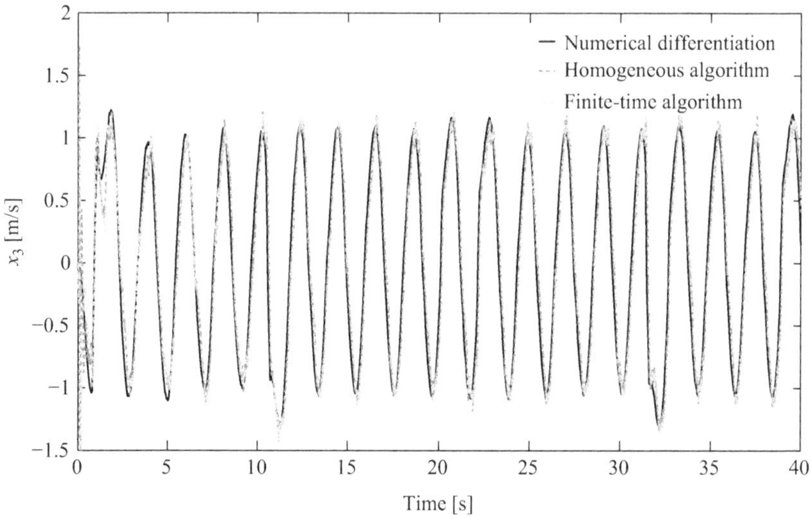

Figure 2.8 Cart velocity estimation

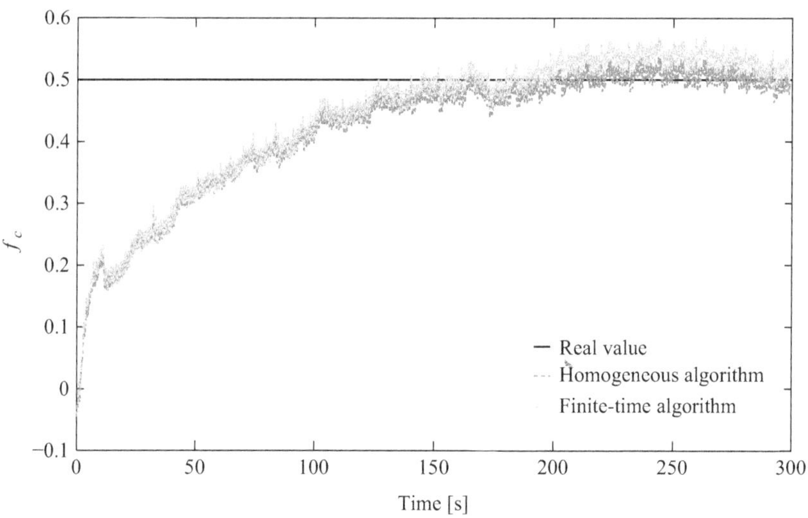

Figure 2.9 Dynamic friction coefficient identification

In order to provide the state estimation $\hat{x}_3$, the fixed-time observer was implemented as in the experimental Section 2.4.3. The results are presented in Figures 2.8–2.11. One can see that both identification algorithms provide a very good identification of the real friction coefficients.

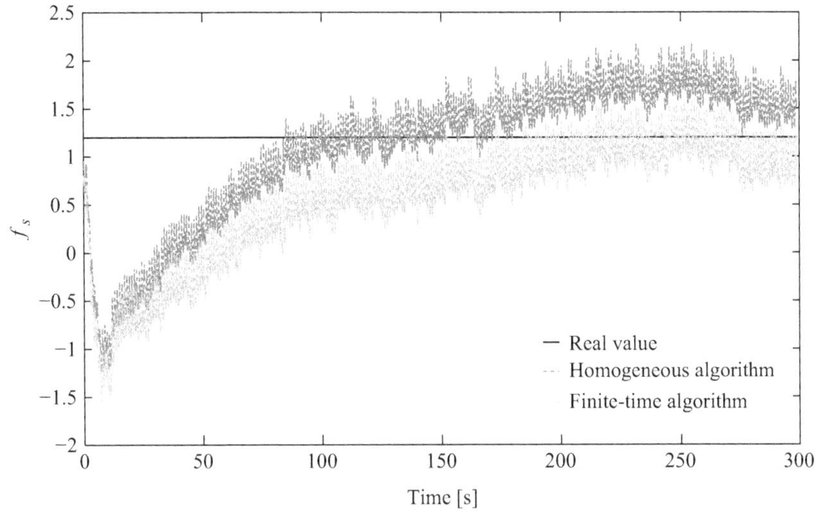

Figure 2.10 Dry friction coefficient identification

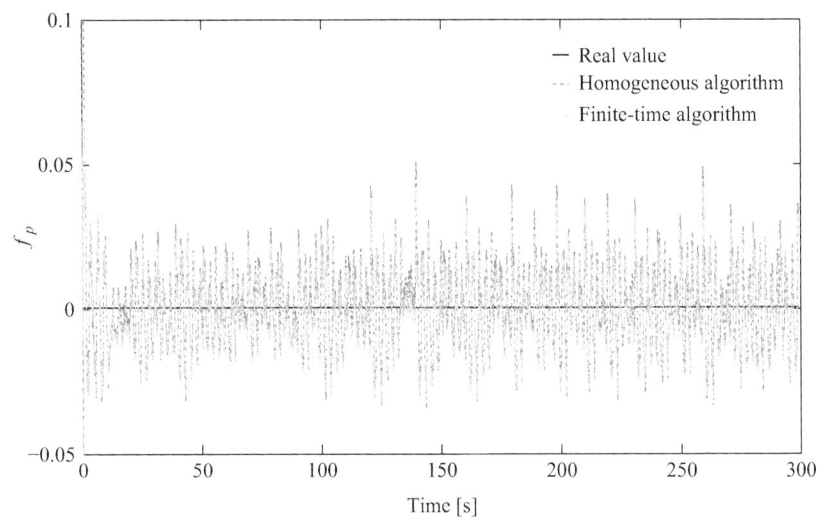

Figure 2.11 Rotational friction coefficient identification

In order to illustrate the performance of both observers the following index is implemented:

$$I = \left(\frac{1}{\Delta T} \int_{t-\Delta T}^{t} |\tilde{\theta}(\tau)|^2 d\tau \right)^{1/2}, \tag{2.125}$$

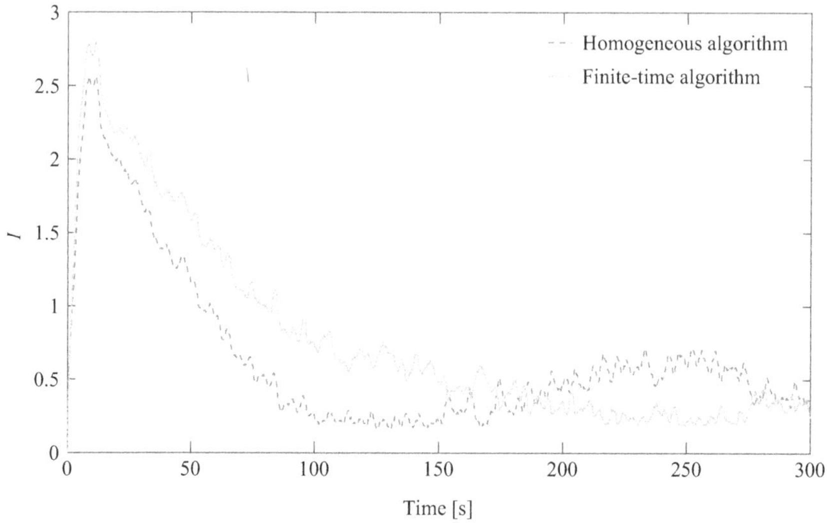

Figure 2.12 Performance index of identification

where $\tilde{\theta}$ is the identification error vector and ΔT is a time window width, which was selected as 2.

The behaviour of the performance index during the experiment is shown in Figure 2.12. It is easy to see that both algorithms provide practically the same index performance. On the other hand, as in the observer algorithms, the Homogeneous algorithm has more parameters than the finite-time algorithm and can be more difficult in its tuning.

2.6 Conclusions

The problems of state estimation and parameter identification for the pendulum–cart system have been addressed. Two HOSM observers, i.e. the super-twisting and the fixed-time, are proposed to estimate the state, in spite of disturbances; exactly and in a finite or fixed time, respectively. Based on the exact state estimation, two parameter identification algorithms, i.e. the Homogeneous and the finite-time, based on sliding-modes techniques are introduced to identify the unknown parameters of the system, i.e. the friction coefficients. The corresponding convergence proofs have been given based on Lyapunov function approach. Some experiments and comparisons have illustrated the effectiveness of the presented algorithms.

References

[1] C. Edwards and S. Spurgeon, *Sliding Mode Control: Theory and Applications*. London: Taylor and Francis, 1998.

[2] Y. Shtessel, C. Edwards, L. Fridman, and A. Levant, *Sliding Mode Control and Observation*. Intuitive theory of sliding mode control. New York, USA: Birkhauser, 2014.
[3] V. Utkin, J. Guldner, and J. Shi, *Sliding Modes in Electromechanical Systems*. London: Taylor and Francis, 1999.
[4] J. Slotine, J. Hedrick, and E. Misawa, 'On sliding observers for nonlinear systems', *Transactions of the ASME: Journal of Dynamic Systems, Measurement and Control,* 1987, **109**, 3, pp. 245–252.
[5] B. Walcott and S. Zak, 'State observation of nonlinear uncertain dynamical systems', *IEEE Transaction on Automatic Control*, 1987, **32**, 2, pp. 166–170.
[6] C. Edwards, S. Spurgeon, and R. Patton, 'Sliding mode observers for fault detection and isolation', *Automatica*, 2000, **36**, 4, pp. 541–553.
[7] C. Tan and C. Edwards, 'Sliding mode observers for detection and reconstruction of sensor faults,' *Automatica*, 2002, **38**, 10, pp. 1815–1821.
[8] M. Hautus, 'Strong detectability and observers', *Linear Algebra and Its Applications*, 1983, **50**, pp. 353–368.
[9] J. Barbot and T. Floquet, 'Super twisting algorithm-based step-by-step sliding mode observers for nonlinear systems with unknown inputs', *International Journal of Systems Science*, 2007, **8**, 10, pp. 803–815.
[10] F. J. Bejarano, L. Fridman, and A. Poznyak, 'Unknown input and state estimation for unobservable systems', *SIAM Journal on Control and Optimization*, 2009, **48**, 2, pp. 1155–1178.
[11] J. Davila, L. Fridman, and A. Levant, 'Second-order sliding-mode observer for mechanical systems', *IEEE Transactions on Automatic Control*, 2005, **50**, 11, pp. 1785–1789.
[12] L. Fridman, A. Levant, and J. Davila, 'Observation of linear systems with unknown inputs via high-order sliding-modes', *International Journal of System Science*, 2007, **38**, 10, pp. 773–791.
[13] L. Fridman, Y. Shtessel, C. Edwards, and X. Yan, 'Higher-order sliding-mode observer for state estimation and input reconstruction in nonlinear systems', *International Journal of Robust and Nonlinear Control*, 2008, **18**, 4–5, pp. 399–413.
[14] H. Du, C. Qian, S. Yang, and S. Li, 'Recursive design of finite-time convergent observers for a class of time-varying nonlinear systems', *Automatica*, 2013, **49**, 2, pp. 601–609.
[15] T. Menard, E. Moulay, and W. Perruquetti, 'A global high-gain finite-time observer', *IEEE Transactions on Automatic Control*, 2010, **55**, 6, pp. 1500–1506.
[16] W. Perruquetti, T. Floquet, and E. Moulay, 'Finite-time observers: Application to secure communication', *IEEE Transactions on Automatic Control*, 2008, **53**, 1, pp. 356–360.
[17] Y. Shen and Y. Huang, 'Uniformly observable and globally Lipschitzian nonlinear systems admit global finite-time observers', *IEEE Transactions on Automatic Control*, 2009, **54**, 11, pp. 2621–2625.

[18] A. Polyakov, D. Efimov, and W. Perruquetti, 'Homogeneous differentiator design using implicit Lyapunov function method', in *The 2014 European Control Conference*, Strasbourg, France, 2014, pp. 288–293.

[19] A. Polyakov, 'Nonlinear feedback design for fixed-time stabilization of linear control systems', *IEEE Transactions on Automatic Control*, 2012, **57**, 8, pp. 2106–2110.

[20] E. Cruz-Zavala, J. Moreno, and L. Fridman, 'Uniform robust exact differentiator', *IEEE Transactions on Automatic Control*, 2011, **56**, 11, pp. 2727–2733.

[21] M. Angulo, J. Moreno, and L. Fridman, 'Robust exact uniformly convergent arbitrary order differentiator', *Automatica*, 2013, **49**, 8, pp. 2489–2495.

[22] H. Huijberts, H. Nijmeijer, and R. Willems, 'System identification in communication with chaotic systems', *IEEE Transactions on Circuits and Systems – I: Fundamental Theory and Applications*, 2000, **47**, 6, pp. 800–808.

[23] Q. Zhongfang, 'Distributed parameter system identification and its application to geophysical parameter identification', in *The 1994 IEEE International Conference on Industrial Technology*, Guangzhou, China, 1994, pp. 799–802.

[24] W. Sun, H. Li, and Q. Han, 'Identification of mechanical parameters of hard-coating materials with strain-dependence', *Journal of Mechanical Science and Technology*, 2014, **28**, 1, pp. 81–92.

[25] J. Hasenauer, S. Waldherr, K. Wagner, and F. Allgöwer, 'Parameter identification, experimental design and model falsification for biological network models using semidefinite programming', *IET Systems Biology*, 2010, **4**, 2, pp. 119–130.

[26] P. Ioannou and J. Sun, *Robust Adaptive Control*. Englewood Cliffs, NJ, USA: Prentice-Hall, Inc., 1996.

[27] R. Isermann and M. Munchhof, *Identification of Dynamic Systems. An Introduction with Applications*. Berlin: Springer-Verlag, 2011.

[28] L. Ljung, *System Identification: Theory for the User*. New Jersey, USA: Prentice Hall Information and System Sciences Series, 1999.

[29] T. Soderstrom and P. Stoica, *System Identification*. Cambridge, Great Britain: Prentice Hall International, 1989.

[30] A. Polyakov and L. Fridman, 'Stability notions and Lyapunov functions for sliding mode control systems', *Journal of the Franklin Institute*, 2014, **351**, 4, pp. 1831–1865.

[31] J. Moreno and E. Guzman, 'A new recursive finite-time convergent parameter estimation algorithm', in *The 18th IFAC World Congress*, Milano, Italy, 2011, pp. 3439–3444.

[32] J. Moreno and M. Osorio, 'Strict Lyapunov functions for the super-twisting algorithm', *IEEE Transactions on Automatic Control*, 2012, **57**, 4, pp. 1035–1040.

[33] J. Davila, L. Fridman, and A. Poznyak, 'Observation and identification of mechanical systems via second order sliding modes', *International Journal of Control*, 2006, **79**, 10, pp. 1251–1262.

[34] N. M'sirdi, A. Rabhi, L. Fridman, J. Davila, and Y. Delanne, 'Second order sliding mode observer for estimation of velocities, wheel sleep, radius and stiffness', in *Proceedings of the 2006 American Control Conference*, Minneapolis, Minnesota USA, 2006, pp. 3316–3321.

[35] V. Adetola and M. Guay, 'Finite-time parameter estimation in adaptive control of nonlinear systems', *IEEE Transactions on Automatic Control*, 2008, **53**, 3, pp. 807–811.

[36] T. Accadia, F. Acernese, P. Astone *et al.*, 'A state observer for the Virgo inverted pendulum', *Review of Scientific Instruments*, 2011, **82**, 9, pp. 34–44.

[37] L. Canete and T. Takahashi, 'Modeling, analysis and compensation of disturbances during task execution of a wheeled inverted pendulum type assistant robot using a unified controller', *Advanced Robotics*, 2015, **29**, 22, pp. 1453–1462.

[38] M. Ramírez-Neria, H. Sira-Ramírez, R. Garrido-Moctezuma, and A. Luviano-Juárez, 'Linear active disturbance rejection control of underactuated systems: The case of the Furuta pendulum', *ISA Transactions*, 2014, **53**, 4, pp. 920–928.

[39] N. Khalid and A. Memon, 'Output feedback control of a class of under-actuated nonlinear systems using extended high gain observer', *Arabian Journal for Science and Engineering*, 2016, **41**, 9, pp. 3531–3542.

[40] J. Lee, R. Mukherjeea, and H. Khalil, 'Output feedback stabilization of inverted pendulum on a cart in the presence of uncertainties', *Automatica*, 2015, **54**, pp. 146–157.

[41] J. Huang, S. Ri, L. Liu, Y. Wang, J. Kim, and G. Pak, 'Nonlinear disturbance observer-based dynamic surface control of mobile wheeled inverted pendulum', *IEEE Transactions on Control Systems Technology*, 2015, **2014**, 6, p. 2400–2407.

[42] S. Ri, J. Huang, Y. Wang, M. Kim, and S. An, 'Terminal sliding mode control of mobile wheeled inverted pendulum system with nonlinear disturbance observer', *Mathematical Problems in Engineering*, 2014, **2014**, pp. 1–8.

[43] A. Roffel and S. Narasimhan, 'Extended Kalman filter for modal identification of structures equipped with a pendulum tuned mass damper', *Journal of Sound and Vibration*, 2014, **333**, 23, pp. 6038–6056.

[44] O. Garcia-Alarcon, S. Puga-Guzman, and J. Moreno-Valenzuela, 'On parameter identification of the Furuta pendulum', *Procedia Engineering*, 2012, **35**, pp. 77–84.

[45] G. Ronquillo-Lomeli, G. Ríos-Moreno, A. Gómez-Espinosa, L. Morales-Hernández, and M. Trejo-Perea, 'Nonlinear identification of inverted pendulum system using Volterra polynomials', *Mechanics Based Design of Structures and Machines*, 2016, **44**, pp. 5–15.

[46] Y. Liang and B. Feeny, 'Parametric identification of a base-excited single pendulum', *Nonlinear Dynamics*, 2006, **46**, 1, pp. 17–29.

[47] Y. Liang and B. Feeny, 'Parametric identification of a chaotic base-excited double pendulum experiment', *Nonlinear Dynamics*, 2008, **52**, 1, pp. 181–197.

[48] INTECO, *Pendulum-Cart System*. User's Manual, U.S.A.: INTECO, 2012.

[49] A. Filippov, *Differential Equations with Discontinuous Right-hand Sides*. Dordrecht, The Netherlands: Kluwer Academic Publishers, 1988.

[50] H. Khalil, *Nonlinear Systems*. New Jersey, USA: Prentice-Hall, 2002.

[51] W. Haddad, S. Nersesov, and L. Du, 'Finite-time stability for time-varying nonlinear dynamical systems', in *The 2008 American Control Conference*, Seattle, USA, 2008, pp. 4135–4139.

[52] H. Ríos, D. Efimov, L. Fridman, J. Moreno, and W. Perruquetti, 'Homogeneity based uniform stability analysis for time-varying systems', *IEEE Transactions on Automatic Control*, 2016, **61**, 3, pp. 725–734.

[53] K. Narendra and A. Annaswamy, *Stable Adaptive Systems*. Mineola, NY, USA: Dover Publications, 2005.

[54] H. Ríos, D. Efimov, A. Polyakov, and W. Perruquetti, 'Homogeneous time-varying systems: Robustness analysis', *IEEE Transactions on Automatic Control*, 2016. DOI:10.1109/TAC.2016.2540806.

[55] A. Poznyak, *Advanced Mathematical Tools for Control Engineers: Volume 1: Deterministic Techniques*. Amsterdam-Boston: Elsevier Science, 2008.

[56] V. I. Bogachev, *Measure Theory, Vol. 1*. Berlin: Springer-Verlag, 2007.

Chapter 3

Higher order sliding-mode stabilization of inverted cart–pendulum

Jesus Mendoza-Avila[1] *and Ismael Castillo*[1]

Abstract

Stabilization of a cart–pendulum system through a generalized higher order surface-controller design is presented. The singular Linear Quadratic (LQ) method presents a natural relationship between order of singularity of a given performance index and the order of sliding-mode controller. Thus, several arbitrary relative degree optimal sliding surfaces and its corresponding higher order sliding-mode controller can be specified for a given system.

Continuous higher order sliding-mode controllers are obtained through a robustification method for arbitrary relative degree nominal controllers based on integral sliding modes and supertwisting algorithm.

An agreement between the accuracy/complexity of the Continuous Higher Order Sliding Mode (CHOSM) controller and the limited accuracy offered by the system's sensors and actuators is obtained through several experiments.

3.1 Introduction

Sliding-mode control (SMC) is one of the main approaches to stabilize uncertain nonlinear systems. One of the main features of SMC is the exact compensation of matched disturbances and the finite-time convergence of the system trajectories to the sliding surface [1].

The conventional sliding-mode design consists of two steps [1–3]:

1. Design of a sliding variable.
2. Sliding-mode control design to enforce the finite-time convergence of system trajectories to the sliding surface.

One advantage of designing a sliding surface, in the first step, is the reduction of the system dynamics to the sliding dynamics. This should exhibit a desired behavior [2]

[1] Departamento de Ingeniería de Control y Robótica, Universidad Nacional Autónoma De México, Mexico

and, recently, a desired order of the reduced system specified by the optimization of a singular performance index [4].

On the other hand, for the second step, higher order sliding modes (HOSM) controllers have been developed [5,6] in order to ensure the finite-time convergence of the system state to a sliding surface of relative degree higher than one (closely related with a higher order of the sliding mode which is indeed the codimension of the sliding surface).

There are some advantages in the realization of HOSM, for example, when the controller's order is equal to the system's relative degree, the design of the sliding variable is not necessary and the controller can drive the system's states to zero in finite time. Furthermore, HOSM controllers provide the ultimate accuracy and can reduce the chattering by means of continuous control signal. Nevertheless, the complexity and computational resources grow exponentially as the order of the HOSM controller. Also, when a sliding variable is designed, there is no theoretical methodology to find the precision on the system's states in spite of the controller order.

In the present chapter, several sliding variables with different relative degrees and its corresponding HOSM controllers are to be tested in order to find a fair compromise between the order reduction and order complexity. Accuracy analysis is presented to show the performance of each case.

3.2 Singular LQ sliding variable design

In [4,7], an extended method to design higher order sliding variables based in optimal singular LQ stabilization [2] is presented.

Consider the system

$$\begin{aligned} \dot{x}(t) &= Ax(t) + Bu(t), \\ x(0) &= x_0, \ \| x_0 \| \leq L, \quad L \in \mathbb{R}^+ \end{aligned} \tag{3.1}$$

where $A \in \mathbb{R}^{n \times n}$ is the system state matrix, $B \in \mathbb{R}^n$ is the input matrix, $x(t) \in \mathbb{R}^n$ is the state vector, $u(t) \in \mathbb{R}$ is the control input. The initial conditions of the system must be in the set defined by L. Pair (A,B) is supposed controllable.

Besides, consider the cost function singular with respect to the control

$$J = \frac{1}{2} \int_{t_1}^{\infty} x^T(t) Q x(t) dt \tag{3.2}$$

where $Q \in R^{n \times n}$ is symmetric and positive semidefinite.

The design algorithm presented in [4] consist of the following steps:

1. Transformation of the system (3.1) to a controllable canonical form, by means of a nonsingular transformation T, such that the transformed system is

   $$\dot{z}(t) = \bar{A}z(t) + \bar{B}u(t), \tag{3.3}$$

   where $z = Tx$, $\bar{A} = TAT^{-1}$, $\bar{B} = TB$, and $\bar{A} \in \mathbb{R}^{n \times n}$, $\bar{B} \in \mathbb{R}^{n \times 1}$.

The matrices $\bar{A}$ and $\bar{B}$ have the *Brunovsky* form:

$$\bar{A} = \begin{bmatrix} 0 & 1 & 0 & \cdots & 0 \\ 0 & 0 & 1 & \cdots & 0 \\ \vdots & \vdots & \vdots & \ddots & \vdots \\ 0 & 0 & 0 & \cdots & 1 \\ -a_n & -a_{n-1} & -a_{n-2} & \cdots & a_1 \end{bmatrix}, \quad \bar{B} = \begin{bmatrix} 0 \\ 0 \\ \vdots \\ 0 \\ 1 \end{bmatrix}, \tag{3.4}$$

where $a_n, \ldots, a_1$ are the coefficients of the characteristic polynomial of (3.1).

The weighing matrix Q in the function (3.2) needs to be transformed to keep its original meaning for the transformed system

$$\bar{Q} = (T^{-1})^T Q T^{-1} \tag{3.5}$$

2. Singularity order of the performance index is defined by means of an analysis of the weighing matrix $\bar{Q}$ where the scalar $\bar{Q}_{22}$ is looked for.

   In higher order case, the matrix $\bar{Q}$ can only have the form

$$\bar{Q} = \begin{pmatrix} \bar{Q}_{11} & \bar{Q}_{12} & 0 & \cdots & 0 \\ \bar{Q}_{21} & \bar{Q}_{22} & 0 & \cdots & 0 \\ 0 & 0 & 0 & \cdots & 0 \\ \vdots & \vdots & \vdots & \ddots & \vdots \\ 0 & 0 & 0 & \cdots & 0 \end{pmatrix}, \tag{3.6}$$

   where $\bar{Q}_{11} \in \mathbb{R}^{n-k-1 \times n-k-1}$ is symmetric positive semidefinite matrix, $\tilde{Q}_{22} > 0$ and $\tilde{Q}_{22} \in R$.

   The singularity order is the number $i = k + 1$, where k is the number of zero columns of the weighing matrix $\bar{Q}$ such that $\tilde{Q}_{22} > 0$, $\tilde{Q}_{22} \in R$.

3. Perform partitions to the system (3.3) according to (3.6). The state vector z is partitioned in three subsets $z = [\bar{z}_1^T \; \bar{z}_2^T \; \bar{z}_3^T]^T$, where $\bar{z}_1$ is the sliding mode reduced dynamics, $\bar{z}_2$ is a virtual control and $\bar{z}_3$ the rest of the variables states.

$$\begin{aligned} \bar{z}_1 &= [z_1 \ldots z_{n-i}]^T, \\ \bar{z}_2 &= z_{n-i+1}, \\ \bar{z}_3 &= [z_{n-i+2} \ldots z_n]^T. \end{aligned} \tag{3.7}$$

   The obtained partitioned system has the form

$$\begin{bmatrix} \dot{\bar{z}}_1 \\ \dot{\bar{z}}_2 \\ \dot{\bar{z}}_3 \end{bmatrix} = \begin{bmatrix} \bar{A}_{11} & \bar{A}_{12} & \bar{A}_{13} \\ \bar{A}_{21} & \bar{A}_{22} & \bar{A}_{23} \\ \bar{A}_{31} & \bar{A}_{32} & \bar{A}_{33} \end{bmatrix} \begin{bmatrix} \bar{z}_1 \\ \bar{z}_2 \\ \bar{z}_3 \end{bmatrix} + \begin{bmatrix} \bar{B}_1 \\ \bar{B}_2 \\ \bar{B}_3 \end{bmatrix} u. \tag{3.8}$$

   Note that subsystem $\bar{z}_1$ is

$$\dot{\bar{z}}_1 = \bar{A}_{11}\bar{z}_1 + \bar{A}_{12}\bar{z}_2, \tag{3.9}$$

where $\bar{A}_{11}$ has the same dimension as $\bar{Q}_{11}$, $\bar{A}_{12}$ is a column matrix with the same rows as $\bar{Q}_{11}$, and the partitions $\bar{A}_{11}$ and $\bar{A}_{12}$ have the Brunovsky form:

$$\bar{A}_{11} = \begin{bmatrix} 0 & 1 & 0 & \cdots & 0 \\ 0 & 0 & 1 & \cdots & 0 \\ 0 & 0 & 0 & \cdots & 0 \\ \vdots & \vdots & \vdots & \ddots & 1 \\ 0 & 0 & 0 & \cdots & 0 \end{bmatrix}, \quad \bar{A}_{12} = \begin{bmatrix} 0 \\ 0 \\ 0 \\ \vdots \\ 1 \end{bmatrix}. \tag{3.10}$$

4. Design the sliding variable by means of the design of an optimal stabilizing control for the subsystem $\bar{z}_1$ solving an optimal control problem with the performance index

$$\bar{J} = \frac{1}{2}\int_{t0}^{\infty} (\bar{z}_1^T(t)\hat{Q}\bar{z}_1(t) + v^T(t)\hat{R}v(t))dt, \tag{3.11}$$

where the variable

$$v(t) = \bar{z}_2(t) + \bar{Q}_{22}^{-1}\bar{Q}_{12}^T\bar{z}_1(t), \tag{3.12}$$

eliminate crossterms of $\bar{z}_1(t)$ and $\bar{z}_2(t)$ due to the partitions in the system (3.3) and the weighing matrix $\bar{Q}$. $\hat{Q} = \bar{Q}_{11} - \bar{Q}_{12}\bar{Q}_{22}^{-1}\bar{Q}_{12}^T$ is the state weighing matrix for the subsystem $\bar{z}_1$ and $\hat{R} = \bar{Q}_{22}$ is the control weighing matrix. Due to the change variable (3.12), the subsystem (3.9) is

$$\dot{\bar{z}}_1 = \hat{A}\bar{z}_1 + \hat{B}v, \tag{3.13}$$

where

$$\begin{aligned} \hat{A} &= \bar{A}_{11} - \bar{A}_{12}\bar{Q}_{22}^{-1}\bar{Q}_{12}^T && \in R^{(n-i)\times(n-i)}, \\ \hat{B} &= \bar{A}_{12} && \in R^{n-i}. \end{aligned} \tag{3.14}$$

This standard problem is solved though

$$\hat{A}^T P + P\hat{A} - P\hat{B}\hat{R}^{-1}\hat{B}^T P + \hat{Q} = 0, \tag{3.15}$$

where P is the only positive definite solution of (3.15) if the pair $(\hat{A}, \bar{D})$ is observable with $\bar{D}$ a matrix such that $\bar{D}^T\bar{D} = \hat{Q}$.

A stabilizing optimal virtual control law with respect to (3.11) for the subsystem (3.13) is given by

$$\begin{aligned} \bar{z}_2 &= -K\bar{z}_1, \\ &= -(\hat{R})^{-1}(\hat{B}^T P + \bar{Q}_{12}^T)\bar{z}_1, \\ &= -(\bar{Q}_{22})^{-1}(\bar{A}_{12}^T P + \bar{Q}_{12}^T)\bar{z}_1. \end{aligned} \tag{3.16}$$

Then, the sliding variable with relative degree $r \geq 1$ can be designed as

$$\sigma = \bar{z}_2 + (\bar{Q}_{22})^{-1}(\bar{A}_{12}^T P + \bar{Q}_{12}^T)\bar{z}_1. \tag{3.17}$$

Note that the sliding variable's relative degree is equal to the performance index singularity order.

There exists a Toolbox for MATLAB® presented in [4] which permit to automatize the sliding variable design method described early.

3.2.1 MATLAB® function: SSLQ

$$[T, S_s, Q_c, U_{eqn}, Algorithm] = SSLQ(A, B, Q_b, AlphaGain) \tag{3.18}$$

Arguments

- A—System matrix of the original linear time invariant system.
- B—Input matrix of the original linear time invariant system.
- Q_b—Is the transformed weighing matrix $\bar{Q}$. For simplicity in the practice is better manipulate and adjust matrix $\bar{Q}$ directly in order to manage the order of singularity, and therefore, the relative degree of the designed sliding variable.
- AlphaGain—Is the gain of the quasicontinuous controller. This must satisfy $AlphaGain > F$, where $| f | \leq F$.

Outputs

- T—Nonsingular transformation matrix to a controllable canonical form, $z = Tx$.
- S_s—Matrix of order $(i-1) \times n$, containing sliding surface and its $i-1$ time derivatives. The sliding variable and its time derivatives are expressed as

$$[\sigma \; \dot{\sigma} \; \cdots \; \sigma^{(i-1)}]^T = S_s z. \tag{3.19}$$

- Q_c—String variable containing the quasicontinuous algorithm of order i.
- U_{neq}—Matrix of order $(1 \times n)$ containing the linear terms of nominal equivalent control.

$$u_{neq} = U_{neq} z \tag{3.20}$$

- Algorithm—String variable which contains all the algorithm in .m code.

3.3 Continuous higher order sliding-mode controllers

The main idea of HOSM is to design an arbitrary order-sliding variable and keep it in zero by means of properly chosen HOSM controller. There exist some HOSM controllers that ensure the convergence to zero of a sliding variable. For example, the sliding variables of relative degree one can be driven to zero by

- Supertwisting algorithm (STA) [8]

$$\begin{aligned} u &= -k_1 |\sigma|^{\frac{1}{2}} \operatorname{sign}(\sigma) + w, \\ \dot{w} &= -k_2 \operatorname{sign}(\sigma), \end{aligned} \tag{3.21}$$

  where k_1 and k_2 are the controller's gains and can be designed as is described in [9].

and the sliding variables of relative degree two by

- Integral discontinuous control (IDC) [10]

$$\begin{aligned} u &= -k_1|\sigma|^{\frac{1}{3}}\text{sign}(\sigma) - k_2|\dot{\sigma}|^{\frac{1}{2}}\text{sign}(\dot{\sigma}) + z, \\ \dot{z} &= -k_3\text{sign}(\sigma), \end{aligned} \tag{3.22}$$

  where k_1, k_2 and k_3 are the controller's gains and must be properly selected to achieve the control aim.

The algorithms STA and IDC presented above can compensate Lipschitz disturbances $|\dot{\phi}(x,t)| < C$; furthermore, these algorithms generate a continuous signal control reducing the chattering effect.

There exist some controllers for sliding variables of relative degree greater than two but in the most of cases those controllers produce discontinuous control signals. However, there exist the so-called integral sliding-mode control (ISMC) which is possible inherit the properties of exact compensation of Lipschitz perturbations to any continuous arbitrary order controller.

3.3.1 Integral sliding-mode control

A way to design a robust control is designing a controller which stabilize the nominal system, later applying a robustify method such that the final controller can stabilize the system with the same nominal convergence features but in presences of disturbances [11].

The sliding-mode-based auxiliary controller that compensates the disturbance from the beginning of the control action, while retaining the order of uncompensated system, is named integral sliding mode (ISM) controller [12,13].

Using ISMC, it is possible to inherit the SMC's properties of exact compensation of disturbances to an arbitrary order nominal controller respecting its convergence features.

3.3.2 Continuous integral sliding-mode control

In [14], a robustification method for arbitrary order controllers via STA is proposed. The main advantage of this method is that the robustified control signal is continuous which reduce significantly the chattering effect.

Considering an arbitrary order dynamic system

$$\begin{aligned} \dot{x}_i &= x_{i+1} \qquad i = 1, 2, \ldots, n-1 \\ \dot{x}_n &= f(x) + g(x)(u + \phi(t)), \end{aligned} \tag{3.23}$$

where $x = [x_1, x_2, \ldots, x_n]^T$ is the state vector, $u \in R$ is the control input, $f(x)$ and $g(x)$ are smooth state functions representing the nominal and known dynamics. In addition, $0 < k_m \leq g(x) \leq k_M$, $\phi(t)$ is a Lipschitz function which represents the matched disturbances acting on the system.

If for a nominal system ($\phi(t) = 0$) there exists a controller $u = u_0(x)$ such that the close loop system origin is stable, then it is possible to design a sliding variable and a controller, such that a disturbed system have a robust behavior with respect to Lipschitz disturbances, keeping all stability features of the nominal close loop system [14].

Defining the virtual sliding variable

$$\varphi(x(t), u_0(x(t))) = x_n(t) - \int_0^t (f(x(\tau)) + g(x(\tau))u_0(x(\tau)))d\tau \tag{3.24}$$

where $x(t)$ are the system's trajectories. When $\sigma(x(t), u_0(x(t))) = 0$, the system trajectories behave with nominal dynamics, even in presence of disturbances.

To ensure the finite-time convergence to zero of (3.24), a possible control law is

$$u = u_0 + v \tag{3.25}$$

where v is an additional control input. The sliding variable dynamics are given by

$$\dot{\varphi} = g(x)(v + \phi(t)). \tag{3.26}$$

Note that siding variable has relative degree one, and STA can be implemented to enforce the system trajectories to zero in finite-time.

Finally, the control law is given by

$$\begin{aligned} u &= u_0 + v \\ v &= -k_{1r}|\varphi|^{\frac{1}{2}}\operatorname{sign}(\varphi) + w \\ \dot{w} &= -k_{2r}\operatorname{sign}(\varphi) \\ \varphi &= x_n(t) - \int_0^t (f(x(\tau)) + g(x(\tau))u_0(x(\tau)))d\tau \end{aligned} \tag{3.27}$$

and stability of the system (3.23) is guaranteed even in presence of Lipschitz matched disturbances. Stability proof of controller (3.27) is presented in [14].

3.3.3 Arbitrary order controllers design

Some controller that can be used as nominal control in (3.27) are

- Exponential convergence controller:
  The LQR control is designed given the weighing matrices of the states Q and of input R then

  $$u(t) = -Kx(t), \tag{3.28}$$

  where K is the gain matrix given by $K = R^{-1}B^TX$, where X is the unique positive semidefinite solution of the algebraic Riccati equation

  $$A^TX + XA - XBR^{-1}B^TX + Q = 0 \tag{3.29}$$

  such that the performance index $J = \frac{1}{2}\int_{t0}^{\infty}(x^T(t)Qx(t) + u^T(t)Ru(t))dt$ is minimized [15].

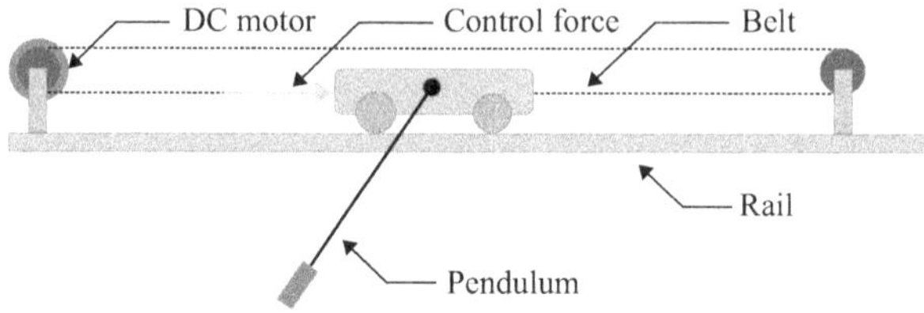

Figure 3.1 The cart–pendulum setup

- Finite-time convergence controller:
  Considering the Hong's controller of orders $r \leq 4$ as

$$u_1 = -k_1||\sigma_1|^{0.9}\text{sign}(\sigma_1)|\text{sign}(|\sigma_1|^{0.9}\text{sign}(\sigma_1)) \tag{3.30}$$

$$u_2 = -k_2||\sigma_2|^{1.11}\text{sign}(\sigma_2) - u_1|^{1.25}\text{sign}(|\sigma_2|^{1.11}\text{sign}(\sigma_2) - u_1) \tag{3.31}$$

$$u_3 = -k_3||\sigma_3|^{1.37}\text{sign}(\sigma_3) - u_2|^{1.57}\text{sign}(|\sigma_3|^{1.37}\text{sign}(\sigma_3) - u_2) \tag{3.32}$$

$$u_4 = -k_4||\sigma_4|^{1.71}\text{sign}(\sigma_4) - u_3|^{2}\text{sign}(|\sigma_4|^{1.71}\text{sign}(\sigma_4) - u_3) \tag{3.33}$$

  where σ is the sliding variable, k_1, k_2, k_3 and k_4 are the controller's gain which must be properly adjusted to achieve the control aim [16].

3.4 Cart–pendulum system

The cart–pendulum consists of a pole mounted on a cart in such a way that the pole can swing freely only on the vertical plane. The cart is driven by a DC flat motor through a belt. In order to swing and to balance the pole, the cart moves back and forward on a rail of 2 m of length (see, Figure 3.1) [17].

The cart–pendulum system can show some nontrivial and complex problems of the control theory as

- subactuated fourth-order system,
- nonlinear system,
- four state variables and one control variable,
- infinite number of equilibrium points,
- unstable equilibrium points,
- the control input is bounded,
- the cart position is bounded by the rail length [17].

Generally, the cart–pendulum control problem is to bring the pole, as fast as possible, to one of the unstable equilibrium points and keep it there despite of disturbances [17]. In order to design a control law for the cart–pendulum, its mathematical model is needed.

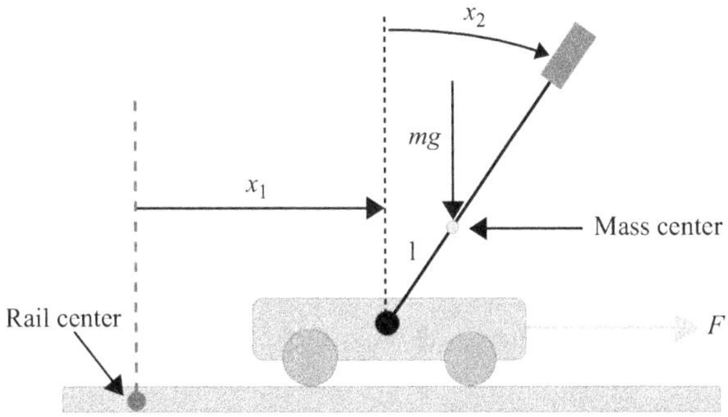

Figure 3.2 Cart–pendulum system

3.4.1 Mathematical model

The dynamic equations which can describe the behavior of the cart–pendulum system are given by four first-order ordinary differential equations.

3.4.1.1 Nonlinear model

The nonlinear mathematical model of the cart–pendulum considering the diagram depicted in Figure 3.2, is given by [17]

$$\begin{aligned}
\dot{x}_1 &= x_3 \\
\dot{x}_2 &= x_4 \\
\dot{x}_3 &= \frac{a_1 w_1(x,u) + w_2(x)\cos(x_2)}{d(x)} \\
\dot{x}_4 &= \frac{w_1(x,u)\cos(x_2) + a_2 w_2(x)}{d(x)}
\end{aligned} \tag{3.34}$$

where the state vector $x = \begin{bmatrix} x_1 & x_2 & x_3 & x_4 \end{bmatrix}^T$ represents: x_1 cart position (m), x_2 pendulum angular position (rad), x_3 cart velocity (m/s) and x_4 pendulum angular velocity (rad/s),

$$w_1(x,u) = k_1(u + \phi_1) + x_4^2 \sin(x_2) - k_2 x_3, \tag{3.35}$$

$$w_2(x,u) = -g\sin(x_2) - k_3(x_4 + \phi_2), \tag{3.36}$$

$$d(x) = b - \cos^2(x_2), \tag{3.37}$$

u is the control input (N), ϕ_1, ϕ_2 represent the external matched and unmatched disturbances (N), respectively.

The system's parameters

$$\begin{aligned}
&a_1 = \tfrac{J_p}{ml},\ a_2 = \tfrac{1}{l}, \quad b = a_1 a_2 = \tfrac{J_p}{ml^2}, \\
&k_1 = \tfrac{p_1}{ml},\ k_2 = \tfrac{f_c - p_2}{ml},\ k_3 = \tfrac{f_p}{ml},
\end{aligned} \tag{3.38}$$

given by the user's manual [17], are given in Table 3.1.

Table 3.1 Table of original system parameters

Symbol	Description	Value
m	Equivalent mass of cart and pendulum	0.872 (kg)
l	Distance from axis of rotation to center of mass of system	0.011 (m)
f_c	Dynamic cart friction coefficient	0.5 (N s/m)
f_s	Cart static friction	1.203 (N)
f_p	Rotational friction coefficient	$6.5 \cdot 10^{-5}$ (N m s/rad)
J_p	Pendulum inertial moment with respect to rotation axis	0.00292 (kg m^2)
g	Gravity acceleration	9.81 (m/s^2)
p_1	Control-force-to-PWM-signal ratio	9.4 (N)
p_2	Control-force-to-cart-velocity ratio	−0.548 (N s/m)
$u_{\max}$	Maximum value of PWM signal	0.5
m_c	Equivalent cart mass	0.768 (kg)
m_{ps}	Pole mass	0.038 (kg)
m_{pw}	Load mass	0.014 (kg)
R_l	Rail length	1.8 (m)
l_p	Pole length	0.5 (m)
l_{po}	Distance between pole mass center and axis rotation	0.107 (m)
l_c	Load length	0.03 (m)
l_{pw}	Distance between load mass center and axis rotation	0.354 (m)
T	Pendulum period	1.17 (s)
J	Inertial moment related to the mass center	0.00282 (kg m^2)

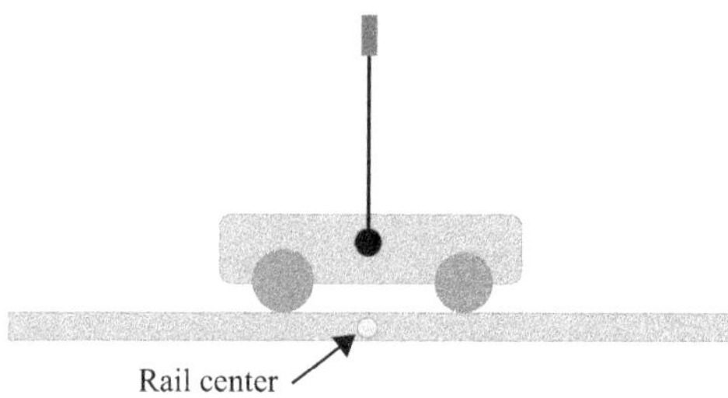

Figure 3.3 Cart–pendulum interesting equilibrium point

3.4.1.2 Equilibrium points

Considering $u = 0$, $\phi = 0$ and making (3.34) equal to zero the equilibrium points of the cart–pendulum are obtained. There exists many solutions for the equations system, however, the equilibrium point of interest is $x_0 = \begin{bmatrix} 0 & 0 & 0 & 0 \end{bmatrix}^T$.

This interesting point represents the upper pendulum position and the cart position at the middle rail as shown in Figure 3.3. The instability of the point x_0 is intuited.

3.4.1.3 Stability analysis

In order to prove the instability of x_0 the Chetaev's instability theorem is applied.

For the system (3.34) with x_0 as an equilibrium point consider the function

$$V(x) = x_1^2 + x_2^2 + x_3^2 + x_4^2 > 0, \tag{3.39}$$

and $V(0) = 0$. The derivative of (3.39) along the trajectories of the system (3.34) is

$$\begin{aligned}
\dot{V}(x) &= \frac{dV}{dx}\dot{x}, \\
&= [2x_1, 2x_2, 2x_3, 2x_4] \begin{bmatrix} x_3 \\ x_4 \\ \frac{0.304w_1(x,u)+w_2(x)\cos(x_2)}{d(x)} \\ \frac{w_1(x,u)\cos(x_2)+90.91w_2(x)}{d(x)} \end{bmatrix}, \\
&= 2x_1x_3 + 2x_2x_4 + 2x_3\frac{0.304w_1(x,u)+w_2(x)\cos(x_2)}{d(x)} \\
&\quad + 2x_4\frac{w_1(x,u)\cos(x_2)+90.91w_2(x)}{d(x)}.
\end{aligned} \tag{3.40}$$

If the trajectories of (3.34) are in $B_r = \{x \in R^n \mid \| x \| \leq r\}$ where $r = 0.01$ and defining the set $U = \{x_1, x_3 > 0, x_2, x_4 < 0, x \in B_r \mid V(x) > 0\}$. In U, (3.40) is positive definite, thus X_{01} is an unstable equilibrium point.

3.4.2 Linear model

The linear model of the system (3.34) can be obtained by Taylor series approximation [11,18] as

$$A = \frac{df}{dx}\Big|_{(x_0,u=0)} = \begin{bmatrix} 0 & 0 & 1 & 0 \\ 0 & 0 & 0 & 1 \\ 0 & \frac{g}{1-b} & \frac{a_1k_2}{1-b} & \frac{k_3}{1-b} \\ 0 & \frac{a_2g}{1-b} & \frac{k_2}{1-b} & \frac{k_3}{1-b} \end{bmatrix}, \quad B = \frac{df}{du}\Big|_{(x_0,u=0)} = \begin{bmatrix} 0 \\ 0 \\ \frac{a_1k_1}{b-1} \\ \frac{k_1}{b-1} \end{bmatrix}, \tag{3.41}$$

then, the linear approximation of (3.34) is given by

$$\dot{x} = \begin{bmatrix} 0 & 0 & 1 & 0 \\ 0 & 0 & 0 & 1 \\ 0 & \frac{g}{1-b} & \frac{a_1k_2}{1-b} & \frac{k_3}{1-b} \\ 0 & \frac{a_2g}{1-b} & \frac{k_2}{1-b} & \frac{k_3}{1-b} \end{bmatrix} x + \begin{bmatrix} 0 \\ 0 \\ \frac{a_1k_1}{b-1} \\ \frac{k_1}{b-1} \end{bmatrix} u. \tag{3.42}$$

The output system is given by

$$y = Cx = \begin{bmatrix} 1 & 0 & 0 & 0 \\ 0 & 1 & 0 & 0 \end{bmatrix} x, \tag{3.43}$$

due to the measurable states are x_1 and x_2.

3.4.2.1 Controllability and observability analysis

The controllability matrix for the system (3.42) is

$$C = [B \mid AB \mid A^2B \mid A^3B], \tag{3.44}$$

substituting the values in Table 3.1

$$C = \begin{bmatrix} 0 & 11.18 & -14 & 3.9 \\ 0 & 36.74 & -45.8 & -1171.1 \\ 11.18 & -13.95 & 3.9 & 12.3 \\ 36.74 & -45.82 & -1171.1 & 1516.2 \end{bmatrix} \tag{3.45}$$

and its determinant is $det(C) = -175{,}320{,}000 \neq 0$ then the matrix (3.45) has complete rank thus the system (3.42) is completely controllable.

The observability matrix for the system (3.42) is

$$O = [C' \mid A'C' \mid A'^2C' \mid A'^3C'], \tag{3.46}$$

substituting the values in Table 3.1

$$O = \begin{bmatrix} 1 & 0 & 0 & 0 \\ 0 & 1 & 0 & 0 \\ 0 & 0 & 1 & 0 \\ 0 & 0 & 0 & 1 \\ 0 & -0.3678 & -1.2469 & -0.0003 \\ 0 & -33.4333 & -4.0959 & -0.0003 \\ 0 & 0.4686 & 1.5560 & -0.3674 \\ 0 & 1.5165 & 5.1084 & -33.43216 \end{bmatrix}. \tag{3.47}$$

and its determinant is $det(C) = -1357.1 \neq 0$ then the matrix (3.47) has complete rank thus the system (3.42) is completely observable. As the system, (3.42) is controllable and observable is possible to design a control law to locally stabilize the equilibrium point x_0.

3.4.2.2 Controllable canonical form

The linear system (3.42) is transformed to the controllable canonical form by means of the transformation $\bar{x} = Tx$ where

$$T = \begin{bmatrix} 0.0028 & -0.0008 & 0 & 0 \\ 0 & 0 & 0.0028 & -0.0008 \\ 0 & 0.0272 & 0 & 0 \\ 0 & 0 & 0 & 0.0272 \end{bmatrix}. \tag{3.48}$$

Applying the transformation $\bar{A} = TAT^{-1}$, $\bar{B} = TB$, $\bar{C} = CT^{-1}$, the system (3.42) in controllable canonical form is

$$\dot{\bar{x}} = \begin{bmatrix} 0 & 1 & 0 & 0 \\ 0 & 0 & 1 & 0 \\ 0 & 0 & 0 & 1 \\ 0 & -40.1813 & -33.4325 & -1.2472 \end{bmatrix} \bar{x} + \begin{bmatrix} 0 \\ 0 \\ 0 \\ 1 \end{bmatrix} u \tag{3.49}$$

$$y = [0.4045 \quad -0.0066 \quad 47.9224 \quad 0]\bar{x} \tag{3.50}$$

The system (3.49) is important in the SMC design.

3.5 Continuous higher order sliding-mode controllers: cart–pendulum system realization

The steps to implement control schemes based on HOSM consist first in the design of the sliding variable and later choose an appropriate HOSM controller to ensure the convergence to the origin.

For the design of sliding variables for the system (3.34), the function (3.18) is used. An interesting output is given by $S_s \in \mathbb{R}^{r \times n}$ which contain the gain vector of the sliding variable given by

$$\begin{bmatrix} \sigma \\ \dot{\sigma} \\ \vdots \\ \sigma^{(r-1)} \end{bmatrix} = S_s \bar{x}. \tag{3.51}$$

The arguments of the function (3.18) are given by the matrices A and B of the system (3.42), the weighting matrix $Q_b = \bar{Q}$ from Table 3.2 and $AphaGain = 1$.

After the sliding variable design, corresponding order HOSM controller is selected and tuned for the experimental results.

Table 3.2 Sliding surfaces designed by SSLQ function

Relative degree	Q_b	S_s
1	$\begin{bmatrix} 560{,}000 & 0 & 138{,}680 & 0 \\ 0 & 100 & 0 & 0 \\ 138{,}680 & 0 & 34{,}609 & 0 \\ 0 & 0 & 0 & 14 \end{bmatrix}$	$\begin{bmatrix} 200 & 14 & 50 & 1 \end{bmatrix}$
2	$\begin{bmatrix} 544{,}000 & 0 & 131{,}750 & 0 \\ 0 & 1 & 0 & 0 \\ 131{,}750 & 0 & 34{,}000 & 0 \\ 0 & 0 & 0 & 0 \end{bmatrix}$	$\begin{bmatrix} 4 & 0.7 & 1 & 0 \\ 0 & 4 & 0.7 & 1 \end{bmatrix}$
2	$\begin{bmatrix} 5{,}299{,}200 & 0 & 1{,}093{,}650 & 0 \\ 0 & 1 & 0 & 0 \\ 1{,}093{,}650 & 0 & 230{,}000 & 0 \\ 0 & 0 & 0 & 0 \end{bmatrix}$	$\begin{bmatrix} 4.8 & 0.5 & 1 & 0 \\ 0 & 4.8 & 0.5 & 1 \end{bmatrix}$
3	$\begin{bmatrix} 49 & 0 & 0 & 0 \\ 0 & 9 & 0 & 0 \\ 0 & 0 & 0 & 0 \\ 0 & 0 & 0 & 0 \end{bmatrix}$	$\begin{bmatrix} 2.333 & 1 & 0 & 0 \\ 0 & 2.333 & 1 & 0 \\ 0 & 0 & 2.333 & 1 \end{bmatrix}$

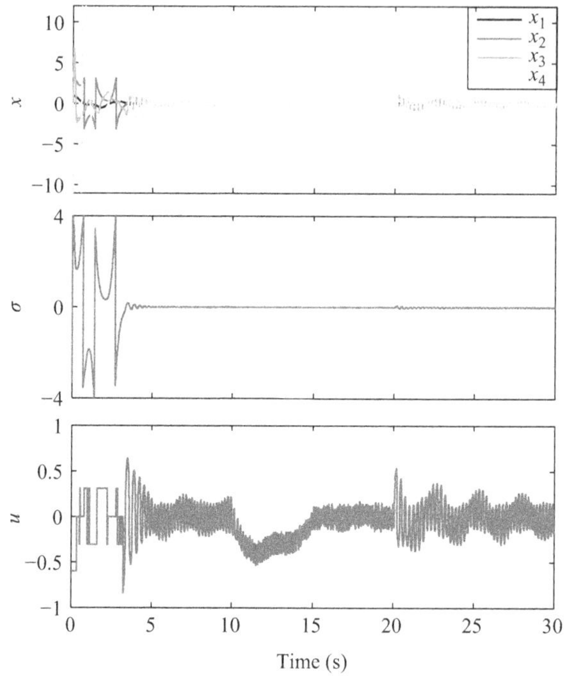

Figure 3.4 Supertwisting algorithm and sliding surface of relative degree one on cart–pendulum system

The experiments are implemented in a cart–pendulum plant developed by INTECO [17]. The control law is programmed in MATLAB with Simulink® and the RTWT (real-time windows target) MathWorks Toolbox is included. Through a data acquisition card RTDAC4/PCI, the communication between the plant and a computer is achieved. The control signal is computed by the PC and sent to the DC flat motor as a PWM signal.

In the experiments, the pendulum is driven near to x_0 by means of a swing-up algorithm which is taken from the test examples of the INTECO cart–pendulum system [17]. When the pendulum angular position is in a region less than 0.2 (rad), i.e., $||x_2|| < 0.2$, the control signal switch to a HOSM controller which stabilize the origin of the system. The initial conditions are different to zero for all states and the sample time is 1 ms. When the system is stable in steady state, two perturbations are exercised: the first one, is a matched perturbation at second 10 of the experiment which consist of applied force to the cart; the second one is an unmatched perturbation (perturbation in subactuated dynamics) at second 20 which consist of hit the pendulum. For the robustified controllers implemented in this work the robustification gains are $k_{1r} = 5$ and $k_{2r} = 5$.

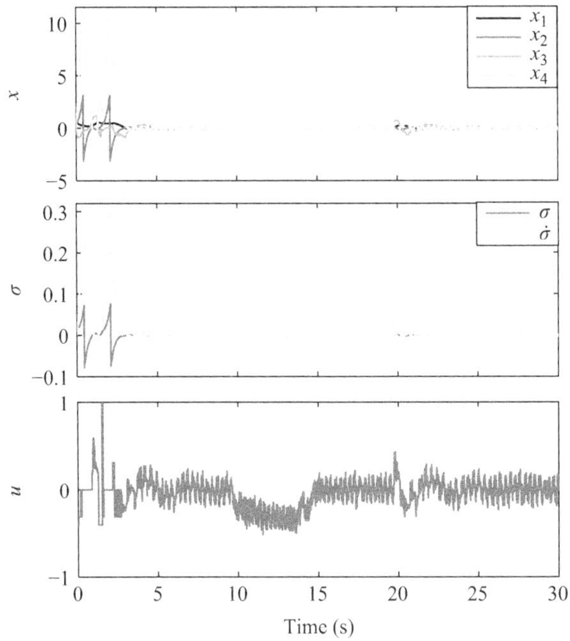

Figure 3.5 Integral discontinuous control and sliding surface of relative degree two on cart–pendulum system

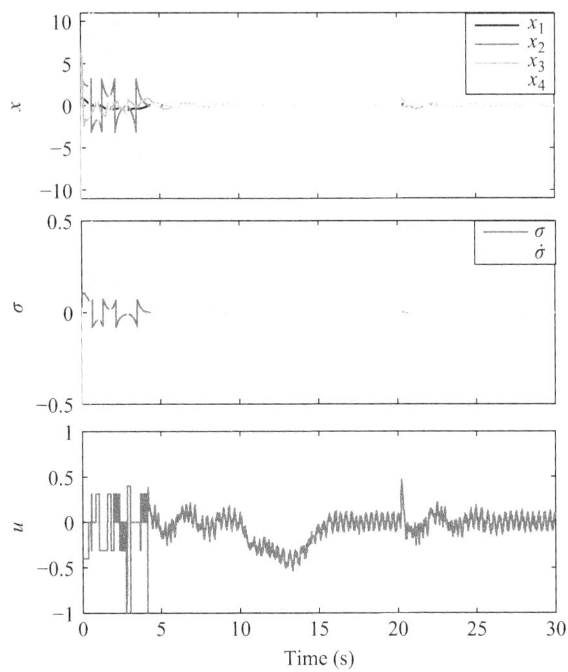

Figure 3.6 Robustified Second-Order Hong Controller and sliding surface of relative degree two on cart–pendulum system

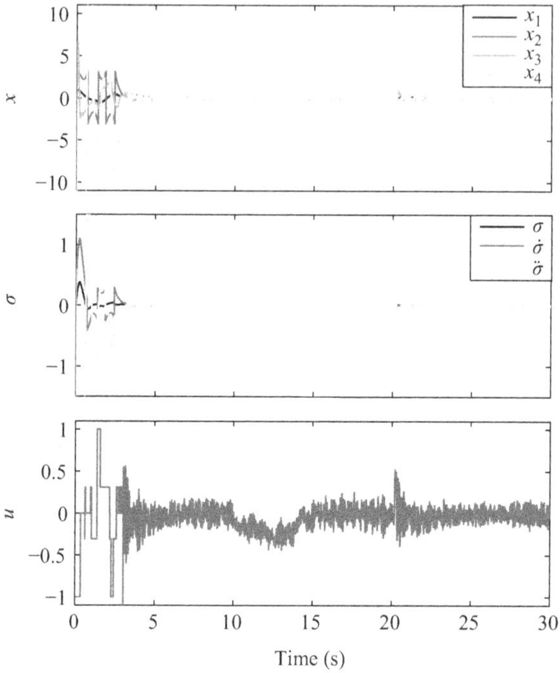

Figure 3.7 Robustified Third-Order Hong Controller and sliding surface of relative degree three on cart–pendulum system

3.5.1 Relative degree one SMC

A sliding variable of relative degree one designed for the system (3.34) is given by $\sigma_1 = [200\ 14\ 50\ 1]\tilde{x}$ (see Table 3.2). To drive the system trajectories to the sliding surface in finite-time, the *Supertwisting algorithm* (3.21) is used with gains $k_1 = 20$ and $k_2 = 13$. The experimental results are shown in Figure 3.4.

3.5.2 Relative degree two SMC

A sliding variable of relative degree two designed for the system (3.34) is given by $\sigma_2 = [4\ 0.7\ 1\ 0]\tilde{x}$ (see Table 3.2). The system trajectories can be carried to the sliding surface in finite-time by *Integral Discontinuous Control* (3.22) with gains $k_1 = 26$, $k_2 = 25$ and $k_3 = 10$. The experimental results are shown in Figure 3.5.

Another controller which can drive to zero sliding variables of relative degree two in finite-time is the *Robustified second-order Hong's controller* (3.31). In this case, the sliding variable is given by $\sigma_2 = [4.8\ 0.5\ 1\ 0]\tilde{x}$ (see Table 3.2) and the controller gains are $k_1 = 4.3$, $k_2 = 120$. The experimental results are shown in Figure 3.6.

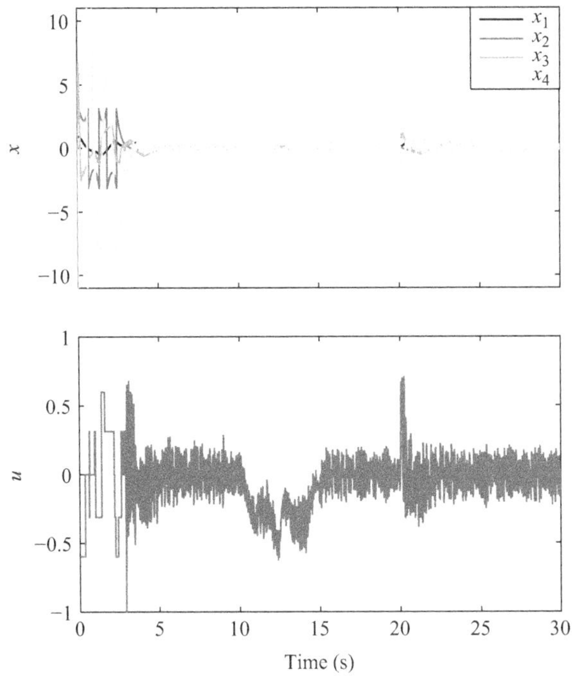

Figure 3.8 Robustified Fourth-Order Hong Controller on cart–pendulum system

3.5.3 Relative degree three SMC

A sliding variable of relative degree three designed for the system (3.34) is given by $\sigma_3 = [2.333\ 1\ 0\ 0]\tilde{x}$ (see Table 3.2). In order to drive the sliding variable to zero in finite-time the *Robustified Third-Order Hong Controller* (3.32) is used with gains $k_1 = 1, k_2 = 1.6, k_3 = 70$. The experimental results are shown in Figure 3.7.

3.5.4 Relative degree four SMC

To stabilize the origin of the system (3.34) in finite time the *Robustified Fourth-Order Hong Controller* (3.33) is used with gains $k_1 = 0.8, k_2 = 0.9, k_3 = 2, k_4 = 140$. The experimental results are shown in Figure 3.8.

Another choice to stabilize the trajectories of the system in the unstable equilibrium point x_0 is the *Robustified LQR controller* (3.28) with gains $K = [7.2\ 42\ 8.4\ 7.2]$. Note that this algorithm gives exponential convergence to the trajectories of the system but with insensitivity with respect to the matched perturbations. The experimental results are shown in Figure 3.9.

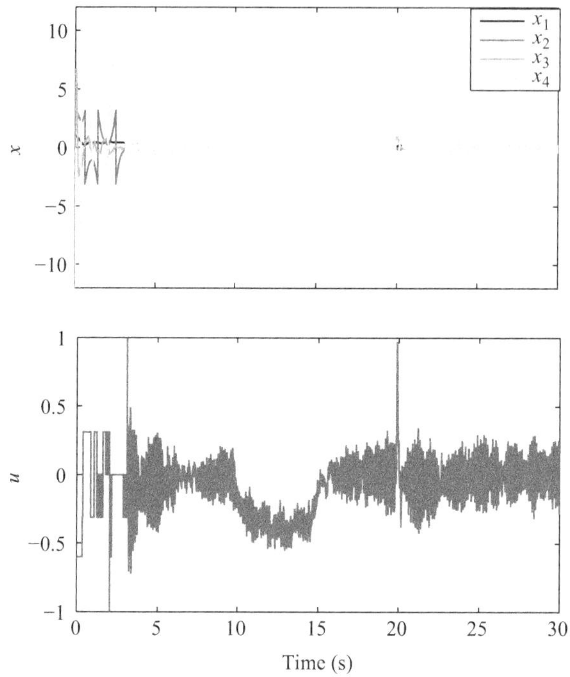

Figure 3.9 Robustified LQR Controller on cart–pendulum system

3.6 Results

The experiments show that HOSM controllers can stabilize the unstable equilibrium point of the system in finite-time (except LQR). Moreover, the controllers provide insensibility to Lipschitz continuous matched disturbances. A quantitative analysis is made to determinate that which controller has the best performance with respect to the accuracy of the sliding variables and states [8].

3.6.1 Accuracy analysis

An accuracy analysis of realized controllers can be done as [8] proposed, let $l = [r]$ be the maximum integer not exceeding r. If the lth derivative $\sigma(t, x(t, \varepsilon))$ is uniformly bounded in ε, for the steady-state part of $x(t, \varepsilon)$, then there exist positive constant $C_1, C_2, \ldots, C_{L-1}$, such that for the steady-state process the following inequalities hold

$$|\sigma| \leq C_0\tau^l,\ |\dot{\sigma}| \leq C_1\tau^{l-1}, |\ddot{\sigma}| \leq C_2\tau^{l-2}, \ldots, |\sigma^{l-1}| \leq C_{l-1}\tau. \tag{3.52}$$

In Table 3.3, it can be seen the accuracy, both in the sliding surface and in the states, obtained in the experiments realized, the accuracy is measured as the maximum value when the system is steady state stable.

Table 3.3 Accuracy of controllers to stabilize the cart–pendulum with sample time 1 ms

Controller	$\lvert\sigma\rvert$	$\lvert\dot{\sigma}\rvert$	$\lvert\ddot{\sigma}\rvert$	$\lvert x_1\rvert$	$\lvert x_2\rvert$	$\lvert x_3\rvert$	$\lvert x_4\rvert$
Super twisting Algorithm	$15 \cdot 10^{-3}$	*	*	0.02	0.025	0.25	0.5
Integral Discontinuous Control	$440 \cdot 10^{-6}$	$7 \cdot 10^{-3}$	*	0.015	0.022	0.08	0.33
Robustified Second-Order Hong Controller	$300 \cdot 10^{-6}$	$5 \cdot 10^{-3}$	*	0.043	0.033	0.12	0.27
Robustified Third-Order Hong Controller	$1.5 \cdot 10^{-3}$	$3 \cdot 10^{-3}$	$28 \cdot 10^{-3}$	0.018	0.018	0.085	0.3
Robustified Fourth-Order Hong Controller	*	*	*	0.015	0.024	0.07	0.28
Robustified LQR	*	*	*	0.021	0.011	0.085	0.35

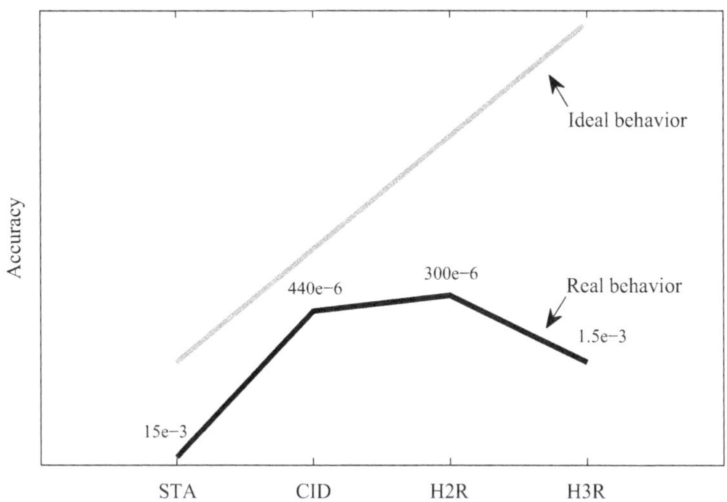

Figure 3.10 Sliding surface accuracy in the experiments

In theory, the accuracy of the sliding surface increases according to the order of the controller. In physical realization, it does not happen. In Figure 3.10, the difference between ideal behavior and real behavior is shown.

In Figure 3.10, it can be seen that in the experiments STA have less accuracy than IDC and robustified second-order Hong controller, but this behavior is no preserve

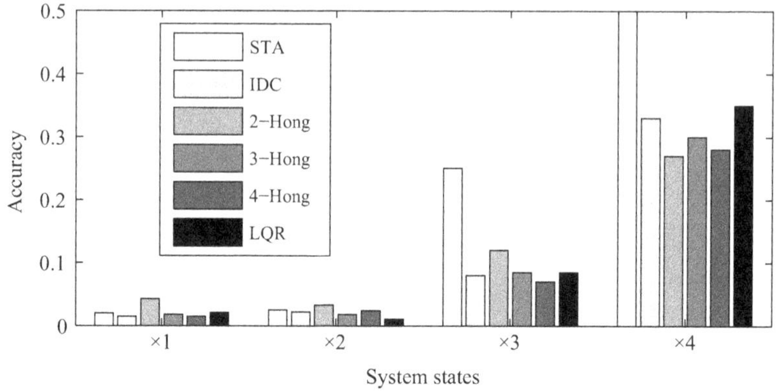

Figure 3.11 System states accuracy in the experiments

in higher order controllers due to the uncertainties in actuators and sensors of the system. However, the accuracy in the system states is similar in all the experiments, as Figure 3.11 shows.

3.7 Conclusions

In this chapter, the generalized higher order two-step sliding-mode methodology to stabilize a cart–pendulum system is implemented. The realization of several Sliding-Mode variables and controllers allows to find a fair compromise between the precision-performance and complexity of the controller that theory is not able to predict.

The sliding variable design is systematically implemented due to the usage of a MATLAB Toolbox for singular LQ design, while HOSM controllers are built by means of a robustification method which allows to inherit the exact compensation of Lipschitz matched perturbations to any arbitrary order nominal controller.

After several experiments, the accuracy analysis shows that the best performance is achieved in a case of order of singularity between the extreme cases, i.e., for this fourth-order system, the best performance is achieved in case of order two.

Moreover, it is shown that HOSM technique is a good option to stabilize uncertain nonlinear systems like the cart–pendulum. The exact compensation of matched disturbances can be appreciated in each experiment.

References

[1] Y. Shtessel, C. Edwards, L. Fridman, and A. Levant. *Sliding Mode Control and Observation*. Series: Control Engineering. Brikhauser: Basel, 2013.

[2] V. I. Utkin, *Sliding Modes in Control and Optimization*, Vol. 116. Berlin: Springer-Verlag Berlin, 1992.

[3] C. Edwards and S. K. Spurgeon, *Sliding Mode Control: Theory and Applications*. Padstow, UK: CRC, 1998.

[4] I. Castillo and L. Fridman. 'Matlab toolbox for singular lq based sliding mode control design', In *Decision and Control (CDC), 2013 IEEE 52nd Annual Conference on*, IEEE, 2013, pp 4242–4247.

[5] L. Fridman and A. Levant. *Higher Order Sliding Modes in Sliding Mode Control in Engineering*. J.P. Barbot, W. Perriguetti (Eds.), New York: Marcel Dekker, 2002.

[6] A. Levant. 'Higher-order sliding modes, differentiation and output–feedback control', *International Journal of Control*, 2003, **76**, pp. 924–941.

[7] I. Castillo, F. Castaños, and L. Fridman. 'Sliding surface design for higher-order sliding modes', *Recent Trends in Sliding Mode Control*, 2016, **102**, p. 29.

[8] A. Levant. 'Sliding order and sliding accuracy in sliding mode control', *International Journal of Control*, 1993, **58**, 6, pp. 1247–1263.

[9] J. A. Moreno and M. Osorio. 'Strict lyapunov functions for the super-twisting algorithm', *IEEE Transactions on Automatic Control*, 2012, **57**, 4, pp. 1035–1040.

[10] C. A. Zamora, J. A. Moreno, and S. Kamal. "Control integral discontinuo para sistemas mecánicos," in 2013 *Congreso Nacional de Control Automático (CNCA AMCA)*. Ensenada, Baja California, Mexico: Asociación de México de Control Automático (AMCA), Oct. 2013, pp. 11–16. [Online]. Available: http://amca.mx/memorias/amca2013/articulos/0155.pdf

[11] H. Khalil. Nonlinear Systems (third ed.). Upper Saddle River, NJ: Prentice-Hall, 2002.

[12] V. Utkin and J. Shi. 'Integral sliding mode in systems operating under uncertainty conditions', *Decision and Control, 1996, Proceedings of the 35th IEEE Conference on*, IEEE, 1996. volume 4, pp. 4591–4596.

[13] L. Fridman, A. S. Poznyak, and F. J. Bejarano. *Robust output LQ Optimal Control via Integral Sliding Modes*. New York: Springer, 2014.

[14] I. Castillo, J. A. Moreno, and L. Fridman. "Método para robustecer controladores para sistemas de orden arbitrario vía algoritmo super-twisting", in *XVI Congreso Latinoamericano de Control Automático (CLCA AMCA)*. Cancun, Quintana Roo, Mexico: Asociación de México de control Automático (AMCA), Oct. 2014, pp. 206–211. [Online]. Available at: http://amca.mx/memorias/amca2014/

[15] S. Skogestad and I. Postlethwaite. *Multivariable Feedback Control: Analysis and Design*, Vol. 2. New York: Wiley, 2007.

[16] Y. Hong, J. Wang, and Z. Xi. 'Stabilization of uncertain chained form systems within finite settling time', *Automatic Control, IEEE Transactions on*, 2005, **50**, 9, pp. 1379–1384.

[17] INTECO. Pendulum–Cart System User's Manual, 2008.

[18] J.-J. E. Slotine and W. Li. *Applied Nonlinear Control*, Vol. 199. Englewood Cliffs, NJ: Prentice-Hall, 1991.

Chapter 4

Stabilization and tracking control of the inverted pendulum on a cart via a modified PSO fractional order PID controller

Haifa Mehdi[1] *and Olfa Boubaker*[1]

Abstract

In this chapter, a fractional-order proportional integral derivative controller, $PI^{\lambda}D^{\mu}$, based super-twisting observer for the cart inverted pendulum system is proposed. The mathematical model of the underactuated robotic system is derived using Lagrange equation and Grunwald–Letnikov fractional calculus with the physical parameters of a commercial device, the Googol Technology experimental Laboratory. $PI^{\lambda}D^{\mu}$ parameters are optimized using a modified Particle Swarm intelligence optimization approach with the help of a multiobjective fitness function. A comparative analysis with the classical Particle Swarm Optimization algorithm shows the superiority of the proposed approach.

4.1 Introduction

In spite of its simple structure, the inverted pendulum on a cart remains one of the most difficult and interesting problems in control theory. Such system is an unstable with highly nonlinear dynamics, a nonminimum phase and an underactuated system, so thus an excellent benchmark for testing different control algorithms. Nevertheless, many standard techniques in control theory are ineffective when tested on the cart inverted pendulum for many reasons [1]. For example, many geometric properties of the system are lost when the pendulum moves through horizontal positions [2]. Moreover, the output-zeroing manifold does not contain any equilibrium points, the relative degree of the system is not constant and the controllability distribution of the system does not have a constant rank. The system is therefore not linearizable [3].

On the other hand, the control problem of the inverted pendulum system can be broadly divided into three sections, swing-up control, stabilization and tracking control. Swing up control is basically used to swing the pendulum rod from pending

[1]National Institute of Applied Sciences and Technology (INSAT), Tunisia

position to stabilization zone. Then, stabilization control is essential to uphold it around the unstable equilibrium point, in upright position, for long interval [4]. So, finding a controller that can render the vertical upward position of the pendulum globally asymptotically stable has been considered a major problem.

In literature, there are many ways to bring the pendulum to the upper half plane [5]. One of the most popular is based on energy control [6,7]. Moreover, there are several different techniques accessible for stabilization and tracking control of inverted pendulum system [8–10]. One of the most used is proportional integral derivative (PID) controller. In fact, PID control is still the most broadly used strategies in the industry to control various real-world processes because of the design simplicity and implementation [11]. Nevertheless, in traditional PID control, there are some limitations. For example, it is a well-known fact that conventional PID controllers cannot effectively control high-order or nonminimum-phase (linear) processes. The quality of control is not also satisfactory when the process is nonlinear or has large dead times [11]. So to improve the performances of PID controllers systems, many researchers have attempted to develop new generations of such compensators. One of the main interesting suggested methods is the fractional-order PID (FOPID) controller approach proposed by Podlubny [12]. This type of PID controller provides us with more tuning parameters compared to the conventional one, and consequently, more control objectives can come into account when it is applied. Fractional-order calculus is a branch of mathematics which generalizes the conventional integer-order calculus to arbitrary orders. FOPID control is currently an emerging technology control with proven performances in numerous applications [13,14]. However, tuning FOPID controllers is a challenging work since there are more parameters to tune than the conventional PID. Indeed, five parameters must be tuned, which means two more parameters than the classical PID controller. More efficient global approach is then required to optimize these parameters automatically. In recent years, one of the most interesting evolutionary optimization techniques used to optimize such controllers for robotic systems is the Particle Swarm Optimization (PSO) [14–17].

The PSO algorithm, introduced by Eberhart and Kennedy in 1995 [18], is a swarm-based metaheuristic technique that models the social behavior of bird flocking. It is well adapted to the optimization of nonlinear functions in multidimensional space. PSO is easy to implement and may outperform other evolutionary algorithms [19]. However, it is liable to trap local optimization and cause premature convergence.

On the other hand, the state estimation problem is one of the most important problems in control theory and for robotic systems. There exist many kinds of observers in the literature in this framework [16,20–22]. The observation problem has been also studied in different types of pendulums. In the context of state estimation, in [23] Kalman filters are introduced to estimate the state of the Virgo inverted pendulum. To estimate the state of the wheeled inverted pendulum, a velocity estimator and linear observers have been presented by [24,25]. An extended high-gain observer and a full-order state observer have been implemented in the inverted pendulum–cart by Lee *et al.* [26] and Hashimoto *et al.* [27].

Motivated by the above discussions, this chapter proposes a FOPID controller based on the Grunwald–Letnikov (GL) fractional calculus and tuned by a modified

PSO (MPSO) intelligence for stabilization and tracking control of the inverted pendulum system. In addition, the problem of state estimation for the pendulum–cart system is addressed in spite of disturbances using a super-twisting observer.

The rest of the paper is divided into the following sections. Section 4.2 gives a description of the inverted pendulum system and derivation of system equations in state space form. Section 4.3 describes the super-twisting observer, the FOPID-based controller design and gives details about MPSO algorithm. Section 4.4 gives a complete description of the control strategy with simulation results and comparative analysis between proposed MPSO-FOPID controllers and PSO-FOPID controllers for different scenarios.

4.2 The cart-inverted pendulum system

4.2.1 System presentation

Consider the cart inverted pendulum system shown by Figure 4.1. The aim is to move the cart along the x-direction to a desired point without the pendulum falling. First, we assume that the pendulum rod is mass-less and that the cart mass and the pendulum mass are denoted as M and m, respectively. The pendulum rod length is l. There is an externally x-directed force on the cart, F, a gravity force mg acts on the point mass at all times and a force of friction F_{fric} between the cart and the track. The coordinate system considered is shown in Figure 4.1, where x represents the cart position, and θ is the tilt angle referenced to the vertically upward direction.

In this chapter, the parameters of the experimental device of the Googol Technology Linear single inverted pendulum (GLIP2001) system is considered for simulation results [28]. The system consists of the following components: the mechanical system, the sensors and actuators, the control hardware and the control software. The block diagram of the setup is shown in Figure 4.2. The system is a single-input and multiple-output system. The voltage acts as a control input for the system and the linear cart displacements and angular position of pendulum are the outputs of the system. Moreover, the mechanical system design, shown in Figure 4.3, consists of a linear servo motion cart with a freely pivoted pendulum above it, a base, a sliding shaft, a timing belt, a timing wheel, an optical encoder and a limit switch.

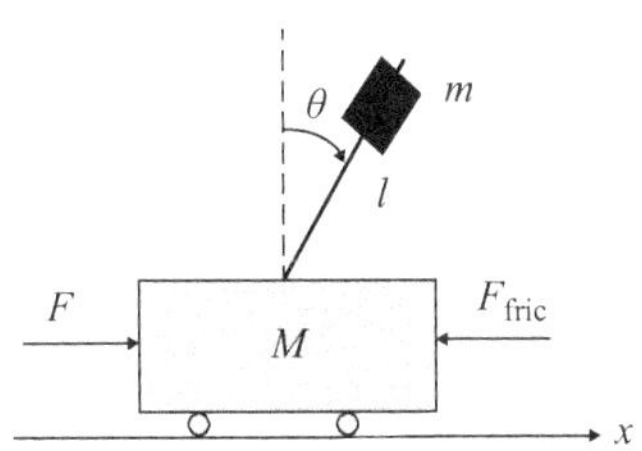

Figure 4.1 Inverted pendulum on a cart

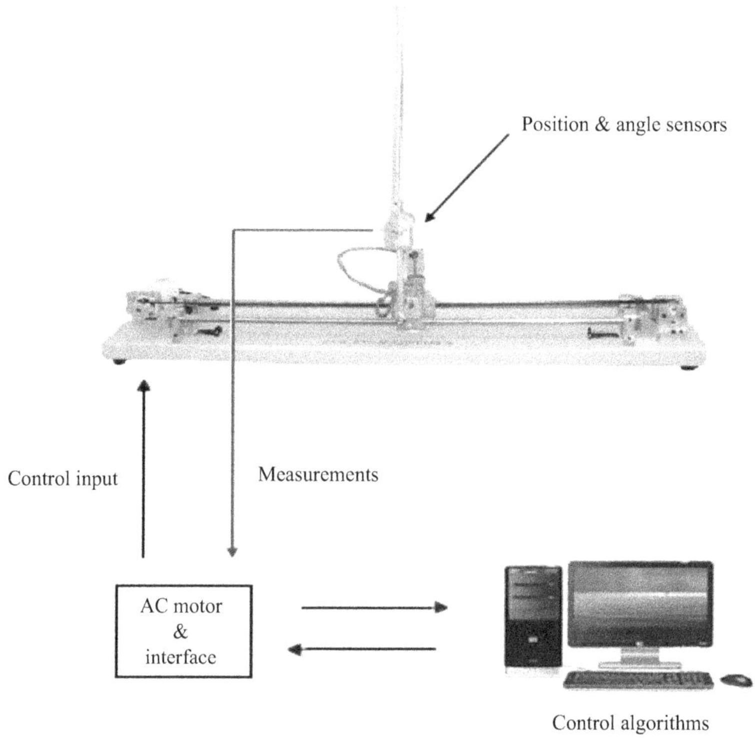

Figure 4.2 Block diagram of the inverted pendulum system

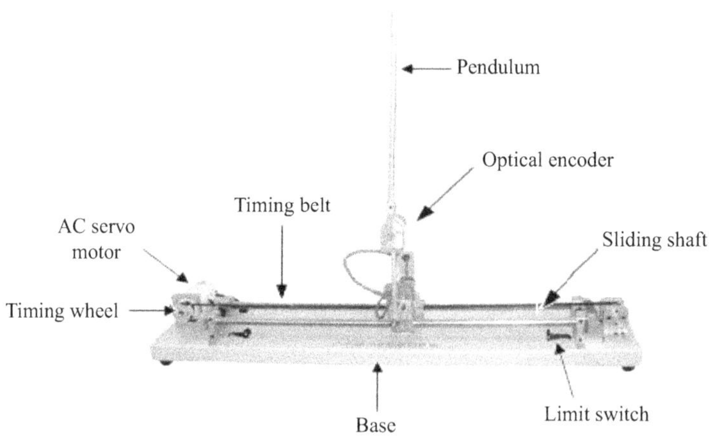

Figure 4.3 Mechanical system of the GoogolTech Linear single inverted pendulum

4.2.2 Dynamic modeling

The Lagrange's equations of the inverted pendulum system are given as follows:

$$\frac{d}{dt}\left(\frac{\partial L}{\partial \dot{x}}\right) - \frac{\partial L}{\partial x} = F + F_{\text{fric}} \tag{4.1}$$

$$\frac{d}{dt}\left(\frac{\partial L}{\partial \dot{\theta}}\right) - \frac{\partial L}{\partial \theta} = 0 \tag{4.2}$$

where L is the Lagrangian defined as

$$L = \frac{1}{2}(M+m)\dot{x}^2 + \frac{1}{2}ml^2\dot{\theta}^2 + ml\dot{x}\dot{\theta}\cos(\theta) - mgl\cos(\theta) \tag{4.3}$$

and F_{fric} is the force of friction between the cart and the track defined as

$$F_{\text{fric}} = -b\dot{x} \tag{4.4}$$

By putting the expressions (4.3) and (4.4) in (4.1) and (4.2) and after solving it, Lagrange's equations of the inverted pendulum system can be expressed as

$$(M+m)\ddot{x} + ml\cos(\theta)\ddot{\theta} - ml\dot{\theta}^2\sin(\theta) + b\dot{x} = F \tag{4.5}$$

$$m\cos(\theta)\ddot{x} + ml\ddot{\theta} - mg\sin(\theta) = 0 \tag{4.6}$$

In the other hand, the experimental setup considered in simulation results uses an AC servo motor, which is controlled by motion controller card via motor drive amplifier. So, in order to develop a more realistic dynamic model, the dynamics of AC servo motor is also taken into account, which is given as

$$B_m\dot{\theta}_m = T_m - T \tag{4.7}$$

where $\dot{\theta}_m$ is the rotor velocity displacement. B_m is the viscous-friction coefficient. T_m and T are the torque generated by a motor and the load torque used to move the cart through a timing pulley and a synchronous belt, respectively.

The torque generated by a motor can be represented as follow [29]:

$$T_m = K_m\left(-\frac{K_b}{R_m}\dot{\theta}_m + \frac{u}{R_m}\right) \tag{4.8}$$

where u is the voltage applied to alternating current (AC) servo motor. K_m, K_b and R_m are the torque and back electromotive force (EFM) constant and the motor armature resistance, respectively.

In this work, we consider a timing pulley and synchronous belt arrangement to convert rotational motion to translation motion. So the relation between those variable can be written as

$$\dot{x} = r\dot{\theta}_m \tag{4.9}$$

where r is the radius of timing pulley.

So the motor torque (4.8) can be written as follows:

$$T_m = -\frac{K_mK_b}{rR_m}\dot{x} + \frac{K_m}{R_m}u \tag{4.10}$$

From (4.7), (4.9) and (4.10), the load torque can be defined as

$$T = \frac{K_m}{R_m}u - \left(\frac{B_m}{r} + \frac{K_m K_b}{rR_m}\right)\dot{x} \tag{4.11}$$

and the force acting on the cart is defined as

$$F = \frac{T}{r} = \frac{K_m}{rR_m}u - \left(\frac{B_m}{r^2} + \frac{K_m K_b}{r^2 R_m}\right)\dot{x} = \alpha u - \beta\dot{x} \tag{4.12}$$

Finally, by substituting the force given in (4.12) into (4.5), the system equations describing the cart position dynamics and the pendulum angle dynamics, respectively, can be described as follows:

$$\ddot{x} = \frac{\alpha u + ml\dot{\theta}^2\sin(\theta) - mg\sin(\theta)\cos(\theta) - (b+\beta)\dot{x}}{(M+m) - m\cos^2(\theta)} \tag{4.13}$$

$$\ddot{\theta} = \frac{u\alpha\cos(\theta) + ml\dot{\theta}^2\sin(\theta)\cos(\theta) - (M+m)g\sin(\theta) - (b\cos(\theta)+\beta)\dot{x}}{ml\cos^2(\theta) - (M+m)l} \tag{4.14}$$

For numerical simulation of the nonlinear model for the inverted pendulum system, it is required to represent the nonlinear equations (4.13) and (4.14) into the standard state space form:

$$\frac{dz}{dt} = f(t, z, u) \tag{4.15}$$

Let, $z_1 = x,\ z_2 = \dot{x} = \dot{z}_1,\ z_3 = \theta,\ z_4 = \dot{\theta} = \dot{z}_3$, the system dynamics may be written as follows:

$$\begin{cases} \dot{z}_1 = z_2 \\ \dot{z}_2 = f_1(t,z,u) + \xi_1(t,z,u) \\ \dot{z}_3 = z_4 \\ \dot{z}_4 = f_2(t,z,u) + \xi_2(t,z,u) \end{cases} \tag{4.16}$$

and the measurable output is given by

$$y = (y_1, y_2)^T = (z_1, z_3)^T \tag{4.17}$$

where the nominal part of the system dynamics is represented by the functions:

$$f_1(t,z,u) = \frac{\alpha u + mlz_4^2\sin(z_3) - mg\sin(z_3)\cos(z_3) - (b+\beta)z_2}{(M+m) - m\cos^2(z_3)} \tag{4.18}$$

$$\begin{aligned} &f_2(t,z,u) \\ &= \frac{u\alpha\cos(z_3) + mlz_4^2\sin(z_3)\cos(z_3) - (M+m)g\sin(z_3) - (b\cos(z_3)+\beta)z_2}{ml\cos^2(z_3) - (M+m)l} \end{aligned} \tag{4.19}$$

while the uncertainties are concentrated in the terms $\xi_1(t,z,u)$ and $\xi_2(t,z,u)$. For the case study of system (4.16) and for $i = 1, 2$, $f_i(t,z,u)$ and $\xi_i(t,z,u)$ are Lebesgue-measurable and uniformly bounded in any region of the state space.

4.3 Control system

4.3.1 Super-twisting observer

For the Googol-inverted pendulum experimental setup, only the cart position, x, and the pendulum angle, θ, are directly measured. This means that the cart velocity and the pendulum angular velocity are not immediately available for use in any control schemes beyond just stabilization. Thus, an observer is relied upon to supply accurate estimation of the states at all cart–pendulum positions.

In [30], Davila *et al.* proposed a super-twisting observer which reconstructs for second-order mechanical systems the velocity from the position measurements with finite time convergence using the principle of the second-order sliding-mode presented in [31]. It is shown that the separation principle theorem is trivial in this case, and the observer can be designed separately from the controller. Thus, if the controller is known to stabilize the process, then the stabilization of the system in closed-loop is assured whenever the super-twisting observer dynamics are fast enough. Furthermore, not only stabilization problem can be achieved but also tracking problem since the time variable is explicit in the observer model [32–34]. Only partial knowledge of the system model is required in such a case.

Theorem 4.1. *The proposed super-twisting observer given by*

$$\begin{cases} \dot{\hat{z}}_1 = \hat{z}_2 + \sigma_1 \left|z_1 - \hat{z}_1\right|^{1/2} \operatorname{sign}(z_1 - \hat{z}_1) \\ \dot{\hat{z}}_2 = f_1(t, z_1, \hat{z}_2, z_3, \hat{z}_4, u) + \kappa_1 \operatorname{sign}(z_1 - \hat{z}_1) \\ \dot{\hat{z}}_3 = \hat{z}_4 + \sigma_2 \left|z_3 - \hat{z}_3\right|^{1/2} \operatorname{sign}(z_3 - \hat{z}_3) \\ \dot{\hat{z}}_4 = f_2(t, z_1, \hat{z}_2, z_3, \hat{z}_4, u) + \kappa_2 \operatorname{sign}(z_3 - \hat{z}_3) \end{cases} \tag{4.20}$$

converges in finite time to the states of system (4.16) if the observer parameters are chosen as:

$$\begin{cases} \kappa_1 > f_1^+ \\ \sigma_1 > \sqrt{\frac{2}{\kappa_1 - f_1^+} \frac{(\kappa_1 + f_1^+)(1+p_1)}{(1-p_1)}} \\ \kappa_2 > f_2^+ \\ \sigma_2 > \sqrt{\frac{2}{\kappa_2 - f_2^+} \frac{(\kappa_2 + f_2^+)(1+p_2)}{(1-p_2)}} \end{cases} \tag{4.21}$$

with known positive constants f_1^+ and f_2^+ and the following conditions:

$$\begin{cases} \left|f_1(t,z,u) - f_1(t, z_1, \hat{z}_2, z_3, \hat{z}_4, u) + \xi_1(t,z,u)\right| < f_1^+ \\ \left|f_2(t,z,u) - f_2(t, z_1, \hat{z}_2, z_3, \hat{z}_4, u) + \xi_2(t,z,u)\right| < f_2^+ \end{cases} \tag{4.22}$$

hold for system (4.16) for any possible t and $\left|\hat{z}_2\right| \leq 2\sup|z_2|$ and $\left|\hat{z}_4\right| \leq 2\sup|z_4|$. p_1 and p_2 are constants chosen as $0 < p_1, p_2 < 1$ and $f_1(t, z_1, \hat{z}_2, z_3, \hat{z}_4, u)$ and $f_2(t, z_1, \hat{z}_2, z_3, \hat{z}_4, u)$ are known nominal functions defined as

$$f_1(t, z_1, \hat{z}_2, z_3, \hat{z}_4, u) = \frac{\alpha u + ml\hat{z}_4^2 \sin(z_3) - mg \sin(z_3)\cos(z_3) - \beta \hat{z}_2}{(M+m) - m\cos^2(z_3)} \tag{4.23}$$

$$f_2(t, z_1, \hat{z}_2, z_3, \hat{z}_4, u) = \frac{u\alpha\cos(z_3) + ml\hat{z}_4^2\sin(z_3)\cos(z_3) - (M+m)g\sin(z_3) - \beta\hat{z}_2}{ml\cos^2(z_3) - (M+m)l} \tag{4.24}$$

Proof. See [30]. □

4.3.2 Fractional controller

4.3.2.1 Fractional order calculus

Fractional-order integration and differentiation are the generalization of the integer-order ones to a real or complex order. Efforts to extend the specific definitions of the traditional integer-order to the more general arbitrary order context led to different definitions for fractional derivatives [35]. Fractional calculus is also three centuries old as is conventional calculus, but is not very popular in the research field. However, since the last few decades, many researchers have done work in different areas of science and engineering (control system, modeling, etc.) using fractional calculus [36,37].

The real order generalization is as follows:

$$D^\alpha = \begin{cases} \dfrac{d^\alpha}{dt^\alpha} & \text{if } \alpha > 0 \\ 1 & \text{if } \alpha = 0 \\ \displaystyle\int_0^t d\tau^{-\alpha} & \text{if } \alpha < 0 \end{cases} \tag{4.25}$$

There are many definitions of fractional calculus but the three most frequently used definitions are *Riemann–Liouville (RL), Caputo* and *GL* [38]. The first one is *Caputo* witch expression is given by

$${}_aD_t^\alpha f(t) = \frac{1}{\Gamma(n-\alpha)} \int_a^t \frac{f^{(n)}(\tau)}{(t-\tau)^{\alpha-n+1}} d\tau \tag{4.26}$$

where α is a real number, a and t are the limits of operator, n is the integer value which satisfies the condition $n - 1 < \alpha < n$ and Γ is the Gamma function defined by

$$\Gamma(z) = \int_0^\infty t^{z-1} e^{-t} dt \tag{4.27}$$

The *Caputo* definition is popular mostly in engineering applications [39], as it directly connection between the type of the initial conditions and the type of the fractional derivative.

The second definition is the RL given as

$${}_aD_t^\alpha f(t) = \frac{1}{\Gamma(n-\alpha)} \left(\frac{d}{dt}\right)^n \int_a^t \frac{f(\tau)}{(t-\tau)^{\alpha-n+1}} d\tau \tag{4.28}$$

The RL definition for the fractional integral and derivative is appropriate for finding the analytical solution of simple functions like $\cos(t)$, e^t [40].

The *Caputo* and *RL* formulation have the same view when the initial conditions are zero.

The third definition is GL given by

$$ {}_aD_t^\alpha f(t) = \lim_{h\to 0}\frac{1}{h^\alpha}\sum_{r=0}^{\lfloor (t-a)/h \rfloor}(-1)^r\frac{\alpha!}{r!\,\Gamma(\alpha-r+1)}f(t-rh) \tag{4.29} $$

The GL definition is used for numerical evaluations. This definition is very beneficial in obtaining a numerical solution of fractional differential equations [41].

In this work, the GL definition will be used. In such a case, the choice of the initial conditions doesn't play an important role in the obtained solutions.

4.3.2.2 Fractional order PID (FOPID) controller

PID controllers have been used for several decades in industries for process control applications [42]. The reason for their popularity is in the simplicity of its design and good performance [42]. A FOPID controller is an extension of the classical PID controller in which the introduction of two extra parameters λ and μ makes it complex as compared to PID controller because of introduction of fractional calculus in it.

The control signal $u(t)$ can then be expressed in the time domain as

$$ u(t) = k_p e(t) + k_i D^{-\lambda} e(t) + k_d D^{\mu} e(t) \tag{4.30} $$

where $e(t)$ is the error signal. k_p, k_i and k_d are proportional, integral and derivative gains, respectively. λ is the order of integration and μ is the order of differentiator.

In the Laplace domain, assuming zero initial conditions, the FOPID controller has the following form:

$$ C(s) = k_p + \frac{k_i}{s^\lambda} + k_d s^\mu \tag{4.31} $$

All the classical PID controllers are particular cases of the fractional controller:

- if $\lambda = 1$ and $\mu = 1$, then it is classical PID controller.
- if $\lambda = 0$ and $\mu = 1$, then it is classical PD controller.
- if $\lambda = 1$ and $\mu = 0$, then it is classical PI controller.
- if $\lambda = 0$ and $\mu = 0$, then it is classical P controller.

A FOPID controller works better in the following situations/applications [43]:

- Five different specifications can be achieved, which is not possible in the case of classical PID controller.
- The FOPID controller can easily attain the property of iso-damping, compared to classical PID controller.
- For higher order systems, the performance of classical PID controller deteriorates, whereas FOPID controller provides better results.
- For a system with long time delay, a FOPID controller provides better results than classical PID controller.
- A classical PID controller provides lower robust stability, whereas FOPID controller has more robust and stable.

- It is difficult to control the system with nonlinearities using classical PID controller, whereas the FOPID controller can perform better.
- The FOPID controller can achieve better response for nonminimum phase system.
- Mostly, nonlinear system is linearized at different operating points, and controllers are designed for different operating points, whereas one FOPID controller is mostly sufficient for nonlinear system.

The fractional controllers have two parameters more than the conventional PID controllers, thus two more specifications can be met, which improves the performance of the overall system. It can be expected that the FOPID controller may enhance the performance of the system. So, finding an optimal set of values for k_p, k_i, k_d, λ and μ in order to meet the user specifications for a given process needs parameter optimization in five-dimensional hyperspace witch remain a complex and time consuming procedure [44]. For those reasons, researchers have used different methods of tuning like Ziegler–Nichols modified method, PSO, neural network, etc. [45,46].

4.3.3 Tuning parameters of fractional-order controller

Tuning controllers is always a challenging task. In case of FOPID controllers, there are five parameters to tune. Because of a larger number of parameters, it is challenging to tune this controller [47]. Various methods are used for tuning of fractional order controllers (Rule-based methods, analytical methods, numerical methods, etc.) [48–50]. In this work, we will use one of the soft computing methods: the PSO approach. Indeed, PSO has been regarded widely as a promising optimization algorithm due to its combination of simplicity, low computational cost and good performance. In the other hand, the optimal problems solved by genetic algorithms can obtain better solutions with PSO in comparison with conventional methods [51]. However, the classic PSO algorithm has a major problem which is premature convergence. So, to address this problem, we will use in this work the MPSO proposed by Zhao *et al.* [52]. In fact, PSO method was used for designing fractional order controllers by Cao and Cao [53]. The optimization performance target is the weighted combination of ITAE and control input. Hu *et al.* [54] have also used the PSO algorithm for tuning of the FOPID implemented with FPGA.

4.3.3.1 Particle swarm optimization

The PSO method is an evolutionary algorithm inspired from study on swarm such as fish schooling and bird flocking [18]. During flight, each particle adjusts its position according to its own experience, and the experience of other particles, making use of the best position encounter by itself and its neighbors. The swarm direction of a particle is defined by the position of particles adjacent to the particle and its history. The best previous position of ith particle is recorded and represented as $pbest_i$. The best particle among all the particles in the group is represented as $gbest$.

The PSO concept consists of, at each time step, changing the velocity (or acceleration) of each particle i toward its $pbest$ and the $gbest$ position. For the ith particle, a new velocity and position are updated such that [19]:

$$V_i^{(k+1)} = w^{(k)} V_i^{(k)} + c_1\, rand()\, (pbest_i - p_i^{(k)}) + c_2\, Rand()\, (gbest^{(k)} - p_i^{(k)}) \quad (4.32)$$

$$p_i^{(k+1)} = p_i^{(k)} + V_i^{(k+1)} \tag{4.33}$$

for $k = 1, 2, \ldots, N$ and $i = 1, 2, \ldots, M$ and where M is the number of particles in a group, N is the maximum number of iterations, $V_i^{(k)}$ is the velocity of ith particle at iteration k, $p_i^{(k)}$ is the position of ith individual at iteration k, $pbest_i^{(k)}$ is the best position of ith individual at iteration k, $gbest^{(k)}$ is the best position of the group until iteration $k, w^{(k)}$ is the inertia weight factor at iteration k, c_1, c_2 is the acceleration factors and *rand ()*, *Rand ()* is the Random numbers between 0 and 1.

In general, the inertia weight is calculated according to the following equation:

$$w^{(k)} = k\frac{w_{\max} - w_{\min}}{N} \tag{4.34}$$

4.3.3.2 Modified particle swarm optimization

The classic PSO algorithm has a major problem which is premature convergence. This problem incurs a rapid loss of diversity during the evolutionary processing. Therefore, to address this problem, Zhao *et al.* [52] propose a MPSO, which adopts an adaptive factor f_a to prevent premature convergence and a perturbation operator g_a help particles flight from local optima.

The adaptive factor is defined as follows:

$$f_a = rand \times \left(1 - \frac{k}{N}\right) \tag{4.35}$$

where rand meets normal distribution within the range [0, 1]. Equation (4.35) means that the adaptive factor f_a tends to decrease with the iterations k. By introducing f_a, the position of particle (4.33) is modified as follows:

$$\begin{aligned} p_i^{(k+1)} &= p_i^{(k)} + f_a \times V_i^{(k+1)} \\ &= p_i^{(k)} + rand \times \left(1 - \frac{k}{N}\right) \times V_i^{(k+1)} \end{aligned} \tag{4.36}$$

It can be seen in relation (4.36) that in the beginning stage of the iteration f_a is large, and the algorithm has strong ability to explore new areas continuously to overcome premature convergence, then as the number of iteration increases, f_a decreases, thus the algorithm converges gradually and can search finely around the optimal solution. The adaptive factor f_a balances the global and local search capability.

In PSO, the global best location *gbest* steers the flight direction for a particle, so it would be more effective to modify *gbest* to maintain particles diversity for jumping out of the potential local optima by using a perturbation operator g_a. So, to better manage diversity and convergence, the perturbation range of *gbest* is limited to $\left[\,0\ \max(p) - \min(p)\,\right]$. $\max(p) - \min(p)$ is the difference of maximum and minimum location of particles in different iteration. At the beginning of the iteration, $\max(p) - \min(p)$ is large, and the extensive perturbation range keeps the diversity of particles. Then as the number of iteration increases, $\max(p) - \min(p)$ decreases gradually, the lower perturbation rate can converges gradually and search finely around the optimal solution. So the best particle among all the particles in the group *gbest* with perturbation operator g_a will be defined as

$$\begin{aligned} gbest^{(k+1)} &= gbest^{(k)} + g_a \\ &= gbest^{(k)} + \left(\max\left(p^{(k)}\right) - \min\left(p^{(k)}\right)\right) \times rand \end{aligned} \tag{4.37}$$

The design steps for implementing the MPSO algorithm are given as follows:

1. Initialize a population of particles with arbitrary positions and velocities.
2. Update the inertial weight w.
3. For each particle, estimate the desired optimization fitness function.
4. Compare particle's fitness with its $pbest_i$. If present value is better than $pbest_i$, then put $pbest_i$ equal to the recent value.
5. Update the velocity and position of the particle according to (4.32) and (4.36), respectively.
6. Identify the particle in the area with the best success so far, and allocate it to *gbest*.
7. If the maximum number of iterations is not reached, perturb the global best location to take new *gbest* as the global best location by using (4.37).
8. If a criterion is meet (a sufficiently good fitness or a maximum number of iterations), go to 9 else go to 2.
9. The particle that generates the latest *gbest* is an optimal controller parameter.

4.3.3.3 Fitness function

For getting good dynamic performance and avoiding large control input, the following fitness function is used in this paper:

$$\begin{aligned} J = \int_0^T \Bigg(& w_1 \times \sqrt{[\theta_d(t) - \theta(t)]^T [\theta_d(t) - \theta(t)]} + w_2 \\ & \times \sqrt{[x_d(t) - x(t)]^T [x_d(t) - x(t)]} + w_3 \times u(t) \Bigg) dt \end{aligned} \tag{4.38}$$

where w_1, w_2 and w_3 are the weights to give equal weightage to all parameters. T is the simulation time. The fitness function given by (4.38) has three terms. The first term is the integral square of pendulum angle error $\theta_d(t) - \theta(t)$ which is used to stabilize pendulum angle. The second term is the integral square of cart position error $x_d(t) - x(t)$ which is used to track cart position. Finally, the third term is the integral square of control input $u(t)$ which is used to minimize the required control voltage.

4.4 Simulation results

Simulation results are carried out based on the system model depicted in Section 4.2. The system parameter's values are listed in Table 4.1.

The FOPID-controller-based super-twisting observer is designed such as

- The cart position FOPID is given by $u_1(t) = k_p e_1(t) + k_i D^{-\lambda} e_1(t) + k_d D^{\mu} e_1(t)$
- The angle position FOPID is given by $u_2(t) = k_p e_2(t) + k_i D^{-\lambda} e_2(t) + k_d D^{\mu} e_2(t)$

Table 4.1 The Googol Technology experimental device parameters

Symbol	Parameter	Value
M	Cart mass	1.096 kg
M	Pendulum mass	0.109 kg
l	Pendulum rod length	0.5 m
b	Coefficient of friction of the cart	0.1 N s/m
g	Acceleration due to gravity	9.8 m/s^2
B_m	Motor viscous friction coefficient	0.03 N m s/rad
K_m	Motor torque constant	2 kg m^2
K_b	Motor back EFM constant	0.1 N s/rad
R_m	Motor resistance	2.5 Ω
r	Radius of timing pulley	0.0195 m

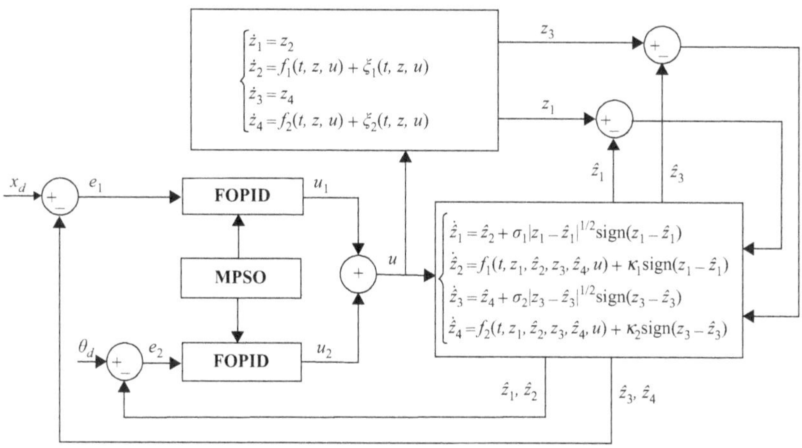

Figure 4.4 FOPID-controller-based super-twisting observer tuned with MPSO

and where $e_1(t) = x_d(t) - x(t)$, $e_2(t) = \theta_d(t) - \theta(t)$ and $u(t) = u_1(t) + u_2(t)$.

Uncertainties are defined as external perturbations satisfying $|\xi_i(t,z,u)| < 1$. They were taken as $\xi_1 = \xi_2 = 0.03\sin(20t)$. In the other hand, to explore the performances of the proposed control approach, we consider two case studies for simulation:

Case study 1: The control input is designed to track a smooth profile for the cart position reference given by $x_d = 0.3\sin(0.1t)$ and, to stabilize the pendulum in its upper position ($\theta_d = 0$).

Case study 2: The control is designed to track a nonsmooth trajectory for the cart position reference given by $x_d = 0.3\sin(0.1t)$ and, to stabilize the pendulum in its upper position.

Table 4.2 MPSO and PSO algorithm parameters

Designation	Variable	Value
Number of particles in a group	M	20
Maximum number of iterations	N	30
Minimum inertia weight factor	$w_{\min}$	0.1
Maximum inertia weight factor	$w_{\max}$	0.9
Search interval for k_p	$\left[k_{p_{\min}}\ k_{p_{\max}}\right]$	$[0\ 300]$
Search interval for k_i	$\left[k_{i_{\min}}\ k_{i_{\max}}\right]$	$[0\ 300]$
Search interval for λ	$\left[\lambda_{\min}\ \lambda_{\max}\right]$	$[0\ 1]$
Search interval for k_d	$\left[k_{d_{\min}}\ k_{d_{\max}}\right]$	$[0\ 300]$
Search interval for μ	$\left[\mu_{\min}\ \mu_{\max}\right]$	$[0\ 1]$
Search interval velocity of k_p	$\left[0\ k_{p_{\max}}/2\right]$	$[0\ 150]$
Search interval velocity of k_i	$\left[0\ k_{i_{\max}}/2\right]$	$[0\ 150]$
Search interval velocity of λ	$\left[0\ \lambda_{\max}/2\right]$	$[0\ 0.5]$
Search interval velocity of k_d	$\left[0\ k_{d_{\max}}/2\right]$	$[0\ 150]$
Search interval velocity of μ	$\left[0\ \mu_{\max}/2\right]$	$[0\ 0.5]$
Acceleration constants	$c_1 c_2$	2

For the two cases, the optimal parameters of the FOPID controllers are determined using the MPSO and classical PSO algorithms.

Figure 4.4 illustrates the block diagram of optimal design of the FOPID-controller-based super-twisting observer for the inverted pendulum on a cart using MPSO algorithm.

The MPSO and PSO algorithm's parameters are given in Table 4.2. The parameters of the fitness function (5) are chosen as $w_1 = 1$, $w_2 = 1$ and $w_3 = 0.1$. The initial conditions are taken as follows: $x = z_1 = \hat{z}_1 = 0.3\,\text{m}$, $\dot{x} = z_2 = -0.1\,\text{m/s}$, $\hat{z}_2 = 0\,\text{m/s}$, $\theta = z_3 = \hat{z}_3 = 0.2\,\text{rad}$, $\dot{\theta} = z_4 = 0.03\,\text{rad/s}$ and $\hat{z}_4 = 0\,\text{rad/s}$.

In the other hand, assume that $|z_1|, \left|\hat{z}_1\right| \le 0.5\,\text{m}$, $|z_2| \le 1.5\,\text{m/s}$, $\left|\hat{z}_2\right| \le 3\,\text{m/s}$, $|z_3|, \left|\hat{z}_3\right| \le 1\,\text{rad}$, $|z_4| \le 2.5\,\text{rad/s}$, $\left|\hat{z}_4\right| \le 5\,\text{rad/s}$ and $|u| \le 7\,\text{V}$, we obtain

$$\begin{cases} \left|f_1(t,z,u) - f_1(t,z_1,\hat{z}_2,z_3,\hat{z}_4,u) + \xi_1(t,z,u)\right| < 0.4 = f_1^+ \\ \left|f_2(t,z,u) - f_2(t,z_1,\hat{z}_2,z_3,\hat{z}_4,u) + \xi_2(t,z,u)\right| < 0.1 = f_2^+ \end{cases} \tag{4.39}$$

Therefore, the observer parameters described by (4.21) are chosen as $\kappa_1 = 2$, $\sigma_1 = 5$, $\kappa_2 = 1$ and $\sigma_2 = 3$.

4.4.1 Simulation results and performance analysis: case study 1

The FOPID controller parameters tuned using the MPSO and the PSO approaches are given in Table 4.3.

Table 4.3 Optimized controller parameters: case study 1

		k_p	k_i	λ	k_d	μ	J
MPSO-FOPID controller	Angle control	167	5	0.37	38	0.29	1.0054
	Position control	59	19	0.23	38	0.18	
PSO-FOPID controller	Angle control	123	10	0.65	19	0.76	1.117
	Position control	89	67	0.12	21	0.32	

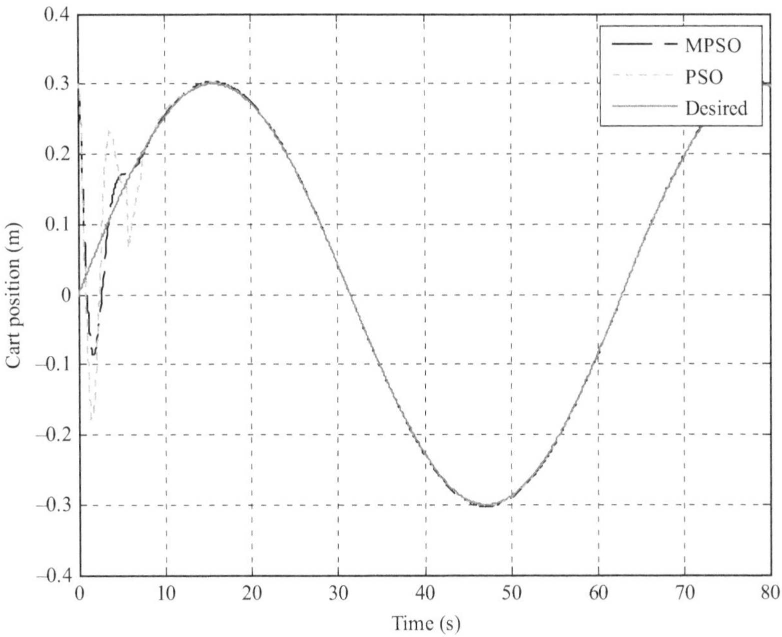

Figure 4.5 Cart's position: case study 1

Figures 4.5–4.8 show the MPSO-FOPID response compared to the PSO-FOPID response for the system under consideration with the objective of controlling the cart position and stabilizing the pendulum angle. From Figure 4.5, we may conclude that there is some oscillations and large settling time for the PSO-FOPID controller regarding the cart's position. Such dynamics are not observed when using the MPSO-FOPID controller. In the other hand, Figure 4.6 shows that cart position error is null after 20 s in the two cases. From Figure 4.7, we can observe oscillations for pendulum's angle when the PSO-FOPID controller is used compared to the MPSO-FOPID controller. In Figure 4.8, both PSO and MPSO FOPID controller provide acceptable control laws but in case of MPSO-FOPID, less control effort is required.

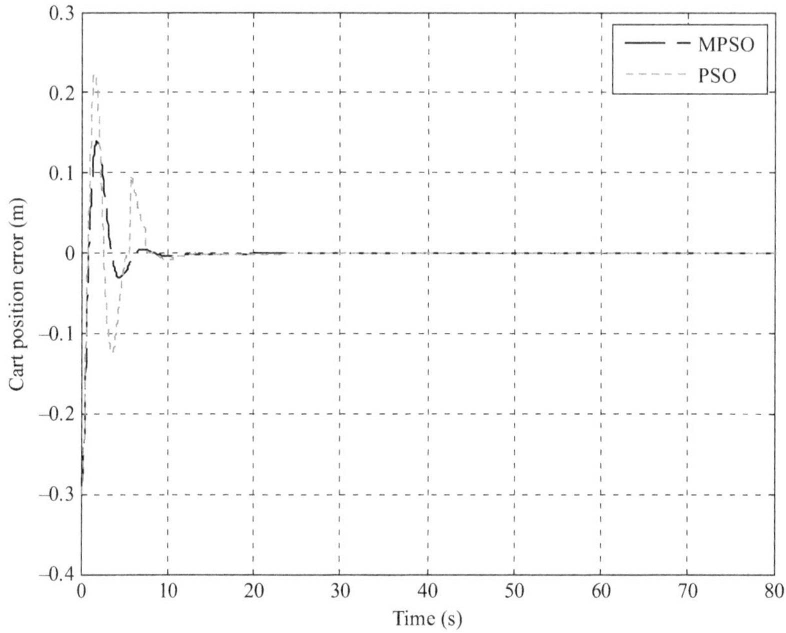

Figure 4.6 Cart's position error: case study 1

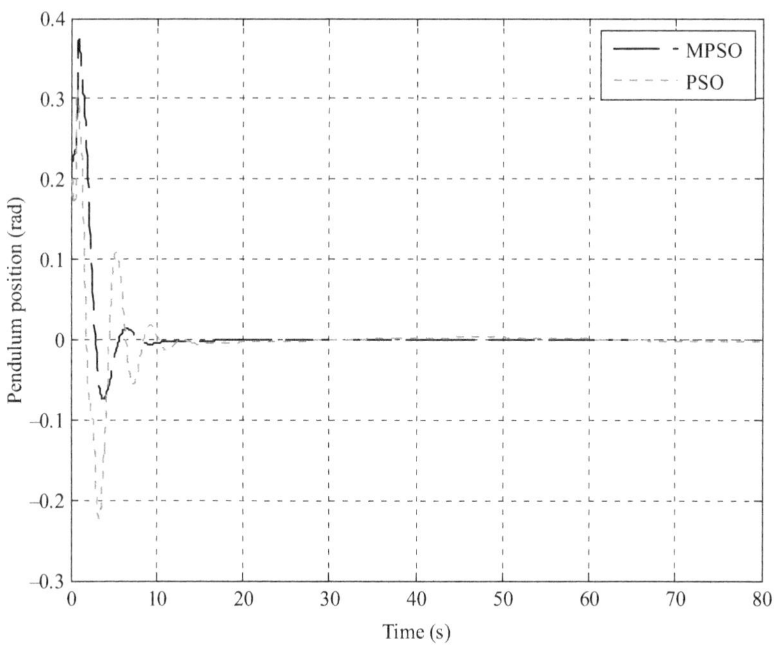

Figure 4.7 Pendulum's position: case study 1

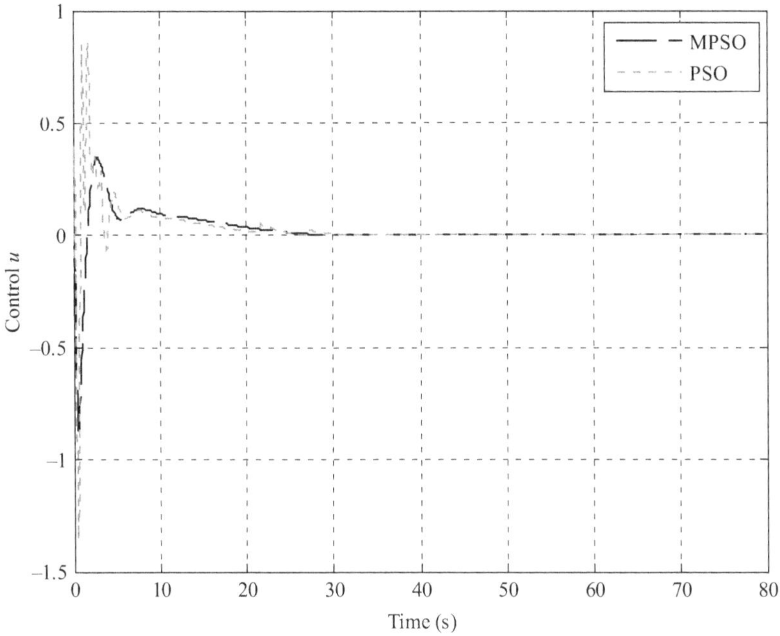

Figure 4.8 Control voltage: case study 1

Table 4.4 Optimized controller parameters: case study 2

		k_p	k_i	λ	k_d	μ	J
MPSO-FOPID controller	Angle control	202	18	0.15	31	0.11	1.4364
	Position control	122	3	0.36	28	0.2	
PSO-FOPID controller	Angle control	281	39	0.59	85	0.71	1.6871
	Position control	159	98	0.46	49	0.13	

4.4.2 Simulation results and performance analysis: case study 2

Table 4.4 gives the best FOPID controller parameters using MPSO and PSO algorithm's for the second case study.

The desired and actual cart trajectories are shown in Figure 4.9. We can conclude that settling time is large for PSO-FOPID controller compared to MPSO-FOPID controller when desired trajectory cart is nonsmooth. From Figure 4.10, we also observe oscillations when parameters of FOPID controller are tuned using PSO algorithm. From Figure 4.11, we can conclude that the profile of the control laws present some oscillations during the deviation of the cart. Furthermore, more overshoots are

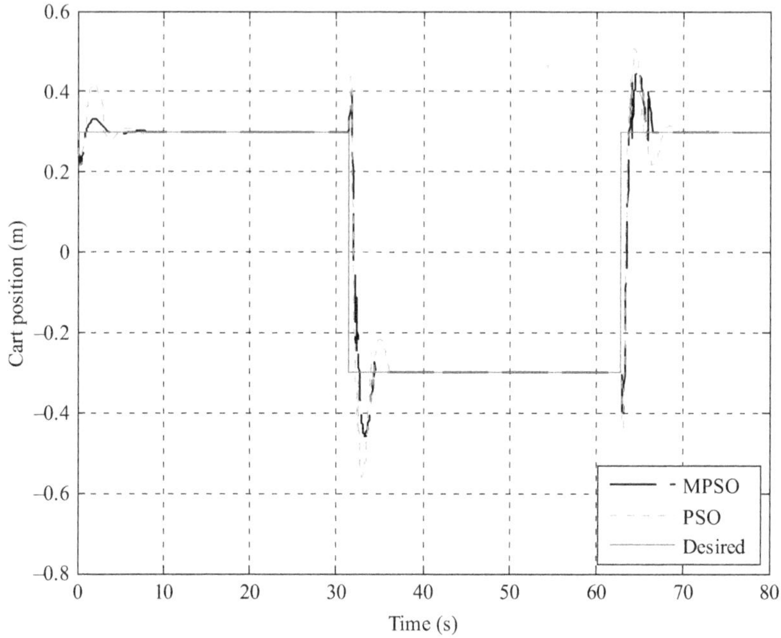

Figure 4.9 Cart's position: case study 2

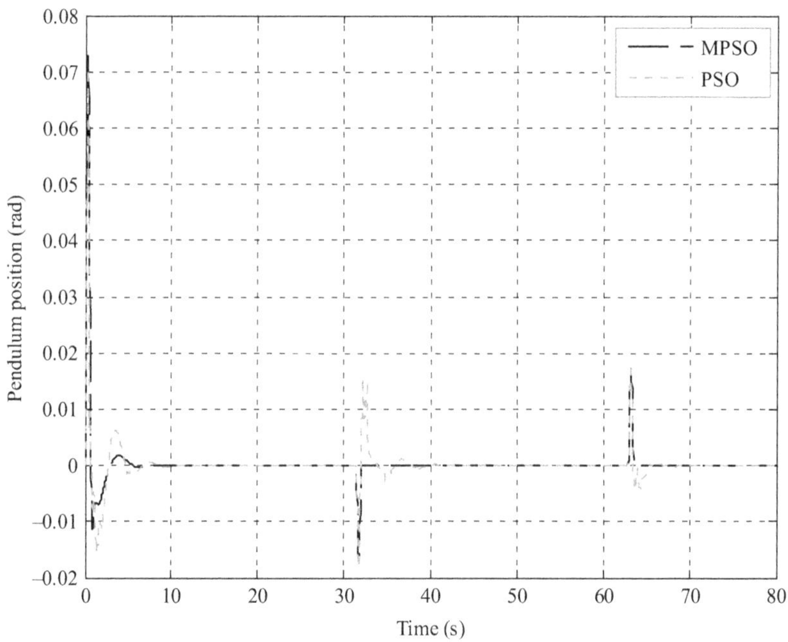

Figure 4.10 Pendulum's position: case study 2

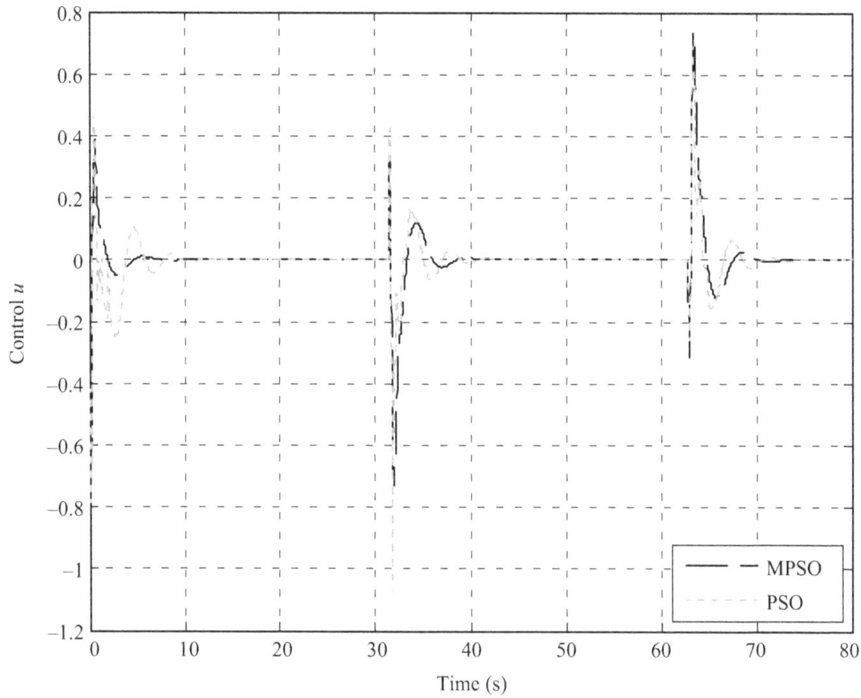

Figure 4.11 Control voltage: case study 2

observed when system is controlled via PSO-PID controllers compared to MPSO-FOPID controllers. Such observations prove that the MPSO algorithm performs well in adjusting the control laws parameter's for nonsmooth motions.

4.5 Conclusion

In this chapter, an MPSO-FOPID-controller-based super-twisting observer is applied for the inverted pendulum on a cart system with the consideration of optimization objectives. The super-twisting observer is proposed to estimate the state vector, in spite of disturbance, exactly and in a finite time. The MPSO algorithm that adopts adaptive factor and perturbation operator conquers the drawback of premature convergence of PSO method, has been improved and applied for simultaneous optimization of several parameters of the controllers. Performances of MPSO-FOPID and PSO-FOPID controllers are studied and compared. Analysis confirms the stability and robustness of the proposed FOPID controller in terms of tracking errors in the case of a smooth or a nonsmooth cart trajectory. The chapter has also shown that MPSO-FOPID gives more flexibility for the control system designing and has better opportunity to adjust system dynamic characteristics, which means that the proposed controller design technique may serve as an efficient alternative for the design of future controllers. The performance investigation of the FOPID control approach using MPSO with a different multiobjective fitness function including linear and angular velocity variables may be done as a future work.

References

[1] Boubaker O. 'The inverted pendulum: A fundamental benchmark in control theory and robotics'. *IEEE International Conference on Education and e-Learning Innovations (ICEELI2012)*; Sousse, Tunisia, Jul. 2012. pp. 1–6.

[2] Zhao J., Spong M.W. 'Hybrid control for global stabilization of the cart-pendulum system'. *Automatica.* 2001; 37(12): 1941–1951.

[3] Chung C.C., Hauser J. 'Nonlinear control of a swinging pendulum'. *Automatica.* 1995; 31(6): 851–862.

[4] Agustinah T., Jazidie A., Nuh M. 'Hybrid fuzzy control for swinging up and stabilizing of the pendulum–cart system'. *IEEE International Conference on Computer Science and Automation Engineering (CSAE '11)*; Shanghai, China, Jun. 2011. pp. 109–113.

[5] Boubaker O. 'The inverted pendulum benchmark in nonlinear control theory: A Survey'. *International Journal of Advanced Robotic Systems.* 2013; 10: 233–242.

[6] Astrom K.J., Furuta K. 'Swinging up a pendulum by energy control'. *Automatica.* 2000; 36: 287–295.

[7] Benrejeb W., Boubaker O. 'FPGA modeling and realtime embedded control design via LabVIEW Software: Application for swinging-up a pendulum'. *International Journal on Smart Sensing and Intelligent Systems.* 2012; 5(3): 576–591.

[8] Sarit K.D., Kaustav K.P. 'Robust compensation of a cart–inverted pendulum system using a periodic controller: Experimental results'. *Automatica.* 2011; 47: 2543–2547.

[9] Zhang Y., Yu X., Yin Y., Peng C., Fan Z. 'Singularity-conquering ZG controllers of z2g1 type for tracking control of the IPC system'. *International Journal of Control.* 2014; 87(9): 1729–1746.

[10] Ngadengon R., Sam Y.M., Osman J.H.S., Tomari R., Wan Zakaria W.N. 'Multirate output feedback with discrete time sliding mode control for inverted pendulum system'. *Procedia Computer Science.* 2015; 76: 290–295.

[11] Astrom K.J., Hagglund T. *Advanced PID control.* Pittsburgh, PA: ISA – The Instrumentation, Systems, and Automation Society; 2006.

[12] Podlubny I. 'Fractional-order systems and $PI^{\lambda}D^{\mu}$ controller'. *IEEE Transactions on Automatic Control.* 1999; 44: 208–214.

[13] Efe M.Ö. 'Neural network assisted computationally simple $PI^{\lambda}D^{\mu}$ control of a quad rotor UAV'. *IEEE Transactions on Industrial Informatics.* 2011; 7: 354–361.

[14] Howimanporn S., Parnichkun M. 'Control of an X–Y planar inverted pendulum using PSO-based SMC'. *International Journal of Robotics & Automation*. 2015; 30(4): 310–321.

[15] Mehdi H., Boubaker O. 'Position/force control optimized by particle swarm intelligence for constrained robotic manipulators'. *IEEE International Conference on Intelligent Systems Design and Applications (ISDA'2011)*; Cordoba, Espagne, Nov. 2011. pp. 190–195.

[16] Mehdi H., Boubaker O. 'Robust impedance control based Lyapunov–Hamiltonian approach for constrained robots'. *International Journal of Advanced Robotic Systems*. 2015; 12: 190–202.

[17] Mehdi H., Boubaker O. 'PSO-Lyapunov motion/force control of robot arms with model uncertainties'. *Robotica*. 2016; 34(3): 634–651.

[18] Kennedy J., Eberhart R. 'Particle swarm optimization'. *Proceeding of the IEEE International Conference of Neural Networks*; Perth, 1995. pp. 1942–1948.

[19] Eberhart R.C., Shi Y. 'Comparison between genetic algorithms and particle swarm optimization'. in Porto V.W., Saravanan N., Waagen D., Eiben A.E. (eds.). *Evolutionary Programming VII*. Berlin, Heidelberg: Springer; 1998. pp. 611–616.

[20] Shtessel Y., Edwards C., Fridman L., Levant A. *Sliding mode control and observation*. New York, NY: Springer; 2014.

[21] Chen L., Edwards C., Alwi H. 'LPV sliding mode observers for sensor fault reconstruction with erroneous scheduling parameter measurements'. *54th IEEE Conference on Decision and Control (CDC)*; Osaka, Japan, Dec. 2015. pp. 4080–4085.

[22] Chan L., Naghdy F., Stirling D.J. 'An improved extended active observer for adaptive control of a n–DOF robot manipulator'. *Journal of Intelligent & Robotic Systems*. 2016. DOI: 10.1007/s10846-016-0402-8.

[23] Accadia T., Acernese F., Astone P., *et al*. 'A state observer for the Virgo inverted pendulum'. *Review of Scientific Instruments*. 2011; 82(9): 34–44.

[24] Maruki Y., Kawano K., Suemitsu H., Matsuo T. 'Adaptive backstepping control of wheeled inverted pendulum with velocity estimator'. *International Journal of Control Automation and Systems*. 2014; 12(5): 1040–1048.

[25] Canete L., Takahashi T. 'Modeling, analysis and compensation of disturbances during task execution of a wheeled inverted pendulum type assistant robot using a unified controller'. *Advanced Robotics*. 2015; 29(22): 1453–1462.

[26] Lee J., Mukherjee R., Khalil H.K. 'Output feedback stabilization of inverted pendulum on a cart in the presence of uncertainties'. *Automatica*. 2015; 54: 146–157.

[27] Hashimoto Y., Otsuki Y., Matsuno T., Yanou A., Minami M. 'Posture control of inverted pendulum cart with balancing mechanism using full-order state observer'. *IEEE/SICE International Symposium on System Integration (SII)*; Tokyo, Japan, Dec. 2014. pp. 398–403.

[28] Googol Technology. Inverted pendulum instruction manual. 2006. Available from http://www.googoltech.com/web/eng/product_details_frame.jsp?module=Educational%20Products&gid=569 [Accessed Dec. 2016].

[29] Kuo B.C., Golnaraghi G. *Automatic control systems*. 8th edn. New York: Wiley; 2003.
[30] Davila J., Fridman L., Levant A. 'Second-order sliding-mode observer for mechanical systems'. *IEEE Transactions on Automatic Control*. 2005; 50(11): 1785–1789.
[31] Levant A. 'Sliding order and sliding accuracy in sliding mode control'. *International Journal of Control*. 1993; 58(6): 1247–1263.
[32] Fang X., Wu A., Shang Y., Du C. 'Multivariable super twisting based robust trajectory tracking control for small unmanned helicopter'. *Mathematical Problems in Engineering*. 2015; 2015: 13 pp. Article ID 620357.
[33] Zong Q., Dong Q., Wang F., Tian B. 'Super twisting sliding mode control for a flexible air-breathing hypersonic vehicle based on disturbance observer'. *Science China Information Sciences*. 2015; 58(7): 1–15.
[34] Hu J., Qiu Y., Liu L. 'High-order sliding-mode observer based output feedback adaptive robust control of a launching platform with backstepping'. *International Journal of Control*. 2016; 89(10): 2029–2039.
[35] Kilbas A.A., Srivastava H.M., Trujillo J.J. *Theory and applications of fractional differential equations*. Amsterdam, The Netherlands: Elsevier B.V; 2006. p. 135.
[36] Monje C.A., Chen Y., Vinagre B.M., Xue D., Feliu-Batlle V. *Fractional-order systems and controls: Fundamentals and applications*. London: Springer; 2010.
[37] Zheng W., Luo Y., Chen Y., Pi Y. 'Fractional-order modeling of permanent magnet synchronous motor speed servo system'. *Journal of Vibration and Control*. 2016; 22(9): 2255–2280.
[38] Podlubny I. *Fractional differential equation*. New York, NY: Academic Press; 1998.
[39] Cafagna D. 'Fractional calculus: A mathematical tool from the past for present engineers'. *IEEE Industrial Electronics Magazine*. 2007; 2(1): 35–40.
[40] Loverro A. *Fractional calculus: History, definitions and applications for the engineer*. Rapport Technique University of Notre Dame, Department of Aerospace and Mechanical Engineering, Indiana; 2004.
[41] Petras I. 'Tuning and implementation methods for fractional-order controllers'. *Fractional Calculus and Applied Analysis*. 2012; 15(2): 282–303.
[42] Anguluri R., Vedurupaka C., Swagatam D. 'Fractional order PID controller design using a modified artificial bee colony algorithm'. In: Panigrahi B.K., Suganthan P.N., Das S., Satapathy S.C. (eds.). *Swarm, Evolutionary, and Memetic Computing*. Berlin, Heidelberg: Springer-Verlag; 2011. pp. 670–678.
[43] Shah P., Agashe S. 'Review of fractional PID controller'. *Mechatronics*. 2016; 38: 29–41.
[44] Chen C., Hu L., Wang L., Gao S., Li L. 'Adaptive particle swarm algorithm for parameters tuning of fractional order PID controller'. *Telkomnika*. 2016; 14(2): 478–488.
[45] Luo Y., Chen Y.Q., Wang C.Y., Pi Y.G. 'Tuning fractional order proportional integral controllers for fractional order systems'. *Journal of Process Control*. 2010; 20(7): 823–831.

[46] Padula F., Visioli A. 'Tuning rules for optimal PID and fractional-order PID controllers'. *Journal of Process Control*. 2011; 21(1): 69–81.

[47] Astrom K.J. *PID controllers: Theory, design and tuning*. Pittsburgh, PA: Instrument Society of America; 1995.

[48] Valério D., Sa da Costa J. 'Tuning of fractional PID controllers with Ziegler–Nichols-type rules.' *Signal Processing*. 2006; 86(10): 2771–2784.

[49] Deepak V.D., Usha Kumari S. 'Modified method of tuning for fractional PID controllers'. *International Conference on Power Signals Control and Computations (EPSCICON)*; Thrissur, India, Jan. 2014. pp. 1–6.

[50] Muresan C.I., De Keyser R., Ionescu C.M. 'Autotuning method for a fractional order controller for a multivariable 13C isotope separation column'. *European Control Conference*; Aalborg, Denmark, Jun. 2016. pp. 358–363.

[51] Song M.P., Gu G.H. 'Research on particle swarm optimization: A review'. *International Conference on Machine Learning and Cybernetics*; Shanghai, China, Aug. 2004. pp. 2236–2241.

[52] Zhao X., Turk M., Li W., Lien K.C., Wang G. 'A multilevel image thresholding segmentation algorithm based on two-dimensional K–L divergence and modified particle swarm optimization'. *Applied Soft Computing*. 2016; 48: 151–159.

[53] Cao J.Y., Cao B.G. 'Design of fractional order controllers based on particle swarm optimization'. *International Journal of Control, Automation, and Systems*. 2006; 4(6): 775–781.

[54] Qu L., Hu H., Huang Y. 'Fractional order PID controller based on particle swarm optimization implemented with FPGA'. *International Conference on Artificial Intelligence and Computational Intelligence (AICI)*; Sanya, China, Oct. 2010. pp. 165–169.

Part II

Controllers for underactuated mechanical systems

Chapter 5

Model-free control of the inertia wheel inverted pendulum with real-time experiments

Ahmed Chemori[1]

5.1 Introduction

The inertia wheel inverted pendulum belongs to the class of underactuated mechanical systems [10,12,22]. These systems are characterized by less control inputs than degrees of freedom, meaning that they have at least one unactuated generalized coordinate. Underactuation in these systems has two main sources, the first one is intentionally, i.e., decided in the design stage to minimize the cost, the weight, consumption, etc. The second is nonintentionally, where a fully actuated mechanical system can become underactuated after the deficiency of one or more actuators. Underactuated mechanical systems are characterized by a high nonlinear coupling between actuated and unactuated coordinates [7], and an internal dynamics which is often unstable (i.e., nonminimum phase systems [2]). In the real world many examples of such systems exist, they include, among others, inverted pendulums [18,21], underactuated robot manipulators [6], gymnast robots [23,31], undersea vehicles [17], aircrafts [26] and some mobile robots [20].

The inertia wheel inverted pendulum [34] is a benchmark for nonlinear control of underactuated mechanical systems. It has attracted the attention of many researchers within control community. Indeed, different control solutions have been proposed in the literature these last decades. In [1], strong damping force on the inertia wheel is taken into account in the design of the controller. The stabilization is achieved via nested saturation based controller. Reference [19] solves the limit cycles generation problem on the inertia wheel pendulum using virtual holonomic constraints. Real-time experiments were carried out showing the robustness of the proposed approach. In [27], collocated partial feedback linearization was performed to exhibit the nonlinear core subsystem which is stabilized using an implicit control. The remaining subsystem is stabilized using multiple sliding mode control technique. In our previous work [4,5], noncollocated partial feedback linearization is used; this gives rise to an unstable internal dynamics, which is stabilized using trajectory optimization and model-based error estimation. In [36], a generalized predictive controller is proposed

[1]Laboratoire d'Informatique, de Robotique et de Microélectronique de Montpellier (LIRMM), France

for the stabilization of the inertia wheel inverted pendulum. In [9], the proposed control solution deals with external disturbance rejection in passivity-based control of the inertia wheel inverted pendulum. All the above-mentioned techniques require knowledge of the system dynamics and parameters, furthermore only [3,4,9,19,36] present real-time experiments as a validation on a prototype of the inertia wheel inverted pendulum.

In this chapter, we propose to design a control scheme based on model-free technique to deal with underactuation in stable limit cycle generation. In order to achieve stable limit cycles on both coordinates of the inertia wheel inverted pendulum, we first design a family of *p-parameterized* periodic trajectories for the pendulum's angle. Those trajectories are then tracked using the control input thanks to a classical model-free controller [8,15]. Since the system is underactuated and nonminimum phase, a second controller is needed to update the parameter p of the above trajectories in order to deal with the convergence of the internal dynamics of the system. To achieve this control, we propose a second model-free controller using actuated coordinate (inertia wheel) as output and trajectories' parameter p as control input. Note that this control scheme can be easily applied to the stabilization case by carefully choosing appropriate trajectories. Numerical simulations as well as real-time experiments are presented to show the effectiveness of the proposed control scheme and its ability for external disturbances rejection.

The rest of this chapter is organized as follows. In Section 5.2, the experimental platform of the inertia wheel inverted pendulum is presented, with a description of its main components. Then, the proposed control solution is described in Section 5.3 for the general case of underactuated mechanical systems. Section 5.4 is devoted to the application of the proposed control scheme to the case of inertia wheel inverted pendulum. Numerical simulations are introduced in Section 5.5, where a first validation of the proposed control solution is illustrated. In Section 5.6, real-time experimental results are presented and discussed. Finally, Section 5.7 is devoted to some concluding remarks.

5.2 Description of the system: inertia wheel inverted pendulum

The testbed of the inertia wheel inverted pendulum system is shown in Figure 5.1, with its different components. It consists of three main parts:

- the mechanical part,
- the electronic part and
- the computer (control PC).

These three main parts will be described in the following sections.

5.2.1 Description of the mechanical part of the system

The mechanical part of the inertia wheel inverted pendulum is shown in Figure 5.2, which consists of an inverted pendulum equipped with an actuated rotating wheel.

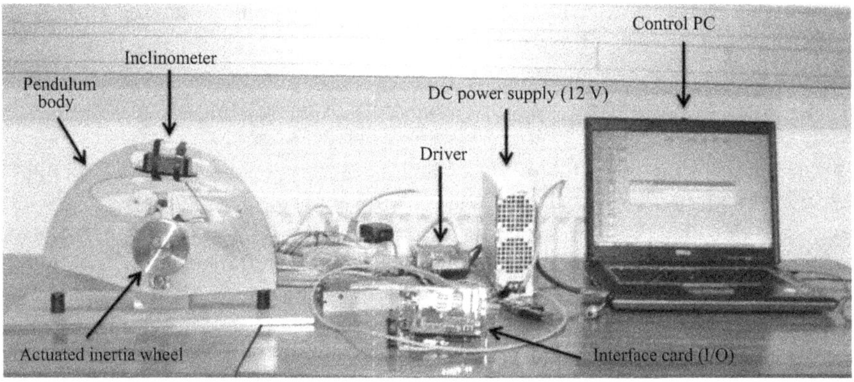

Figure 5.1 View of the testbed of the inertia wheel inverted pendulum system

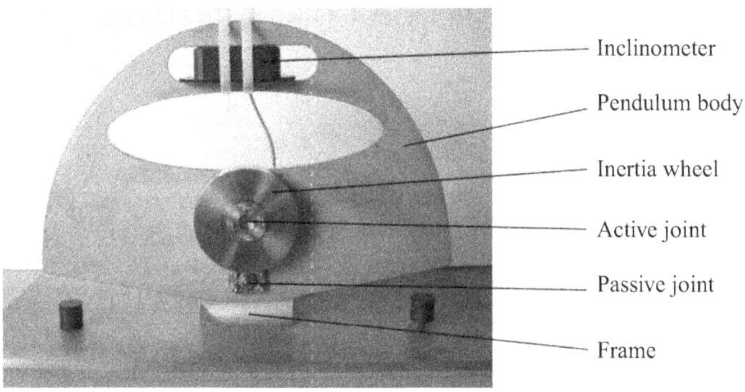

Figure 5.2 View of the mechanical part of the inertia wheel inverted pendulum

The joint between the pendulum body and the frame is unactuated whereas the joint between the pendulum body and the wheel is actuated by a Maxon EC-powermax 30 DC motor.

The pendulum angle q_{na} (with respect to the vertical) is constrained to remain within the interval $[-10°, +10°]$ due to mechanical stops (cf. Figure 5.2).

5.2.2 Description of the electronic part of the system

The actuator of the system is a `Maxon EC-powermax 30` DC motor, equipped with an integrated incremental encoder allowing the measurement in real-time of the inertia wheel angular position. To measure the angle of the pendulum with respect to the vertical, the system is equipped with an inclinometer `FAS-G` of `Micro strain`. The electronic part of the testbed includes also a `MAGMA` driver to control the DC motor with a power up to 200 W, a DC power supply and an I/O interface card. These components are visible on Figure 5.1 of the testbed.

5.2.3 Description of the control PC and software

As illustrated in Figure 5.1, the whole system is controlled with a computer equipped with a 2.4-GHz microprocessor. The control software is implemented using `C++` language, and the whole system is running under `Ardence` RTX (Real-Time Extension for Windows) operating system. The computer is the heart of the control system, it is connected to the I/O interface card to communicate with the physical system.

5.3 A proposed control solution for underactuated systems

In this section, the proposed control scheme is detailed. The main idea is to develop a control architecture based on model-free control for tracking of reference periodic trajectories while keeping stable the internal dynamics of underactuated mechanical systems.

5.3.1 Brief background on model-free control

Model-free control strategy has been initially proposed in [8,15] resulting in a breakthrough in nonlinear control. This technique is based on previous results on fast estimation and identification of nonlinear signals [16,24]. The control scheme is based on local linear approximation of the controlled system dynamics which is valid for a small time window. This approximation is updated in an online fashion thanks to a fast estimator. The control law proposed consists in a PID controller augmented with compensating terms provided by online estimation of the system dynamics. The overall controller is also called *i*-PID (standing for intelligent PID). Comparison of such a controller with classical PID controller can be found in [11]. The main advantage of this control strategy is that it doesn't require neither prior knowledge of the system dynamics nor complex parameters tuning. Consequently, it can be easily applied to controller unknown systems. Some efforts have been made for two particular cases of underactuated systems: the ball and beam [14] (where the dynamics of the beam has not been taken into account) and the Planar Vertical Take Off and Landing (PVTOL) aircraft [28] where the addressed control problem is limited to stabilization.

For the sake of simplicity, we present in the following model-free control for a single-input–single-output nonlinear systems.

5.3.1.1 Nonlinear dynamics

Consider a nonlinear system with unknown dynamics. The input–output behavior of the system can be expressed in the following general form as an ordinary differential equation (ODE):

$$E(y, \dot{y}, \ldots, y^{(a)}, u, \dot{u}, \ldots, u^{(b)}) = 0 \tag{5.1}$$

Given that this finite dimensional ODE is smooth enough, it can be approximated for a short time-window by the following simplified model:

$$y^{(\nu)} = F + \alpha u \tag{5.2}$$

The derivation order ν and the constant parameter $\alpha \in \mathbb{R}$ can be arbitrarily chosen by the designer. In model-free control literature, ν is generally chosen to be 1 or 2. The nonphysical constant α is a design parameter. $F \in \mathbb{R}$ captures all the unknown nonlinearities in the input–output behavior and can be compensated in the control law. Since (5.2) is valid for a short time window, it must be updated at each sample time. Therefore, the value of F is updated from the measurement of u and $y^{(\nu)}$ in the following manner:

$$[F(k)]_e = [y^{(\nu)}(k)]_e - \alpha u(k-1) \tag{5.3}$$

where $[F(k)]_e$ is the estimated value of F at sample time k which will be used for the computation of the control input $u(k)$. $[y^{(\nu)}(k)]_e$ is the estimated value of the νth derivative of the output y at discrete time k and $u(k-1)$ is the control input previously applied at discrete time $k-1$. The value of F can be initialized to 0 at initial time $k = 0$.

5.3.1.2 The control law

Given numerical knowledge of F (based on estimation) expressed by (5.15), the control input may cancel the unknown nonlinearities and add compensating terms corresponding to a closed-loop tracking of a given reference trajectory $y^*(t)$ using a conventional PID controller resulting in an *intelligent*-PID (often referred to as *i*-PID):

$$u = \frac{1}{\alpha}\left(-F + y^{*(\nu)} + K_p e + K_i \int e + K_d \dot{e}\right) \tag{5.4}$$

where $y^{*(\nu)}$ is the νth derivative of the reference trajectory y^*. K_p, K_i, K_d are the PID feedback gains, $e = y^* - y$ is the output tracking error, and $\dot{e}$ is its first time-derivative. The tuning of the PID feedback gains can be performed using poles placement techniques since all nonlinearities are supposed to be canceled. It is no longer necessary to perform complex system identification [13,14]. If $\nu = 1$, the PID controller reduces to a PI controller since the first derivative of output $\dot{y}$ is taken into account in the estimation of F in (5.15).

5.3.2 Proposed control solution

In our case, we are interested in tracking of periodic reference trajectories for stable limit cycles generation on underactuated mechanical systems (mainly known as nonlinear and nonminimum phase) and particularly the inertia wheel inverted pendulum. In order to simplify our presentation of the control problem as well as the proposed solution, let as focus on 1-input–2-degree-of-freedom underactuated mechanical systems (these are the minimum dimensions for a system to be underactuated) without loss of generality. The dynamics of such systems takes the following Lagrangian matrix form [29,33]:

$$M(q)\ddot{q} + H(q,\dot{q}) + G(q) = Ru \tag{5.5}$$

where $M \in \mathbb{R}^{2\times 2}$ is the inertia matrix of the system, $q \in \mathbb{R}^2$ is the vector of generalized coordinates. $\dot{q}$, $\ddot{q} \in \mathbb{R}^2$ are, respectively, their first and second time-derivatives. $H \in \mathbb{R}^2$ is a vector of centrifugal and Coriolis forces and $G \in \mathbb{R}^2$ is a vector of gravitational terms. $u \in \mathbb{R}$ is the control input and $R \in \mathbb{R}^{2\times 1}$ is a matrix distributing the effects of u on the generalized coordinates of the system. Consider the following partition $q = [q_a, q_{na}]^T$ of the vector of generalized coordinates, where q_a is the actuated coordinate and q_{na} is the unactuated one, then (5.5) may be rewritten as:

$$m_{11}(q)\ddot{q}_a + m_{12}(q)\ddot{q}_{na} + h_1(q,\dot{q}) + g_1(q) = u \tag{5.6}$$

$$m_{21}(q)\ddot{q}_a + m_{22}(q)\ddot{q}_{na} + h_2(q,\dot{q}) + g_2(q) = 0 \tag{5.7}$$

with:

$$M = \begin{bmatrix} m_{11} & m_{12} \\ m_{21} & m_{22} \end{bmatrix}; \quad H = \begin{bmatrix} h_1 \\ h_2 \end{bmatrix}; \quad G = \begin{bmatrix} g_1 \\ g_2 \end{bmatrix} \tag{5.8}$$

We suppose that the state of the system is the vector $[q_a\ q_{na}\ \dot{q}_a\ \dot{q}_{na}]^T$.

Our goal is to generate stable limit cycles on both actuated and unactuated coordinates of the underactuated mechanical system described by (5.6) and (5.7).

5.3.2.1 Basic principle of the proposed control scheme

A family of p-parameterized τ-periodic reference trajectories $q^*_{na}(p,\tau,t)$ should be defined for the unactuated coordinate. Those trajectories should have the same boundary conditions for all p values, allowing the controller to switch from one trajectory to another while keeping the overall trajectory smooth. Thanks to the dynamic coupling existing between the actuated and unactuated coordinates of the system, it is possible to control the unactuated coordinate using the control input u (i.e., the torque on the actuated coordinate) which allows those trajectories to be tracked on the unactuated coordinate q_{na} using the control input u.

Indeed, the dynamics (5.6) and (5.7) can be rewritten in a form with a relation between unactuated coordinate and control input [30]. First, (5.6) is solved for $\ddot{q}_a$[1]:

$$\ddot{q}_a = m_{11}^{-1}(-m_{12}\ddot{q}_{na} - h_1 - g_1 + u)\,. \tag{5.9}$$

Injecting this solution in (5.7) leads to:

$$\overline{m}_2\ddot{q}_{na} + \overline{h}_2 + \overline{g}_2 = -m_{21}m_{11}^{-1}u \tag{5.10}$$

where $\overline{m}_2 = m_{22} - m_{21}m_{11}^{-1}m_{12}$, $\overline{h}_2 = h_2 - m_{21}m_{11}^{-1}h_1$ and $\overline{g}_2 = g_2 - m_{21}m_{11}^{-1}g_1$.

Based on the resulting dynamics (5.10), involving a relationship between $\ddot{q}_{na}$ and u, a model-free controller can then be designed to perform the tracking of reference trajectories on unactuated coordinate q_{na} using the control input u. However, since

[1]For clarity reasons, the dependency in q and $\dot{q}$ of the involved terms is omitted in the adopted notation.

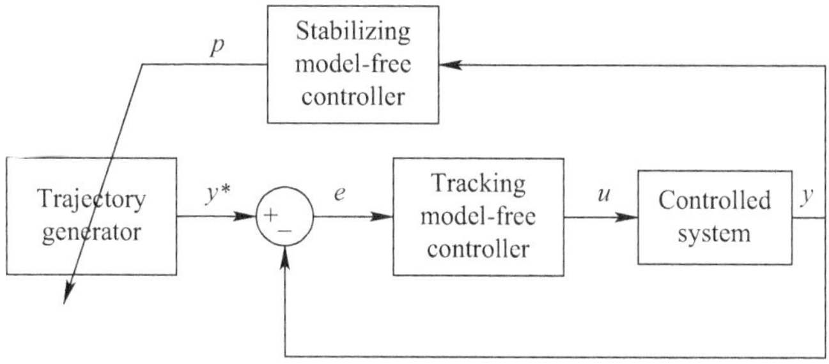

Figure 5.3 Schematic view of the proposed dual model-free controller. The controlled system has as output y and as input u. The reference output trajectories y^* are generated given a parameter p, and tracked by the first model-free controller (called also tracking controller) based on the evaluation of the error $e = y^* - y$. The parameter p is updated at each period by the second model-free controller (called also stabilizing controller)

the internal dynamics of the system is unstable (the inertia wheel is a nonminimum phase underactuated mechanical system), even a perfect tracking of these trajectories cannot guarantee the stability of the internal dynamics of the system (i.e., the inertia wheel dynamics). Consequently, a second controller is needed to manage the stability of the system's internal dynamics.

Indeed, the closed-loop system dynamics can be obtained by substituting u in (5.6) by its value taken from (5.7). This requires the invertibility of $-m_{21}m_{11}^{-1}$ and particularly of m_{21}. This invertibility constraint is also known as *Strong Inertial Coupling* [35]. The resulting internal dynamics is then the following:

$$\ddot{q}_a + m_{12}^{-1}m_{22}\ddot{q}_{na} + m_{12}^{-1}h_2 + m_{12}^{-1}g_2 = 0. \tag{5.11}$$

Note that $\ddot{q}_{na}$ depends on the control input u and therefore on the reference parameterized trajectory and its design parameter p. This second controller therefore will acts on the reference trajectory's parameter p (this parameter is considered as the control input to be generated by this second controller) in the aim of keeping the stability of the internal dynamics of the system (i.e., the actuated coordinate is considered as the output to be controlled). In other words, a periodic parameterized reference trajectory is tracked on the unactuated coordinate by the first controller, and at the end of each period, the second controller chooses the new trajectory parameter p in order to stabilize the actuated coordinate. The chosen parameter p is used in the reference trajectory of the forthcoming cycle. The overall control scheme is illustrated in the block-diagram of Figure 5.3.

5.3.2.2 Cyclic reference trajectories generation

The proposed control framework needs a trajectory generator to provide parameterized reference trajectories $q_{na}^*(p, \tau, t)$ to be tracked on the unactuated coordinate of the system. The proposed trajectories must fulfill some necessary conditions. First of all, they have to be continuous, time-derivable and periodic in order to generate limit cycles. This consideration leads to design oscillatory shaped trajectories which are split up in half period, where we will use symmetry to generate the whole cycle. The parametrization of these trajectories must allow the controller to update the parameter p (which corresponds to the time at which the trajectory q^*_{na} crosses zero) during tracking, while the overall trajectory remains smooth. This leads to some initial and final conditions of each half period part. That is for a given period τ and amplitude $2A$:

$$\forall p \in \mathscr{P}, \begin{cases} q^*_{na}(p,\tau,0) = q^*_{na}(p,\tau,\tau) = A \\ q^*_{na}(p,\tau,\frac{\tau}{2}) = -A \\ \dot{q}^*_{na}(p,\tau,0) = \dot{q}^*_{na}(p,\tau,\frac{\tau}{2}) = \dot{q}^*_{na}(p,\tau,\tau) = 0 \end{cases} \tag{5.12}$$

for some domain $\mathscr{P} \subset \mathbb{R}$ (chosen to keep an oscillatory shape). We propose to use a six-degree polynomial function parameterized with p such that:

$$\forall p \in \mathscr{P}, \; q^*_{na}(p,\tau,t=p) = 0 \tag{5.13}$$

Figure 5.4 shows normalized reference trajectories (for $\tau = 2$ and $A = 1$) for different values of the parameter p during half a period. The domain $\mathscr{P}$ is restricted to interval $[p_{\min}, p_{\max}]$ guaranteeing an oscillatory shape trajectory. It is worth to note, due to the symmetry property of half period trajectory parts, $p_{\min} = 1 - p_{\max}$.

5.3.2.3 The dual model-free controller

Consider the design of a model-free tracking controller (the first controller in the control framework of Figure 5.3). The first step is to consider the unactuated coordinate

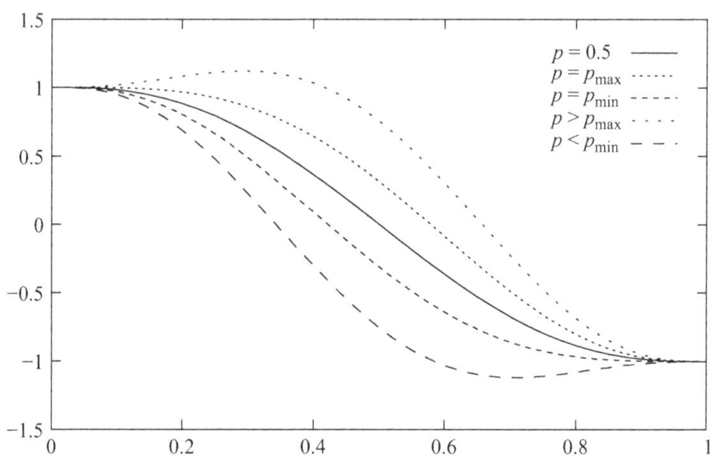

Figure 5.4 Examples of normalized reference trajectories during a half-period for different values of p, the period τ is 2 s and the amplitude 2A is 2

nonlinear dynamics (5.10) to be replaced by a local model according to model-free control principle:

$$\ddot{q}_{na} = F_1 + \alpha_1 u \tag{5.14}$$

where the constant parameter α_1 is a control design parameter. The term F_1 is supposed to capture the nonlinearities in the unactuated coordinate dynamics and is updated at each sample time according to the following equation:

$$[F_1(k)]_e = [\ddot{q}_{na}(k)]_e - \alpha_1 u(k-1) \tag{5.15}$$

where the notation $[.]_e$ states for the estimated value.

The tracking controller is obtained using numerically computed value of F_1 in (5.15) using the first model-free controller (*i*-PID) as follows:

$$u = \frac{1}{\alpha_1}\left(-F_1 + \ddot{q}^*_{na}(p,\tau,t) + K_{p1}e + K_{i1}\int e + K_{d1}\dot{e}\right) \tag{5.16}$$

where the feedback gains K_{p1}, K_{i1}, K_{d1} can be chosen using poles placement technique based on the assumption that the system's nonlinearities are compensated by the term F_1. The unactuated coordinate q_{na} should follow therefore the desired periodic trajectory $q^*_{na}(p,\tau,t)$.

The parameter p, in the tracking control input (5.16), is constant over half a period, i.e., $\forall t \in [k\frac{\tau}{2}\ (k+1)\frac{\tau}{2}]$, $k \in \mathbb{N}$ and is updated at the end of each half period, i.e., at time $k\frac{\tau}{2}$, $k \in \mathbb{N}$ by a second model-free controller to be designed.

For that purpose, let us consider now the dynamics of the actuated coordinate of the system. This dynamics may be approximated by the following local discrete-time model:

$$\Delta_\tau v_a = F_2 + \alpha_2 p \tag{5.17}$$

where $\Delta_\tau v_a = \dot{q}_a(k\frac{\tau}{2}) - \dot{q}_a((k-1)\frac{\tau}{2})$ is the variation of the velocity of the actuated joint (inertia wheel) $v_a = \dot{q}_a$ measured between half periods and the constant α_2 is a control design parameter. According the basic principle of model-free control the value of F_2 is updated at the end of each half cycle ($t = k\frac{\tau}{2}$ for $k \in \mathbb{N}$) as follows:

$$\left[F_2\left(k\frac{\tau}{2}\right)\right]_e = \left[\Delta_\tau \dot{q}_a\left(k\frac{\tau}{2}\right)\right]_e - \alpha_2 p\left((k-1)\frac{\tau}{2}\right) \tag{5.18}$$

Note that continuous dynamics of the actuated coordinate within the half cycle $t = [k\frac{\tau}{2}\ (k+1)\frac{\tau}{2}]$ is not taken into account in this local model. The idea is that we only aim at limit cycle generation on this coordinate; therefore, it is only required that its evolution should be periodic. In other words, the aim of the second controller is to bring the actuated coordinate as close as possible to a fixed desired state $(q_a^d, \dot{q}_a^d)$ at the end of each half period, ensuring a certain periodicity of its evolution, leading therefore to limit cycles generation. The resulting second model-free controller should updates the trajectory parameter p according to the following control law:

$$p = \frac{1}{\alpha_2}\left(-F_2 + K_{p2}(q_a^d - q_a) + K_{i2}\int (q_a^d - q_a) + K_{d2}(\dot{q}_a^d - \dot{q}_a)\right) \tag{5.19}$$

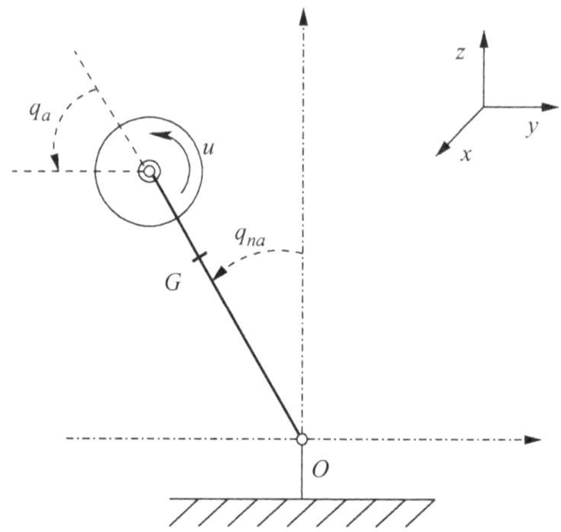

Figure 5.5 Principle schematic view of the inertia wheel inverted pendulum: the joint between the frame and the beam is unactuated (passive joint), while the one between the beam and the inertia wheel is actuated (active joint)

where K_{p2}, K_{i2}, K_{d2} are the PID feedback gains. It is worth to note that since the desired state for actuated coordinate $(q_a^d, \dot{q}_a^d)$ is constant, the $(\Delta_\tau \dot{q}_a)^d$ term should be zero and is then omitted.

5.4 Application: the inertia wheel inverted pendulum system

The testbed of the inertia wheel inverted pendulum is presented in Section 5.2. The basis of its mechanical structure is illustrated in Figure 5.5. The actuator produces an angular acceleration of the rotating inertia wheel which generates, thanks to the dynamic coupling between coordinates, a torque acting on the pendulum's passive joint; therefore, this passive joint can be controlled through the acceleration of the inertia wheel. The goal of the proposed control solution (as introduced in Section 5.3) is to generate stable limit cycles on both coordinates (actuated and unactuated) using only one control input (namely the torque applied to the inertia wheel). The idea is to control the angular position q_{na} of the pendulum with respect to the vertical by the first controller (tracking controller in Figure 5.3), while the angular velocity of the inertia wheel (system's internal dynamics) is controlled by the second controller (stabilizing controller in Figure 5.3).

5.4.1 Dynamic modeling of the system

The mechanical structure of the inertia wheel inverted pendulum is sketched in Figure 5.5. The nonlinear dynamic model [4] of the plant is obtained using Lagrange formulation [32] and is given by:

$$(I_1 + I_2)\ddot{q}_{na} + I_2\ddot{q}_a - \overline{ml}g \sin q_{na} = 0 \tag{5.20}$$

$$I_2(\ddot{q}_{na} + \ddot{q}_a) = u \tag{5.21}$$

Table 5.1 *Description of dynamic parameters of the inertia wheel inverted pendulum*

Parameter	Description	Value	Unit
m_1	Body mass	3.228	kg
m_2	Wheel mass	0.86422	kg
I_1	Body inertia	3.042×10^{-2}	kg m^2
I_2	Wheel inertia	7.986×10^{-4}	kg m^2
l_1	Body CoM position	6.354×10^{-4}	m
l_2	Wheel CoM position	52×10^{-3}	m

where I_1, I_2 are, respectively, the moments of inertia of the beam and the wheel. u is the torque generated by the motor acting on the inertia wheel. $\overline{ml} = m_1 l_1 + m_2 l_2$ with m_1 and m_2 being the masses of the pendulum and the inertia wheel. l_1, l_2 are distances from origin O (cf. Figure 5.5) to the gravity centers of the pendulum and the rotating mass (respectively). This model has the general form (5.6) to (5.7) of dynamics of underactuated mechanical systems. The dynamic parameters of the inertia wheel inverted pendulum system are summarized in Table 5.1.

5.5 Numerical simulations

Reference trajectories are generated for the unactuated coordinate q_{na} as presented in Section 5.3.2.2. The trajectories amplitude is $2A = 6°$ and period is $\tau = 2$ s. Numerical simulations have been performed using MATLAB®/Simulink® software of MathWorks. It is worth noting that the dynamic model (5.20) and (5.21) is used to simulate the dynamic behavior of the system, it is not used in the controller design.

The following design parameters were used in the implemented control scheme: $\alpha_1 = -100$, $\alpha_2 = 50$, the feedback gains of the first controller (tracking controller) are as follows: $K_{p1} = 200$, $K_{i1} = 0$ and $K_{d1} = 100$. The feedback gains of the second controller (stabilizing controller) are given by: $K_{p2} = K_{i2} = 0$ and $K_{d2} = 1$. The sampling frequency is set to 150 Hz which is a reasonable value for real-time implementation. The desired state for the inertia wheel velocity (controlled by the second controller) is set to $\dot{q}_a^d = 0$.

The obtained simulation results are depicted in Figure 5.6. It is worth to note that a disturbing torque of intensity 0.5 N m is applied to the pendulum beam at time $t = 15$ s. The convergence to a stable limit cycle can clearly be observed on the phase portrait of the pendulum angle q_{na}. The controller reacts immediately to the disturbance as we can see a spike in the control signal (cf. Figure 5.6(e)). This disturbance induces a deviation from the reference trajectories of the pendulum angle position and velocity which is immediately compensated. Despite the big deviation of

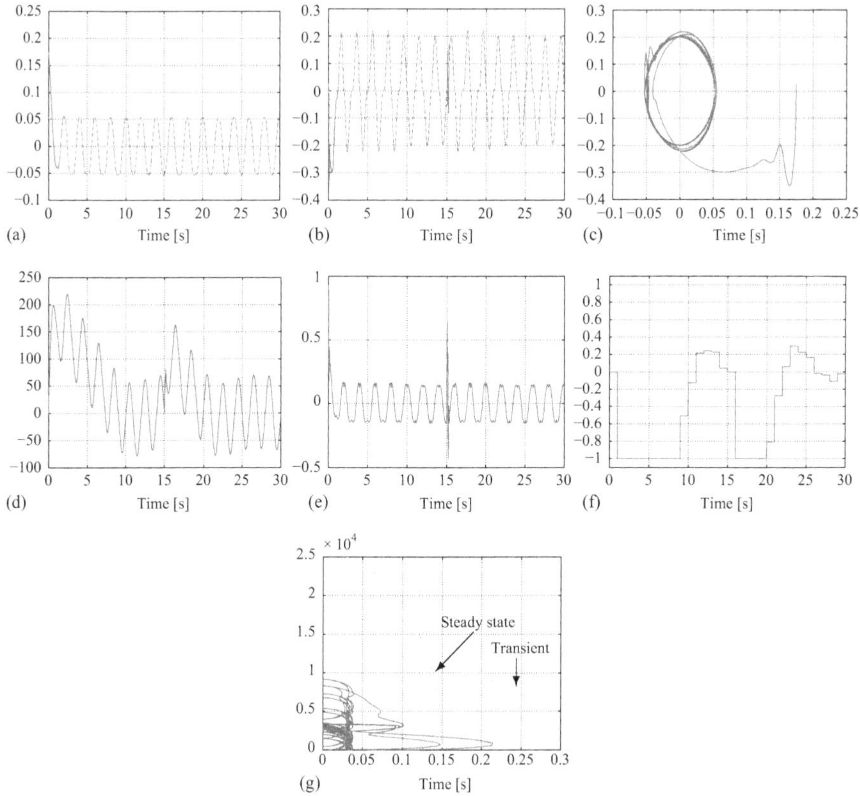

Figure 5.6 Simulation results. A punctual external disturbance is introduced as a torque applied at time $t = 15$ s to the pendulum beam. (a) Pendulum position q_{na} [rad] real position (solid line) and desired position (dashed line). (b) Pendulum velocity $\dot{q}_{na}$ [rad $\cdot$ s^{-1}] real velocity (solid line) and desired velocity (dashed line). (c) Phase portrait of the pendulum (q_{na} [rad], $\dot{q}_{na}$ [rad $\cdot$ s^{-1}]). (d) Inertia wheel velocity $\dot{q}_a$ [rad $\cdot$ s^{-1}]. (e) Control input u[N $\cdot$ m] versus time. (f) evolution of parameter p [−]. (g) Motor rotationnal speed [rpm] vs Absolute motor torque [N $\cdot$ m]

the inertia wheel velocity oscillations at the beginning of the simulation due to nonzero initial conditions and just after the disturbance, the second controller successfully brings back the inertia wheel velocity trajectory to a periodic trajectory. Indeed, the internal dynamics of the systems converges also to a stable limit cycle. To check the admissibility of the scenario to be implemented on the real system, the motor power admissibility is depicted in Figure 5.6(g), where it can be seen clearly that the trajectories are inside the admissible region.

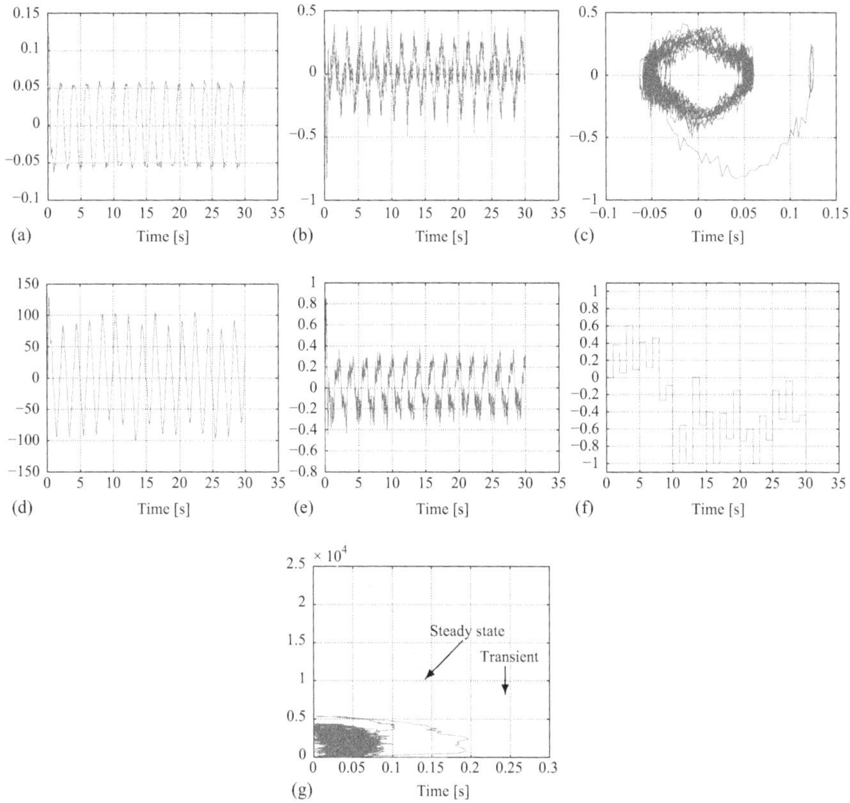

Figure 5.7 Real-time experimental results in nominal case, without external disturbances. (a) Pendulum position q_{na} real position (solid line) and desired position (dashed line). (b) Pendulum velocity $\dot{q}_{na}$ real velocity (solid line) and desired velocity (dashed line). (c) Phase portrait of the pendulum (q_{na}, $\dot{q}_{na}$). (d) Inertia wheel velocity $\dot{q}_a$. (e) Control input u versus time. (f) Evolution of parameter p. (g) Check of power admissible-region of the actuator

5.6 Real-time experiments

Real-time experiments are achieved thanks to the experimental testbed described in Section 5.2. It is worth to note that the control design parameters used for these experiments are different from those used in numerical simulation. The first controller parameters are chosen as $\alpha_1 = -150$, $K_{p1} = 70$, $K_{i1} = 0$ and $K_{d1} = 12$. For the Second controller the following parameters are used $\alpha_2 = 70$, $K_{p2} = 0$, $K_{i2} = 0$ and $K_{d2} = 1$. Noise filtering was performed through the use of an *alpha-beta* filter [25] to correctly estimate the pendulum acceleration used in (5.15). The following two experimental scenarios have been considered:

- *control in nominal case,*
- *external disturbance rejection.*

5.6.1 Scenario 1: Control in nominal case

In this first scenario, no external disturbances have been considered. The obtained results for this scenario are depicted in Figure 5.7. The experiment was started by an initial angular position corresponding to a pendulum body at the mechanical stop, introducing an initial position error of 0.125 rad (i.e., 7.16°). Despite this initial error, the controller was able to steer rapidly the system toward the periodic reference trajectory as it can be observed on Figure 5.7(a). The convergence to a stable limit cycle is clearly visible in Figure 5.7(c). The control inputs are shown in Figure 5.7(e) for the first controller and in Figure 5.7(f) for the second one. The motor power admissibility is illustrated through the plot of Figure 5.7(g) showing the evolution of the motor speed (absolute value) versus its torque (absolute value).

5.6.2 Scenario 2: External disturbance rejection

In this experiment, external disturbances are introduced by pushing the pendulum body in a punctual manner as illustrated in Figure 5.8. Two external disturbances have been introduced at approximately $t_1 = 12$ s and $t_2 = 23$ s. The obtained experimental results are plotted in Figure 5.9. Noise in measurement is observable on all plots, particularly amplified on the pendulum velocity $\dot{q}_{na}$ in Figure 5.9(b). Despite this fact, the first controller was able to track the reference trajectory on the pendulum (cf. Figure 5.9(a)). The effect of the introduced external disturbances can clearly be observed on the pendulum trajectories in Figure 5.9(a) and the control input in Figure 5.9(e). The tracking controller brings back rapidly the pendulum position and velocity to their respective reference trajectories. The convergence to a stable limit cycle is clearly visible on the phase portrait of the pendulum shown on Figure 5.9(c).

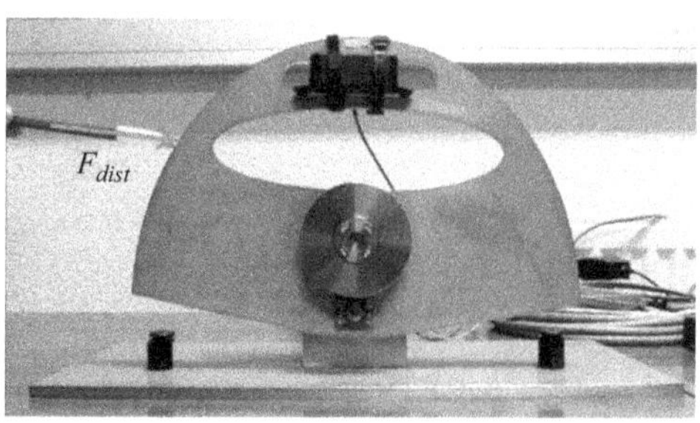

Figure 5.8 Illustration of the external disturbance action applied on the pendulum body during experiments

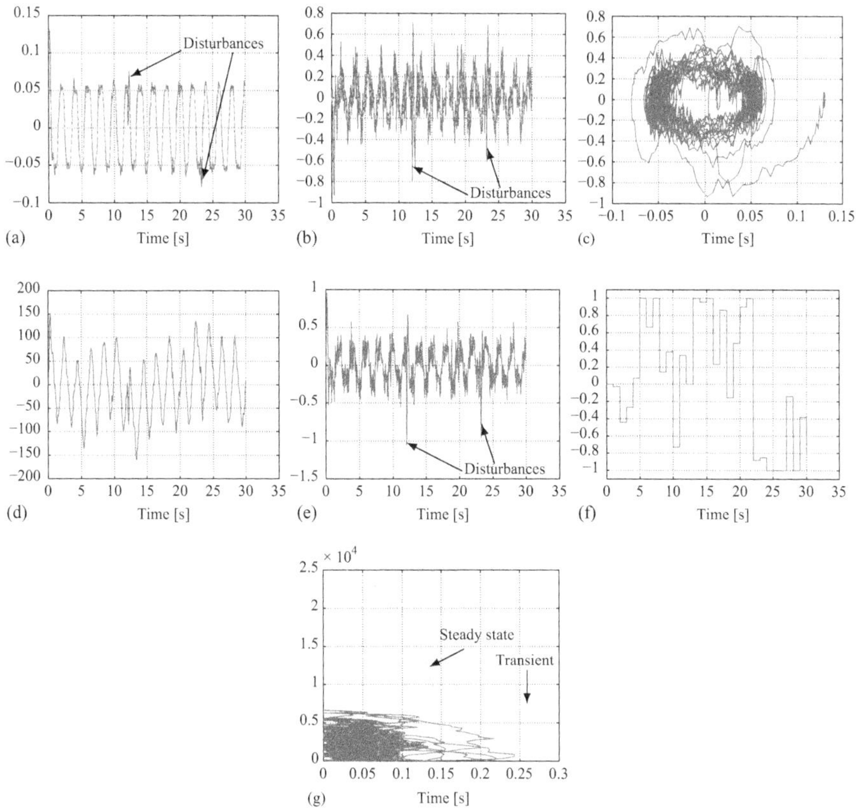

Figure 5.9 Real-time experimental results. Two external disturbances have been introduced at approximately $t_1 = 12$ s and $t_2 = 23$ s on the pendulum body. (a) Pendulum position q_{na} [rad] real position (solid line) and desired position (dashed line). (b) Pendulum velocity $\dot{q}_{na}$ [rad $\cdot$ s^{-1}] real velocity (solid line) and desired velocity (dashed line). (c) Phase portrait of the pendulum (q_{na} [rad], $\dot{q}_{na}$ [rad $\cdot$ s^{-1}]). (d) Inertia wheel velocity $\dot{q}_a$ [rad $\cdot$ s^{-1}]. (e) Control input u [N $\cdot$ m] versus time. (f) Evolution of parameter p [−]. (g) Motor rotational speed [rpm] vs Absolute motor torque [N $\cdot$ m]

The evolution of the inertia wheel velocity is depicted in Figure 5.9(d). Indeed, the introduced disturbances induce a shift in the inertia wheel velocity trajectory cycle. However, thanks to the second controller reaction, which is observable in Figure 5.9(f), the inertia wheel velocity converges back to the limit cycle in few periods. The absolute value of motor velocity versus the generated torque are displayed in Figure 5.9(g) to check the admissible-region of the actuator. Notice that the obtained trajectory remains, throughout the experiments, within the admissible region of the actuator's power.

5.7 Conclusion

This chapter deals with the control problem of the inertia wheel inverted pendulum. This benchmark belongs to the class of underactuated mechanical systems. The proposed control solution, based on model-free control technique, includes two controllers to deal with stable limit cycle generation. The first controller (called *tracking controller*) is used to control the angular position of the pendulum (unactuated coordinate), while the angular velocity of the inertia wheel (actuated coordinate) is controlled by the second controller (*called stabilizing controller*). Indeed, to stabilize the resulting internal dynamics of the system, a family of *p-parameterized* periodic trajectories is generated and tracked by the first controller, while their parameter p is on-line updated by the second controller in the aim of improving the behavior the internal dynamics (i.e., *convergence to a stable limit cycle*). To validate the proposed control scheme, numerical simulations as well as real-time experiments are introduced. They show the effectiveness of the proposed control scheme as well as its ability to reject external disturbances. Future work may include the extension of the proposed control scheme to endow it with an automatic adaptive tuning of the parameters of the controllers, as well as its generalization to the case of underactuated systems with more than one degree of underactuation.

References

[1] C.F. Aguilar-Ibanez, O.O. Gutierrez F, and M.S. Suarez Castanon. 'Stabilization of the strongly damping inertia wheel pendulum by a nested saturation functions', In *Proc. of American Control Conference*, 2008, pp. 3434–3439.

[2] S. Andary and A. Chemori. 'A dual model-free control of non-minimum phase systems for generation of stable limit cycles', In *IEEE CDC–ECC*, Orlando, FL, USA, 2011, pp. 1387–1392.

[3] S. Andary, A. Chemori, M. Benoit, and J. Sallantin. 'A dual model-free control of underactuated mechanical systems, application to the inertia wheel inverted pendulum', In *American Control Conference (ACC'12)*, Montréal, Canada, 2012, pp. 1029–1034.

[4] S. Andary, A. Chemori, and S. Krut. 'Control of the underactuated inertia wheel inverted pendulum for stable limit cycle generation', *Advanced Robotics*, 2009, **23**, pp. 1999–2014.

[5] S. Andary, A. Chemori, and S. Krut. 'Estimation-based disturbance rejection in control for limit cycle generation on inertia wheel inverted pendulum testbed', In *IEEE/RSJ International Conference on Intelligent Robots and Systems (IROS'09)*, St. Louis, MO, USA, 2009, pp. 1302–1307.

[6] M. Bergerman. *Dynamics and Control of Underactuated Manipulators*. PhD Thesis, Carnegie Mellon University, 1996.

[7] W. Blajer and K. Kolodziejczyk. 'Modeling of underactuated mechanical systems in partly specified motion', *Journal of Theoretical and Applied Mechanics*, 2008, **46**, 2, pp. 383–394.

[8] R. Bourdais, M. Fliess, C. Join, and W. Perruquetti. 'Towards a model-free output tracking of switched nonlinear systems', In *NOLCOS 2007—7th IFAC Symposium on Nonlinear Control Systems*, Pretoria, South Africa, 2007.

[9] N. Khraief A. Chemori and S. Belghith. External disturbance rejection in IDA-PBC controller for underactuated mechanical systems: from theory to real-time experiments. In *IEEE Multi-Conference on Systems and Control (MSC'14)*, Antibes, France, 2014, pp. 1747–1752.

[10] A. Choukchou-Braham, B. Cherki, M. Djemai, and K. Busawon. *Analysis and Control of Underacuated Mechanical Systems*. London, UK: Springer, 2014.

[11] B. D'Andrea Novel, M. Fliess, C. Join, H. Mounier, and B. Steux. 'A mathematical explanation via "intelligent" PID controllers of the strange ubiquity of PIDs'. In *18th Mediterranean Conference on Control and Automation, MED'10*, Marrakech Morocco, 2010.

[12] I. Fantoni and R. Lozano. *Non-linear Control for Underactuated Mechanical Systems*. New York, USA: Springer-Verlag, 2001.

[13] M. Fliess and C. Join. 'Intelligent PID controllers', In *16th Mediterranean Conference on Control and Automation*, Ajaccio, France, 2008.

[14] M. Fliess and C. Join. 'Model-free control and intelligent PID controllers: towards a possible trivialization of nonlinear control?', In *15th IFAC Symposium on System Identification (SYSID 2009)*, Saint-Malo, France, 2009.

[15] M. Fliess, C. Join, M. Mboup, and H. S. Ramirez. 'Vers une commande multivariable sans modèle', In *Conférence internationale francophone d'automatique (CIFA 2006)*, Bordeaux, France, 2006.

[16] M. Fliess, C. Join, and H. S. Ramirez. 'Non-linear estimation is easy', *International Journal of Modelling Identification and Control*, 2008, **4**, pp. 12–27.

[17] T. I. Fossen. *Nonlinear Modelling and Control of Underwater Vehicles*. PhD Thesis, Norwegian Institute of Technology, 1991.

[18] L. B. Freidovich, P. La Hera, U. Mettin, A. Robertsson, and A. S. Shiriaev. 'Shaping stable periodic motions of inertia wheel pendulum: theory and experiments', *Asian Journal of Control*, 2009, **11**, 5, pp. 548–556.

[19] L. Freidovich, P. La Hera, U. Mettin, A. Robertsson, A. Shiriaev, and R. Johansson. 'Shaping stable periodic motions of inertia wheel pendulum: theory and experiments', *Asian Journal of Control*, 2009, **11**, 5, pp. 548–556.

[20] S. Ge and F. Lewis. *Autonomous Mobile Robots*. Boca Raton, FL, USA: CRC Press, 2006.

[21] S. Jadlovska and J. Sarnovsky. 'A complex overview of modeling and control of the rotary single inverted pendulum system', *Power Engineering and Electrical Engineering*, 2013, **11**, 2, pp. 73–85.

[22] S. Jadlovska and J. Sarnovsky. 'Modeling and optimal control of nonlinear underactuated mechanical systems – a survey', In *13th Scientific Conference of Young Researchers – FEI TU of Kosice*, Herlany, Slovakia, 2013, pp. 32–35.

[23] A. De Luca, R. Mattone, and G. Oriolo. 'Stabilization of underactuated planar 2r manipulator', *International Journal of Robust and Nonlinear Control*, 2000, **10**, 4, pp. 181–198.

[24] M. Mboup, C. Join, and M. Fliess. 'Numerical differentiation with annihilators in noisy environment', *Numerical Algorithms*, 2009, **50**, pp. 439–467.

[25] R. Penoyer. 'The alpha-beta filter', *The C Users Journal*, 1993, **11**, 7, pp. 73–86.

[26] G. Poulin, A. Chemori, and N. Marchand. 'Minimum energy oriented global stabilizing control of the PVTOL aircraft', *International Journal of Control*, 2007, **80**, 3, pp. 430–442.

[27] N. Qaiser, S. B. Tariq, N. Haq, and T. Aziz. 'A novel nonlinear implicit sliding surface controller design for inertia wheel pendulum', In *Proceedings of the 10th WSEAS International Conference on Robotics, Control and Manufacturing Technology*, ROCOM'10, Stevens Point, Wisconsin, USA, 2010, pages 21–25. World Scientific and Engineering Academy and Society (WSEAS).

[28] S. Riachy, M. Fliess, C. Join, and J.-P. Barbot. 'Vers une simplification de la commande non linéaire: l'exemple d'un avion à décollage vertical', In *Sixième Conférence Internationale Francophone d'Automatique, CIFA 2010*, Nancy, France, 2010.

[29] L. Sciavicco and B. Siciliano. *Modeling and Control of Robot Manipulators*. New York: McGraw-Hill, 1996.

[30] M. Spong. 'Partial feedback linearization of underactuated mechanical systems', In *EEE/RSJ/GI International Conference on Intelligent Robots and Systems*, Munich, Germany, 1994, pp. 314–321.

[31] M. Spong. 'The swing up control problem for the acrobot', *IEEE Control Systems Magazine*, February 1995, **15**, 1, pp. 49–55.

[32] M. Spong and M. Vidyasagar. *Robot Dynamics and Control*. New York, USA: John Wiley and Sons, 1989.

[33] Mark W. Spong. Underactuated mechanical systems. In *Control Problems in Robotics and Automation*. London, UK: Springer-Verlag, 1998.

[34] M. W. Spong, P. Corke, and R. Lozano. Nonlinear control of the inertia wheel pendulum. *Automatica*, 1999, **37**, pp. 1845–1851.

[35] M.W. Spong. 'Partial feedback linearization of underactuated mechanical systems', In *Proceedings of the IEEE/RSJ/GI International Conference on Intelligent Robots and Systems '94*, 1994.

[36] N. Touati and A. Chemori. 'Predictive control for the stabilization of a fast mechatronic system: from simulation to real-time experiments', In *International IFAC Symposium on Mechatronic Systems (Mechatronics'13)*, Hangzhou, China, 2013, pp. 237–242.

Chapter 6

Output feedback second-order sliding-mode tracking control for perturbed inertia wheel pendulum

Luis T. Aguilar[1] *and Rafael Iriarte*[2]

We address the robust output feedback tracking control problem for an inertia wheel pendulum in the presence of uniformly bounded matched disturbances. The periodic motion of the pendulum will be at the upright position which corresponds to the unstable equilibrium point of the unforced system. A two-relay-controller-based reference model was developed for generating the desired trajectories to be tracked by the unactuated link of the inertia wheel pendulum and then design an output feedback robust tracking controller. The desired amplitude and frequency were tuned by choosing the two-relay control gains properly. A second-order sliding-mode tracking controller interconnected with a second-order sliding-mode observer was capable to track the prescribed reference trajectory rejecting matched external disturbances. Performance issues of the constructed controller-observer were illustrated in a numerical study.

6.1 Introduction

For nonlinear mechanical systems, tracking control problem is known to be more difficult than stabilization mainly for underactuated systems whose initial conditions are close to an unstable equilibrium point. The tracking problem for nonminimum-phase systems is a challenging topic in control since that further research applications in the motion control of underactuated systems have gone in many directions. Orbital stabilization problems where underactuated systems are evolved are the coordinated motion of biped robots [8,11,20], gymnastic robots, electrical converters [1,18], and others (see, e.g., [10,22] and references therein).

A strategy addressing the self-oscillation and motion planning for a class of underactuated systems is presented in [2] based on the two-relay controller methodology. In previous works [3–5], the oscillations at the output are generated for an inertia wheel pendulum using such methodology without a tracking controller, reason that we called

[1]Instituto Politécnico Nacional, México
[2]Department of Control and Robotics, National University of Mexico, UNAM, Mexico

self-oscillation. The drawback of the method is that the closed-loop system turns out to be high sensitive to disturbances and the formulas of the two-relay controller are function of parameters of the plant. In other words, the deviation of the frequency and amplitude of the periodic trajectory at the output of the closed-loop structure proposed in [3–5] with respect to the desired ones, is related with the uncertainties of the parameters of the plant because formulas to compute the two values of the two-relay controller depends on the exact knowledge of the physical parameters (e.g., inertias, length of the links, and masses), while neglected dynamics, as friction, were not considered as part of the formulas. Iriarte *et al.* [12] develop the two-relay controller robustified with a state-feedback tracking controller, complemented with experimental verification, dealing with such inaccuracies. However, the output feedback case has remained unaddressed in the above papers.

The aim of this chapter is to solve the output feedback-tracking control problem for the inertia wheel pendulum under the presence of undesired disturbances. The periodic motion of the pendulum will be at the upright position which corresponds to the unstable equilibrium point of the unforced system. A two-relay-controller-based reference model was used for generating the desired trajectories to be tracked by the unactuated link of the inertia wheel pendulum and then, design an output feedback robust tracking controller. The second-order sliding-mode observer was taken from [16]. The desired amplitude and frequency were set by choosing the two-relay control gains according to the formulas obtained from the describing function method [13]. This chapter also refines the results of [12] where the asymptotical stability proof of the closed-loop system is revisited.

The chapter is organized as follows. In Section 6.2, we give the dynamic model of the plant and define the problem statement. Moreover, the trajectory generation, via two-relay controller, to be tracked by the output of the plant is revised. In Section 6.3, the variable structure observer is designed to estimate the velocity of the motor and the pendulum. In Section 6.4, a second-order sliding-mode control is synthesized to guarantee robust tracking of the reference trajectory. Numerical results illustrate the performance of the proposed output feedback controller in Section 6.5. Finally, conclusions are provided in Section 6.6.

6.2 Dynamic model and problem statement

Dynamics of an inertia wheel pendulum, taken from [7], augmented with viscous friction and disturbances can be described as follows:

$$\begin{bmatrix} J_1 & J_2 \\ J_2 & J_2 \end{bmatrix} \begin{bmatrix} \dot{v}_1 \\ \dot{v}_2 \end{bmatrix} + \begin{bmatrix} h \sin q_1 \\ f_s v_2 \end{bmatrix} = \begin{bmatrix} 0 \\ 1 \end{bmatrix} \tau + w, \tag{6.1}$$

$$y = [q_1, q_2]^T. \tag{6.2}$$

In the above equation, $q_1(t) \in \mathbb{R}$ is the absolute angle of the pendulum, counted clockwise from the vertical downward position, $q_2(t) \in \mathbb{R}$ is the absolute angle of the disk, $v_1(t) = \dot{q}_1(t)$ and $v_2(t) = \dot{q}_2(t)$ are the angular velocity of the pendulum and the

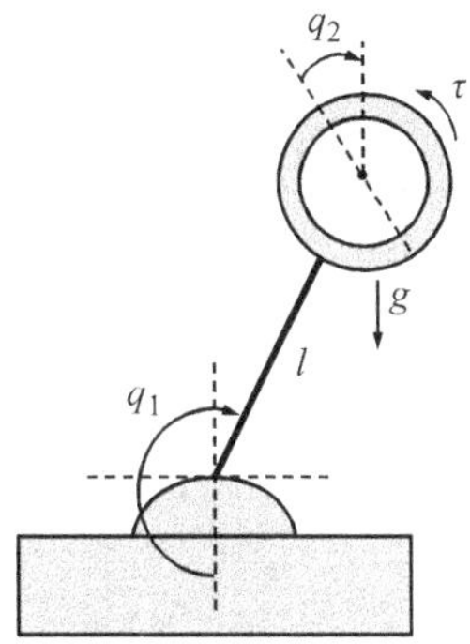

Figure 6.1 Inertia wheel pendulum

wheel, respectively; $t \in \mathbb{R}^+$ is the time, $y(t) \in \mathbb{R}^2$ is the measurement output, $\tau \in \mathbb{R}$ is the controlled torque applied to the disk (see Figure 6.1), $f_s v_2$ represents the viscous friction force affecting the actuator where $f_s > 0$ is the viscous friction coefficient, J_1 and J_2 are positive parameters denoting the inertia of the pendulum and that of the wheel, respectively; the factor $h = mgl$ depends on the combined mass m of the overall wheel pendulum, on the gravitational constant g, and on the length l of the link, and finally, $w = (w_1, w_2)^T$ are the external disturbances affecting the system. An upper bound $M_i > 0$, $(i = 1, 2)$ for the magnitude of the disturbances is normally known a priori

$$\sup_t |w_i(t)| \leq M_i, \qquad i = 1, 2. \tag{6.3}$$

It should be noted that the system (6.1) is nonlinear and underactuated.

The control objective is to find τ such that

$$\lim_{t \to \infty} \|q_{1r}(t) - q_1(t)\| = 0 \tag{6.4}$$

where $q_{1r}(t)$ is the desired trajectory of the pendulum. The desired trajectory is designed such that $q_r(t)$, $\dot{q}_r(t)$, and $\ddot{q}_r(t) \in \mathbb{R}^2$ are uniformly bounded in t.

6.2.1 Generating desired trajectories centered at the upright equilibrium point

Let us recall from [12], the procedure to generate the set of trajectories for the inertia wheel pendulum to follow. First, let us consider the dynamics of the wheel pendulum in terms of the reference positions and velocities $(q_r, \dot{q}_r)$ without viscous friction

$$\begin{bmatrix} J_1 & J_2 \\ J_2 & J_2 \end{bmatrix} \begin{bmatrix} \ddot{q}_{1r} \\ \ddot{q}_{2r} \end{bmatrix} + \begin{bmatrix} h \sin q_{1r} \\ 0 \end{bmatrix} = \begin{bmatrix} 0 \\ 1 \end{bmatrix} \tau_r. \tag{6.5}$$

Second, we need to find the reference torque $\tau_r \in \mathbb{R}$ to produce a set of desired periodic motion of the underactuated link $(y = q_{1r})$ such that the output has a periodic motion with desired frequency Ω and amplitude A_p. Throughout, we confine our research interest in desired oscillations around the upright position of the pendulum which correspond to a challenging problem.

The inertia wheel pendulum has underactuation degree one satisfying certain structural property provided in [10] (see details in Appendix A). As a result, it is possible to make exact linearization thus achieving local stability of zero dynamics, that is,

$$p_1 = q_{1r} + J_1^{-1} J_2 q_{2r} - \pi \tag{6.6}$$

$$\eta = J_1 \dot{q}_{1r} + J_2 \dot{q}_{2r} + K p_1 \tag{6.7}$$

where $K > 0$ is a constant. The last term in (6.6) ensures that the desired trajectories will be generated around the equilibrium point $q_{1r}^{\star} = \pi$ rad (see Figure 6.1). From the latter set of equations, it is easy to verify that

$$J_1 \dot{p}_1 = \eta - K p_1 \tag{6.8}$$

while

$$\dot{\eta} = K J_1^{-1} J_2 \dot{q}_{2r} - h \sin(q_{1r}) + K \dot{q}_{1r}, \tag{6.9}$$

$$\ddot{\eta} = -h \cos(q_{1r}) \dot{q}_{1r} - K J_1^{-1} h \sin(q_{1r}), \tag{6.10}$$

$$\dddot{\eta} = R(q_{1r}, \dot{q}_{1r}) + H(q_{1r}) \tau_r \tag{6.11}$$

where

$$H(q_{1r}) = \frac{h \cos(q_{1r})}{J_1 - J_2}, \tag{6.12}$$

$$R(q_{1r}, \dot{q}_{1r}) = h\big(\dot{q}_{1r}^2 + H(q_{1r})\big) \sin(q_{1r}) - \frac{hK}{J_1} \dot{q}_{1r} \cos(q_{1r}). \tag{6.13}$$

Hence, we can take

$$\tau_r = H^{-1}(q_{1r}) \left(u_r - a_0 \eta - a_1 \dot{\eta} - a_2 \ddot{\eta} - R(q_{1r}, \dot{q}_{1r})\right), \tag{6.14}$$

where $H(q_r)$ is nonsingular around the equilibrium point $(q_{1r}^{\star}, \dot{q}_{1r}^{\star}) = (\pi, 0)$, a_0, a_1, and a_2 are positive constants. Introducing the new state coordinates $x = (x_1, x_2, x_3)^T = (\eta, \dot{\eta}, \ddot{\eta})^T$, we obtain

$$\begin{bmatrix} \dot{x}_1 \\ \dot{x}_2 \\ \dot{x}_3 \end{bmatrix} = \underbrace{\begin{bmatrix} 0 & 1 & 0 \\ 0 & 0 & 1 \\ -a_0 & -a_1 & -a_2 \end{bmatrix}}_{A} \begin{bmatrix} x_1 \\ x_2 \\ x_3 \end{bmatrix} + \underbrace{\begin{bmatrix} 0 \\ 0 \\ 1 \end{bmatrix}}_{B} u_r, \tag{6.15}$$

$$\dot{p}_1 = -\frac{K}{J_1} p_1 + \frac{1}{J_1} y_r, \tag{6.16}$$

$$y_r = \underbrace{\begin{bmatrix} 1 & 0 & 0 \end{bmatrix}}_{C} x. \tag{6.17}$$

Equation (6.16) constitutes the zero dynamics which is orbitally asymptotically stable [23].

The following *two-relay controller* is proposed for the purpose of exciting a periodic motion in (6.15):

$$u_r = -c_1 \operatorname{sign}(y_r) - c_2 \operatorname{sign}(\dot{y}_r) \tag{6.18}$$

where c_1 and c_2 are scalars parameters designed such that the scalar-valued function output $y_r(t)$ has a periodic motion with the desired frequency Ω and amplitude A_p. The formulas to compute c_1 and c_2, derived from the describing function method, are [2]

$$c_1 = \begin{cases} \dfrac{\pi}{4} \dfrac{A_p}{|W(j\Omega)|} \left(\sqrt{1+\xi^2}\right)^{-1} & \text{if } \Omega \in Q_2 \cup Q_3 \\ -\dfrac{\pi}{4} \dfrac{A_p}{|W(j\Omega)|} \left(\sqrt{1+\xi^2}\right)^{-1} & \text{elsewhere} \end{cases} \tag{6.19}$$

$$c_2 = \xi \cdot c_1 \tag{6.20}$$

where $W(j\omega) = C(j\omega I - A)^{-1}B$ is the corresponding transfer function of (6.15) and (6.17), $\xi = -\mathrm{Im}\{W(j\Omega)\}/\mathrm{Re}\{W(j\Omega)\}$, and the quadrants Q_i $(i = 1, \ldots, 4)$ are defined as

$$Q_1 = \{\omega \in \mathbb{R} : \mathrm{Re}\{W(j\omega)\} > 0, \mathrm{Im}\{W(j\omega)\} \geq 0\} \tag{6.21}$$

$$Q_2 = \{\omega \in \mathbb{R} : \mathrm{Re}\{W(j\omega)\} \leq 0, \mathrm{Im}\{W(j\omega)\} \geq 0\} \tag{6.22}$$

$$Q_3 = \{\omega \in \mathbb{R} : \mathrm{Re}\{W(j\omega)\} \leq 0, \mathrm{Im}\{W(j\omega)\} < 0\} \tag{6.23}$$

$$Q_4 = \{\omega \in \mathbb{R} : \mathrm{Re}\{W(j\omega)\} > 0, \mathrm{Im}\{W(j\omega)\} < 0\}. \tag{6.24}$$

The difference between (6.18) and the second-order sliding-mode controller given, for example, in [15] is that c_1 is not constrained to be positive and greater than c_2. The orbital stability analysis of (6.5), (6.18) is detailed in [2, Ch. 3] and [12], and it is therefore omitted.

6.3 Observer design

In this work, the velocity estimation is carried out by a family of finite-time velocity observers [16]

$$\text{Pendulum} \begin{cases} \dot{\hat{q}}_1 = \hat{v}_1 - l_{1p}|\tilde{q}_1|^{1/2} \operatorname{sign}(\tilde{q}_1) + \mu_p \tilde{q}_1 \\ \dot{\hat{v}}_1 = -\frac{1}{2} l_{2p} \operatorname{sign}(\tilde{q}_1) + \frac{3}{2}|\tilde{q}_1|^{1/2} \operatorname{sign}(\tilde{q}_1) + \mu_p^2 \tilde{q}_1 + b_p \tau \end{cases} \tag{6.25}$$

$$\text{Wheel} \begin{cases} \dot{\hat{q}}_2 = \hat{v}_2 - l_{1w}|\tilde{q}_2|^{1/2} \operatorname{sign}(\tilde{q}_2) + \mu_w \tilde{q}_2 \\ \dot{\hat{v}}_2 = -\frac{1}{2} l_{2w} \operatorname{sign}(\tilde{q}_2) + \frac{3}{2}|\tilde{q}_2|^{1/2} \operatorname{sign}(\tilde{q}_2) + \mu_w^2 \tilde{q}_2 + b_w \tau \end{cases} \tag{6.26}$$

where $\tilde{q}_1 = \hat{q}_1 - q_1$ and $\tilde{q}_2 = \hat{q}_2 - q_2$ denote the estimation error of the position of the pendulum and the wheel, respectively; $\hat{v}_1$ and $\hat{v}_2$ are the estimated angular velocity of the pendulum and the wheel, respectively; $\mu_p \geq 0$ and $\mu_w \geq 0$ are nonnegative constants; $b_p = 1$, $b_w = J_1/J_2$ were chosen according to [16], and l_{1p}, l_{2p}, l_{1w}, and l_{2w} are positive constants.

6.4 Control synthesis

The control objective is to stabilize the pendulum around the desired trajectory and to keep the wheel velocity bounded. This last can be achieved by introducing a feedback from the rotor velocity. In this section, we will design an output feedback controller which ensures (6.4) while providing boundedness of $v_2(t)$ and attenuating external disturbances where the reference signal $q_r(t) \in \mathbb{R}^2$ is computed on line from (6.5), (6.14), and (6.18). The control law is based on the assumption that $q(t) \in \mathbb{R}^2$ is the only measured variable.

Let the pendulum tracking error and the wheel velocity error be defined, respectively, as

$$e_1(t) \triangleq q_{1r}(t) - q_1(t), \tag{6.27}$$

$$\dot{e}_1(t) \triangleq \dot{q}_{1r}(t) - \hat{v}_1(t), \tag{6.28}$$

$$\dot{e}_2(t) \triangleq \dot{q}_{2r}(t) - \hat{v}_2(t). \tag{6.29}$$

Due to (6.1), (6.5)–(6.18), the errors dynamics are governed by

$$\ddot{e}_1 = \ddot{q}_{1r} + \frac{J_2}{\Delta}(h\sin(q_{1r} - e_1) - f_s(\dot{q}_{2r} - \dot{e}_2)) - \frac{J_2}{\Delta}(w_1 - w_2) + \frac{J_2}{\Delta}\tau \tag{6.30}$$

$$\ddot{e}_2 = \ddot{q}_{2r} - \frac{J_2}{\Delta}h\sin(q_{1r} - e_1) + \frac{J_1}{\Delta}f_s(\dot{q}_{2r} - \dot{e}_2) - \frac{1}{\Delta}(J_1 w_2 - J_2 w_1) - \frac{J_1}{\Delta}\tau \tag{6.31}$$

where $\Delta = (J_1 - J_2)J_2$. Under the following control law

$$\tau = -\frac{\Delta}{J_2}(\alpha_1 \operatorname{sign}(e_1) + \alpha_2 \operatorname{sign}(\dot{e}_1) + \beta_1 e_1 + \beta_2 \dot{e}_1 - \gamma \operatorname{sign}(\dot{e}_2)) - h\sin(q_{1r} - e_1) + f_s(\dot{q}_{2r} - \dot{e}_2) - \frac{\Delta}{J_2}\ddot{q}_{1r} \tag{6.32}$$

with parameters such that

$$\alpha_1 > \alpha_2 > 2MJ_2\Delta^{-1} + \gamma, \qquad \beta_1, \beta_2, \gamma > 0, \tag{6.33}$$

and using the fact that $J_1\ddot{q}_{1r} + J_2\ddot{q}_{2r} = -h\sin(q_{1r})$, the errors dynamics (6.31) result in

$$\ddot{e}_1 = -\alpha_1 \operatorname{sign}(e_1) - \alpha_2 \operatorname{sign}(\dot{e}_1) - \beta_1 e_1 - \beta_2 \dot{e}_1 + \gamma \operatorname{sign}(\dot{e}_2) - \frac{J_2}{\Delta}(w_1 - w_2) \tag{6.34}$$

$$\ddot{e}_2 = \frac{h}{J_2}\sin(q_{1r} - e_1) - \frac{h}{J_2}\sin(q_{1r}) - \frac{1}{\Delta}(J_1 w_2 - J_2 w_1) + \frac{J_1}{J_2}(\alpha_1 \operatorname{sign}(e_1) + \alpha_2 \operatorname{sign}(\dot{e}_1) + \beta_1 e_1 + \beta_2 \dot{e}_1 - \gamma \operatorname{sign}(\dot{e}_2)). \tag{6.35}$$

Throughout, solutions of above system are defined in the sense of Filippov [9] as that of a certain differential inclusion with a multivalued right-hand side. The discontinuous controller (6.32) consists of the linear gain $-\beta_1 e_1 - \beta_2 \dot{e}_1$, the twisting control part $-\alpha_1 \operatorname{sign}(e_1) - \alpha_2 \operatorname{sign}(\dot{e}_1)$ which ensure robustness and finite-time stability, and the rest are the nonlinear compensation terms.

To verify that $(e_1^e, \dot{e}_1^e, \dot{e}_2^e)^T = 0 \in \mathbb{R}^3$ is an equilibrium point of the unperturbed closed-loop system, notice from (6.34) that

$$\gamma \operatorname{sign}(\dot{e}_2) = \alpha_1 \operatorname{sign}(e_1) + \alpha_2 \operatorname{sign}(0) + \beta_1 e_1, \tag{6.36}$$

since $\dot{e}_1^e = 0$. Substituting the above equation into the last term of (6.35) together with $\ddot{e}_2 = 0$, we find that $e_1^e = 0$.

Theorem 6.1. *The controller introduced in (6.32) subject to parameters (6.33) ensures asymptotic stability of the equilibrium point* $\varphi^e = (e_1^e, \dot{e}_1^e, \dot{e}_2^e)^T = 0$.

Proof. We break the proof in two steps.
1. Finite-time convergence of $(e_1, \dot{e}_1)$ *to the origin.* First, let us assume that $v_1(t) \equiv \hat{v}_1(t)$ and $v_2(t) \equiv \hat{v}_2(t)$. Now, we will demonstrate that $(e_1, \dot{e}_1)$ reach the origin asymptotically. To this end, we consider the Lyapunov function

$$V_1 = \frac{1}{2} \begin{bmatrix} e_1 \\ \dot{e}_1 \end{bmatrix}^T \underbrace{\begin{bmatrix} \beta_1 + \varepsilon\beta_2 & \varepsilon \\ \varepsilon & 1 \end{bmatrix}}_{\Gamma} \begin{bmatrix} e_1 \\ \dot{e}_1 \end{bmatrix} + \alpha_1 |e_1| \tag{6.37}$$

which is radially unbounded almost everywhere. The positivity of the Lyapunov function V_1 is verified if and only if $\Gamma = \Gamma^T$ is positive definite and if α_1 is also positive. In particular, matrix Γ will be positive definite if $\beta_1 + \varepsilon\beta_2 > 0$ and if

$$\beta_1 + \varepsilon\beta_2 - \varepsilon^2 > 0 \tag{6.38}$$

are satisfied. By solving the latter quadratic inequality with respect to ε, we have

$$\frac{1}{2}\beta_2 - \frac{1}{2}\sqrt{\beta_2^2 + 4\beta_1} < \varepsilon < \frac{1}{2}\beta_2 + \frac{1}{2}\sqrt{\beta_2^2 + 4\beta_1}. \tag{6.39}$$

It is straightforward to show that the time derivative of V_1 along the solution of the closed-loop system (6.34) and (6.35) satisfies

$$\begin{aligned} \dot{V}_1 \le & -\varepsilon\beta_1 e_1^2 - (\beta_2 - \varepsilon)\dot{e}_1^2 - (\alpha_2 - \gamma - 2J_2\Delta^{-1}M)|\dot{e}_1| \\ & - \varepsilon(\alpha_1 - \alpha_2 - \gamma - 2J_2\Delta^{-1}M)|e_1|, \end{aligned} \tag{6.40}$$

where inequalities, due to (6.3),

$$\frac{J_2}{\Delta}(w_1 - w_2) \le \frac{J_2}{\Delta}\|w_1 - w_2\| \le 2\frac{J_2}{\Delta}M \tag{6.41}$$

$$\frac{1}{\Delta}(J_1 w_2 - J_2 w_1) \le \frac{1}{\Delta}\|J_1 w_2 - J_2 w_1\| \le 2\frac{J_1}{\Delta}M, \tag{6.42}$$

has been used.

The function $\dot{V}_1$ will be negative-definite provided that (6.33) and $\beta_2 - \varepsilon > 0$ are satisfied. Then, by applying [19, Th. 4.4] with parameters given in (6.33), it is concluded that (6.4) is reached in finite-time t_s.

Now, we will proof that $\dot{e}_2$ will tends to the origin from t_s to infinity. To this end, note that the remaining dynamics of the velocity error of the wheel (6.35) is

$$\ddot{e}_2 = -J_1 J_2^{-1} \gamma \operatorname{sign}(\dot{e}_2). \tag{6.43}$$

Using the Lyapunov function $V_2(\dot{e}_2) = J_1^{-1} J_2 |\dot{e}_2|$ we have that

$$\dot{V}_2(\dot{e}_2) = -\gamma, \tag{6.44}$$

therefore $\dot{V}_2(\dot{e}_2)$ will be negative definite for all $t \geq t_s$.
2. Boundedness of wheel velocity. Finally, we need to prove that $\dot{e}_2$ does not escape to infinity in finite time, that is, $\lim_{t \to t_s} \|\dot{e}_2(t)\| < \infty$. To this end, let us rewrite (6.35) as

$$\ddot{e}_2 = -J_1 J_2^{-1} \gamma \operatorname{sign}(\dot{e}_2) + g(e_1, \dot{e}_1, w) \tag{6.45}$$

where

$$\begin{aligned} g(e_1, \dot{e}_1, w) = {} & \frac{h}{J_2} \sin(q_{1r} - e_1) - \frac{h}{J_2} \sin(q_{1r}) - \frac{1}{\Delta}(J_1 w_2 - J_2 w_1) \\ & + \frac{J_1}{J_2}(\alpha_1 \operatorname{sign}(e_1) + \alpha_2 \operatorname{sign}(\dot{e}_1) + \beta_1 e_1 + \beta_2 \dot{e}_1). \end{aligned} \tag{6.46}$$

Since (6.41)–(6.42) and the fact that e_1 and $\dot{e}_1$ decreases asymptotically to the origin (i.e., $|e_1(t)| \leq |e_1(0)|$ and $|\dot{e}_1(t)| \leq |\dot{e}_1(0)|$), the following upper bound for $g(e_1, \dot{e}_1, w)$ is obtained

$$\|g(e_1, \dot{e}_1, w)\| \leq \underbrace{\frac{2h}{J_2} + \frac{J_1}{J_2}(\alpha_1 + \alpha_2 + \beta_1 |e_1(0)| + \beta_2 |\dot{e}_1(0)|) + \frac{2J_1}{\Delta} M}_{=\varepsilon_0}. \tag{6.47}$$

Consider now the following Lyapunov function

$$V(\dot{e}_2) = J_1^{-1} J_2 |\dot{e}_2|. \tag{6.48}$$

The time derivative of $V(\dot{e}_2)$ along the solution of the closed-loop system (6.45)–(6.46) is

$$\begin{aligned} \dot{V}(\dot{e}_2) &= -J_1^{-1} J_2 \operatorname{sign}(\dot{e}_2) \ddot{e}_2 \\ &= -J_1^{-1} J_2 \operatorname{sign}(\dot{e}_2)(-J_1 J_2^{-1} \gamma \operatorname{sign}(\dot{e}_2) + g(e_1, \dot{e}_1, w)) \\ &= -\gamma - J_1^{-1} J_2 \operatorname{sign}(\dot{e}_2) g(e_1, \dot{e}_1, w) \\ &\leq -\gamma + J_1^{-1} J_2 \|g(e_1, \dot{e}_1, w)\| \\ &\leq -\gamma + J_1^{-1} J_2 \varepsilon_0 \end{aligned} \tag{6.49}$$

where inequality (6.47) has been used. Due to inequality (6.33), the time derivative of the Lyapunov function $\dot{V}(\dot{e}_2)$ is not negative definite. However, integrating $\dot{V}(\dot{e}_2(t))$ from 0 to t_s, we get

$$V(t_s) \leq (J_1^{-1} J_2 \varepsilon_0 - \gamma) t_s + V(0) \tag{6.50}$$

which implies

$$|\dot{e}_2(t_s)| \leq (\varepsilon_0 - J_1 J_2^{-1}\gamma)t_s + |\dot{e}_2(0)|. \tag{6.51}$$

Therefore, we conclude that $\dot{e}_2$ is finite for $t_s < \infty$. □

6.5 Numerical results

In this section, we present simulation results using the laboratory inertial wheel pendulum manufactured by Quanser Inc., depicted in Figure 6.1 where $J_1 = 4.572 \times 10^{-3}$ kg m^2, $J_2 = 2.495 \times 10^{-5}$ kg m^2, and $h = 0.3544$ kg m^2 s^{-2} (see [6]). The viscous friction coefficient $f_s = 8.80 \times 10^{-5}$ N m s was identified by applying the procedure from [14]. Simulations were carried out to achieve the orbital stabilization of the unactuated link (pendulum) q_1 around the equilibrium point $q^\star = (\pi, 0)$.

Setting $\Omega = \pi/2$ rad/s and $A_p = 0.07$ as desired frequency and amplitude, respectively, we have $c_1 = 2$ and $c_2 = -2$ which have been computed from (6.19) and (6.20). Invoking [2, Th. 2.1], the system (6.15) was orbitally asymptotically stable under $K = 1$, $a_0 = 1$, $a_1 = 2$, and $a_2 = 1$.

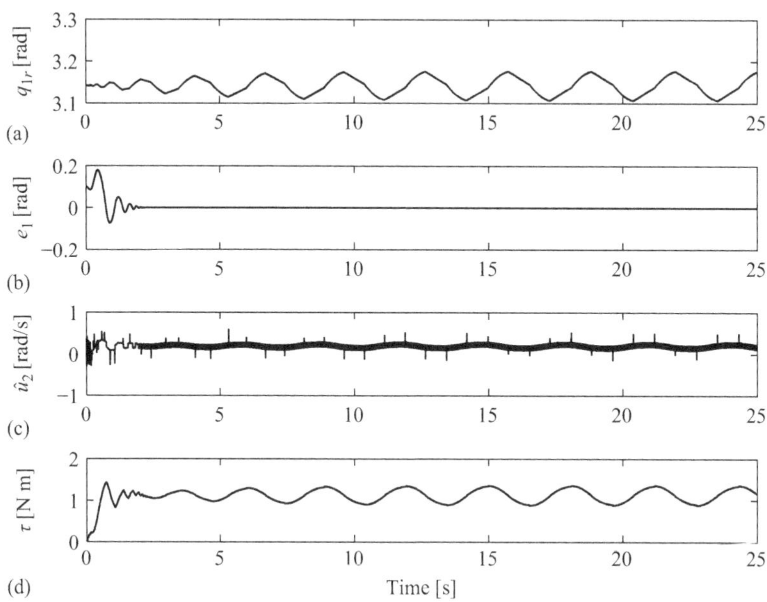

Figure 6.2 Disturbance-free closed-loop system results: (a) desired trajectory to follow, (b) tracking pendulum position error, (c) wheel velocity, and (d) input torque

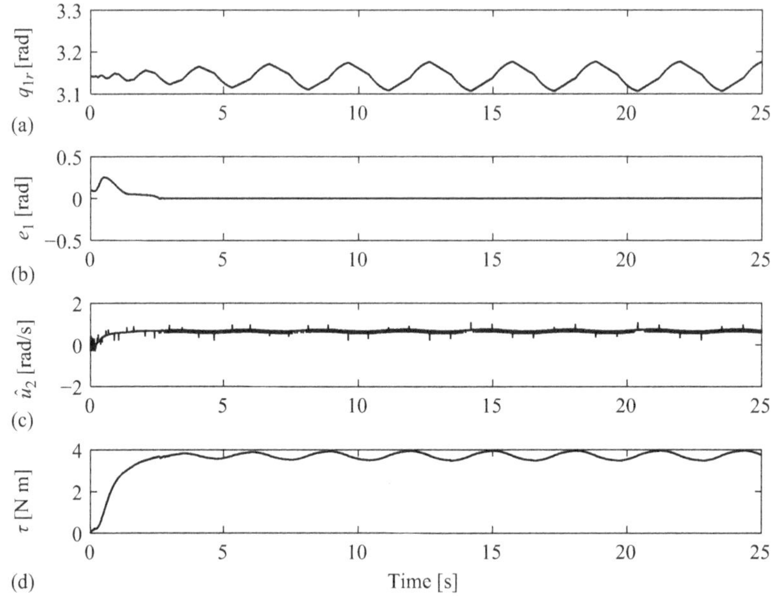

Figure 6.3 Perturbed closed-loop system results: (a) desired trajectory to follow, (b) tracking pendulum position error, (c) wheel velocity, and (d) input torque

The initial conditions for the inertia wheel pendulum, selected for the simulations, were close to $q_1(0) = 3$ rad and $q_2(0) = 0$ rad, whereas all the velocity initial conditions were set to $v_1(0) = v_2(0) = 0$ rad/s. The gains of the controller (6.32) were set to $\alpha_1 = 5$, $\alpha_2 = 2$, $\beta_1 = 2$, $\beta_2 = 1$, and $\gamma = 0.1$. The gains of the observer (6.25)–(6.26) were set to $\mu_p = 0.1$, $\mu_w = 0.1$, $b_p = 1$, $b_w = 183.24$, $l_{1p} = 10$, $l_{2p} = 100$, $l_{1w} = 100$, and $l_{2w} = 10 \times 10^3$.

Figure 6.2 shows the desired trajectory to follow, the tracking pendulum position error, the wheel velocity, and the input torque for the disturbance-free case. This figure corroborates that second-order sliding-mode controller guarantee finite-time stabilization of the pendulum around the desired trajectory. In order to test the controller robustness, a harmonic external disturbance $w_i = 0.02\sin(t)\,(i = 1, 2)$ is then applied to the closed-loop system. Good performance and desired robustness properties of the controller are concluded from Figure 6.3.

6.6 Conclusions

In this paper, we have synthesized an output feedback second-order sliding-mode controller to solve the tracking control problem of the inertia wheel pendulum. Similar to [2], the periodic motion of the pendulum will be at the upright position around the unstable equilibrium point of the unforced system where the two-relay controller was also used to induce of oscillations at the scalar output of the inertia wheel pendulum reference model and then injected as desired trajectory to the closed-loop system. Such configuration allows to generate oscillations without depend on the neglected dynamics. Therefore, the resulting control law ensures rejection against bounded matched disturbances, while numerical study verifies the performance of the synthesized controller. The methodology allows to extend the results to a class of underactuated systems of relative degree one.

Appendix A

A.1 Normal form transformation of the inertia wheel pendulum reference model

In this Appendix we present the normal form transformation for the inertia wheel pendulum. To begin with, let us introduce the following Theorem extracted from [17, Th. 4.2.1].

Theorem 6.2. *Consider an underactuated system with kinetic symmetry with respect to the unactuated variable q_1 and fully actuated shape variables q_2,*

$$m_{11}(q_2)\ddot{q}_1 + m_{12}(q_2)\ddot{q}_2 + h_1(q,\dot{q}) = 0 \tag{A.1}$$

$$m_{21}(q_2)\ddot{q}_1 + m_{22}(q_2)\ddot{q}_2 + h_2(q,\dot{q}) = \tau. \tag{A.2}$$

Let $\tau = \alpha(q_2)u + \beta(q,\dot{q})$ be the collocated partially linearizing change of control for (A.2). Assume all the elements of

$$\omega = m_{11}^{-1}(q_2)m_{12}(q_2)dq_2 \tag{A.3}$$

are exact one-forms and let $\omega = d\varpi(q_2)$. Then, there exists a global change of coordinates (i.e., diffeomorphism) obtained from the Lagrangian of the system

$$p_1 = q_1 + \varpi(q_2) \tag{A.4}$$

$$\sigma = m_{11}(q_2)\dot{q}_1 + m_{12}(q_2)\dot{q}_2 = \frac{\partial \mathcal{L}}{\partial \dot{q}_1} \tag{A.5}$$

that transforms the dynamics of the underactuated system in (A.2) into a cascade normal form in strict feedback form

$$\dot{p}_1 = m_{11}^{-1}(q_2)\sigma \tag{A.6}$$

$$\dot{\sigma} = -\frac{V(p_1,q_2)}{\partial p_1} \tag{A.7}$$

$$\dot{q}_2 = p_s \tag{A.8}$$

$$\dot{p}_2 = u, \tag{A.9}$$

where

$$\varpi(q_2) = \int_0^{q_2} m_{11}^{-1} m_{12} ds. \tag{A.10}$$

In addition, the (q_t, σ)*-subsystem is a Lagrangian system with configuration vector* q_t *that belongs to the reduced manifold* $Qr = \Phi(Q_1 \times Q_2)$ *with reduced Lagrangian*

$$\mathcal{L}_t = \frac{1}{2}\dot{p}_1^T m_r(q_2)\dot{p}_1 - V_t(p_1 - \gamma(q_2), q_2) \tag{A.11}$$

that satisfies the unforced Euler–Lagrange equation

$$\frac{d}{dt}\frac{\partial \mathcal{L}_t}{\partial \dot{p}_1} - \frac{\partial \mathcal{L}_t}{\partial p_1} = 0. \tag{A.12}$$

Let us consider the decomposition of (6.5) into its actuated and unactuated parts (A.2) with $m_{11} = J_1$, $m_{12} = m_{22} = J_2$, $h_1(q_r, \dot{q}_r) = h\sin(q_{1r})$, and $h_2(q_r, \dot{q}_r) = 0$. So, by Theorem 6.2, after applying the change of coordinates (A.5), that is

$$p_1 = q_{1r} - \pi + J_1^{-1} J_2 q_{2r} \tag{A.13}$$

$$\sigma = J_1\dot{q}_{1r} + J_2\dot{q}_{2r} = \frac{\partial \mathcal{L}}{\partial \dot{q}_{1r}}, \tag{A.14}$$

the dynamics of the inertia wheel pendulum reference model into cascade system in strict feedback form is given by

$$\dot{q}_{1r} = J_1^{-1}\sigma - J_1^{-1} J_2 \dot{q}_{2r} \tag{A.15}$$

$$\dot{\sigma} = -h\sin(q_{1r}) \tag{A.16}$$

$$\ddot{q}_{2r} = u_r \tag{A.17}$$

where

$$\varpi(q_{2r}) = J_1^{-1} J_2 q_{2r}. \tag{A.18}$$

In the latter theorem, u_r is an input which orbitally asymptotically stabilizes the system, σ denote the generalized momentum conjugate to q_{1r}, and $\mathcal{L}(q_r, \dot{q}_r) = K(q_r, \dot{q}_r) - V(q_r)$ is the Lagrangian where $K(q_r, \dot{q}_r)$ and $V(q_r)$ denotes the kinetic and potential energy, respectively.

Defining the output as

$$\eta = Kp_1 + \sigma \tag{A.19}$$

along with (6.14) and (A.13)–(A.14), the system (A.15)–(A.17) in the coordinates $(\eta, \dot{\eta}, \ddot{\eta}, p_1)^T$ becomes (6.15)–(6.17). By applying Theorem 6.1 from Grizzle *et al.* [10], it follows that the zero dynamics (6.16) is exponentially stable.

Acknowledgments

L. Aguilar gratefully acknowledges the financial support from Consejo Nacional de Ciencia y Tecnología (CONACYT) under Grant number 268364. R. Iriarte acknowledges the financial support from UNAM-DGAPA PAPIIT Grant number IN112915.

References

[1] Albea, C., Canudas-de-Wit, C., and Gordillo, F. (2007). Adaptive control of the boost DC-AC converter. *Proceedings of the IEEE International Conference on Control Applications*, 611–616.

[2] Aguilar, L., Boiko, I., Fridman, L., and Iriarte, R. (2015). *Self-Oscillations in Dynamic Systems—A New Methodology via Two-Relay Controllers*. Birkhäuser, London, UK.

[3] Aguilar, L., Boiko, I., Fridman, L., and Freidovich, L. (2012). Generating oscillations in inertia wheel pendulum via two-relay controller. *International Journal of Robust and Nonlinear Control*, **22**, 318–330.

[4] Aguilar, L., Boiko, I., Fridman, L., and Iriarte, R. (2009). Generating self-excited oscillations for underactuated mechanical systems via two-relay controller. *International Journal of Control*, **82**(9), 1678–1691.

[5] Aguilar, L., Boiko, I., Fridman, L., and Iriarte, R. (2009). Generating self-excited oscillations via two-relay controller. *IEEE Transactions on Automatic Control*, **54**(2), 416–420.

[6] Åstrom, K., Block, D., and Spong, M. (2001). *The Reaction Wheel Pendulum*. Lecture Notes for the Reaction Wheel Pendulum (Part of the Mechatronics Control Kit). Quanser: Urbana-Champaign, IL.

[7] Block, D., Åstrom, K., and Spong, M. (2007). *The Reaction Wheel Pendulum*. Synthesis Lectures on Control and Mechatronics #1. Williston, VT: Morgan & Claypool Publisher.

[8] Chevallereau, C., Abba, G., Aoustin, Y., Plestan, E., Canudas-de-Wit, C., and Grizzle, J. (2003). Rabbit: a testbed for advanced control theory. *IEEE Control Systems Magazine*, **23**(5), 57–79.

[9] Filippov, A. (1988). *Differential Equations with Discontinuous Right-hand Sides*. Kluwer Academic Publisher, Dordrecht.

[10] Grizzle, J., Moog, C., and Chevallereau, C. (2005). Nonlinear control of mechanical systems with an unactuated cyclic variable. *IEEE Transactions on Automatic Control*, **50**(5), 559–576.

[11] Hurmuzlu, Y., Génot, F., and Brogliato, B. (2004). Modeling, stability, and control of biped robots—a general framework. *Automatica*, **40**, 1647–1664.

[12] Iriarte, R., Aguilar, L., and Fridman, L. (2013). Second order sliding mode tracking controller for inertia wheel pendulum. *Journal of the Franklin Institute*, **350**, 92–106.

[13] Khalil, H. (2002). *Nonlinear Systems*. Third edition, Prentice-Hall, Upper Saddle River.

[14] Kelly, R., Llamas, J., and Campa, R. (2000). A measurement procedure for viscous and Coulomb friction. *IEEE Transactions on Instrumentation and Measurements*, **49**(4), 857–861.

[15] Levant, A. (2003). High-order sliding modes: differentiation and output-feedback control. *International Journal of Control*, **76**, 924–941.

[16] Moreno, J. (2012). A Lyapunov approach to output feedback control using second-order sliding modes. *IMA Journal of Mathematical Control and Information*, **29**(3), 291–308.

[17] Olfati-Saber, R. (2001). *Nonlinear control of underactuated mechanical systems with applications to robotics and aerospace vehicles*. Ph.D. dissertation, California Inst. Technol., Pasadena, CA.

[18] Olivier, J.C., Le Claire, J.C., and Loron, L. (2008). An efficient switching frequency limitation process applied to high dynamic voltage supply. *IEEE Transactions on Power Electronics*, **23**(1), 153–162.

[19] Orlov, Y. (2009). *Discontinuous Systems: Lyapunov Analysis and Robust Synthesis under Uncertain Conditions*. Springer, London.

[20] Plestan, F., Grizzle, J.,Westervelt, E., and Abba, G. (2003). Stable walking of a 7-DOF biped robot. *IEEE Transactions on Robotics and Automation*, **19**(4), 653–668.

[21] Reyhanoglu, M., van der Schaft, A., McClamroch, N., and Kolmanovsky, I. (1999). Dynamics and control of a class of underactuated mechanical systems. *IEEE Transactions on Automatic Control*, **44**(9), 1663–1671.

[22] Shiriaev, A., Freidovich, L., Robertsson, A., and Sandberg, A. (2007). Virtual-holonomic-constraints based design stable oscillations of Furuta pendulum: theory and experiments. *IEEE Transactions on Robotics*, **23**(4), 827–832.

[23] Wiggins, S. (2000). *Introduction to Applied Nonlinear Dynamical Systems and Chaos*. Text in Applied Mathematics. Second edition, Springer, New York.

Chapter 7

Switched integral sliding mode control for robust generation of self-oscillation in pendulum systems

Juan Eduardo Velázquez-Velázquez[1], *Rosalba Galván-Guerra*[1], *Rafael Iriarte*[2]

Abstract

Self-oscillations for pendulum systems generated by a two-relay controller are robustified by using a switched integral sliding mode control. The robustification is theoretically exact, assuring the frequency and amplitude of the oscillations remain despite the presence of matched uncertainties/perturbations. To illustrate the efficacy of the proposed robustifying strategy, an inertial wheel pendulum is considered along the chapter.

7.1 Introduction

Self-oscillation (SO) is the generation and maintenance of a periodic motion by a source of power that lacks a corresponding periodicity: the oscillation itself controls the phase with which the power source acts on it (see [1] for more details about this physical phenomenon). This phenomenon appears when the trajectories of the system are attracted to a limit cycle [2].

The SO is characterized by its frequency and amplitude, which can be analyzed in the frequency or in the state-space domains. In the frequency domain, the describing function method [2–4] can be used to obtain an approximation of the oscillating characteristics (frequency and amplitude) of the periodic motion. In state-space domain, the Poincaré maps [2,3,5] give the necessary and sufficient conditions for the existence and stability of the periodic motion. Moreover, when a discontinuous controller is used to generate a SO, the locus of a perturbed relay system (LPRS) [5] tool can be used to find theoretically exactly the gains of the controller guaranteeing the desired oscillating characteristics of the periodic motion.

[1]Unidad Profesional Interdisciplinaria de Ingeniería campus Hidalgo, Instituto Politécnico Nacional, San Agustín Tlaxiaca, Mexico

[2]Facultad de Ingeniería, Universidad Nacional Autónoma de México, Ciudad Universitaria, Mexico

A common method for producing periodic motions is through tracking, that gives traceability of different shapes, but if the desired frequency is high, it requires a lot of energy. While by using the principle of SO, the shape is hardly controllable, but it is possible to achieve high frequencies [6].

To generate SO, it is factible to introduce virtual holonomic constraints to the nonlinear system [7], generating an exponentially orbitally stable periodic solution of the system. To assure finite-time convergence to the limit cycle, a relay controller [5] can be used. However, there are some systems (e.g., a double integrator) in which the use of feedback having a single-valued nonlinearity (e.g., relay) may not produce a desirable result. For systems with relative degree two, the two-relay controller (TRC) [6] is capable to generate SO with a desired frequency and amplitude. But the main disadvantage with the SO generation methodologies is that they are not robust, and any uncertainty/perturbation may change the frequency and amplitude of the oscillations.

To robustify SO, the sliding modes methodologies are a good option when the uncertainties/perturbations are matched, i.e., they affect directly the channels of the controller. One way to robustify SO is to identify and to compensate the uncertainties/perturbations [8], to track an asymptotic harmonic generator based on a Van der Pol equation [9] using a twisting controller or to track a nominal trajectory generated by the TRC by using a high-order sliding mode control or a quasicontinuous one [10,11]. But during the reaching phase, those approaches are incapable to compensate the uncertainties/perturbations. Furthermore, by the intrinsic characteristics of the TRC, the closed-loop system has a switched behavior, and none of the mentioned approaches contemplate this issue.

The integral sliding modes (ISM) [12] allow theoretically exact compensation of the matched uncertainties/perturbations without explicit computation of the nominal trajectory. For switched linear systems, a switched ISM (SISM) approach has been recently extended in [13]. This approach allows to compensate theoretically exactly the matched uncertainties/perturbations just after the initial time despite the presence of switchings.

In this chapter, a robustifying methodology for SO with a given frequency and amplitude generated by a TRC is proposed. A SISM controller is used to compensate the matched uncertainties/perturbations right after the initial time.

The chapter is organized as follows: Section 7.2 presents some preliminaries concepts for stable SO generation with the TRC together with a motivational example of an inertial wheel pendulum (IWP) that shows the necessity of a robustifying procedure; this section ends with the problem formulation. Then, in Section 7.3, the robustifying strategy is presented, and a computational algorithm resumes the proposed strategy. In Section 7.4, robust SO are generated in the IWP by applying the robustifying strategy. Finally, in Section 7.5, some comments conclude the chapter.

Acronyms

- SO – self-oscillation
- LPRS – locus of a perturbed relay system

- TRC – two-relay controller
- ISM – integral sliding modes
- SISM – switched integral sliding modes
- IWP – inertia wheel pendulum
- ULTIS – uncertain linear time-invariant system
- NLTIS – nominal linear time-invariant system
- ASNLTIS – affine switched nominal linear time-invariant system
- ASULTIS – affine switched uncertain linear time-invariant system

7.2 Preliminaries and problem formulation

7.2.1 Uncertain system

Consider the following uncertain nonlinear system

$$\dot{x}(t) = f(x,t) + g(x,t)(u(t) + w(t)) \tag{7.1}$$

where $f : \mathbb{R}^n \times \mathbb{R} \to \mathbb{R}^n$, $g : \mathbb{R}^n \times \mathbb{R} \to \mathbb{R}^n$ are known continuous functions, $x(t) \in \mathbb{R}^n$ and $u(t) \in \mathbb{R}$ are the states and control vector, respectively. Note that we are assuming the full state vector is available. $w(t) \in \mathbb{R}$ denotes the uncertainties/perturbations present in the system, and it is assumed unknown but bounded, i.e., $|w(t)| \leq w_{\max}$. By using any linearization method, the above system is represented by an uncertain linear time-invariant system (ULTIS)

$$\dot{x}(t) = Ax(t) + B(u(t) + \phi(t)) \tag{7.2}$$

we assume that the system (7.2) is controllable and $B\phi(t) = f(x,t) + g(x,t)(u(t) + w(t)) - Ax(t) - Bu(t)$. The last assumption assures the uncertainties/perturbations ϕ is matched. Moreover, if the system is stable, the uncertainties/perturbations remain bounded, i.e., $|\phi(t)| \leq \phi_{\max}$.

Without loss of generality, the control $u(t)$ is designed such that it generates SO on an arbitrary scalar output $y(t) = Cx(t)$, around a given reference $y_{\text{ref}} \in \mathbb{R}$ with a desired frequency Ω and amplitude $\mathscr{A}$.

7.2.2 Generation of self-oscillations: two-relay controller

Let's define a nominal linear time-invariant system (NLTIS) by assuming the absence of uncertainties/perturbations, i.e., $\phi(t) = 0$,

$$\dot{x}_{\text{Nom}}(t) = Ax_{\text{Nom}}(t) + B\bar{u}(t) \tag{7.3}$$

where the control vector $\bar{u}(t) = u_{\text{Nom}}(t) + u_{\text{TRC}}(t)$ is composed by a stabilizer controller $u_{\text{Nom}}(t) = -\mathscr{K}x_{\text{Nom}}(t)$ and a TRC [6] u_{TRC} given by

$$u_{\text{TRC}}(t) = -c_1 \operatorname{sign}(\varsigma(t)) - c_2 \operatorname{sign}(\dot{\varsigma}(t)) \tag{7.4}$$

where $\varsigma(t) = y_{\text{ref}} - y_{\text{Nom}}(t)$, and c_1 and c_2 are design parameters chosen in such a way that the scalar output y has a steady periodic motion with frequency Ω and amplitude

$\mathscr{A}$ around the reference y_{ref}. Under these conditions, the nominal system takes the form

$$\dot{x}_{\text{Nom}}(t) = \tilde{A}x_{\text{Nom}}(t) + Bu_{\text{TRC}}(t) \tag{7.5}$$

where $\tilde{A} = A - B\mathscr{K}$ is Hurwitz.

Assume the parameters c_1 and c_2 are designed using the LPRS method (see [6] for more details about this result)

$$\begin{aligned} c_1 &\approx \frac{\pi}{4}\frac{\mathscr{A}}{|W(j\Omega)|}\frac{1}{\sqrt{1+2\xi\cos(2\pi\gamma)+\xi^2}} \\ c_2 &\approx \frac{\pi}{4}\frac{\mathscr{A}}{|W(j\Omega)|}\frac{\xi}{\sqrt{1+2\xi\cos(2\pi\gamma)+\xi^2}} \end{aligned} \tag{7.6}$$

where

$$W(j\omega) = C(j\omega I - \tilde{A})^{-1}B, \quad \xi = -\frac{L(\Omega,0)}{L(\Omega,\gamma)} \tag{7.7}$$

and γ is solution of the equation

$$L(\omega,0)L_1(\omega,-\gamma) - L(\omega,\gamma)L_1(\omega,0) = 0 \tag{7.8}$$

with

$$L(\omega,\gamma) = \frac{\pi}{4}C\left(e^{\tilde{A}\gamma\frac{2\pi}{\omega}}\left(I + e^{\tilde{A}\frac{\pi}{\omega}}\right)^{-1}\tilde{A}^{-1}\left(I - e^{\tilde{A}\frac{\pi}{\omega}}\right) + \tilde{A}^{-1}\left(e^{\tilde{A}\gamma\frac{2\pi}{\omega}} - I\right)\right)B \tag{7.9}$$

and

$$L_1(\omega,\gamma) = \frac{\pi}{4}C\tilde{A}\left(e^{\tilde{A}\gamma\frac{2\pi}{\omega}}\left(I + e^{\tilde{A}\frac{\pi}{\omega}}\right)^{-1}\tilde{A}^{-1}\left(I - e^{\tilde{A}\frac{\pi}{\omega}}\right) + \tilde{A}^{-1}\left(e^{\tilde{A}\gamma\frac{2\pi}{\omega}} - I\right)\right)B. \tag{7.10}$$

To assure the orbital asymptotic stability of the SO periodic motion of the closed-loop system (7.3), the TRC gains must fulfill the following theorem.

Theorem 7.1. *[6] Suppose that the selected parameters c_1 and c_2 of the TRC algorithm (7.4) result in a closed-loop that generates a periodic motion of the output y of the plant (7.3). This solution is orbitally exponentially stable, if and only if all the eigenvalues of the matrix $\Phi = (\Phi_1\Phi_2)^2$, where*

$$\begin{aligned} \Phi_1 &= \left(I - \tfrac{v_1 C\tilde{A}}{C\tilde{A}v_1}\right)e^{\tilde{A}\frac{T}{2}}, & \Phi_2 &= \left(I - \tfrac{v_2 C}{Cv_2}\right)e^{\tilde{A}T}, \\ v_1 &= \tilde{A}\eta_p + B(c_1+c_2), & v_2 &= \tilde{A}\rho_p + B(c_1-c_2), \\ \eta_p &= e^{\tilde{A}\frac{T}{2}}\rho_p + \tilde{A}^{-1}(e^{\tilde{A}\frac{T}{2}} - I)B(c_1+c_2), & \rho_p &= -e^{\tilde{A}T}\eta_p - \tilde{A}^{-1}(e^{\tilde{A}T} - I)B(c_1-c_2); \end{aligned} \tag{7.11}$$

are located inside the unit circle.

Note that the TRC can be seen as a switched controller of the form

$$u_{\mathrm{TRC}}(t) = \begin{cases} u_{\mathrm{TRC},1} = -c_1 - c_2 & \text{if } \varsigma(t) \geq 0 \text{ and } \dot{\varsigma}(t) \geq 0 \\ u_{\mathrm{TRC},2} = -c_1 + c_2 & \text{if } \varsigma(t) \geq 0 \text{ and } \dot{\varsigma}(t) < 0 \\ u_{\mathrm{TRC},3} = c_1 - c_2 & \text{if } \varsigma(t) < 0 \text{ and } \dot{\varsigma}(t) \geq 0 \\ u_{\mathrm{TRC},4} = c_1 + c_2 & \text{if } \varsigma(t) < 0 \text{ and } \dot{\varsigma}(t) < 0 \end{cases} \tag{7.12}$$

Hence, (7.5) can be restated as an affine switched NLTIS (ASNLTIS)

$$\dot{x}_{\mathrm{Nom}}(t) = \tilde{A}x_{\mathrm{Nom}}(t) + F_{\sigma(t)} \tag{7.13}$$

$$\sigma(t) = \begin{cases} 1 & \text{if } \varsigma(t) \geq 0 \text{ and } \dot{\varsigma}(t) \geq 0 \\ 2 & \text{if } \varsigma(t) \geq 0 \text{ and } \dot{\varsigma}(t) < 0 \\ 3 & \text{if } \varsigma(t) < 0 \text{ and } \dot{\varsigma}(t) \geq 0 \\ 4 & \text{if } \varsigma(t) < 0 \text{ and } \dot{\varsigma}(t) < 0 \end{cases} \tag{7.14}$$

and $F_i = Bu_{\mathrm{TRC},i}$, $i \in \{1, 2, 3, 4\}$.

The nominal trajectory can be used to detect when the switching variables ς and $\dot{\varsigma}$ change their values, i.e., it can be used to detect the switching moments t_j, defined as

$$t_j = \min\{t : t > t_{j-1}, \sigma(t^-) \neq \sigma(t), \ j = 0, 1, \ldots\} \tag{7.15}$$

Observe that since the nominal trajectory is designed to assure a steady periodic motion with a desired frequency and amplitude, we can assume that the switching moments conform an ordered sequence

$$0 = t_0 < t_1 < \cdots < t_{j-1} < t_j < \cdots \tag{7.16}$$

7.2.3 Motivational example

Consider the dynamic model of an IWP with friction effects and exogenous perturbations [6] (see Figure 7.1)

$$\begin{bmatrix} J_1 & J_2 \\ J_2 & J_2 \end{bmatrix} \begin{bmatrix} \ddot{q}_1(t) \\ \ddot{q}_2(t) \end{bmatrix} + \begin{bmatrix} h\sin(q_1(t)) \\ f_s\dot{q}_2(t) \end{bmatrix} = \begin{bmatrix} 0 \\ 1 \end{bmatrix} (\tau(t) + w(t)) \tag{7.17}$$

where $q_1(t), q_2(t) \in \mathbb{R}$ are the absolute angle of the pendulum and the disk, respectively, counted clockwise from the vertical downward position. $\tau(t) \in \mathbb{R}$ is the controlled torque applied to the disk, while $w(t)$ denotes external perturbations. J_1, J_2, and h are positive physical parameters, which depend on the geometric dimensions, and the inertia–mass distribution, while f_s is the viscous friction coefficient, and it is considered unknown.

Consider the change of variable [14]

$$\begin{aligned} p_1(t) &= q_1(t) - \pi + \frac{J_2}{J_1} q_2(t) \\ p_2(t) &= J_1\dot{q}_1(t) + J_2\dot{q}_2 + Kp_1(t) \end{aligned} \tag{7.18}$$

and take $x(t) = \begin{bmatrix} p_2(t) & \dot{p}_2(t) & \ddot{p}_2(t) \end{bmatrix}^T$ as the state variables, and

$$\tau(t) = H^{-1}(q_1(t))\,(\bar{u}(t) - R(q_1(t), \dot{q}_1(t))) \tag{7.19}$$

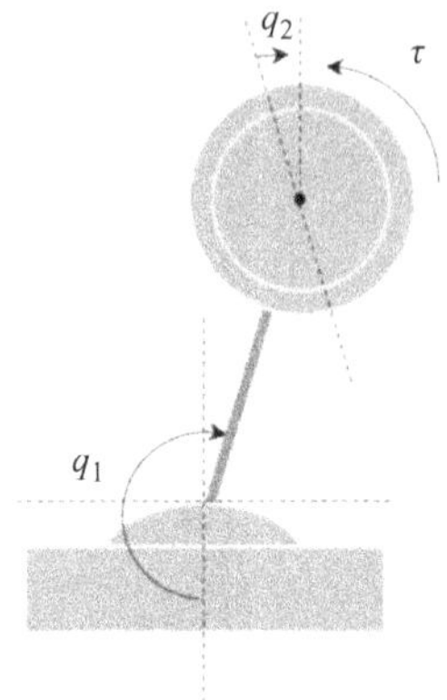

Figure 7.1 IWP schematic

with

$$
\begin{aligned}
H(q_1(t)) &= \frac{h\cos(q_1(t))}{J_1 - J_2}, \\
R(q_1(t), \dot{q}_1(t)) &= (\dot{q}_1^2(t) + H(q_1(t)))h\sin(q_1(t)) - \frac{hK}{J_1}\dot{q}_1(t)\cos(q_1(t))
\end{aligned}
\tag{7.20}
$$

and $\mathscr{K} = [-a_0 \quad -a_1 \quad -a_2]$.

Hence, the obtained linearized IWP model is given by

$$
\begin{aligned}
\dot{x}(t) &= \begin{bmatrix} 0 & 1 & 0 \\ 0 & 0 & 1 \\ -a_0 & -a_1 & -a_2 \end{bmatrix} x(t) + \underbrace{\begin{bmatrix} 0 \\ 0 \\ 1 \end{bmatrix} u_{\mathrm{TRC}}(t) + \begin{bmatrix} 0 \\ 0 \\ 1 \end{bmatrix}}_{F_{\sigma(t)}} \phi(t), \\
\dot{p}_1(t) &= -\frac{K}{J_1}p_1(t) + \frac{1}{J_1}x_1(t);
\end{aligned}
\tag{7.21}
$$

where $\phi(t) = H(q_1(t))\,(w(t) - f_s\dot{q}_2(t))$.

The control objective is to achieve orbital stabilization of the unactuated link q_1 around the unstable equilibrium point $q^* = [\pi \quad 0]^T$ with a desired frequency $\Omega = 2\pi$ rad/s and amplitude $\mathscr{A} = 0.07$ rad. For simulation purposes, it is assumed that $q_1(0) = 3$ rad, $q_2(0) = 0$ rad, $\dot{q}_1(0) = \dot{q}_2(0) = 0$ rad/s, $J_1 = 4.572 \times 10^{-3}$, $J_2 = 2.495 \times 10^{-5}$, $f_s = 8.80 \times 10^{-5}$, $h = 0.3544$, $K = 1 \times 10^{-4}$, $a_0 = 350$, $a_1 = 155$ and $a_2 = 22$. Here, we assume $y(t) = q_1(t)$.

Under the assumption that $\phi(t) = 0$, to achieve the desired frequency and amplitude, the LPRS method was applied to obtain the TRC gains $c_1 = 2$ and $c_2 = -2.5$. In Figure 7.2, the behavior of the IWP is obtained with the TRC. Observe that q_1 has frequency of 2π rad/s and an amplitude of 0.07 rad. Recall that due to the switched characteristic of the TRC, the IWP model is seen as a switched system whose switching

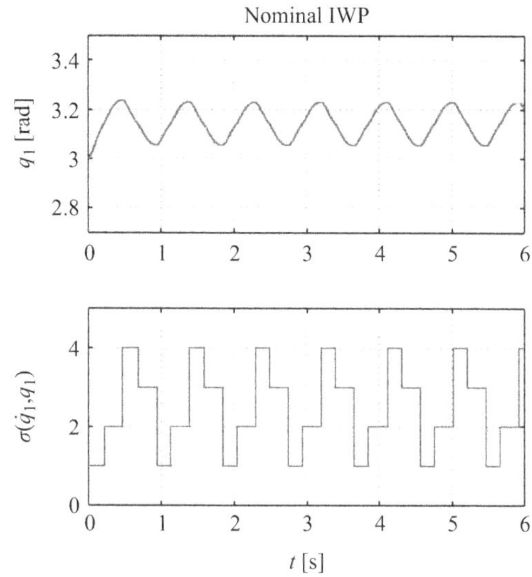

Figure 7.2 Nominal behavior of the IWP: TRC

moments are ruled by (7.14), the detected switching moments are depicted in the second graph of Figure 7.2.

Now, let's apply the same TRC designed previously to the IWP model (7.21). For simulations purposes,

$$w(t) = 0.2\sin\left(\frac{\pi t}{2\cos(\pi t)}\right) + 0.1. \tag{7.22}$$

The behavior of the pendulum angle is shown in Figure 7.3. Note that the nominal controller and the TRC are incapable to maintain the desired frequency and amplitude of the pendulum angle. Furthermore, the uncertainties/perturbations affect the trajectory of the system, and hence, the switching moments are quite different from the nominal ones (see second plot of Figure 7.3).

7.2.4 Problem statement

Considering a control input $u(t) = \bar{u}(t) + u_{\mathrm{Int}}$, the linearized uncertain system (7.2) can be restated as an affine switched ULTIS (ASULTIS)

$$\dot{x}(t) = \tilde{A}x(t) + F_{\sigma(t)} + B(u_{\mathrm{Int}}(t) + \phi(t)), \;\; y(t) = Cx(t). \tag{7.23}$$

In the rest of the chapter, we will design a robustifying controller u_{Int} such that the ASULTIS (7.23) has SO with a desired frequency Ω and amplitude $\mathscr{A}$ despite the presence of matched uncertainties/perturbations, i.e., the ASULTIS behaves as the ASNLTIS (7.13) theoretically exactly right after the initial time.

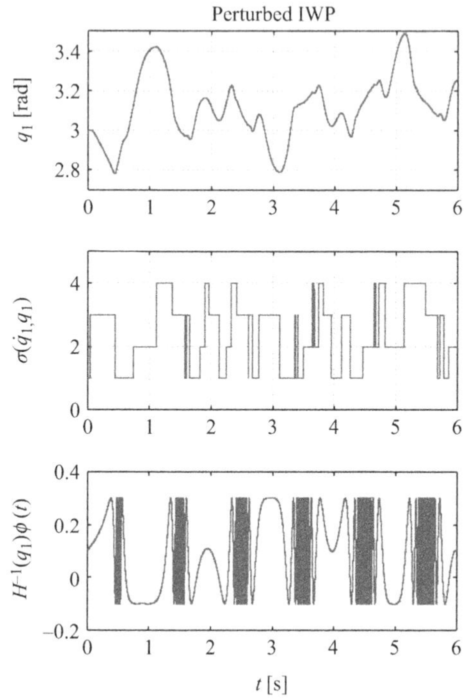

Figure 7.3 Behavior of the IWP: TRC

7.3 Switched integral sliding modes

7.3.1 Sliding variable design

To compensate the uncertainties/perturbations, we use a SISM strategy [13] that compensates the effects of the matched uncertainties/perturbations theoretically exactly right after the initial time. A block diagram with the overall control strategy is shown in Figure 7.4.

Let's define the following integral sliding variable

$$s(x,t) = G(x(t) - x(t_j)) - G\int_{t_j}^{t} \left(\tilde{A}x(\eta) + F_{\sigma(\eta)}\right) d\eta. \tag{7.24}$$

with the projection matrix $G = B^+$. Observe that this variable is in the origin at every switching moment, i.e., $s(x(t_j), t_j) = 0$. Now, taking the first derivative with respect to time along the trajectories of (7.23),

$$\dot{s}(x,t) = u_{\text{Int}}(t) + \phi(t) \tag{7.25}$$

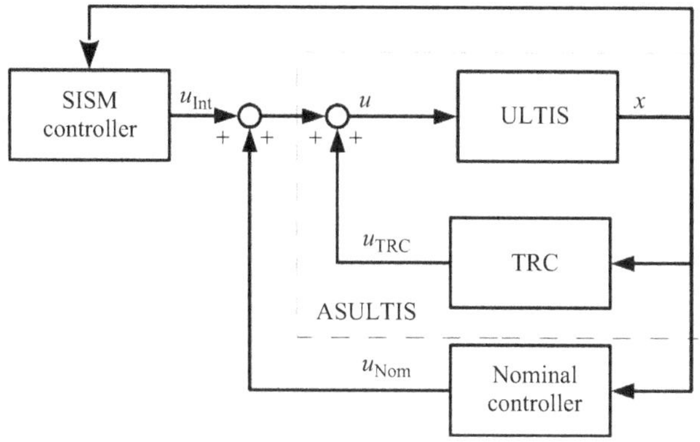

Figure 7.4 Block diagram of the proposed control strategy

and the equivalent control has the form

$$u_{\text{Int,eq}}(t) = -\phi(t). \tag{7.26}$$

Hence, the sliding modes dynamic of the system (7.23) takes the form

$$\dot{x}(t) = \tilde{A}x(t) + F_{\sigma(t)} \tag{7.27}$$

that is equivalent to the nominal system (7.13), if the establishment of the sliding mode since the initial time is assured.

7.3.2 Controller design

Let the SISM controller in the unitary form

$$u_{\text{Int}} = -\beta \frac{s}{|s|} \tag{7.28}$$

where β is a design constant. The next theorem gives sufficient conditions to assure the sliding variable (7.24) remains in the origin since the initial time.

Theorem 7.2. *[13]*
If

$$\beta > \|B\|\phi_{\max} \tag{7.29}$$

the sliding mode is established since the initial time, and the matched uncertainties/perturbations are compensated right after the initial time.

The robust SO control strategy based on the SISM controller can be summarized in the following algorithm (see Figure 7.5 for a flowchart of the algorithm).

SISM robustifying algorithm

1. Define the desired frequency Ω and amplitude $\mathscr{A}$ for the output $y(t)$
2. Linearize the plant if necessary

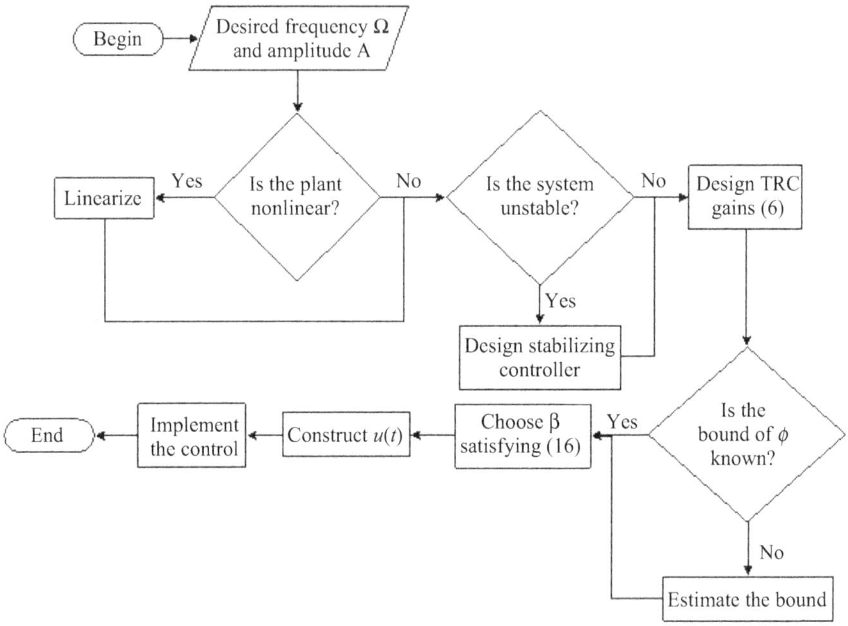

Figure 7.5 Flowchart of the SISM algorithm

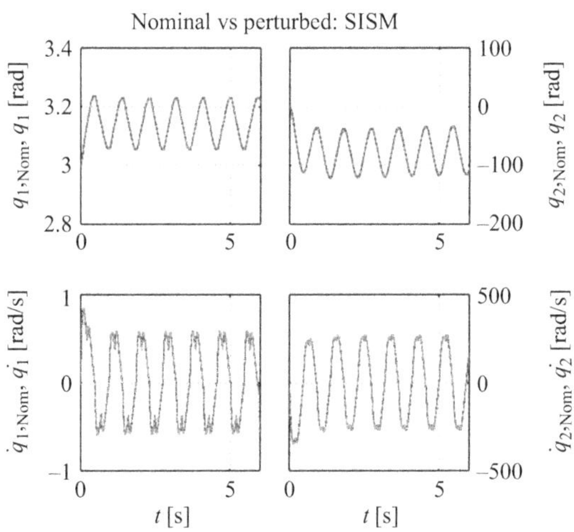

Figure 7.6 Nominal IWP vs perturbed IWP: SISM

3. If the system is unstable design the stabilizing nominal controller u_{Nom}
4. Design the TRC gains (7.6)
5. Determine the bound of the matched uncertainties/perturbations $\phi(t)$
6. Choose a suitable constant β satisfying (7.29)
7. Construct the control signal $u(t)$

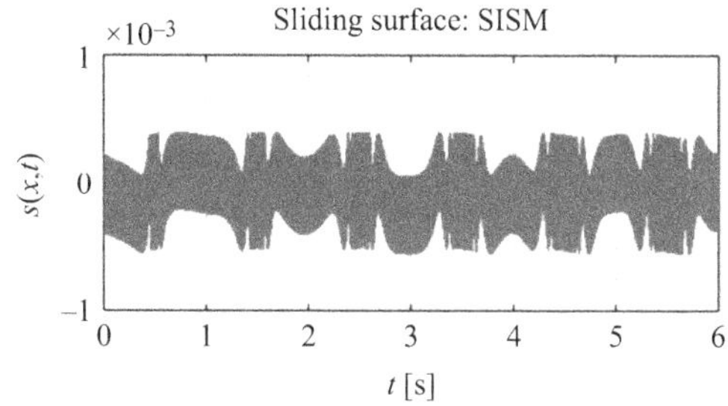

Figure 7.7 Sliding variable: SISM

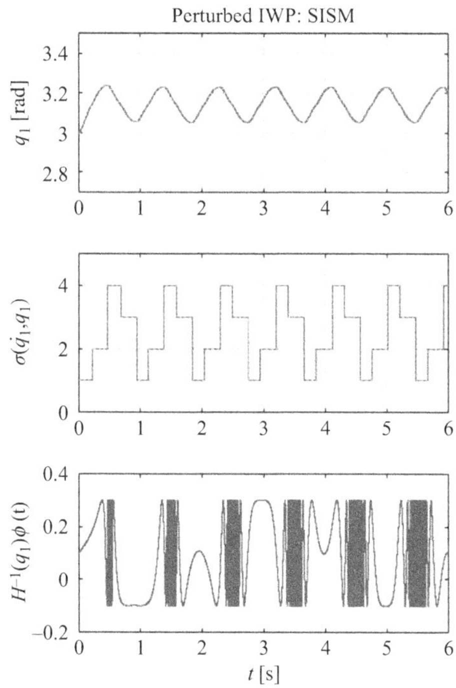

Figure 7.8 Behavior of the IWP: SISM

7.4 Simulation example

To show the applicability of the proposed approach, let's robustify the IWP motivational example with the same control objective: to generate SO around the unstable equilibrium point $q^* = [\pi \quad 0]$, i.e., $y_{\text{ref}} = \pi$; with a desired frequency $\Omega = 2\pi$ rad/s and amplitude $\mathscr{A} = 0.07$ rad. Let's recall the linearized model (7.21), and by using the nominal model, it can be shown that the matched uncertainties/perturbations are bounded, i.e., $|\phi(t)| \leq 30$.

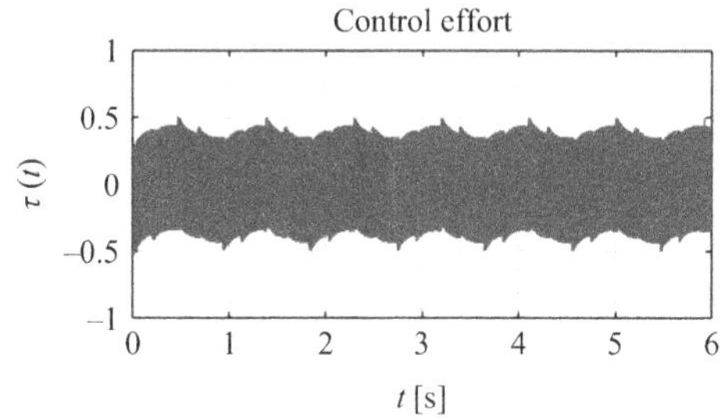

Figure 7.9 Applied torque: SISM

The behavior of the IWP with the SISM controller is illustrated in Figure 7.6, and the sliding mode variable is given in Figure 7.7. Observe that the perturbed IWP is in the sliding mode since the initial time and behaves as the nominal one right after the initial time; this assures the switching moments coincide, and as it is shown in Figure 7.8, the desired frequency and amplitude for the pendulum angle is achieved despite the presence of the matched uncertainties/perturbations. Finally, the applied torque is given in Figure 7.9.

7.5 Conclusion

For pendulum systems with SO generated by the TRC, a robustifying strategy based on an ISM approach is proposed. The TRC generates SO with a desired frequency and amplitude, and due to the intrinsic switching characteristic of the TRC, the closed-loop system is considered a switched affine one. The considered approach robustifies the SO against bounded matched uncertainties/perturbations. By using MATLAB® simulations, for the IWP example, it is shown that the SISM compensates right after the initial time the matched uncertainties/perturbations, assuring the switching moments and the desired frequency and amplitude of the SO remain.

Acknowledgment

The authors gratefully acknowledge the financial support from PAPIIT, UNAM, grants IN112915 and IN113216.

References

[1] A. Jenkins. "Self-oscillation", *Physics Reports*, **525(2)**, pp. 167–222, (2013), [Online]. Available: http://www.sciencedirect.com/science/article/pii/S0370157312004073, self-oscillation.

[2] H. K. Khalil. *Nonlinear Systems*, third edition, (Prentice-Hall, Upper Saddle River, New Jersey 07458, 2002).

[3] A. Gelb, W. Velde. *Multiple-Input Describing Functions and Nonlinear Systems Design*, (McGraw-Hill, New York, 1968).

[4] D. Atherton. *Nonlinear Control Engineering–Describing Function Analysis and Design*, (Van Nostrand, Wokingham, 1975).

[5] I. Boiko. *Discontinuous Control Systems: Frequency-domain Analysis and Design*, (Birkhäuser Boston, 2009), [Online]. Available: http://books.google.com.mx/books?id=P0palNA5BIMC.

[6] L. T. Aguilar, I. Boiko, L. Fridman, R. Iriarte. *Self-Oscillations in Dynamic Systems: A New Methodology via Two-Relay Controllers*, first edition, (Basel, Switzerland: Birkhäuser Basel, 2015).

[7] A. Shiriaev, J. W. Perram, C. C. de Wit. "Constructive tool for orbital stabilization of underactuated nonlinear systems: Virtual constraints approach", *IEEE Transactions on Automatic Control*, **50(8)**, pp. 1164–1176, (Aug 2005).

[8] L. T. Aguilar, I. Boiko, L. Fridman, A. Ferreira. "Identification based generation of self-excited oscillations for underactuated mechanical systems via two-relay algorithm", *2008 International Workshop on Variable Structure Systems*, pp. 41–46, (2008).

[9] Y. Orlov, L. T. Aguilar, L. Acho, A. Ortiz. "Asymptotic harmonic generator and its application to finite time orbital stabilization of a friction pendulum with experimental verification", *International Journal of Control*, **81(2)**, pp. 227–234, (2008), [Online]. Available: http://dx.doi.org/10.1080/00207170701516371.

[10] A. Ferreira, F. J. Bejarano, L. M. Fridman. "Robust control with exact uncertainties compensation: With or without chattering?", *IEEE Transactions on Control Systems Technology*, **19(5)**, pp. 969–975, (Sept 2011).

[11] A. Estrada, L. T. Aguilar, R. Iriarte, L. Fridman. "Two relay controller for real time trajectory generation and its application to inverted orbital stabilization of inertia wheel pendulum via quasi-continuous hosm", *Asian Journal of Control*, **14(1)**, pp. 58–66, (2012), [Online]. Available: http://dx.doi.org/10.1002/asjc.339.

[12] V. I. Utkin. *Sliding Modes in Control and Optimization*, (Berlin, Heidelberg: Springer-Verlag, 1992).

[13] R. Galván-Guerra, L. Fridman. "Output integral sliding mode based lq control for switched uncertain systems", *13th International Workshop on Variable Structure Systems*, (2013).

[14] J. W. Grizzle, C. H. Moog, C. Chevallereau. "Nonlinear control of mechanical systems with an unactuated cyclic variable", *IEEE Transactions on Automatic Control*, **50(5)**, pp. 559–576, (May 2005).

Chapter 8
Finite-time stabilization of underactuated mechanical systems in the presence of uncertainties: application to the cart–pole system

Jawhar Ghommam[1], Ahmed Chemori[2], and Faiçal Mnif[3]

8.1 Introduction

Underactuated mechanical systems [13,14,27] are those systems with less control inputs than generalized coordinates (called also degrees of freedom), i.e., they have unactuated generalized coordinates. For such systems, the unactuated generalized coordinates may indirectly be controlled by the actuated coordinates through the dynamic coupling, inherent to these systems [8]. This coupling is often nonlinear, resulting in generally nonintegrable dynamic constraints and therefore second-order nonholonomic.

Many examples of such systems exist, and mainly in robotics. They include, among others, inverted pendulums [16,25,26], underactuated robot manipulators [6], gymnast robots and particularly the acrobot [41] and the pendubot [31], undersea vehicles [15], PVTOL (Planar Vertical Takeoff and Landing) aircrafts [35] and mobile robots [18].

Underactuation in these mechanical systems is generally introduced intentionally at the design level to reduce the manufacturing cost, the weight and/or failure rate; consequently, the obtained systems may be able to perform complex tasks with a reduced number of actuators. The source of underactuation may also be unintentionally due to failure of one or more actuators in a fully actuated system.

Underactuated systems may require new approaches/techniques to design effective control strategies; therefore, they constitute a good framework for nonlinear control problems regarding both theoretical and practical aspects. For these reasons, they are attracting more and more attention of researchers from nonlinear control community as well as from many research fields in robotics.

In the literature, many research efforts have been made in control aspects [30–32,39,41–43], giving rise to multiple control solutions; however, control problem of such systems is still open.

[1]National Institute of Applied Sciences and Technology (INSAT), Tunisia
[2]Laboratoire d'Informatique, de Robotique et de Microélectronique de Montpellier (LIRMM), France
[3]College of Engineering, Sultan Qaboos University, Sultanate of Oman

For instance in [9], a technique of pseudolinearization using spline functions is proposed to control the acrobot underactuated system. In [42], a design technique combining partial feedback linearization and Lyapunov methods based on saturation functions, switching and energy shaping has been proposed. In [43], the author discuss a survey on some existing results such as geometric nonlinear control and passivity-based control for stabilization and tracking control of such systems. Three control algorithms have been proposed in [30] for an underactuated two-link robot, namely an optimal LQ controller, a partial feedback-linearizing regulator and a sliding mode controller. In [31], the testbed is also a two-link underactuated robot but the actuator is located at the base (in contrast to the one proposed in [30] where the actuator is located at the joint between the two links), and the authors propose a method based on partial feedback linearization [40], nilpotent approximation and an iterative stabilization procedure. Roughly speaking, they propose an open-loop controller able to steer the system closer to the desired equilibrium point in finite time, and apply it iteratively, which provides a robust exponential convergence to the equilibrium. In [32], the authors propose to study the nonlinear behavior of the same robot proposed in [31] through a graphical tool based on Poincaré map. The study in [39] concerns underactuated mechanical systems with several actuated degrees of freedom and a single unactuated degree of freedom, where a discontinuous nonlinear feedback controller is proposed, allowing the closed-loop equilibrium at the origin to be globally attractive. In [45], a generalized predictive controller has been proposed, based on the linearized dynamics of the system. In [10], a nonlinear predictive controller is proposed for both stabilization and stable-limit cycle generation with an application to the ECP 505 inverted pendulum. In [44], a technique based on switching and saturation control is proposed. In [11], a passivity-based approach has been proposed to control the inertia wheel inverted pendulum. In [17], the technique of virtual holonomic constraints has been proposed with an application to the pendubot for generation of periodic motions, and to the Inertia Wheel Pendulum in [16]. In [3], the proposed control techniques are based on partial feedback linearization, combined with the optimization of the reference trajectories to stabilize the internal dynamics of the system. In [4] and [5], a dual-model-free control technique is proposed. The control architecture includes two model-free controllers, the first one is used to track the reference trajectories on the actuated degrees of freedom of the system, whereas the second one is used to stabilize the internal dynamics. In [25], a technique based on Riccati and nonlinear control is proposed for inverted pendulums. Output feedback control with an extended high gain observer was recently proposed in [28].

To deal with model uncertainties, a time-scale approach along with the Lyapunov design has been proposed in [38]. The authors in [34] used a sliding mode technique for the cart–pole system to stabilize the system in presence of disturbances. Model uncertainties however have not been tackled in this work. In [1], disturbances and model uncertainties have been fully considered in a technique that involves a backstepping procedure combined with sliding mode, applied to the inverted pendulum system after the system was converted into a normal form. Integral sliding mode control was also applied in [46] to deal with uncertainties in the two-wheeled mobile inverted pendulum. The control law being designed is based on the linearized system dynamics which resulted in a small region of attraction for the equilibrium. The problem becomes more challenging whenever nonparametric uncertainties and unknown time-varying disturbances are considered in the design of effective and robust control law for the underactuated system.

An important research issue which has attracted the attention of the control community is the finite-time convergence and finite-time stabilization [2,7,19–21,24], that ensures faster convergence rate, better disturbance rejection and robustness against uncertainty. Although myriad approaches have been proposed in the literature that tackle the problem of finite-time stabilization, they are not effective in dealing with time-varying and state-depending uncertainties. Few attempts have been concentrated in the design of adaptive schemes [22,23] to resolve the problem of state dependent and time-varying uncertainties. However, the adaptive scheme has only been applied to a class of p-normal forms [12]. To the best of the authors' knowledge, few have dealt with finite-time stabilization of underactuated systems subject to time-varying and state-dependent uncertainties. This chapter focuses on the design of a novel nonlinear controller for a class of underactuated system; we would like to introduce a new strategy that addresses an adaptive robust finite-time stabilization law derived from adding power integrator (API) [36] technique to a class of single input dynamics described as a normal form system [33]. The idea is to generate coordinate transformations that decouple the underactuated dynamics into an upper-triangular form allowing for only one control input governing the whole system to be designed, whereby a continuous finite-time stabilizing control law is derived from the API technique, and an adaptive compensator is proposed to compensate for the matched and unmatched uncertainties collected throughout the steps of the design. As a consequence, the resulting control law is able to stabilize the underactuated mechanical system in a finite time.

The reminder of this chapter is organized as follows. Section 8.2 discusses the Lagrangian model of mechanical underactuated systems and the coordinate transformation that transforms this model into a normal-form representation. Section 8.3 formulates the finite-time stabilization problem for the nth-order underactuated system. Section 8.4 gives a background on finite-time stability and provides some useful lemmas needed throughout the paper. The control design is developed in Section 8.5. Simulation results with an application to the cart–pole system are presented and discussed in Section 8.6. Finally, Section 8.7 draws a conclusion for this chapter.

8.2 Dynamic modeling of underactuated mechanical systems: Lagrangian formalism

The Lagrangian formalism is one among the powerful mathematical tools based on the vibrational method to model a large class of mechanical systems. Underactuated mechanical systems are branches of this class. They can be modeled using the Euler–Lagrange equations. Let Q be an n-dimensional configuration manifold, and $q \in Q$ is a vector of generalized coordinates. The Euler–Lagrange equations for any mechanical system of n degrees of freedom (DOF) are given by [33]

$$\frac{d}{dt}\frac{\partial L(q,\dot{q})}{\partial \dot{q}} - \frac{\partial L(q,\dot{q})}{\partial q} = B(q)u \tag{8.1}$$

where $u \in \mathbb{R}^m$ and $B(q)$ is the external forces matrix. $L(q,\dot{q})$ is the Lagrangian associated to the mechanical system and expressed by the difference between kinetic and potential energies:

$$L(q,\dot{q}) = K - V = \frac{1}{2}\dot{q}^\top M(q)\dot{q} - V(q) \tag{8.2}$$

where K is the kinetic energy, $V(q)$ denotes the potential energy and $M(q) = M^\top(q) \in \mathbb{R}^{n\times n}$ is a positive definite inertia matrix. Let $m_{ik} \in \mathbb{R}^+$ denote the entries of the inertia matrix M with $i = 1,\dots,n$ and $k = 1,\dots,n$. Based on (8.1) and (8.2), the equations of motion can be deduced as follows:

$$\sum_k m_{ik}(q)\ddot{q}_k + \sum_{i,k} \Gamma^\ell_{ik}(q)\dot{q}_i\dot{q}_k + g_i(q) = e_\ell^\top B(q)u, \quad \ell = 1,\dots,n \tag{8.3}$$

where e_ℓ is the ℓth standard basis in R^n, $g_i = \partial_{qi}V(q)$, and $\Gamma^\ell_{ik}(q)$ is the Christoffel symbol defined by

$$\Gamma^\ell_{ik}(q) = \frac{1}{2}\left(\frac{\partial m_{\ell k}(q)}{\partial q_i} + \frac{\partial m_{\ell i}(q)}{\partial q_k} - \frac{\partial m_{ik}(q)}{\partial q_\ell}\right) \tag{8.4}$$

In a matrix form, (8.3) can be rewritten as

$$M(q)\ddot{q} + C(q,\dot{q})\dot{q} + G(q) = B(q)u \tag{8.5}$$

The term $C(q,\dot{q})$ includes two distinct terms, in particular, the first term $q_i\dot{q}_k$ involves the centrifugal forces ($i = k$), and the second is the Coriolis forces for ($i \neq k$), and $G(q)$ represents the gravity vector. It is interesting to point out that the mechanical system described by (8.5) satisfies the skew symmetric property of the matrix $\dot{M}(q) - 2C(q,\dot{q})$.

It is important to distinguish two categories of mechanical systems governed by the equation of motion (8.5). A mechanical system described by (8.5) is said to be *fully actuated* if $m = n$, which is equivalent to say that the matrix $B(q) \in R^{n\times m}$ is invertible. In other words, it can be roughly inferred that the mechanical system is fully actuated, if the number of actuators is equal to the dimension of its configuration manifold. On the other hand, if $m < n$, the mechanical system is said to be *underactuated*, which is equivalent to have fewer independent inputs than the number of degree of freedom to be controlled. This implies that the matrix $B(q)$ becomes noninvertible and consequently the underactuated mechanical system as opposed to fully actuated systems is not exact feedback linearizable. Because the number of inputs is less than the number of degree of freedom, the controllability of the mechanical underactuated system is not always satisfied. In this chapter, however, we will assume that the considered mechanical systems are controllable. Furthermore, we suppose that the equation of motion (8.5) can be rearranged so that the matrix $B(q)$ can be written as:

$$B(q) = \begin{bmatrix} \mathbf{0}_{(n-m)\times m} \\ \mathbf{I}_m \end{bmatrix} \tag{8.6}$$

where the first $n - m$ equations of (8.5) represent the unactuated dynamics while the last m equations are the actuated dynamics. The generalized coordinates q can further be partitioned as follows:

$$q = \begin{bmatrix} q_u \\ q_a \end{bmatrix} \tag{8.7}$$

where $q_u \in \mathbb{R}^{(n-m)}$ is the vector of unactuated configuration variables and $q_a \in \mathbb{R}^m$ is the vector of actuated configuration variables. Then, it is straightforward to rewrite the equation of motion (8.5) as follows:

$$\begin{bmatrix} m_{11}(q) & m_{12}(q) \\ m_{21}(q) & m_{22}(q) \end{bmatrix} \begin{bmatrix} \ddot{q}_u \\ \ddot{q}_a \end{bmatrix} + \begin{bmatrix} c_{11}(q) & c_{12}(q) \\ c_{21}(q) & c_{22}(q) \end{bmatrix} \begin{bmatrix} \dot{q}_u \\ \dot{q}_a \end{bmatrix} + \begin{bmatrix} g_1(q) \\ g_2(q) \end{bmatrix} = \begin{bmatrix} \mathbf{0} \\ \mathbf{u}_2 \end{bmatrix} \tag{8.8}$$

The partitioned equation of motion (8.8) can also be represented in component form as follows:

$$m_{11}(q)\ddot{q}_u + m_{12}(q)\ddot{q}_a + h_1(q,\dot{q}) = \mathbf{0} \tag{8.9}$$

$$m_{21}(q)\ddot{q}_u + m_{22}(q)\ddot{q}_a + h_2(q,\dot{q}) = \mathbf{u}_2 \tag{8.10}$$

where $h_1(q,\dot{q}) = c_{11}(q)\dot{q}_u + c_{12}(q)\dot{q}_a + g_1(q)$ and $h_2(q,\dot{q}) = c_{21}(q)\dot{q}_u + c_{22}(q)\dot{q}_a + g_2(q)$. It is worth to notice that there is no control input in (8.9); therefore, it is not possible to feedback linearize the system (8.9) and (8.10). It is however possible to partially feedback linearize the system using a change of control. The procedure of linearization of the actuated dynamics is called *collocated partial linearization* which is credited to Spong [43] and consists in finding a global invertible change of control of the form $u_2 = \alpha(q)v + \beta(q,\dot{q})$ that partially linearizes the dynamics (8.8) such as, after partial linearization, the new control v appears in the nonlinear-subsystem dynamics as well as in the linearized subsystem dynamics as follows:

$$\ddot{q}_u = f_1(q,\dot{q}) + g_1(q,\dot{q})v \tag{8.11}$$

$$\ddot{q}_a = v \tag{8.12}$$

It can be observed that the unactuated part (8.11) is still complex, while the actuated part (8.12) is linearized into a double-integrator dynamics. The new control input v is present in both subsystems, which largely increases the difficulty in the control design. The author in [33] introduced a global change of coordinates that decouples these two subsystems and transforms the dynamics of the system into a cascaded form called the *normal form*, where the control input of the original actuated subsystem does not appear in the unactuated subsystem. This significantly simplifies the control design for the underactuated mechanical system. However, with this approach, there is an inherent requirement that the model parameters of the mechanical system must be perfectly known to be able to design controller for the transformed normal form.

Due to lack of knowledge on the system parameters and the inability to measure the whole system state, the controller has to cope with these uncertainties. This situation justifies the need for the development of a global transformation that decouples the underactuated mechanical system without passing by the partial feedback linearization of the system.

Consider the dynamic (8.9), resolve for $\ddot{q}_u$, it gives

$$\ddot{q}_u = -m_{11}(q)^{-1}(m_{12}(q)\ddot{q}_a + h_1(q, \dot{q})) \tag{8.13}$$

Substituting (8.13) into (8.10), after some manipulation, we obtain

$$\ddot{q}_a = f_3(q, \dot{q}) + b_3(q)\mathbf{u}_2 \tag{8.14}$$

Again substituting the resulting dynamics (8.14) into the dynamics of (8.10), then solving for $\ddot{q}_u$ leads to the following dynamics:

$$\ddot{q}_u = f_4(q, \dot{q}) + b_4(q)\mathbf{u}_2 \tag{8.15}$$

where$f_3(q, \dot{q}) = (m_{22} - m_{21}m_{11}^{-1}m_{12})^{-1}[m_{21}m_{11}^{-1}h_1(q, \dot{q}) - h_2(q, \dot{q})] =, b_3(q) = (m_{22} - m_{21}m_{11}^{-1}m_{12})^{-1}, f_4(q, \dot{q}) = -m_{11}^{-1}m_{12}f_4(q, \dot{q} - m_{11}^{-1}h_1(q, \dot{q}))$ and $b_4(q) = -m_{11}^{-1}m_{12}b_3(q)$. Regrouping the dynamic (8.14) and (8.15), one can obtain a state-space representation of the system as follows:

$$\begin{aligned} \dot{x}_1 &= x_2 \\ \dot{x}_2 &= f_3(q, \dot{q}) + b_3(q)\mathbf{u}_2 \\ \dot{x}_3 &= x_4 \\ \dot{x}_4 &= f_4(q, \dot{q}) + b_4(q)\mathbf{u}_2 \end{aligned} \tag{8.16}$$

where the state space vector x has been defined as $[x_1, x_2, x_3, x_4]^\top = [q_u, \dot{q}_u, q_a, \dot{q}_a]^\top$. Note the structural difference between (8.12), (8.11) and (8.16) is that the actuated dynamics in (8.12) is a double integrator, while in (8.16), the actuated dynamics has some nonlinearities.

To decouple the nonlinear system represented by (8.16), we may use the global transformation proposed in [33] as follows:

$$\begin{aligned} z_1 &= x_1 - \int_0^{x_3} \frac{b_3(s)}{b_4(s)} ds \\ z_2 &= x_2 - \frac{b_3(x)}{b_4(x)} \\ z_3 &= x_3 \\ z_4 &= x_4 \end{aligned} \tag{8.17}$$

The equation dynamics in the Z-space is found by taking the time derivative of (8.17), leading to

$$\begin{aligned}
\dot{z}_1 &= z_2 + \frac{b_3(z)}{b_4(z)}z_4 - \frac{d}{dt}\int_0^{z_3}\frac{b_3(s)}{b_4(s)}ds \\
\dot{z}_2 &= f_3(z) - \frac{b_3(z)}{b_4(z)}f_4(z) - \frac{d}{dt}\left(\frac{b_3(z)}{b_4(z)}\right)z_4 \\
\dot{z}_3 &= z_4 \\
\dot{z}_4 &= f_4(z) + b_4(z)\mathbf{u}_2
\end{aligned} \tag{8.18}$$

Clearly, the dynamics of the underactuated system in Z-space, is in a special cascade form with only one control input appearing in the last subsystem. To ease the representation, let us define $d_1(z) = \frac{b_3(z)}{b_4(z)}z_4 - \frac{d}{dt}\int_0^{z_3}\frac{b_3(s)}{b_4(s)}ds$, $d_2(z) = -z_3 + f_3(z) - \frac{b_3(z)}{b_4(z)}f_4(z) - \frac{d}{dt}\left(\frac{b_3(z)}{b_4(z)}\right)z_4$, $d_3(z) = 0$, then the system (8.18) can be rewritten as

$$\begin{aligned}
\dot{z}_1 &= z_2 + d_1(z) \\
\dot{z}_2 &= z_3 + d_2(z) \\
\dot{z}_3 &= z_4 + d_3(z) \\
\dot{z}_4 &= f_4(z) + b_4(z)\mathbf{u}_2
\end{aligned} \tag{8.19}$$

where $d_1(z), d_2(z)$ and $d_3(z)$ are regarded as unmatched uncertainties to be compensated by the control $\mathbf{u}_2$. In a similar manner, an nth-order underactuated system can be represented in the state space form as

$$\begin{aligned}
\dot{x}_1 &= x_2 \\
\dot{x}_2 &= f_3(q,\dot{q}) + b_3(q)\mathbf{u} \\
\dot{x}_3 &= x_4 \\
\dot{x}_4 &= f_4(q,\dot{q}) + b_4(q)\mathbf{u} \\
&\vdots \\
\dot{x}_{n-1} &= x_n \\
\dot{x}_n &= f_n(q,\dot{q}) + b_n(q)\mathbf{u}
\end{aligned} \tag{8.20}$$

To decouple the nth order system (8.20), the global transformation as proposed in [33] can be extended to nth order recursively as follows:

$$\begin{aligned}
z_1 &= x_1 - \int_0^{x_3}\frac{b_3(s)}{b_4(s)}ds, \\
z_2 &= x_2 - \frac{b_3(x)}{b_4(x)}, \\
z_3 &= x_3 - \int_0^{x_5}\frac{b_4(s)}{b_5(s)}ds
\end{aligned}$$

$$
\begin{aligned}
z_4 &= x_4 - \frac{b_4(x)}{b_5(x)} \\
&\vdots \\
z_{n-3} &= x_{n-3} - \int_0^{x_{n-1}} \frac{b_{n-2}(s)}{b_n(s)} ds \\
z_{n-2} &= x_{n-2} - \frac{b_{n-1}(x)}{b_n(x)} \\
z_{n-1} &= x_{n-1} \\
z_n &= x_n
\end{aligned} \tag{8.21}
$$

Hence, the system dynamics in Z-space can similarly be obtained, taking the time derivative of (8.21), we have

$$
\begin{aligned}
\dot{z}_1 &= z_2 + \frac{b_3(z)}{b_4(z)} z_4 - \frac{d}{dt} \int_0^{z_3} \frac{b_3(s)}{b_4(s)} ds \\
\dot{z}_2 &= f_3(z) - \frac{b_3(z)}{b_4(z)} f_4 - \frac{d}{dt} \left(\frac{b_3(z)}{b_4(z)} \right) z_4 \\
\dot{z}_3 &= z_4 + \frac{b_4(z)}{b_6(z)} z_6 - \frac{d}{dt} \int_0^{z_5} \frac{b_4(s)}{b_6(s)} ds \\
\dot{z}_4 &= f_4(z) - \frac{b_4(z)}{b_6(z)} f_6(z) - \frac{d}{dt} \left(\frac{b_4(z)}{b_6(z)} \right) z_6 \\
&\vdots \\
\dot{z}_{n-3} &= z_{n-2} + \frac{b_{n-2}(z)}{b_n(z)} z_n - \frac{d}{dt} \int_0^{z_{n-1}} \frac{b_{n-2}(s)}{b_n(s)} ds \\
\dot{z}_{n-2} &= f_{n-2}(z) - \frac{b_{n-2}(z)}{b_n(z)} f_n - \frac{d}{dt} \left(\frac{b_{n-2}(z)}{b_n(z)} \right) z_n \\
\dot{z}_{n-1} &= z_n \\
\dot{z}_n &= f_n(z) + b_n(z)\mathbf{u}
\end{aligned} \tag{8.22}
$$

Let us now define the following transformation:

$$
\begin{aligned}
d_1(z) &= \frac{b_3(z)}{b_4(z)} z_4 - \frac{d}{dt} \int_0^{z_3} \frac{b_3(s)}{b_4(s)} ds \\
d_2(z) &= -z_3 + f_3(z) - \frac{b_3(z)}{b_4(z)} f_4 - \frac{d}{dt} \left(\frac{b_3(z)}{b_4(z)} \right) z_4 \\
&\vdots
\end{aligned} \tag{8.23}
$$

$$
\begin{aligned}
d_{n-3} &= \frac{b_{n-2}(z)}{b_n(z)} z_n - \frac{d}{dt}\int_0^{z_{n-1}} \frac{b_{n-2}(s)}{b_n(s)} ds \\
d_{n-2} &= -z_{n-1} + f_{n-2}(z) - \frac{b_{n-2}(z)}{b_n(z)} f_n - \frac{d}{dt}\left(\frac{b_{n-2}(z)}{b_n(z)}\right) z_n \\
d_{n-1} &= 0 \\
d_n &= f_n(z)
\end{aligned}
$$

The dynamics in (8.23) can be similarly written in a standard form like in (8.18) as follows:

$$
\begin{aligned}
\dot{z}_i &= z_{i+1} + d_i(z), \quad 1 \le i \le n-1 \\
\dot{z}_n &= b_n(z)\mathbf{u} + d_n(z) \\
y &= h(z)
\end{aligned}
\tag{8.24}
$$

where $d_i(z) \in \mathbb{R}, i = 1, 2, \ldots, n$ are regarded as matched and unmatched uncertainties. It can be noticed that by construction, the variation bounds of the uncertainties $d_i(z), i = 1, \ldots, n$ are not assumed to be known and also cannot be assumed linearly parameterizable into a multiplication of a known regressor and a vector of unknown constant parameters. Also for the simplicity of the control design, let us consider the following two assumptions:

Assumption 8.1. *The terms $d_i(z), i = 1, \ldots, n$ are all unknown functions, whose bounds are also unknown.*

Assumption 8.2. *The uncertain control coefficient $b_n(z)$ is bounded as $b_{\min} \le b_n(z) \le b_{\max}$ and $b_n(z)$ is sign-definite such that $sign(b_n(z)) = +1$. Further, it is convenient to assume that $b_n(z) \neq 0$.*

8.3 Control problem formulation

In this chapter, we consider a class of underactuated systems of the form (8.20), which can be transformed without partial feedback linearization and through the global change of coordinates (8.21) into the cascaded form (8.24). Where $z = [z_1, z_2, \ldots, z_n]^\top \in \Omega$, where Ω is a compact set of $\mathbb{R}^n$, $u \in \mathbb{R}$ is the control input and $d_i(z) \in \mathbb{R}, i = 1, \ldots, n-1$ are unmatched uncertainties, while d_n is the matched uncertainty. Under Assumptions 8.1 and 8.2, the idea is to design the controller $\mathbf{u}$ such that the system state z tracks in finite time the desired trajectory $z_d^* = [z_{1d}^*, z_{2d}^*, \ldots, z_{nd}^*]^\top \in \Omega_d$ where Ω_d is a compact set of Ω. This means that for any finite initial condition $z_i(0), i = 1, \ldots, n$, there exists a finite time T such that

$$
\lim_{t \to T} \left| z_i - z_{id}^* \right| = 0, \quad i = 1, \ldots, n \tag{8.25}
$$

Remark 8.1. *The control design satisfying the limit (8.25) guarantees stabilization in finite time of the underactuated mechanical system of the form (8.20) to a desired reference input x_d. In most of the cases, stabilization around the origin of the first state*

x_1 is desired. The reference input for the Z-subsystem $z_{1d}^* = 0$ implies stabilization of the state variable x_1 to zero.

8.4 Preliminaries

In this section, a review of some basic concepts and lemma related to the concept of finite-time stability and the corresponding Lyapunov theorem are presented; they will be useful for most of the stability analysis of this chapter.

8.4.1 Notions of finite stability

Theorem 8.1. *[7]: Consider the non-Lipschitz continuous autonomous system $\dot{x} = f(x), f(0) \in \mathbb{R}$. Assume, there are C^1 function $V(x)$ defined on a neighborhood $D \subset \mathbb{R}$ of the origin, and real numbers $c > 0$ and $0 < \alpha < 1$ such that*

1. *$V(x)$ is positive definite on D*
2. *$\dot{V}(x) + cV^{\alpha} \leq 0, \quad \forall x \in D$*

Then, the origin x is locally finite-time stable. If $D = \mathbb{R}$, and $V(x)$ is radially unbounded, then the origin $x = 0$ is globally finite-time stable. Moreover, it can be verified that the settling time being dependent on the initial state $x(0) = x_0$ satisfies $T_x(x_0) \leq \frac{V(x_0)^{1-\alpha}}{c(1-\alpha)}$ for all x_0 in some open neighborhood of the origin.

Lemma 8.1. *[36] For any $x_i \in \mathbb{R}, i = 1, \ldots, n$ and $0 < p \leq 1$, the following inequality holds $(|x_1| + \cdots + |x_n|)^p \leq |x_1|^p + \cdots + |x_n|^p \leq n^{1-p}(|x_1| + \cdots + |x_n|)^p$. When $0 < p = \frac{p_1}{p_2} \leq 1$, where p_1 and p_2 are positive odd integers, then the following holds $|x^p - y^p| \leq 2^{1-p}|x - y|^p$, for any $x, y \in \mathbb{R}$.*

Lemma 8.2. *[36] For any $x \in \mathbb{R}, y \in \mathbb{R}, c > 0, d > 0$ and $\gamma(x,y) > 0$, a real-valued function, the following holds: $|x|^c|y|^d \leq (c\gamma(x,y)/(c+d))|x|^{c+d} + (d\gamma(x,y)^{-c/d}/(c+d))|y|^{c+d}$.*

8.5 Proposed control design

The purpose of this section is to devise a control law that stabilizes the system (8.24) in finite time. Given the fact that the Z-system (8.24) contains unmatched uncertainty, traditional adaptive techniques like the backstepping technique [29] and the multiple-surface sliding [37] cannot be used, because the variation bounds of $d_i, i = 1, \ldots, n$ are not assumed to be known. To solve the problem of finite-time stabilization with unmatched uncertainties, we propose a novel robust finite-time control scheme whereby a continuous recursive finite-time stabilizing control law is derived from the API technique [36] and an adaptive compensator.

To develop the finite-time controller for (8.24), we first introduce the following tracking error e_1 and the virtual errors $e_m, m = 2, \ldots, n$ as follows:

$$e_1 = z_1 - z^*_{1d} \tag{8.26}$$

and

$$e_m = z_m^{1/q_m} - z_{md}^{*1/q_m}, \quad m = 2, \ldots, n \tag{8.27}$$

where $q_m = (4\ell n + 3 - 2m)/(4\ell n + 1) > 0, (\ell \in \mathbb{Z}_+); z^*_m, m = 2, \ldots, n$ is the virtual control given by $z^*_{md} = -\beta_{m-1} e_{m-1}^{q_m}$ with β_{m-1} being positive constant to be determined later. One main feature for the use of the parameter ℓ in the fraction power q_m is that it allows adjusting the control precision and finite-time convergence of the state variables.

The design procedure consists in n steps, throughout the $n - 1$ steps, the virtual controllers $z^*_{(m-1)d}$ will be designed. Upon the completion of step n, the fast control term as well the adaptive compensator are designed.

Step 1: Consider the following Lyapunov candidate function:

$$V_1(z) = \frac{1}{1+S} e_1^{1+S} \tag{8.28}$$

where $S = (4\ell n - 1)/(4\ell n + 1)$. The time derivative of (8.28) gives

$$\begin{aligned} \dot{V}_1(z) &= e_1^S \dot{e}_1 \\ &= e_1^S z_2^* + e_1^S (z_2 - z^*_{2d}) + e_1^S d_1(z) \end{aligned} \tag{8.29}$$

By introducing the virtual control $z^*_{2d} = -\beta_1 e_1^S$, where β_1 is a design parameter, $\dot{V}_1(z)$ rewrites:

$$\dot{V}_1(z) = -\beta_1 e_1^{2S} + e_1^S (z_2 - z^*_{2d}) + e_1^S d_1(z) \tag{8.30}$$

Using Lemmas 8.1 and 8.2, the second term of (8.30) can be upper-bounded as follows:

$$e_1^S (z_2 - z^*_{2d}) \leq 2^{1-S} |e_2|^S |e_1|^S \leq \frac{1}{2} |e_1|^{2S} + C_2 |e_2|^{2S} \tag{8.31}$$

where C_2 is a positive constant. In light of (8.31), the time derivative of $V_1(z)$ can be written as follows:

$$\dot{V}_1(z) \leq -\beta_1 e_1^{2S} + \frac{1}{2} |e_1|^{2S} + C_2 |e_2|^{2S} + e_1^S d_1(z) \tag{8.32}$$

Step 2: According to [24], a C^1 and positive definite Lyapunov candidate function is constructed as

$$V_2((z)) = V_1(z) + W_2(z_1, z_2) \tag{8.33}$$

with

$$W_2(z_1, z_2) = \int_{z^*_{2d}}^{z_2} \left(\chi^{\frac{1}{q_2}} - z_{2d}^{*\frac{1}{q_2}} \right)^{1+S-q_2} d\chi \tag{8.34}$$

then it is easy to show that the Lyapunov function candidate $V_2(z)$ is positive definite and satisfies $V_2(z) \le \max\{\frac{1}{1+S}, 2\}(e_1^{1+S} + e_2^{1+S})$. Note that the function $W_2(z_1, z_2)$ has the following properties:

$$\frac{\partial W_2(.)}{\partial z_2} = e_2^{1+S-q_2} \tag{8.35}$$

$$\begin{aligned} \frac{\partial W_2(.)}{\partial z_1} &= -(1+S-q_2)\frac{\partial z_{2d}^{*\frac{1}{q_2}}}{\partial z_1}\int_{z_{2d}^*}^{z_2}\left(\chi^{\frac{1}{q_2}} - z_{2d}^{*\frac{1}{q_2}}\right)^{S-q_2} d\chi \\ &= -\frac{\partial z_{2d}^{*\frac{1}{q_2}}}{\partial z_1}(z_2 - z_{2d}^*) \end{aligned} \tag{8.36}$$

Hence, the derivative of $W_2(.)$ can be computed as follows:

$$\begin{aligned} \dot{W}_2(.) &= e_2\dot{z}_2 - \left((z_2 - z_{2d}^*)\frac{\partial z_{2d}^{*\frac{1}{q_2}}}{\partial z_1}\dot{z}_1\right) \\ &= e_2(z_3 - z_{3d}^*) + e_2 z_3^* + e_2 d_2(z) - (z_2 - z_{2d}^*)\frac{\partial z_{2d}^{*\frac{1}{q_2}}}{\partial z_1}\dot{z}_1 \end{aligned} \tag{8.37}$$

using the fact that

$$\left|(z_2 - z_{2d}^*)\frac{\partial z_{2d}^{*\frac{1}{q_2}}}{\partial z_1}\dot{z}_1\right| \le 2^{1-q_2} e_2^{q_2}\left|\frac{\partial z_{2d}^{*\frac{1}{q_2}}}{\partial z_1} z_2\right| + 2^{1-q_2} e_2^{q_2}\left|\frac{\partial z_{2d}^{*\frac{1}{q_2}}}{\partial z_1} d_1(z)\right| \tag{8.38}$$

Let the first lumped unmatched uncertainty be $\tau_1 = e_2 d_2(z) + 2^{1-q_2}\beta_1^{\frac{1}{q_2}} e_2^{q_2} d_1(z)$, it is then straightforward to obtain

$$\dot{W}_2 \le e_2(z_3 - z_{3d}^*) + e_2 z_{3d}^* + 2^{1-q_2}\beta^{\frac{1}{q_2}} e_2^{q_2}|z_2| + \tau_1 \tag{8.39}$$

According to Lemma 8.1, the third term of the right hand side of (8.39) can be bounded such that

$$|z_2| \le |e_2 + z_{2d}^{*\frac{1}{q_2}}|^{q_2} \le |e_2|^{q_2} + \beta_1|e_1|^{q_2} \tag{8.40}$$

which implies that

$$\begin{aligned} |e|_2^{q_2} 2^{1-q_2}\beta^{\frac{1}{q_2}}|z_2| &\le 2^{1-q_2}\beta^{\frac{1}{q_2}}|e_2|^{2q_2} + 2^{1-q_2}\beta^{1+\frac{1}{q_2}}|e_2|^{q_2}|e_1|^{q_2} \\ &\le 2^{2(1-q_2)-1}\beta^{2(1+\frac{1}{q_2})}|e_2|^{2q_2} + \frac{1}{2}|e_1|^{2q_2} + 2^{1-q_2}\beta^{\frac{1}{q_2}}|e_2|^{2q_2} \\ &= \frac{1}{2}|e_1|^{2q_2} + \bar{C}_2|e_2|^{2q_2} \end{aligned} \tag{8.41}$$

where $\bar{C}_2 = 2^{2(1-q_2)}\beta^{2(1+\frac{1}{q_2})} + 2^{1-q_2}\beta^{\frac{1}{q_2}}$.

Combining (8.32), (8.39) and (8.41) yields the following:

$$\dot{V}_2(.) \le -(\beta_1 - 1)|e_1|^{2S} + (C_1 + \bar{C}_2)|e_2|^{2S} + e_3 z_{3d}^* + e_2(z_3 - z_{3d}^*) + \tau_1 \tag{8.42}$$

At this stage, it is worth to note that $q_3 = q_2 - \frac{2}{4\ell n+1}$, which implies that $q_3 = 1 + 2q_2 = 1 + 2S$. Therefore, by selecting $\beta_1 > n - 1 - \kappa$, where $\kappa > 0$ is a design parameter, and the virtual control $z_{3d}^* = -\beta_2 z_{2d}^{q_3}$ with $\beta_2 \geq n - 2 + \kappa + C_2 + \bar{C}_2$, we obtain

$$\dot{V}_2 \leq -(n-2+\kappa)|e_1|^{2S} - (n-2+\kappa)|e_2|^{2S} + e_2(z_3 - z_{3d}^*) + \tau_1 \tag{8.43}$$

Step *k* (*k*: = 3 . . . *n* − 1): We proceed to the derivation of the virtual control by using an inductive argument. Suppose at step $k-1$, that there exists a C^1 Lyapunov candidate function $V_{k-1}(z_1, \ldots, z_{k-1})$, positive definite and verifies

$$V_{k-1}(.) \leq \max\{\frac{1}{1+S}, 2\} \sum_{m=1}^{k-1} e_m^{1+S} \tag{8.44}$$

such that

$$\dot{V}_{k-1}(.) \leq -(n-k+1+\kappa) \sum_{m=1}^{k-1} e_m^{2S} + e_{k-1}^{1+S-q_{k-1}}(z_k - z_{kd}^*) + \sum_{m=1}^{k-2} \tau_m \tag{8.45}$$

Now let us claim that (8.44) and (8.45) hold as step k. To prove this claim, consider the following Lyapunov candidate function:

$$V_k(z_1, \ldots, z_k) = V_{k-1}(.) + W_k(z_1, \ldots, z_k) \tag{8.46}$$

with

$$W_k(.) = \int_{z_{dk}^*}^{z_k} (\chi^{\frac{1}{q_k}} - z_{dk}^{*\frac{1}{q_k}})^{1+S-q_k} d\chi \tag{8.47}$$

From the previous step, it can be observed that $W_k(.)$ has the following properties:

$$\frac{\partial W_k(.)}{\partial z_k} = e^{1+S-q_k} \tag{8.48}$$

$$\frac{\partial W_k(.)}{\partial z_m} = -(1+S-q_k)\frac{\partial z_{dk}^{*\frac{1}{q_k}}}{\partial z_m} \int_{z_{dk}^*}^{z_k} (\chi^{\frac{1}{q_k}} - z_{dk}^{\frac{1}{q_k}})^{S-q_k} d\chi \tag{8.49}$$

also, it is easy to show that $V_k(.)$ is C^1 and positive definite, which verifies $V_k(.) =\leq \max\left\{\frac{1}{1+S}, 2\right\} \sum_{m=1}^{k} e_m^{1+S}$. The time derivative of $V_k(.)$ satisfies

$$\dot{V}_k(.) \leq -(n-k+1+\kappa)\sum_{m=1}^{k-1} e_m^{2S} + e_{k-1}^{1+S-q_{k-1}}(z_k - z_{dk}^*) + e_k^{1+S-q_k} z_{k+1}$$
$$+ e_k^{1+S-q_k} d_k(z) + \sum_{m=2}^{k-1} \frac{\partial W_k(.)}{\partial z_m}\dot{z}_m + \sum_{m=1}^{k-1} \tau_m$$
$$\leq -(n-k+1+\kappa)\sum_{m=1}^{k-1} e_m^{2S} + e_{k-1}^{1+S-q_{k-1}}(z_k - z_{dk}^*) + e_k^{1+S-q_k} z_{k+1}$$
$$+ \sum_{m=2}^{k-1} \frac{\partial W_k(.)}{\partial z_m} z_{m+1} + \left[e_k^{1+S-q_k} d_k(z) + \sum_{m=2}^{k-1} \frac{\partial W_k(.)}{\partial z_m} d_m(z)\right] + \sum_{m=1}^{k-1} \tau_m \quad (8.50)$$

If the kth lumped unmatched uncertainty is defined as $\tau_k = e_k^{1+S-q_k} d_k(z) + \sum_{m=2}^{k-1} \frac{\partial W_k(.)}{\partial z_m} d_m(z)$, then (8.50) rewrites:

$$\dot{V}_k(.) \leq -(n-k+1+\kappa)\sum_{m=1}^{k-1} e_m^{2S} + e_{k-1}^{1+S-q_{k-1}}(z_k - z_{dk}^*) + e_k^{1+S-q_k} z_{k+1}$$
$$+ \sum_{m=2}^{k-1} \frac{\partial W_k(.)}{\partial z_m} z_{m+1} + \sum_{m=1}^{k} \tau_m \quad (8.51)$$

Next, we bound the second and the fourth terms of the right hand side of (8.51). First according to Lemma 8.2, it holds that

$$|e_{k-1}^{1+S-q_{k-1}}(z_k - z_{dk}^*)| \leq 2^{1+S-q_{k-1}} |e_{k-1}|^{1+S-q_{k-1}} |e_k|^{q_k}$$
$$\leq \frac{|e_{k-1}|^{2S}}{2} + C_k |e_k^{2S}| \quad (8.52)$$

with C_k a positive constant. As for the fourth term, it is easy to obtain the following:

$$\left|\sum_{m=2}^{k-1} \frac{\partial W_k(.)}{\partial z_m} z_{m+1}\right| \leq (1+S-q_k)2^{1-q_k}|e_k|^S \left|\sum_{m=2}^{k-1} \frac{\partial z_{dk}^*}{\partial z_m} z_{m+1}\right| \quad (8.53)$$

To further bound the fourth term in (8.51), we need to conduct the analysis by inductive argument and assume that at step $k-1$, the following holds

$$\left|\sum_{m=2}^{k-2} \frac{\partial z_{dk}^*}{\partial z_m} z_{m+1}\right| \leq \sum_{m=1}^{k-1} \gamma_{(k-1)m} e_m^S \quad (8.54)$$

where $\gamma_{(k-1)m} \geq 0$, then show the inequality also holds for k. Therefore, we have

$$\left|\sum_{m=2}^{k-1} \frac{\partial z_{dk}^*}{\partial z_m} z_{m+1}\right| \leq \left|-\beta_{k-1}^{\frac{1}{q_k}} \sum_{m=2}^{k-1} \frac{\partial e_{k-1}}{\partial z_m} z_{m+1}\right|$$
$$\leq \beta_{k-1}^{\frac{1}{q_k}} \left|\frac{z_{k-1}^{\frac{1}{q_k}-1}}{q_{k-1}} z_k + \sum_{m=2}^{k-2} \frac{\partial z_{k-1}^{*\frac{1}{q_k-1}}}{\partial z_m} z_{m+1}\right| \quad (8.55)$$

In light of the definition of the tracking error, it is worth recalling that $e_m = z_m^{\frac{1}{q_m}} - z_{md}^{*\frac{1}{q_m}}$ and $z_m^* = -\beta_{m-1} e_{m-1}^{q_m}$, it can then be inferred that $z_m^{*\frac{1}{q_m}} = -\beta_{m-1}^{\frac{1}{q_m}} e_{m-1}$, and therefore, the following inequality holds

$$|z_m| \leq |e_m + z_m^{*\frac{1}{q_m}}|^{q_m} \leq |e_m|^{q_m} + \beta_{m-1}|e_{m-1}|^{q_m} \tag{8.56}$$

Applying (8.56) to the inequality (8.55) and using the assumption of the inequality (8.54) yields

$$\begin{aligned} \left|\sum_{m=2}^{k-1} \frac{\partial z_{dk}^*}{\partial z_m} z_{m+1}\right| &\leq \beta_{k-1}^{\frac{1}{q_k}} \left[\frac{1}{q_{k-1}} \left(|e_{k-1}|^{1-q_{k-1}} + \beta_{k-2}^{\frac{1}{q_{k-1}}-1} e_{k-2}^{1-q_{k-1}} \right) \left(|e_k|^{q_k} \right.\right. \\ &\quad \left.\left. + \beta_{k-1}|e_{k-1}|^{q_k} \right) + \sum_{m=2}^{k-2} \frac{\partial z_{dk}^*}{\partial z_m} z_{m+1} \right] \\ &\leq \sum_{m=1}^{k} \gamma_{km} |e_m|^S \end{aligned} \tag{8.57}$$

with γ_{km} being a positive constant. Therefore,

$$\begin{aligned} \left|\sum_{m=2}^{k-1} \frac{\partial W_k(.)}{\partial z_m} z_{m+1}\right| &\leq (1 + S - q_k) 2^{1-q_k} |e_k|^S \left(\sum_{m=1}^{k} \gamma_{km} |e_m|^S \right) \\ &\leq \frac{1}{2} \sum_{m=1}^{k-1} e_m^{2S} + \bar{C}_k |e_k|^{2S} \end{aligned} \tag{8.58}$$

with $\bar{C}_k \geq 0$ being a positive constant.

Substituting (8.52) and (8.58) into (8.51) leads to

$$\begin{aligned} \dot{V}_k(.) &\leq -(n - k + 1 + \kappa) \sum_{m=1}^{k-1} e_m^{2S} + (C_k + \bar{C}_k)|e_k|^{2S} + \frac{|e_{k-1}|}{2} + \frac{1}{2} \sum_{m=1}^{k-1} e_m^{2S} \\ &\quad + \sum_{m=1}^{k} \tau_k + e_k^{1+S-q_k} (z_{k+1} - z_{d(k+1)})^* + e_k^{1+S-q_k} z_{k+1}^* \\ &\leq -(n - k + \kappa) \sum_{m=1}^{k-1} e_m^{2S} + (C_k + \bar{C}_k)|e_k|^{2S} + e_k^{1+S-q_k} (z_{k+1} - z_{d(k+1)}^*) \\ &\quad + e_k^{1+S-q_k} z_{k+1}^* + \sum_{m=1}^{k} \tau_k \end{aligned} \tag{8.59}$$

By introducing the virtual control $z_{(k+1)d}^* = -\beta_k e_k^{q_{k+1}}$ with β_k selected such that $\beta_k \geq n - k + \kappa + C_k + \bar{C}_k > 0$, we get

$$\dot{V}_k(.) \leq -(n-k+\kappa)\sum_{m=1}^{k} e_m^{2S} + e_k^{1+S-q_k}(z_{k+1} - z^*_{(k+1)d}) + \sum_{m=1}^{k} \tau_k \tag{8.60}$$

Step n: This is the final stage of the design where the real control input appears in the dynamics. For this step, consider the nth part of the Lyapunov candidate function

$$V_n(z_1,\ldots,z_n) = V_{n-1}(.) + W_n(z_1,\ldots,z_n) \tag{8.61}$$

where

$$W_n(.) = \int_{z^*_{nd}}^{z_n} (\chi^{\frac{1}{q_n}} - z_n^{\frac{1}{q_n}})^{1+S-q_n} d\chi \tag{8.62}$$

then, it is obvious to conclude that $V_n(.)$ is C^1, positive definite and satisfies $V_n(.) \leq \max\{\frac{1}{1+S},2\}\sum_{m=1}^{n} e_m^{1+S}$. Furthermore, let $\widetilde{G} = G - \widehat{G}$ be the estimate error of the matched and unmatched uncertainty and G is the total lumped uncertainty to be defined later. From the above inductive argument, one can conclude that

$$\begin{aligned}
\dot{V}_n &\leq -\kappa\sum_{m=1}^{n-1} e_m^{2S} + e_n^{1+S-q_n}\dot{z}_n + (C_n + \bar{C}_n)e_n^{2S} + \sum_{m=1}^{n-1}\tau_m \\
&\leq -\kappa\sum_{m=1}^{n-1} e_m^{2S} + e_n^{1+S-q_n}(b_n u + d_n(z)) + (C_n + \bar{C}_n)e_n^{2S} \\
&\quad + e_n^{1+S-q_n}\left(e_n^{q_n-1-S}\sum_{m=1}^{n-1}\tau_m\right) \\
&\leq -\kappa\sum_{m=1}^{n-1} e_m^{2S} + b_n e_n^{1+S-q_n} u + (C_n + \bar{C}_n)e_n^{2S} + b_n e_n^{1+S-q_n}\left(\frac{d_n(z)}{b_n}\right. \\
&\quad \left. + \frac{e_n^{q_n-1-S}}{b_n}\sum_{m=1}^{n-1}\tau_m\right)
\end{aligned} \tag{8.63}$$

where $C_n > 0$ and $\bar{C}_n > 0$ are positive constants. The actual control law can therefore be designed as follows:

$$u = -\beta_n e_n^{q_{n+1}} - \widehat{G}\,\mathrm{sign}(\sigma) \tag{8.64}$$

with the compensator being designed as follows:

$$\dot{\widehat{G}} = |\sigma|, \quad \sigma = e^{1+S-q_n} \tag{8.65}$$

where $\widehat{G}$ is the estimation of the total uncertainty function G defined as $G = \frac{d_n(z)}{b_n} + \frac{e_n^{q_n-1-S}}{b_n}\sum_{m=1}^{n-1}\tau_m$ and β_n is a positive design parameter chosen arbitrarily by the designer.

The main result of this chapter can be summarized in the following theorem.

Theorem 8.2

Consider the nth order underactuated system in the X-space represented by (8.20), through the coordinate transformation (8.21) and under Assumptions 8.1 and 8.2, the finite-time convergence objective specified in (8.25) is achieved by the control input u consisting of (8.64) and the compensator (8.65) with the virtual control $z^*_{md}, m = 1, \ldots, n$ being applied.

Proof. Consider the Lyapunov function $V = V_n(.)$. Since $\frac{b_n}{2}\widetilde{G}^2$ is positive definite, then $V \leq V_n(.) + \frac{b_n}{2}\widetilde{G}^2$. Taking the derivative of V along (8.63), yields

$$\begin{aligned}\dot{V} &\leq \dot{V}_n - b_n\widetilde{G}\dot{\widehat{G}}\\ &\leq -\kappa\sum_{m=1}^{n-1} e_m^{2S} + b_n e_n^{1+S-q_n}u + (C_n + \bar{C}_n)e_n^{2S} + b_n e_n^{1+S-q_n}\left(\frac{d_n(z)}{b_n}\right.\\ &\quad\left. + \frac{e_n^{q_n-1-S}}{b_n}\sum_{m=1}^{n-1}\tau_m\right) - b_n\widetilde{G}\dot{\widehat{G}}\\ &\leq -\kappa\sum_{m=1}^{n-1} e_m^{2S} + b_n e_n^{1+S-q_n}u + (C_n + \bar{C}_n)e_n^{2S} + b_n e_n^{1+S-q_n}G - b_n\widetilde{G}\dot{\widehat{G}}\end{aligned} \tag{8.66}$$

Substituting the control law (8.64) and the update law (8.65) in (8.66) leads to

$$\begin{aligned}\dot{V} &\leq -\kappa\sum_{m=1}^{n-1} e_m^{2S} - b_n\beta_n e_n^{1+S-q_n+q_{n+1}} + (C_n + \bar{C}_n)e_n^{2S} - b_n\sigma\widehat{G}\mathrm{sign}(\sigma)\\ &\quad + b_n\sigma G - b_n G\dot{\widehat{G}}\\ &\leq -\kappa\sum_{m=1}^{n-1} e_m^{2S} - (b_n\beta_n - C_n - \bar{C}_n)e_n^{2S} + b_n(G\sigma - \widehat{G}|\sigma|) - b_n G|\sigma|\end{aligned} \tag{8.67}$$

If β_n is chosen such that $\beta_n b_{\max} - C_n - \bar{C}_n > \kappa$, it follows from (8.67) that

$$\dot{V} \leq -\kappa\sum_{m=1}^{n} e_m^{2S} \tag{8.68}$$

Noting that $V = V_n(.) \leq \max\{\frac{1}{1+S}, 2\}\sum_{m=1}^{n} e_m^{1+S}$ and using Lemma 8.1, we have

$$V_n(.)^{\frac{2S}{1+S}} \leq \max\left\{\frac{1}{1+S}, 2\right\}\sum_{m=1}^{n} e_m^{2S} \tag{8.69}$$

Let $\alpha := \frac{2S}{1+S}$, then it is easy to conclude that

$$\frac{dV(.)}{dt} \leq \mu V^{\alpha} \tag{8.70}$$

with $\mu = \frac{\kappa}{\max\{\frac{1}{1+S}, 2\}}$. From (8.70), it is straightforward to conclude from Theorem 8.1 that the solutions of the closed-loop system consisting of (8.24) and the control input (8.64) along with the compensator (8.65) are finite-time stable, and therefore, the limit in (8.25) is satisfied. This completes the proof. □

8.6 Numerical simulations: the case of the cart–pole system

The cart–pole inverted pendulum depicted schematically in Figure 8.1 consists of a pole mounted on a cart by means of a pivot in such a way that the pole can freely swing in the x–y plane. The equations of motion can be obtained by neglecting friction in the pivot and by applying the Euler–Lagrange formulation. The system model can be represented as follows [34]:

$$\begin{aligned} (M + m\sin^2\theta)\ddot{x} - m\sin\theta(l\dot{\theta}^2 - g\cos\theta) &= u \\ (M + m\sin^2\theta)l\ddot{\theta} + ml\dot{\theta}^2\sin\theta\cos\theta - (M+m)g\sin\theta &= -\cos\theta(u) \end{aligned} \tag{8.71}$$

where M is the mass of the cart, m is the mass of the pendulum, l is half the length of the pendulum, i.e., the distance from the pivot to the center of mass of the pendulum, g is the acceleration due to gravity. The cart–pole system has two equilibrium points, one of which is known as the stable vertically downward position where $\theta = \pi$, and the other one being the unstable vertically upward position where $\theta = 0$. $F(t)$ is the horizontal force being applied to the cart to swing the pendulum until it reaches the desired unstable position where $\theta = 0, \dot{\theta} = 0$ and $\dot{x} = 0$. As can be seen from the equations of motion (8.71), only one control u appears in both dynamic equations of the cart–pole system; therefore, the whole system is underactuated. Meaning that it has fewer control inputs than degrees of freedom. Let $\mathbf{x} = [x, \dot{x}, \theta, \dot{\theta}]^\top = [x_1, x_2, x_3, x_4]^\top \in \mathbb{R}^4$, then the state-space representation of the cart–pole system becomes

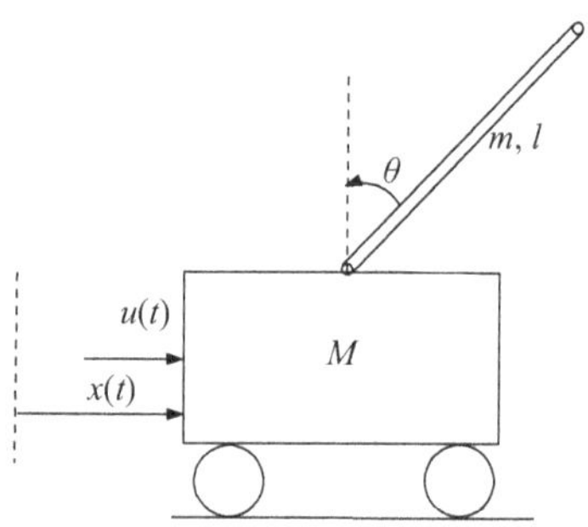

Figure 8.1 Schematic view of the cart–pole inverted pendulum

$$\begin{aligned}\dot{x}_1 &= x_2\\ \dot{x}_2 &= f_2(\mathbf{x}) + b_2(\mathbf{x})u\\ \dot{x}_3 &= x_4\\ \dot{x}_4 &= f_4(\mathbf{x}) + b_4(\mathbf{x})u\end{aligned} \tag{8.72}$$

where $f_2(\mathbf{x}) = \frac{mlx_4^2 \sin x_3 - mg \sin x_3 \cos x_3}{M + m \sin^2 x_3}$, $b_2(\mathbf{x}) = \frac{1}{M + m \sin^2 x_3}$, $f_4(\mathbf{x}) = \frac{(M+m)g \sin x_3 - mlx_4^2 \sin x_3 \cos x_3}{Ml + ml \sin^2 x_3}$ and $b_4(\mathbf{x}) = \frac{-\cos x_3}{Ml + ml \sin^2 x_3}$. The state-space representation (8.72) is in the form (8.20) with $n = 4$. The system can be readily transformed from the X-space into Z-space using the coordinate transformation (8.21) as follows:

$$\begin{aligned}z_1 &= x_1 - \int_0^{x_3} \frac{b_2(s)}{b_4(s)} ds = x_1 + l \ln(\sec x_3 + \tan x_3)\\ z_2 &= x_2 - \frac{b_2(\mathbf{x})}{b_4(\mathbf{x})} x_4 = x_2 + l x_4 \sec x_3\\ z_3 &= x_3\\ z_4 &= x_4\end{aligned} \tag{8.73}$$

with the change of coordinate (8.73), the system dynamics in the Z-space becomes

$$\begin{aligned}\dot{z}_1 &= z_2 + d_1(z)\\ \dot{z}_2 &= z_3 + d_2(z)\\ \dot{z}_3 &= z_4 + d_3(z)\\ \dot{z}_4 &= b_4(z)\mathbf{u} + d_4(z)\end{aligned} \tag{8.74}$$

where $d_1(z) = d_3(z) = 0$ and

$$\begin{aligned}d_2(z) &= -z_3 - lz_4^2 \sec z_3 \tan z_3 + \frac{(M+m)g \tan z_3 - mg \sin z_3 \cos z_3}{M + m \sin^2 z_3}\\ d_4(z) &= \frac{(M+m)g \sin z_3 - mlz_4^2 \sin z_3 \cos z_3}{Ml + ml \sin^2 z_3}\\ b_4(z) &= \frac{-\cos z_3}{Ml + ml \sin^2 z_3}\end{aligned} \tag{8.75}$$

The objective is then to control the transformed system (8.74) such that all states z_1, z_2, z_3 and z_4 converge in finite time to zero regardless of the internal uncertainties and external disturbances. For such system, the virtual controllers and the actual

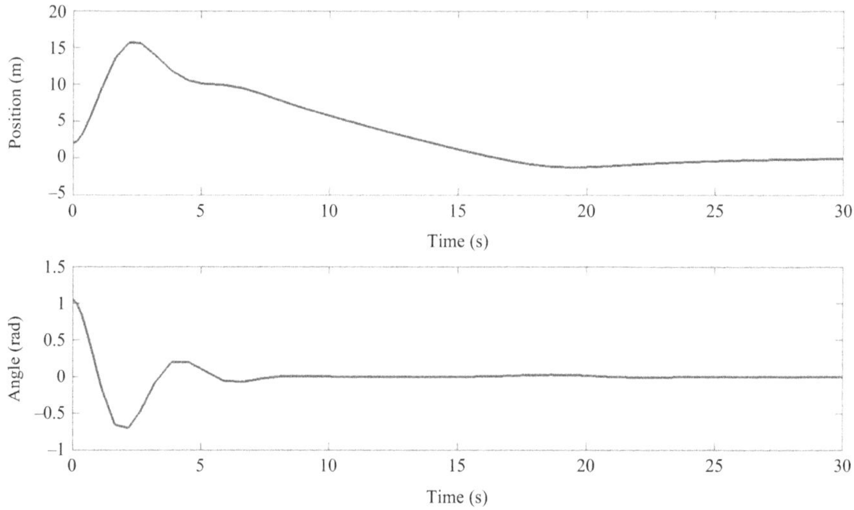

Figure 8.2 Trajectories of the cart position and pendulum angle versus time

control law are designed according to the development of Theorem 8.2 and are as follows:

$$\begin{cases} z_{1d}^{*} = 0, & e_1 = z_1^{\frac{1}{q_1}} - z_{1d}^{*\frac{1}{q_1}} \\ z_{2d}^{*} = -\beta_1 e_1^{q_2}, & e_2 = z_2^{\frac{1}{q_2}} - z_{2d}^{*\frac{1}{q_2}} \\ z_{3d}^{*} = -\beta_2 e_2^{q_3}, & e_3 = z_3^{\frac{1}{q_2}} - z_{3d}^{*\frac{1}{q_2}} \\ z_{4d}^{*} = -\beta_3 e_3^{q_4}, & e_4 = z_4^{\frac{1}{q_2}} - z_{4d}^{*\frac{1}{q_2}} \end{cases} \tag{8.76}$$

and

$$\begin{cases} u = -\beta_4 e_4^{q_5} - \widehat{G}\operatorname{sign}(\sigma) \\ \dot{\widehat{G}} = |\sigma|, \quad \sigma = e^{1+S-q_4} \end{cases} \tag{8.77}$$

To demonstrate the performance of the proposed robust finite control law, we have performed numerical simulations on MATLAB®/Simulink® framework, considering the following system parameters $m = 1$ kg, $M = 2$ kg and $l = 0.5$ m. The initial conditions are chosen as $\mathbf{x} = [1, 0, \frac{\pi}{3}, 0]^{\top}$, the desired state is $\mathbf{x}_d = [0, 0, 0, 0]^{\top}$. The control gains are chosen to be $\kappa = 10, \beta_1 = 13, \beta_2 = 12, \beta_3 = 11$ and $\beta_4 = 20$. Different simulation scenarios have been considered. The first simulation scenario is the control of uncertain cart–pole system without external disturbances. The second one considers the case of an uncertain cart–pole system with the presence of external disturbances.

1. **Scenario 1:** Uncertain cart–pole system without external disturbances
   In this scenario, we assume that the system dynamics is completely unknown except for the sign condition in Assumption 8.2. The obtained simulation results

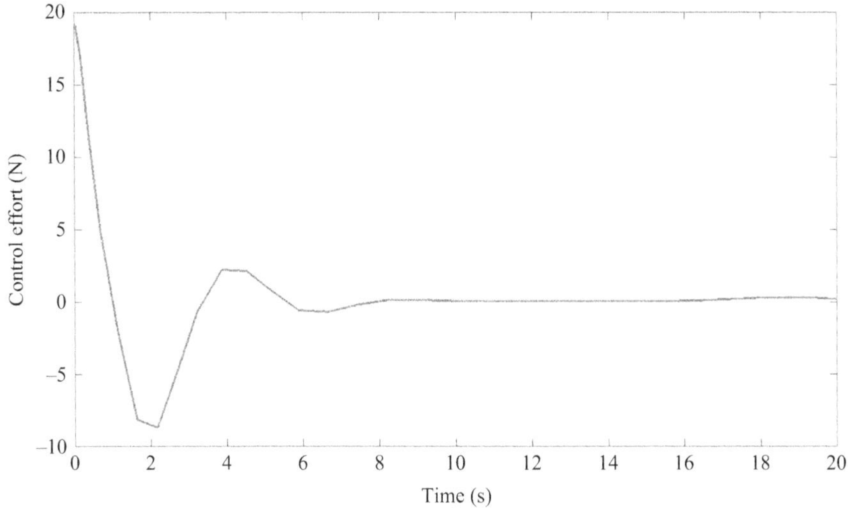

Figure 8.3 Evolution of the control effort versus time

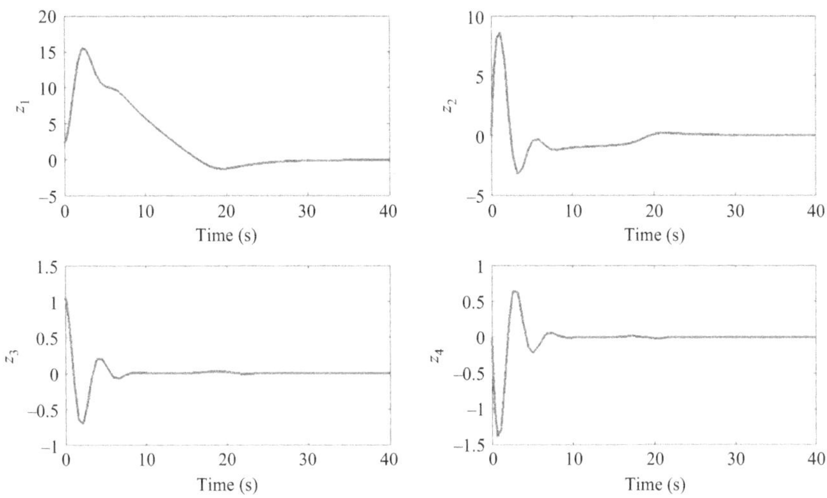

Figure 8.4 Evolution of the trajectories of Z-space state variables versus time

are shown in Figures 8.2–8.4. It is clear in Figure 8.2 how the cart displacement and the pendulum angle converge in finite time to zero despite the uncertainties present in the system dynamics. The control effort being deployed is shown in Figure 8.3. Figure 8.4 shows the convergence to zero of the state variables in the Z-space. Clearly, the algorithm performs well in the presence of model parameters' uncertainties.

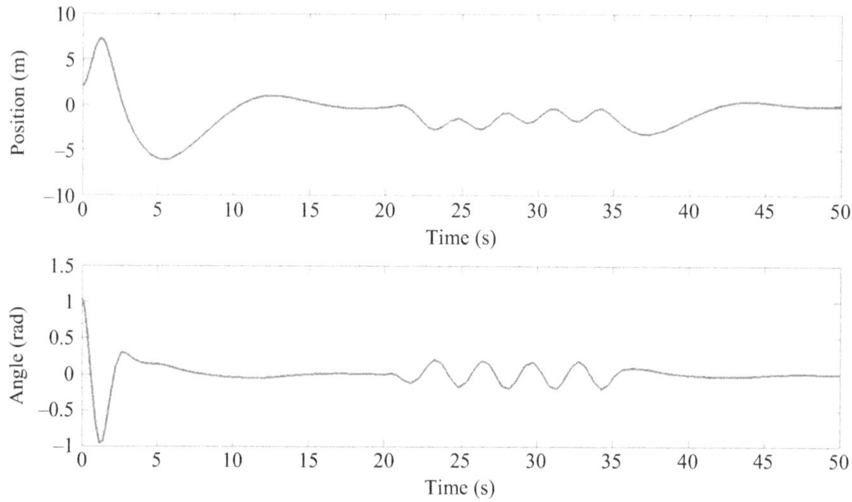

Figure 8.5 Evolution of the trajectories of the cart position and pendulum angle accounting for disturbances

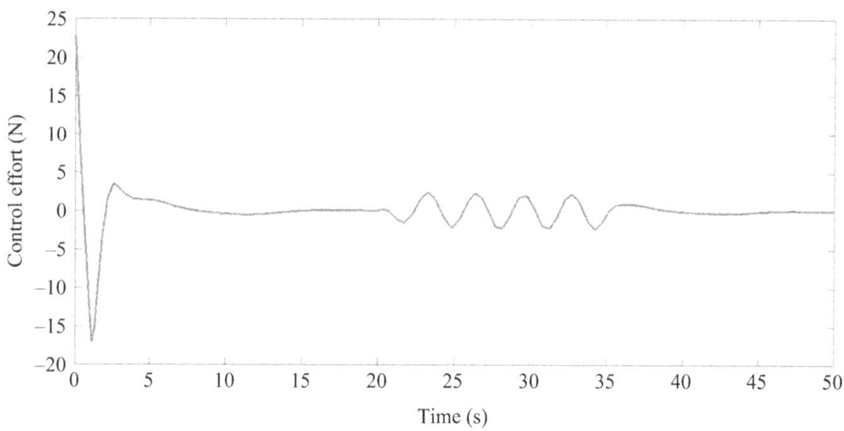

Figure 8.6 Evolution of the control effort accounting for disturbances

2. **Scenario 2:** Uncertain cart–pole system with external disturbances
In this simulation scenario, on top of the uncertainties that the system dynamics contains, a periodic perturbation $d(t) = \sin(2t)$ is added as an external excitation to the cart–pole system during the time interval $t \in [20, 35]$ s in order to test the robustness of the proposed control approach. The obtained simulation results are depicted in Figures 8.5–8.7. From Figure 8.5, it can be noticed that despite the existing uncertainties and the considered external disturbance,

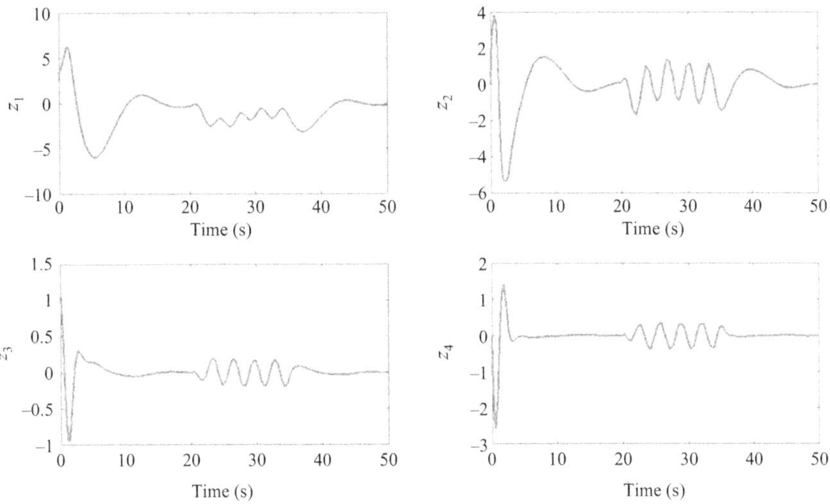

Figure 8.7 Evolution of the trajectories of Z-space state variables accounting for disturbances

the controller is able to steer the system to its desired values. The control effort after the application of the disturbances is shown in Figure 8.6, clearly the amplitude of the control input even with the external perturbation remains with reasonable size. The convergence of the state variables in Z-space is shown in Figure 8.7. From these figures, it can be seen that our control design is robust to tolerate significant variation of the system parameters as well as the external disturbances.

8.7 Conclusion

In this paper, the problem of global finite-time stabilization was addressed for the control of a class of uncertain underactuated mechanical systems. By integrating a fractional power feedback control method with a compensator, the uncertainties in the system can be effectively handled and finite-time stabilization is achieved. Future work is to extend the current design technique to a class of chained form systems perturbed by external disturbances as well as experiments on the inertia wheel inverted pendulum.

References

[1] N. Adhikary and C. Mahanta. 'Integral backstepping sliding mode control for underactuated systems: swing-up and stabilization of the cart–pendulum system', *ISA Transactions*, 2013, **58**, pp. 870–880.

[2] F. Amato, M. Ariola, and P. Dorato. 'Finite-time control of linear systems subject to parametric uncertainties and disturbances', *Automatica*, 2001, **37**, 9, pp. 1459–1463.

[3] S. Andary, A. Chemori, and S. Krut. 'Control of the underactuated inertia wheel inverted pendulum for stable limit cycle generation', *RSJ Advanced Robotics*, 2009, **23**, 15, pp. 1999–2014.

[4] S. Andary and A. Chemori. 'A dual model-free control of non-minimum phase systems for generation of stable limit cycles', In *Proc. 50th IEEE Conf. Decision Contr. (CDC'11)*, Montreal, Canada, 2011, pp. 1387–1392.

[5] S. Andary, A. Chemori, M. Benoit, and J. Sallantin. 'A dual model-free control of underactuated mechanical systems, application to the inertia wheel inverted pendulum', In *American Control Conference (ACC'12)*, Montreal, Canada, 2012, pp. 1029–1034.

[6] M. Bergerman. *Dynamics and Control of Underactuated Manipulators*. PhD Thesis, Carnegie Mellon University, 1996.

[7] S. P. Bhat and D. S. Bernstein. 'Finite-time stability of continuous autonomous systems', *SIAM Journal on Control and Optimization*, 2000, **38**, 3, pp. 751–766.

[8] W. Blajer and K. Kolodziejczyk. 'Modeling of underactuated mechanical systems in partly specified motion', *Journal of Theoretical and Applied Mechanics*, 2008, **46**, 2, pp. 383–394.

[9] S. A. Bortoff and M. W. Spong. 'Pseudolinearization of the acrobot using spline functions', In *Proc. 31st IEEE Conf. Decision Contr.*, Tucson, Arizona, USA, 1992, pp. 593–598.

[10] A. Chemori and M. Alamir. 'Nonlinear predictive control of under-actuated mechanical systems, application: the ECP 505 inverted pendulum', In *MTNS'06 Proc. Mathematical Theory of Networks and Systems*, Leven, Belgium, 2004.

[11] N. Khraief, A. Chemori and S. Belghith. ' External disturbance rejection in IDA-PBC controller for underactuated mechanical systems: from theory to real-time experiments', In *IEEE Multi-Conference on Systems and Control (MSC'14)*, Antibes, France, 2014, pp. 1747–1752.

[12] D. Cheng and W. Lin. 'On normal form of nonlinear systems', *IEEE Transactions on Automatic Control*, 2003, **48**, 11, pp. 1242–1248.

[13] A. Choukchou-Braham, B. Cherki, M. Djemai, and K. Busawon. *Analysis and Control of Underwatcuated Mechanical Systems*. London, UK: Springer, 2014.

[14] I. Fantoni and R. Lozano. *Non-linear Control for Underactuated Mechanical Systems*. New York, USA: Springer-Verlag, 2001.

[15] T. I. Fossen. *Nonlinear Modelling and Control of Underwater Vehicles*. PhD Thesis, Norwegian Institute of Technology, 1991.

[16] L. B. Freidovich, P. La Hera, U. Mettin, A. Robertsson, and A. S. Shiriaev. 'Shaping stable periodic motions of inertia wheel pendulum: theory and experiments', *Asian Journal of Control*, 2009, **11**, 5, pp. 548–556.

[17] L. Freidovicha, A. Robertssona, A. Shiriaeva, and R. Johanssonb. 'Periodic motions of the pendubot via virtual holonomic constraints: theory and experiments', *Automatica*, 2008, **44**, pp. 785–791.

[18] S. Ge and F. Lewis. *Autonomous Mobile Robots*. Boca Raton, FL: CRC Press, Taylor & Francis Group, 2006.

[19] V. T. Haimo. 'Finite-time controllers', *SIAM Journal on Control and Optimization*, 1986, **24**, pp. 760–770.

[20] Y. Hong. 'Finite-time stabilization and stabilizability of a class of controllable systems', *System & Control Letters*, 2002, **46**, pp. 231–236.

[21] Y. Hong, J. Wang, and Z. Xi. 'Stabilization of uncertain chained form systems within finite settling time', *IEEE Transactions on Automatic Control*, 2005, **50**, 9, pp. 1379–1384.

[22] Y. G. Hong and Z. P. Jiang. 'Finite-time stabilization of nonlinear systems with parametric and dynamic uncertainties', *IEEE Transactions on Automatic Control*, 2006, **51**, pp. 1950–1956.

[23] Y. G. Hong, K. Wang, and D. Cheng. 'Adaptive finite-time control of nonlinear systems with parametric uncertainty', *IEEE Transactions on Automatic Control*, 2006, **51**, pp. 858–862.

[24] X. Huang, W. Lin, and B. Yang. 'Global finite-time stabilization of a class of uncertain nonlinear systems', *Automatica*, 2005, **41**, pp. 881–888.

[25] S. Jadlovska and J. Sarnovsky. 'Application of the state-dependent riccati equation method in nonlinear control design for inverted pendulum systems', In *IEEE 11th International Symposium on Intelligent Systems and Informatics (SISY'13)*, pages 209–214, Subotica, Serbia, 2013.

[26] S. Jadlovska and J. Sarnovsky. 'A complex overview of modeling and control of the rotary single inverted pendulum system', *Power Engineering and Electrical Engineering*, 2013, **11**, 2, pp. 73–85.

[27] S. Jadlovska and J. Sarnovsky. 'Modeling and optimal control of nonlinear underactuated mechanical systems a survey', In *13th Scientific Conference of Young Researchers FEI TU of Kosice*, Herlany, Slovakia, 2013, pp. 32–35.

[28] N. Khalid and A. Y. Memon. 'Output feedback control of a class of underactuated nonlinear systems using extended high gain observer', *Arabian Journal for Science and Engineering*, 2016, **41**, 9, pp. 3531–3542.

[29] H. Khalil. *Nonlinear Systems*. Prentice-Hall, Upper Saddle River, Second ed., NJ, USA, 1996.

[30] K. Lee and V. 'Coverstone-Carroll. Control algorithms for stabilizing underactuated robots'. *Journal of Robotic Systems*, 1998, **15**, 12, pp. 681–697.

[31] A. De Luca, R. Mattone, and G. Oriolo. 'Stabilization of underactuated planar 2r manipulator', *International Journal of Robust and Nonlinear Control*, 2000, **10**, 4, pp. 181–198.

[32] Y. Nakamura, T. Suzuki, and M. Koinuma. 'Nonlinear behavior and control of nonholonomic free-joint manipulator', *IEEE Transactions on Robotics and Automation*, 1997, **13**, 6, pp. 853–862.

[33] R. Reza Olfati-Saber. *Nonlinear Control of Underactuated Mechanical Systems with Application to Robotics and Aerospace Vehicles.* PhD Thesis, Massachusetts Institute of Technology, 2001.

[34] M. Park and D. Chwa. 'Swing-up and stabilization control of inverted pendulum systems via coupled sliding-mode control method', *IEEE Transactions on Industrial Electronics*, 2009, **56**, pp. 3541–3555.

[35] G. Poulin, A. Chemori, and N. Marchand. 'Minimum energy oriented global stabilizing control of the PVTOL aircraft', *International Journal of Control*, 2007, **80**, 3, pp. 430–442.

[36] C. Qian and W. Lin. 'A continuous feedback approach to global strong stabilization of nonlinear systems', *IEEE Transactions on Automatic Control*, 2001, **46**, 7, pp. 1061–1079.

[37] D. Qian, J. Yi, and D. Zhao. 'Hierarchical sliding mode control for a class of SIMO under-actuated systems', *Control and Cybernetics*, 2008, **37**, 1, pp. 1–18.

[38] M. T. Ravichandran and A. D. Mahindrakar. 'Robust stabilization of a class of underactuated mechanical systems using time scaling and Lyapunov redesign', *IEEE Transactions on Industrial Electronics*, 2011, **58**, pp. 4299–4313.

[39] M. Reyhanoglu, S. Ludvigsen, and O. Egeland. 'Discontinuous feedback control of a special class of underactuated mechanical systems', *International Journal of Robust and Nonlinear Control*, 2000, **10**, 4, pp. 265–282.

[40] M. Spong. 'Partial feedback linearization of underactuated mechanical systems', In *EEE/RSJ/GI International Conference on Intelligent Robots and Systems*, Munich, Germany, 1994, pp. 314–321.

[41] M. Spong. 'The swing up control problem for the acrobot', *IEEE Control Systems Magazine*, February 1995, **15**, 1, pp. 49–55.

[42] M. Spong. 'Energy based control of a class of underactuated mechanical system', In *13th IFAC World Congress*, San Francisco, USA, 1996, pp. 431–436.

[43] M. Spong. Underactuated mechanical systems. *Control Problems in Robotics and Automation*, B. Siciliano and K. P. Valavanis Eds., LNCIS, Springer Verlag, 230:135–150, 1998.

[44] M. W. Spong and L. Praly. *Control Using Logic-Based Switching*, volume 222 of *Lecture Notes in Control and Information Sciences*, chapter Control of underactuated mechanical systems using switching and saturation, pages 162–172. Springer, Heidelberg, Germany, 2006.

[45] N. Touati and A. Chemori. 'Predictive control for the stabilization of a fast mechatronic system: from simulation to real-time experiments', In, Hangzhou, China, 2013 *International IFAC Symposium on Mechatronic Systems (Mechatronics'13)*, pp. 237–242.

[46] J. Xu, Z. Guo, and T. H. Lee. 'Design and implementation of integral sliding mode control on an underactuated two-wheeled mobile robot', *ISA Transactions*, 2013, **58**, pp. 870–880.

Part III

Nonlinear controllers for mobile inverted pendulum systems

Chapter 9
Advances in robust control of mobile wheeled inverted pendulum

Jian Huang[1] and Toshio Fukuda[2]

Abstract

There has been increasing interest in a type of underactuated mechanical systems, mobile-wheeled inverted-pendulum (MWIP) models, which are widely used in autonomous robotics and intelligent vehicles. To cope with the model uncertainties and external disturbances, several robust controllers are designed for the MWIP models. For the velocity-tracking problem of the MWIP systems, we proposed two sliding-mode-control (SMC) methods. There is still a steady tracking error when the first SMC method is used. By assuming a novel sliding surface, the second SMC method is designed to solve this problem. Using a coordinate transformation, the non-"Class-I" type underactuated MWIP system is presented as a semistrict feedback form which is convenient for controller design. A dynamic surface controller with a nonlinear disturbance observer (DSCNDO) is then designed to solve the balance control problem of the MWIP systems. The proposed DSCNDO can compensate the external disturbances and the model uncertainties to improve the system performance significantly. The stabilities of the closed-loop MWIP systems using the proposed methods are proved by Lyapunov theorem. The effectiveness of all the methods is verified by numerical simulations.

9.1 Introduction

In the past decades, many efforts have been devoted to controlling problems of mobile wheeled inverted pendulum (MWIP) models, which are widely applied in the field of autonomous robotics and intelligent vehicles [1–10]. The MWIP is a typical underactuated system, which is nonlinear and inherently unstable. This makes the control of MWIP to be a very challengeable problem. On the other hand, there are a lot of

[1]School of Automation, Huazhong University of Science and Technology, China
[2]Faculty of Science and Engineering, Meijo University, Nagoya, Japan

practical systems implemented based on MWIP models, including the JOE [5], the Nbot [6], the Legway [7], the B2 [8], the Segway [9], the UW-Car [11] and so on. These systems are characterized by the ability to balance on its two wheels and spin on the spot. This additional maneuverability allows easy navigation on various terrains, turning sharp corners and traversing small steps or curbs. In addition, the compact structure design allows drivers to access most places that can only be accessed by walkers in the past. Moreover, people can travel in short distances in a small area by such kind of vehicles instead of by cars or by buggies that are more polluted.

Due to the nonlinear, underactuated and inherent unstable dynamics, the MWIP models have attracted much attention in the field of control theory and engineering. It should be noted that there are a lot of control methods have been applied in benchmark underactuated systems including the inverted pendulum, the acrobot and the rotating pendulum [12–15], whereas most of these methods are based on exact mathematical models. Therefore, they are lack of robustness to model errors and external disturbances. In the practical applications, there are unavoidable uncertainties caused by both internal and external effects. Therefore, the robust control of MWIP has to be dealt with.

It is not easy to design robust controllers for nonlinear underactuated systems, whose control problems have been proved to be challenging because the techniques of fully actuated systems cannot be used directly. With the great efforts of researchers, several robust control methods have been proposed for the underactuated system in the recent years. The first common-used robust control strategy is the sliding mode control (SMC) technique. Sankaranarayanan [16] proposed a SMC algorithm to robustly stabilize a class of underactuated mechanical systems that are not linearly controllable and violate Brockett's necessary condition for smooth asymptotic stabilization of the equilibrium, with parametric uncertainties. Park *et al.* [17] proposed an adaptive neural SMC method for trajectory tracking control of nonholonomic wheeled mobile robots with model uncertainties and external disturbances. The second robust control strategy discussed in this chapter is the dynamic surface control (DSC). In order to avoid the "explosion of terms" problem of Integrator Backstepping controller and the "chattering problem" of SMC controller, the DSC method was developed by Swaroop *et al.* [18]. Recently, the DSC technique has been applied into the control of different underactuated mechanical systems, including the underactuated marine vessels [19], the inertia wheel pendulum, the translational oscillator with rotational actuator, the acrobot [20] and so on. For the MWIP model, we investigated both the SMC and DSC strategies in our previous studies [11,21,22].

To reduce the effect from uncertainties to the MWIP system, the nonlinear disturbance observer (NDO) might be a candidate solution. It is found that using a disturbance observer can further improve the robustness of controller. A NDO was proposed by Mohammadi to manage the disturbance of nonlinear system, which is applied for a four-degree of freedom selective compliance assembly robot arm (SCARA) manipulator [23]. Chen proposed a NDO to deal with the disturbance of nonlinear system [24], which is applied to tracking control of pneumatic artificial muscle actuator by using DSC control method [25]. For an uncertain structural systems, Wei *et al.* proposed a new type of composite control scheme of disturbance observer-based control and terminal SMC [26]. Yang *et al.* [27] for systems with mismatched uncertainties proposed a SMC approach by using a novel sliding surface based on a disturbance observer.

In this chapter, we first introduce the mathematical model of MWIP using Lagrange motion equations in Section 9.2. The velocity SMC controller of the MWIP with modeling uncertainties and external is designed in Section 9.3. In Section 9.4, a new dynamic surface controller with nonlinear disturbance observer (DSCNDO) is developed by combining conventional DSC and NDO techniques for the balancing control of MWIP system. For all these controllers, the stabilities of closed-loop MWIP system are proved. Finally, numerical simulations are presented to verify the effectiveness of the proposed approaches.

9.2 Modeling of mobile wheeled inverted pendulum

9.2.1 Derivation of dynamic model

The MWIP system is modeled as a one dimensional inverse pendulum which rotates about the wheels' axles. Hence, the body's motion on a plane is determined by the inclination and translational motion. Figure 9.1 shows the structure of an MWIP system, where θ_w and θ_b are the wheel's rotation angle and the inclination angle of the body, respectively. The coordinates of the wheel and body are denoted by (x_w, y_w) and (x_b, y_b), respectively. We suppose that the system moves on a slope with constant but uncertain angle α, which is hard to be measured by sensors. To describe the parameters of the MWIP system, some notations should be clarified first (see also Figure 9.1):

1. m_b, m_w—masses of the body and the wheel.
2. I_b, I_w—moments of inertia of the body and the wheel.
3. l—length between the wheel axle and the center of gravity of the body.

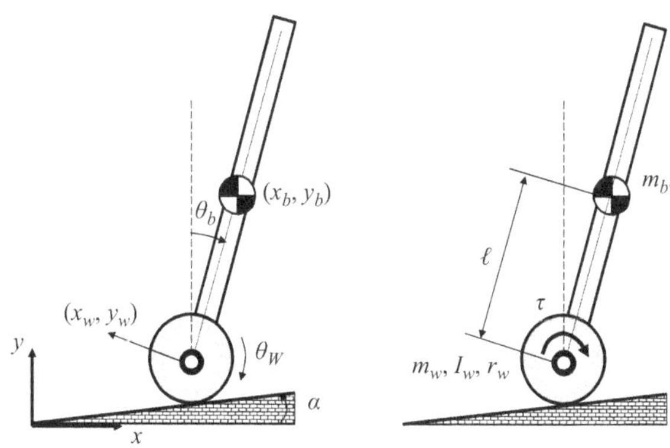

Figure 9.1 The mobile wheeled inverted pendulum system model

4. r_w—radius of the wheel.
5. D_b—viscous resistance in the driving system.
6. D_w—viscous resistance of the ground.

The positions and velocities of the wheel and the body can be given by

$$\begin{cases} x_b = l\sin\theta_b + r_w\theta_w\cos\alpha \\ y_b = l\cos\theta_b + r_w\theta_w\sin\alpha \end{cases}, \quad \begin{cases} x_w = r_w\theta_w\cos\alpha \\ y_w = r_w\theta_w\sin\alpha \end{cases} \tag{9.1}$$

$$\begin{cases} \dot{x}_b = l\dot{\theta}_b\cos\theta_b + r_w\dot{\theta}_w\cos\alpha \\ \dot{y}_b = -l\dot{\theta}_b\sin\theta_b + r_w\dot{\theta}_w\sin\alpha \end{cases}, \quad \begin{cases} \dot{x}_w = r_w\dot{\theta}_w\cos\alpha \\ \dot{y}_w = r_w\dot{\theta}_w\sin\alpha \end{cases} \tag{9.2}$$

The Lagrangian equation of motion is used for the derivation of the dynamic equation. The kinetic, potential and dissipated energy and their contributions to the dynamic equation are computed as follows.

The kinetic energies of the body and the wheel can then be computed as

$$V_b = \frac{1}{2}m_b\left(\dot{x}_b^2 + \dot{y}_b^2\right) + \frac{1}{2}I_b\dot{\theta}_b^2 = \frac{1}{2}m_b\left(r_w^2\dot{\theta}_w^2 + 2r_w l\dot{\theta}_b\dot{\theta}_w\cos(\theta_b + \alpha) + l^2\dot{\theta}_b^2\right) + \frac{1}{2}I_b\dot{\theta}_b^2 \tag{9.3}$$

$$V_w = \frac{1}{2}m_w\left(\dot{x}_w^2 + \dot{y}_w^2\right) + \frac{1}{2}I_w\dot{\theta}_w^2 = \frac{1}{2}m_w r_w^2\dot{\theta}_w^2 + \frac{1}{2}I_w\dot{\theta}_w^2. \tag{9.4}$$

The gravitational potential energy of the body and the wheel can be computed as

$$U_b = m_b g(l\cos\theta_b + r_w\theta_w\sin\alpha) \tag{9.5}$$

$$U_w = m_w g r_w\theta_w\sin\alpha \tag{9.6}$$

Therefore, the Lagrange function L is given by

$$L = V_b + V_w - U_b - U_w \tag{9.7}$$

According to Lagrange's motion equation, the dynamic model of an MWIP system can be represented by

$$\begin{aligned} \frac{d}{dt}\left(\frac{\partial L}{\partial\dot{\theta}_b}\right) - \frac{\partial L}{\partial\theta_b} &= -\tau - D_b\left(\dot{\theta}_b - \dot{\theta}_w\right) \\ \frac{d}{dt}\left(\frac{\partial L}{\partial\dot{\theta}_w}\right) - \frac{\partial L}{\partial\theta_w} &= \tau + D_b\left(\dot{\theta}_b - \dot{\theta}_w\right) - D_w\dot{\theta}_w \end{aligned} \tag{9.8}$$

It follows from (9.8) that the final nonlinear model can be

$$\begin{cases} m_{11}\ddot{\theta}_w + (m_{12}\cos(\theta_b + \alpha))\ddot{\theta}_b = \tau - D_w\dot{\theta}_w + D_b\left(\dot{\theta}_b - \dot{\theta}_w\right) \\ \quad + m_{12}\dot{\theta}_b^2\sin(\theta_b + \alpha) - \tilde{M} \\ (m_{12}\cos(\theta_b + \alpha))\ddot{\theta}_w + m_{22}\ddot{\theta}_b = -\tau - D_b\left(\dot{\theta}_b - \dot{\theta}_w\right) + G_b\sin\theta_b \end{cases} \tag{9.9}$$

where parameters m_{11}, m_{12}, m_{22} and G_b satisfy

$$\begin{cases} m_{11} = (m_b + m_w)r_w^2 + I_w \\ m_{12} = m_b l r_w \\ m_{22} = m_b l^2 + I_b \\ G_b = m_b g l \\ \tilde{M} = (m_b + m_w) r_w g \sin\alpha. \end{cases} \tag{9.10}$$

9.2.2 Analysis of equilibriums

By choosing the state variable as $\mathbf{x} = \begin{bmatrix} x_1 & x_2 & x_3 \end{bmatrix}^{\mathrm{T}} = \begin{bmatrix} \theta_b & \dot{\theta}_w & \dot{\theta}_b \end{bmatrix}^{\mathrm{T}}$, the state model of an MWIP system can be represented by

$$\dot{x}_1 = x_3 \tag{9.11a}$$

$$\begin{aligned} m_{11}\dot{x}_2 + (m_{12}\cos(x_1+\alpha))\,\dot{x}_3 = \tau - D_w x_2 + D_b(x_3 - x_2) \\ + m_{12}{x_3}^2 \sin(x_1+\alpha) - \tilde{M} \end{aligned} \tag{9.11b}$$

$$(m_{12}\cos(x_1+\alpha))\dot{x}_2 + m_{22}\dot{x}_3 = -\tau - D_b(x_3 - x_2) + G_b \sin x_1 \tag{9.11c}$$

Then we can get the vector form of this dynamic model:

$$\mathbf{M(x)\dot{x}} = \mathbf{F(x)} + \mathbf{u} \tag{9.12}$$

where

$$\mathbf{M(x)} = \begin{bmatrix} 1 & 0 & 0 \\ 0 & m_{11} & m_{12}\cos(x_1+\alpha) \\ 0 & m_{12}\cos(x_1+\alpha) & m_{22} \end{bmatrix} \tag{9.13}$$

$$\mathbf{F(x)} = \begin{bmatrix} x_3 \\ -D_w x_2 + D_b(x_3 - x_2) + m_{12}{x_3}^2 \sin(x_1+\alpha) - \tilde{M} \\ -D_b(x_3 - x_2) + G_b \sin x_1 \end{bmatrix} \tag{9.14}$$

$$\mathbf{u} = \begin{bmatrix} 0 & \tau & -\tau \end{bmatrix}^{\mathrm{T}}. \tag{9.15}$$

Suppose $\mathbf{x}^* = \begin{bmatrix} {x_1}^* & {x_2}^* & {x_3}^* \end{bmatrix}^{\mathrm{T}}$ is the desired equilibrium of system (9.12), satisfying

$${x_3}^* = 0, \quad G_b \sin {x_1}^* = D_w {x_2}^* + \tilde{M} \tag{9.16}$$

In the case of velocity control problem, the desired velocity ${\theta_w}^* = {x_2}^*$ is always given in advance. The desired inclination angle of the body, ${x_1}^*$, is then determined by (9.16), whereas different body parameters and unknown slope angle result in different inclination angle ${x_1}^*$, which makes it impossible to compute the exact equilibrium before the controller design. Normally, it is difficult to design a controller if we do not know the exact equilibrium in advance. Especially in the case of using a linearized model, we even cannot obtain such model because it is generated by linearizing the nonlinear system around the equilibrium. This problem can be partially solved by estimating the equilibrium using nominal system parameters without considering any uncertainties, whereas the controller based on the estimated equilibrium cannot guarantee that the system finally moves at the desired velocity, even if it is asymptotic stable. This will be further discussed in the rest of this chapter.

In the rest of this chapter, "$\hat{a}$" denotes that the term a is evaluated based on corresponding parameter of the nominal system moving on a flat ground without any uncertainties and disturbances.

9.3 Sliding mode controller design for mobile wheeled inverted pendulum

9.3.1 Type-I sliding mode controller for MWIP

Two types of SMC controllers, Type-I and Type-II, will be discussed for the MWIP system. We first introduce the design procedure of Type-I SMC controller for the MWIP system. This controller design is based on full state feedback. The sliding surface is defined as a weighted combination of position tracking error and velocity tracking error, which is given by

$$s = \lambda_1 \tilde{x}_1 + \lambda_2 \tilde{x}_2 + \tilde{x}_3 \tag{9.17}$$

where $\tilde{x}_i = x_i - \hat{x}_i^*$ are the tracking errors with $i = 1, 2, 3$ and coefficients λ_1, λ_2 are constants to be determined. As pointed out in the previous section, exact system equilibriums cannot be obtained in advance. Hence we use nominal system parameters to calculate the estimation $\hat{x}_i^*$ of the actual desired $x_i{}^*$. Thus the estimation $\hat{x}_i^*$ is given by

$$\hat{x}_3^* = 0, \quad \hat{x}_2^* = x_2{}^*, \quad \hat{G}_b \sin \hat{x}_1^* = D_w \hat{x}_2^* \tag{9.18}$$

The state trajectories can be driven to the sliding surface by choosing the control law τ such that

$$s\dot{s} \leq -\eta \, |s| \tag{9.19}$$

where η is a positive constant.

From (9.11), it follows that

$$M_1(\mathbf{x})\dot{x}_2 + M_2(\mathbf{x})\dot{x}_3 = -D_w x_2 + G_b \sin x_1 + m_{12} x_3{}^2 \sin(x_1 + \alpha) - \tilde{M} \tag{9.20}$$

where

$$M_1(\mathbf{x}) = m_{11} + m_{12}\cos(x_1 + \alpha), \quad M_2(\mathbf{x}) = m_{22} + m_{12}\cos(x_1 + \alpha) \tag{9.21}$$

From (9.20), we can rewrite (9.11) as

$$\begin{cases} \dot{x}_1 = x_3 \\ \bar{M}(\mathbf{x})\dot{x}_2 + b(\mathbf{x}) = M_2(\mathbf{x})\tau \\ \bar{M}(\mathbf{x})\dot{x}_3 + a(\mathbf{x}) = -M_1(\mathbf{x})\tau \end{cases} \tag{9.22}$$

where

$$\bar{M}(\mathbf{x}) = m_{11}m_{22} - (m_{12}\cos(x_1 + \alpha))^2, \tag{9.23}$$

$$\begin{aligned} a(\mathbf{x}) &= -[m_{12}(D_b + D_w)\cos(x_1 + \alpha) + m_{11}D_b]x_2 \\ &\quad + (m_{11} + m_{12}\cos(x_1 + \alpha))D_b x_3 + (m_{12}x_3)^2\cos(x_1 + \alpha)\sin(x_1 + \alpha) \\ &\quad - m_{11}G_b \sin x_1 - m_{12}\cos(x_1 + \alpha)\tilde{M}, \end{aligned} \tag{9.24}$$

$$\begin{aligned} b(\mathbf{x}) &= [m_{22}(D_b + D_w) + m_{12}D_b\cos(x_1 + \alpha)]x_2 \\ &\quad - (m_{22} + m_{12}\cos(x_1 + \alpha))D_b x_3 - m_{12}m_{22}{x_3}^2\sin(x_1 + \alpha) \\ &\quad + m_{12}G_b \sin x_1 \cos(x_1 + \alpha) + m_{22}\tilde{M}. \end{aligned} \tag{9.25}$$

Note that variable $\bar{M}$ is positive because it is the determinant of the positive definite matrix of the kinetic energy [28].

Define

$$f_1(\mathbf{x}) = \frac{a(\mathbf{x}) + \lambda_2 b(\mathbf{x})}{\bar{M}(\mathbf{x})}, \quad \hat{f}_1(\mathbf{x}) = \frac{\hat{a}(\mathbf{x}) + \lambda_2 \hat{b}(\mathbf{x})}{\hat{\bar{M}}(\mathbf{x})} \tag{9.26}$$

and the estimation error caused by modeling uncertainty and disturbances is assumed to be bounded by some known function $F_1(x)$,

$$\left| f_1(\mathbf{x}) - \hat{f}_1(\mathbf{x}) \right| \le F_1 \tag{9.27}$$

Further, we assume that

$$\frac{\lambda_2 M_2(\mathbf{x}) - M_1(\mathbf{x})}{\bar{M}(\mathbf{x})} = (1 + \Delta)\frac{\lambda_2 \hat{M}_2(\mathbf{x}) - \hat{M}_1(\mathbf{x})}{\hat{\bar{M}}(\mathbf{x})}, \qquad |\Delta| \le \beta_1 < 1 \tag{9.28}$$

Note that the assumptions (9.27) and (9.28) are common-used in the SMC, which can be found in [29].

Theorem 9.1. *The achievement of a sliding motion on the surface (9.17) can be guaranteed by selecting the control law*

$$\begin{aligned} \tau &= \frac{1}{\lambda_2 \hat{M}_2(\mathbf{x}) - \hat{M}_1(\mathbf{x})}\left(\hat{a}(\mathbf{x}) + \lambda_2 \hat{b}(\mathbf{x})\right) - \frac{\lambda_1 \hat{\bar{M}}(\mathbf{x})}{\lambda_2 \hat{M}_2(\mathbf{x}) - \hat{M}_1(\mathbf{x})}x_3 \\ &\quad + \frac{\hat{\bar{M}}(\mathbf{x})}{\lambda_2 \hat{M}_2(\mathbf{x}) - \hat{M}_1(\mathbf{x})}k(\mathbf{x})\mathrm{sgn}(s) \end{aligned} \tag{9.29}$$

where

$$k(\mathbf{x}) \le -\frac{F_1(\mathbf{x}) + \beta_1 |r(\mathbf{x})| + \eta}{1 - \beta_1}, r(\mathbf{x}) = \hat{f}_1(\mathbf{x}) - \lambda_1 \dot{x}_1 \tag{9.30}$$

Proof. Using the equivalent control method, the equivalent control law $\hat{\tau}$ can be easily obtained as

$$\hat{\tau} = \frac{1}{\lambda_2 \hat{M}_2(\mathbf{x}) - \hat{M}_1(\mathbf{x})}\left(\hat{a}(\mathbf{x}) + \lambda_2 \cdot \hat{b}(\mathbf{x})\right) - \frac{\lambda_1 \hat{\bar{M}}(\mathbf{x})}{\lambda_2 \hat{M}_2(\mathbf{x}) - \hat{M}_1(\mathbf{x})}x_3 \tag{9.31}$$

Choose the sliding control law τ as $\tau = \hat{\tau} + \Delta\tau$, the rest work is to find appropriate switching control law $\Delta\tau$.

From (9.17), (9.22) and (9.31), the derivative of s is given by

$$
\begin{aligned}
\dot{s} &= \lambda_1\dot{x}_1 + \lambda_2\dot{x}_2 + \dot{x}_3 \\
&= \lambda_1\dot{x}_1 + \lambda_2\left(-\frac{b(\mathbf{x})}{\bar{M}(\mathbf{x})} + \frac{M_2(\mathbf{x})(\hat{\tau}+\Delta\tau)}{\bar{M}(\mathbf{x})}\right) + \left(-\frac{a(\mathbf{x})}{\bar{M}(\mathbf{x})} - \frac{M_1(\mathbf{x})(\hat{\tau}+\Delta\tau)}{\bar{M}(\mathbf{x})}\right) \\
&= \lambda_1\dot{x}_1 + \frac{1}{\bar{M}(\mathbf{x})}\left[-a(\mathbf{x}) - \lambda_2 b(\mathbf{x}) + \lambda_2 M_2(\mathbf{x})\hat{\tau} - M_1(\mathbf{x})\hat{\tau}\right] \\
&\quad + \left(\frac{\lambda_2 M_2(\mathbf{x}) - M_1(\mathbf{x})}{\bar{M}(\mathbf{x})}\right)\Delta\tau = \frac{1}{\bar{M}(\mathbf{x})}\left\{\frac{\lambda_2 M_2(\mathbf{x}) - M_1(\mathbf{x})}{\lambda_2\hat{M}_2(\mathbf{x}) - \hat{M}_1(\mathbf{x})}\left[\hat{a}(\mathbf{x}) + \lambda_2\hat{b}(\mathbf{x})\right]\right. \\
&\quad \left. - \left[a(\mathbf{x}) + \lambda_2 b(\mathbf{x})\right]\right\} - \left(\frac{\hat{\bar{M}}(\mathbf{x})}{\bar{M}(\mathbf{x})}\cdot\frac{\lambda_2 M_2(\mathbf{x}) - M_1(\mathbf{x})}{\lambda_2\hat{M}_2(\mathbf{x}) - \hat{M}_1(\mathbf{x})} - 1\right)\lambda_1\dot{x}_1 \\
&\quad + \left(\frac{\lambda_2 M_2(\mathbf{x}) - M_1(\mathbf{x})}{\bar{M}(\mathbf{x})}\right)\Delta\tau = \left(\hat{f}_1(\mathbf{x}) - f_1(\mathbf{x})\right) \\
&\quad + \Delta\left(\hat{f}_1(\mathbf{x}) - \lambda_1\dot{x}_1\right) + \left(\frac{\lambda_2 M_2(\mathbf{x}) - M_1(\mathbf{x})}{\bar{M}(\mathbf{x})}\right)\Delta\tau
\end{aligned}
\tag{9.32}
$$

Choose

$$
\Delta\tau = \frac{\hat{\bar{M}}(\mathbf{x})}{\lambda_2\hat{M}_2(\mathbf{x}) - \hat{M}_1(\mathbf{x})}k(\mathbf{x})\mathrm{sgn}(s) \tag{9.33}
$$

where $k = k(x)$ is a state-dependent function and $sgn(\cdot)$ is the signum function. It follows that

$$
\begin{aligned}
s\dot{s} &= s\left[\left(\hat{f}_1(\mathbf{x}) - f_1(\mathbf{x})\right) + r(\mathbf{x})\Delta + (\Delta+1)\,k(\mathbf{x})\mathrm{sgn}(s)\right] \\
&\le |s|\,(F_1(\mathbf{x}) + \beta_1\,|r(\mathbf{x})| + (1-\beta_1)\,k(\mathbf{x})).
\end{aligned}
\tag{9.34}
$$

If k has the bound defined by (9.30), then the sliding condition (9.19) is satisfied. □

The controller (9.29) ensures that all system trajectories will reach the surface and remain there in the presence of model uncertainties and disturbances. The next step is to find appropriate coefficients λ_1, λ_2 that guarantee the trajectories are asymptotically stable during the sliding phase. On the sliding surface, the closed-loop system dynamics is represented by the combination of the sliding surface in the form $s = 0$ and (9.20):

$$\dot{x}_1 = x_3 \tag{9.35a}$$

$$M_1(\mathbf{x})\dot{x}_2 + M_2(\mathbf{x})\dot{x}_3 = -D_w x_2 + G_b \sin x_1 + m_{12}{x_3}^2\sin(x_1+\alpha) - \tilde{M} \tag{9.35b}$$

$$\lambda_1\left(x_1 - \hat{x}_1^*\right) + \lambda_2\left(x_2 - \hat{x}_2^*\right) + x_3 = 0 \tag{9.35c}$$

To avoid the confusion of notations, we use $\bar{\mathbf{x}}^* = \left[\,\bar{x}_1^*\ \bar{x}_2^*\ \bar{x}_3^*\,\right]^{\mathrm{T}}$ to denote the real equilibrium of system (9.35), which is given by

$$
\bar{x}_3^* = 0, \quad \lambda_1\left(\bar{x}_1^* - \hat{x}_1^*\right) + \lambda_2\left(\bar{x}_2^* - \hat{x}_2^*\right) = 0, \quad G_b\sin\bar{x}_1^* = D_w\bar{x}_2^* + \tilde{M}. \tag{9.36}
$$

If the closed-loop system is asymptotically stable, it will finally converge to $\bar{\mathbf{x}}^*$. It should be noted that usually the values of three kinds of equilibriums $\mathbf{x}^*$, $\hat{\mathbf{x}}^*$ and

$\bar{\mathbf{x}}^*$ are different because they are determined by different (9.16), (9.18) and (9.36), respectively.

Theorem 9.2. *For the MWIP system (9.11), design sliding surface (9.17) and control law (9.29). Assume inequalities* $M_1(\bar{\mathbf{x}}^*), M_2(\bar{\mathbf{x}}^*) > 0$ *and* $\cos(\bar{x}_1^*) > 0$ *hold, if coefficients* λ_1, λ_2 *satisfy*

$$\lambda_1,\ \lambda_2 > 0, \quad \lambda_2 M_2(\bar{\mathbf{x}}^*) - M_1(\bar{\mathbf{x}}^*) < 0 \tag{9.37}$$

then the closed-loop system is locally asymptotically stable around equilibrium $\bar{\mathbf{x}}^*$.

Proof. From (9.35a) and (9.35c), we have

$$\begin{cases} x_2 = \left[-\lambda_1 x_1 - \dot{x}_1 + \lambda_1 \hat{x}_1^* + \lambda_2 \hat{x}_2^*\right]/\lambda_2 \\ \dot{x}_2 = (-\lambda_1 \dot{x}_1 - \ddot{x}_1)/\lambda_2 \end{cases} \tag{9.38}$$

By substituting (9.38) into (9.35b) and introducing a new state vector

$$\mathbf{y} = \left[\, y_1 \ y_2 \,\right]^{\mathrm{T}} = \left[\, x_1 \ \dot{x}_1 \,\right]^{\mathrm{T}} \tag{9.39}$$

a new system is derived as

$$\dot{\mathbf{y}} = \Phi(\mathbf{y}) \tag{9.40}$$

where

$$\Phi(\mathbf{y}) = \begin{bmatrix} \Phi_1(\mathbf{y}) \\ \Phi_2(\mathbf{y}) \end{bmatrix}, \quad \Phi_1(\mathbf{y}) = y_2, \tag{9.41}$$

$$\begin{aligned} \Phi_2(\mathbf{y}) = \frac{1}{\lambda_2 M_2 - M_1} \Big[& (\lambda_1 M_1 + D_w) y_2 + \lambda_2 M_{12} y_2{}^2 \sin(y_1 + \alpha) \\ & + \lambda_1 D_w y_1 + \lambda_2 G_b \sin y_1 - D_w\left(\lambda_1 \hat{x}_1^* + \lambda_2 \hat{x}_2^*\right) - \lambda_2 \tilde{M} \Big] \end{aligned} \tag{9.42}$$

It is apparent that system (9.35) and (9.40) are equivalent in the sense of stability. The equilibrium of (9.40) is denoted by $\mathbf{y}^* = \left[\, y_1{}^* \ y_2{}^* \,\right]^{\mathrm{T}} = \left[\, \bar{x}_1^* \ \bar{x}_3^* \,\right]^{\mathrm{T}}$. As Ashrafiuon did in [30], we linearize (9.40) about the equilibrium $\mathbf{y}^*$ to establish linear stability criteria that guarantee local exponential stability of the nonlinear system. The linearized system is described by

$$\dot{\mathbf{y}} = \mathbf{A}\mathbf{y} \tag{9.43}$$

where

$$\mathbf{A} = \left.\frac{\partial \Phi}{\partial \mathbf{y}}\right|_{\mathbf{y}^*} = \begin{bmatrix} 0 & 1 \\ \Delta_1 & \Delta_2 \end{bmatrix}, \quad \Delta_1 = \left.\frac{\partial \Phi_2}{\partial y_1}\right|_{\mathbf{y}^*}, \quad \Delta_2 = \left.\frac{\partial \Phi_2}{\partial y_2}\right|_{\mathbf{y}^*} \tag{9.44}$$

The characteristic polynomial for the eigenvalues of matrix $\mathbf{A}$ is represented in Laplace domain as

$$|S\mathbf{I} - \mathbf{A}| = S^2 - \Delta_2 S - \Delta_1 = 0 \tag{9.45}$$

where S is the Laplace variable. According to the Hurwitz stability criteria, the equilibrium of linearized system (9.43) is asymptotically stable if we have

$$\Delta_1, \Delta_2 < 0 \tag{9.46}$$

After computing the two derivatives in (9.44), we have

$$\Delta_1 = \frac{\lambda_1 D_w + \lambda_2 G_b \cos y_1{}^*}{\lambda_2 M_2(\mathbf{y}^*) - M_1(\mathbf{y}^*)}, \qquad \Delta_2 = \frac{\lambda_1 M_1(\mathbf{y}^*) + D_w}{\lambda_2 M_2(\mathbf{y}^*) - M_1(\mathbf{y}^*)} \tag{9.47}$$

If we choose λ_1, λ_2 that satisfy (9.37), inequality (9.46) is then satisfied. □

Because the MWIP system is required to move with a nearly upright body, the values of $M_1(\bar{\mathbf{x}}^*)$ and $M_2(\bar{\mathbf{x}}^*)$ are usually positive if the slope angle is small enough. The value of $\cos(\bar{x}_1^*)$ is positive for the same reason. Hence, the assumptions of Theorem 9.2 are naturally satisfied. It is found that if the coefficients λ_1 and λ_2 satisfy $\lambda_1 > 0, 0 < \lambda_2 < \min(1, m_{11}/m_{22})$, then the assumptions (9.37) will hold.

Note that the real equilibrium $\bar{\mathbf{x}}^*$ is the intersection point of curve $x_2 = g(x_1) = (G_b/D_w)\sin(x_1) - \tilde{M}/D_w$ and a straight line crossing equilibrium $\hat{\mathbf{x}}^*$ with a slope $-\lambda_1/\lambda_2$ in the $x_1 - x_2$ coordinate frame. Although the steady tracking error $|\bar{x}_2^* - x_2^*|$ cannot be eliminated, it can be reduced to some extent by decreasing $|\lambda_1/\lambda_2|$.

9.3.2 *Type-II sliding mode controller for MWIP*

When using the Type-I SMC controller, the real equilibrium $\bar{\mathbf{x}}^*$ is usually different from the desired equilibrium $\mathbf{x}^*$, even if the closed-loop system is asymptotically stable. This leads to a steady tracking error in the case of velocity control for MWIP systems. To ensure an MWIP system can move at a desired velocity $\dot{\theta}_w^*$, a novel sliding surface and the corresponding SMC controller design scheme is proposed in this section.

We start by introducing a new state vector

$$\mathbf{z} = \begin{bmatrix} z_1 & z_2 & z_3 & z_4 \end{bmatrix}^{\mathrm{T}} = \begin{bmatrix} \theta_1 & \dot{\theta}_1 & \theta_w - \dot{\theta}_w^* t & \dot{\theta}_w - \dot{\theta}_w^* \end{bmatrix}^{\mathrm{T}} \tag{9.48}$$

The new sliding surface is given by

$$s = \mu_1 z_1 + z_2 + \mu_2 z_3 + \mu_3 z_4 \tag{9.49}$$

where coefficients μ_1, μ_2, μ_3 are constants to be determined.

Define

$$f_2(\mathbf{x}) = \frac{a(\mathbf{x}) + \mu_3 b(\mathbf{x})}{\bar{M}}, \quad \hat{f}_2(\mathbf{x}) = \frac{\hat{a}(\mathbf{x}) + \mu_3 \hat{b}(\mathbf{x})}{\hat{\bar{M}}} \tag{9.50}$$

and the estimation error caused by modeling uncertainty and disturbances is assumed to be bounded by some known function $F_2(\mathbf{x})$,

$$\left| f_2(\mathbf{x}) - \hat{f}_2(\mathbf{x}) \right| \le F_2(\mathbf{x}). \tag{9.51}$$

Further, we assume that

$$\frac{\mu_3 M_2(\mathbf{x}) - M_1(\mathbf{x})}{\bar{M}(\mathbf{x})} = (1 + \tilde{\Delta}) \frac{\mu_3 \hat{M}_2(\mathbf{x}) - \hat{M}_1(\mathbf{x})}{\hat{\bar{M}}(\mathbf{x})}, \qquad |\tilde{\Delta}| \le \beta_2 < 1 \tag{9.52}$$

The Type-II SMC controller is proposed by the following theorem.

Theorem 9.3. *The achievement of a sliding motion on the surface (9.49) can be guaranteed by selecting the control law*

$$\tau = \frac{1}{\mu_3\hat{M}_2(\mathbf{x}) - \hat{M}_1(\mathbf{x})}\left(\hat{a}(\mathbf{x}) + \mu_3\hat{b}(\mathbf{x})\right) - \frac{\mu_1\hat{\tilde{M}}(\mathbf{x})}{\mu_3\hat{M}_2(\mathbf{x}) - \hat{M}_1(\mathbf{x})}z_2$$
$$- \frac{\mu_2\hat{\tilde{M}}(\mathbf{x})}{\mu_3\hat{M}_2(\mathbf{x}) - \hat{M}_1(\mathbf{x})}z_4 + \frac{\hat{\tilde{M}}(\mathbf{x})}{\mu_3\hat{M}_2(\mathbf{x}) - \hat{M}_1(\mathbf{x})}\tilde{k}(\mathbf{x})\mathrm{sgn}(s) \tag{9.53}$$

where $\hat{a}(\mathbf{x})$ *and* $\hat{b}(\mathbf{x})$ *are the same as described in Section 9.2.1, and* $\tilde{k}$, $\tilde{r}$ *satisfy*

$$\tilde{k}(\mathbf{x}) \le -\frac{F_2 + \beta_2\,|\tilde{r}(\mathbf{x})| + \eta}{1 - \beta_2}, \quad \tilde{r}(\mathbf{x}) = \hat{f}_2(\mathbf{x}) - \mu_1 z_2 - \mu_2 z_4 \tag{9.54}$$

Proof. This theorem can be easily proven using the similar method in the proof of Theorem 9.1. □

A similar approach as used in Section 9.2.1 is assumed to analyze the stability of the sliding surface (9.49). Choosing a new state vector $\mathbf{q} = \begin{bmatrix} q_1 & q_2 & q_3 \end{bmatrix}^{\mathrm{T}} = \begin{bmatrix} z_1 & z_2 & z_4 \end{bmatrix}^{\mathrm{T}}$, the closed-loop system dynamics on sliding surface (9.49) is represented by

$$\dot{q}_1 = q_2, \tag{9.55a}$$

$$M_1\dot{q}_3 + M_2\dot{q}_2 = -D_w\left(q_3 + \dot{\theta}_w^*\right) + G_b\sin q_1 + m_{12}{q_2}^2\sin(q_1 + \alpha) - \tilde{M}, \tag{9.55b}$$

$$\mu_1 q_2 + \dot{q}_2 + \mu_2 q_3 + \mu_3\dot{q}_3 = 0 \tag{9.55c}$$

Equations (9.55a)–(9.55c) can be further transformed into the following vector form:

$$\dot{\mathbf{q}} = \Psi(\mathbf{q}) \tag{9.56}$$

where

$$\begin{aligned}
\Psi(\mathbf{q}) &= \begin{bmatrix} \Psi_1(\mathbf{q}) & \Psi_2(\mathbf{q}) & \Psi_3(\mathbf{q}) \end{bmatrix}^{\mathrm{T}}, \\
\Psi_1(\mathbf{q}) &= q_2, \\
\Psi_2(\mathbf{q}) &= \frac{1}{\mu_3 M_2 - M_1}\Big[\mu_1 M_1 q_2 + (\mu_2 M_1 - \mu_3 D_w)\,q_3 \\
&\quad + \mu_3 m_{12}{q_2}^2\sin(q_1 + \alpha) + \mu_3 G_b\sin q_1 - D_w\mu_3\dot{\theta}_w^* - \mu_3\tilde{M}\Big], \\
\Psi_3(\mathbf{q}) &= \frac{1}{\mu_3 M_2 - M_1}\Big[-\mu_1 M_2 q_2 - (\mu_2 M_2 - D_w)\,q_3 \\
&\quad - M_{12}{q_2}^2\sin(q_1 + \alpha) - G_b\sin q_1 + D_w\dot{\theta}_w^* + \tilde{M}\Big].
\end{aligned} \tag{9.57}$$

Obviously point $\mathbf{q}^* = \begin{bmatrix} {x_1}^* & 0 & 0 \end{bmatrix}$ is one of the equilibriums of the nonlinear system (9.56), where ${x_1}^*$ is the real desired inclination body angle determined by (9.16).

Our aim is to choose suitable coefficients μ_1, μ_2, μ_3 that guarantee the asymptotical stability of this equilibrium.

Theorem 9.4. *For the MWIP system (9.12), design sliding surface (9.49) and control law (9.53). Assume inequalities* $M_1(\mathbf{q}^*)$, $M_2(\mathbf{q}^*) > 0$ *and* $\cos(x_1{}^*) > 0$ *hold, if coefficients* μ_1, μ_2, μ_3 *satisfy*

$$\begin{aligned} &\mu_1 M_1(\mathbf{q}^*) + D_w > \mu_2 M_2(\mathbf{q}^*), \\ &\mu_3 M_2(\mathbf{q}^*) - M_1(\mathbf{q}^*) < 0, \\ &\mu_1, \mu_2, \mu_3 > 0, \quad \mu_1\mu_3 > \mu_2 \end{aligned} \tag{9.58}$$

then the closed-loop system is locally asymptotically stable around equilibrium $\mathbf{q}^*$.

Proof. We linearize (9.57) about the equilibrium $\mathbf{q}^*$ to establish linear stability criteria that guarantee local exponential stability of the nonlinear system. The linearized system is described by

$$\dot{\mathbf{q}} = \mathbf{B}(\mathbf{q} - \mathbf{q}^*) \tag{9.59}$$

where

$$\mathbf{B} = \left.\frac{\partial \Psi}{\partial \mathbf{q}}\right|_{\mathbf{q}^*} = \begin{bmatrix} 0 & 1 & 0 \\ \Delta_1 & \Delta_2 & \Delta_3 \\ \Delta_4 & \Delta_5 & \Delta_6 \end{bmatrix} \tag{9.60}$$

$$\begin{aligned} &\Delta_1 = \left.\frac{\partial \Psi_2}{\partial q_1}\right|_{\mathbf{q}^*}, \quad \Delta_2 = \left.\frac{\partial \Psi_2}{\partial q_2}\right|_{\mathbf{q}^*}, \quad \Delta_3 = \left.\frac{\partial \Psi_2}{\partial q_3}\right|_{\mathbf{q}^*}, \quad \Delta_4 = \left.\frac{\partial \Psi_3}{\partial q_1}\right|_{\mathbf{q}^*}, \\ &\Delta_5 = \left.\frac{\partial \Psi_3}{\partial q_2}\right|_{\mathbf{q}^*}, \quad \Delta_6 = \left.\frac{\partial \Psi_3}{\partial q_3}\right|_{\mathbf{q}^*} \end{aligned} \tag{9.61}$$

According to the Hurwitz stability criteria, the equilibrium of linearized system (9.59) is asymptotically stable if we have

$$\begin{aligned} &\Delta_2 + \Delta_6 < 0, \quad \Delta_1\Delta_6 - \Delta_3\Delta_4 > 0, \quad (\Delta_2 + \Delta_6)(\Delta_3\Delta_5 - \Delta_2\Delta_6) \\ &\quad + \Delta_1\Delta_2 + \Delta_3\Delta_4 > 0 \end{aligned} \tag{9.62}$$

It follows from (9.57) and (9.61) that

$$\begin{aligned} &\Delta_2 + \Delta_6 = -\mu_1\delta^* M_1(\mathbf{q}^*) + \mu_2\delta^* M_2(\mathbf{q}^*) - \delta^* D_w, \\ &\Delta_1\Delta_6 - \Delta_3\Delta_4 = \mu_2\delta^* G_b \cos x_1{}^*, \\ &\Delta_3\Delta_5 - \Delta_2\Delta_6 = (\delta^*)^2 \mu_1 D_w(\mu_3 M_2(\mathbf{q}^*) - M_1(\mathbf{q}^*)), \\ &\Delta_1\Delta_2 + \Delta_3\Delta_4 = (\delta^*)^2 G_b \cos x_1{}^*[(\mu_1\mu_3 - \mu_2) M_1(\mathbf{q}^*) + \mu_3 D_w], \end{aligned} \tag{9.63}$$

where δ^* is given by

$$\delta^* = \frac{1}{M_1(\mathbf{q}^*) - \mu_3 M_2(\mathbf{q}^*)} \tag{9.64}$$

Apparently inequalities (9.62) hold if coefficients μ_1, μ_2, μ_3 are chosen to satisfy (9.58). This completes the proof. □

It is found that if inequalities $\mu_1 \min(m_{11}/m_{22}, 1) > \mu_2$ and $\mu_3 < \min(m_{11}/m_{22}, 1)$ hold, then inequalities (9.58) are always satisfied. It is also should be noted that the problem of uncertain equilibriums is solved by using a special sliding surface (9.49). The third item of (9.49) plays an important role in avoiding uncertain equilibriums. In fact, state z_3 is the integration of state z_4, which can eliminate the static tracking error like an integral part in a common-used PID controller.

9.4 Dynamic surface controller with nonlinear disturbance observer for MWIP

In this section, we proposed a DSCNDO for the balance control of an MWIP system. Compared with the linear controllers, larger stability region can be achieved by applying the DSC strategy. The introduction of NDO enhances the robustness of closed-loop MWIP system to model errors and external disturbances. Moreover, the "explosion-term" problem is also avoided in the controller design.

Here we rewrite the dynamic model (9.9) as

$$\begin{cases} m_{11}\ddot{\theta}_w + m_{12}\cos(\theta_b)\,\ddot{\theta}_b \\ = u - (D_w + D_b)\,\dot{\theta}_w + D_b\dot{\theta}_b + m_{12}\dot{\theta}_b^2\sin(\theta_b) + \tau_{\text{ext1}}, \\ (m_{11} + m_{12}\cos(\theta_b))\,\ddot{\theta}_w + (m_{22} + m_{12}\cos(\theta_b))\,\ddot{\theta}_b \\ = -D_w\dot{\theta}_w + m_{12}\dot{\theta}_b^2\sin(\theta_b) + G_b\sin(\theta_b) + \tau_{\text{ext2}} \end{cases} \tag{9.65}$$

where all model uncertainties and external disturbances are lumped to τ_{ext1} and τ_{ext2}.

9.4.1 Nonlinear disturbance observer design for MWIP

To improve the robustness and control performance of the closed-loop MWIP control system, it is necessary to apply a NDO estimating model uncertainties, frictions and external disturbances. This subsection illustrates the design procedure of a NDO in the MWIP system.

Considering the nonlinear underactuated system with disturbance, to simplify the denotation, we rewrite (9.65) as vector form:

$$\mathbf{M}(\mathbf{q})\,\ddot{\mathbf{q}} + \mathbf{N}(\mathbf{q}, \dot{\mathbf{q}}) + \mathbf{F}(\dot{\mathbf{q}}) = \tau + \tau_{\text{ext}} \tag{9.66}$$

where

$$\mathbf{q} = \begin{bmatrix} q_1 & q_2 \end{bmatrix}^T = \begin{bmatrix} \theta_w & \theta_b \end{bmatrix}^T, \tag{9.67}$$

$$\mathbf{M}(\mathbf{q}) = \begin{bmatrix} m_{11} & m_{12}\cos(q_2) \\ m_{11} + m_{12}\cos(q_2) & m_{22} + m_{12}\cos(q_2) \end{bmatrix}, \tag{9.68}$$

$$\mathbf{N}(\mathbf{q}, \dot{\mathbf{q}}) = \begin{bmatrix} -m_{12}\dot{q}_2^2\sin(q_2) \\ -G_b\sin(q_2) - m_{12}\dot{q}_2^2\sin(q_2) \end{bmatrix}, \tag{9.69}$$

$$\mathbf{F}(\dot{\mathbf{q}}) = \begin{bmatrix} (D_w + D_b)\dot{q}_1 - D_b\dot{q}_2 \\ D_w\dot{q}_1 \end{bmatrix}, \tag{9.70}$$

$$\tau = \begin{bmatrix} u \\ 0 \end{bmatrix}, \tau_{\text{ext}} = \begin{bmatrix} \tau_{\text{ext}1} \\ \tau_{\text{ext}1} + \tau_{\text{ext}2} \end{bmatrix} \tag{9.71}$$

Let us assume that $\hat{\mathbf{M}}(\mathbf{q})$ and $\hat{\mathbf{N}}(\mathbf{q}, \dot{\mathbf{q}})$ are the estimates of the actual $\mathbf{M}(\mathbf{q})$ and $\mathbf{N}(\mathbf{q}, \dot{\mathbf{q}})$, and that $\Delta\mathbf{M}(\mathbf{q})$ and $\Delta\mathbf{N}(\mathbf{q}, \dot{\mathbf{q}})$ are the corresponding additive uncertainties presented in the model of the MWIP. That is we have

$$\begin{cases} \mathbf{M}(\mathbf{q}) = \hat{\mathbf{M}}(\mathbf{q}) + \Delta\mathbf{M}(\mathbf{q}) \\ \mathbf{N}(\mathbf{q}, \dot{\mathbf{q}}) = \hat{\mathbf{N}}(\mathbf{q}, \dot{\mathbf{q}}) + \Delta\mathbf{N}(\mathbf{q}, \dot{\mathbf{q}}) \end{cases} \tag{9.72}$$

The lumped disturbance vector τ_d is defined as

$$\tau_d = \begin{bmatrix} \tau_{d1} & \tau_{d2} \end{bmatrix}^T = \tau_{\text{ext}} - \Delta\mathbf{M}(\mathbf{q})\,\ddot{\mathbf{q}} - \Delta\mathbf{N}(\mathbf{q}, \dot{\mathbf{q}}) - \mathbf{F}(\dot{\mathbf{q}}) \tag{9.73}$$

By this definition, the effect of all dynamic uncertainties, joint frictions and external disturbances is lumped into a single disturbance vector τ_d. From (9.66), it is seen that

$$\hat{\mathbf{M}}(\mathbf{q})\,\ddot{\mathbf{q}} + \hat{\mathbf{N}}(\mathbf{q}, \dot{\mathbf{q}}) = \tau + \tau_d \tag{9.74}$$

To estimate the lumped disturbance τ_d, the NDO is designed as

$$\dot{\hat{\tau}}_d = -\mathbf{L}(\mathbf{q}, \dot{\mathbf{q}})\,\hat{\tau}_d + \mathbf{L}(\mathbf{q}, \dot{\mathbf{q}})\left(\hat{\mathbf{M}}(\mathbf{q})\,\ddot{\mathbf{q}} + \hat{\mathbf{N}}(\mathbf{q}, \dot{\mathbf{q}}) - \tau\right) \tag{9.75}$$

Define $\tilde{\tau}_d = \tau_d - \hat{\tau}_d$ as the disturbance tracking error and using (9.75), it is observed that

$$\dot{\hat{\tau}}_d = \mathbf{L}(\mathbf{q}, \dot{\mathbf{q}})\,\tilde{\tau}_d \tag{9.76}$$

or, equivalently

$$\dot{\tilde{\tau}}_d = \dot{\tau}_d - \mathbf{L}(\mathbf{q}, \dot{\mathbf{q}})\,\tilde{\tau}_d \tag{9.77}$$

In general, there is no prior information about the derivative of the disturbance τ_d. When the disturbance varies slowly relative to the observer dynamics, it is reasonable to suppose that $\dot{\tau}_d = 0$. Therefore, we have

$$\dot{\tilde{\tau}}_d = -\dot{\hat{\tau}}_d = -\mathbf{L}(\mathbf{q}, \dot{\mathbf{q}})\,\tilde{\tau}_d \tag{9.78}$$

Let us define an auxiliary variable $\mathbf{z} = \begin{bmatrix} z_1 & z_2 \end{bmatrix}^T = \hat{\tau}_d - \mathbf{p}(\mathbf{q}, \dot{\mathbf{q}})$, where

$$\frac{d}{dt}\mathbf{p}(\mathbf{q}, \dot{\mathbf{q}}) = \mathbf{L}(\mathbf{q}, \dot{\mathbf{q}})\,\hat{\mathbf{M}}(\mathbf{q})\,\ddot{\mathbf{q}}. \tag{9.79}$$

And substitute it to (9.75), the observer can be designed as

$$\begin{cases} \dot{\mathbf{z}} = \mathbf{L}(\mathbf{q}, \dot{\mathbf{q}})\left[\hat{\mathbf{N}}(\mathbf{q}, \dot{\mathbf{q}}) - \tau - \mathbf{p}(\mathbf{q}, \dot{\mathbf{q}}) - \mathbf{z}\right] \\ \hat{\tau}_d = \mathbf{z} + \mathbf{p}(\mathbf{q}, \dot{\mathbf{q}}) \end{cases} \tag{9.80}$$

In this study, the following disturbance observer gain matrix $\mathbf{L}(\mathbf{q}, \dot{\mathbf{q}})$ and vector $\mathbf{p}(\mathbf{q}, \dot{\mathbf{q}})$ is used as

$$\begin{cases} \mathbf{L}(\mathbf{q}, \dot{\mathbf{q}}) = \mathbf{L}(\mathbf{q}) = \mathbf{X}^{-1}\hat{\mathbf{M}}^{-1}(\mathbf{q}) \\ \mathbf{p}(\mathbf{q}, \dot{\mathbf{q}}) = \mathbf{p}(\dot{\mathbf{q}}) = \mathbf{X}^{-1}\dot{\mathbf{q}} \end{cases} \tag{9.81}$$

where $\mathbf{X}$ is a constant invertible matrix to be determined, that is

$$\mathbf{X} = \begin{bmatrix} c_1 & c_2 \\ c_3 & c_4 \end{bmatrix}, \qquad c_i \geq 0, \quad i = 1, 2, 3, 4 \tag{9.82}$$

Substituting (9.81) and (9.82) into (9.80), and using (9.66) we have

$$\begin{aligned} \dot{\mathbf{z}} &= \begin{bmatrix} \dot{z}_1 \\ \dot{z}_2 \end{bmatrix} \\ &= A^{-1}\begin{bmatrix} c_1D_1 - c_2D_2 & c_2\hat{m}_{11} - c_1\hat{m}_{12}\cos(\theta_b) \\ c_3D_1 - c_4D_2 & c_4\hat{m}_{11} - c_3\hat{m}_{12}\cos(\theta_b) \end{bmatrix} \\ &\quad \times \begin{bmatrix} -\hat{m}_{12}\dot{\theta}_b^2\sin(\theta_b) - u - c_1\dot{\theta}_w - c_2\dot{\theta}_b - z_1 \\ -\hat{G}_b\sin(\theta_b) - \hat{m}_{12}\dot{\theta}_b^2\sin(\theta_b) - c_3\dot{\theta}_w - c_4\dot{\theta}_b - z_2 \end{bmatrix} \end{aligned} \tag{9.83}$$

and

$$\begin{bmatrix} \hat{\tau}_{d1} \\ \hat{\tau}_{d2} \end{bmatrix} = \begin{bmatrix} z_1 \\ z_2 \end{bmatrix} + \begin{bmatrix} c_1 & c_2 \\ c_3 & c_4 \end{bmatrix}\begin{bmatrix} \dot{\theta}_w \\ \dot{\theta}_b \end{bmatrix} \tag{9.84}$$

where

$$\begin{cases} D_1 = \hat{m}_{22} + \hat{m}_{12}\cos(\theta_b) \\ D_2 = \hat{m}_{11} + \hat{m}_{12}\cos(\theta_b) \\ A = \det\left(\hat{\mathbf{M}}(\mathbf{q})\right) = \hat{m}_{11}\hat{m}_{22} - \left(\hat{m}_{12}\cos(\theta_b)\right)^2 \end{cases} \tag{9.85}$$

Therefore, the disturbance observer can be designed as follows

$$\begin{cases} \dot{z}_1 = A^{-1}\big[(c_2D_2 - c_1D_1)\left(\hat{m}_{12}\dot{\theta}_b^2\sin(\theta_b) + u + c_1\dot{\theta}_w + c_2\dot{\theta}_b + z_1\right) \\ -\left(c_2\hat{m}_{11} - c_1\hat{m}_{12}\cos(\theta_b)\right)\left(\hat{G}_b\sin(\theta_b) + \hat{m}_{12}\dot{\theta}_b^2\sin(\theta_b) + c_3\dot{\theta}_w + c_4\dot{\theta}_b + z_2\right)\big] \\ \dot{z}_2 = A^{-1}\big[(c_4D_2 - c_3D_1)\left(\hat{m}_{12}\dot{\theta}_b^2\sin(\theta_b) + u + c_1\dot{\theta}_w + c_2\dot{\theta}_b + z_1\right) \\ -\left(c_4\hat{m}_{11} - c_3\hat{m}_{12}\cos(\theta_b)\right)\left(\hat{G}_b\sin(\theta_b) + \hat{m}_{12}\dot{\theta}_b^2\sin(\theta_b) + c_3\dot{\theta}_w + c_4\dot{\theta}_b + z_2\right)\big] \\ \hat{\tau}_{d1} = z_1 + c_1\dot{\theta}_w + c_2\dot{\theta}_b \\ \hat{\tau}_{d2} = z_2 + c_3\dot{\theta}_w + c_4\dot{\theta}_b \end{cases} \tag{9.86}$$

9.4.2 Controller design

To analyze the type and features of the underactuated systems, we first introduce the following definition and lemma.

Definition 9.5. *([31], Definition 3.9.1.) (Class-I, II underactuated systems) We call an underactuated system (9.87) Class-I underactuated system, if q_2 is actuated. We call it a Class-II underactuated system, if q_2 is unactuated. In order words, for a Class-I system $\tau_1 = 0$ and $\tau_2 = 0$ for a Class-II system.*

$$\begin{cases} m_{11}(q_2)\ddot{q}_1 + m_{12}(q_2)\ddot{q}_2 + m'_{11}(q_2)\dot{q}_1\dot{q}_2 + m'_{12}(q_2)\dot{q}_2^2 - g_1(q_1, q_2) = \tau_1 \\ m_{21}(q_2)\ddot{q}_1 + m_{22}(q_2)\ddot{q}_2 - \frac{1}{2}m'_{11}(q_2)\dot{q}_1^2 + \frac{1}{2}m'_{22}(q_2)\dot{q}_2^2 - g_2(q_1, q_2) = \tau_2 \end{cases} \tag{9.87}$$

Lemma 9.1. *([31], Proposition 3.9.1.) (Class-I normal form) Consider a Class-I underactuated system with two DOF. Then, the following global change of coordinates obtained from the Lagrangian of the system*

$$\begin{cases} z_1 = q_1 + \gamma(q_2) \\ z_2 = m_{11}(q_2)\,p_1 + m_{12}(q_2)\,p_2 = \dfrac{\partial L}{\partial \dot{q}_1} \\ \xi_1 = q_2 \\ \xi_1 = p_2 \end{cases} \tag{9.88}$$

where

$$\gamma(q_2) = \int_0^{q_2} m_{11}^{-1}(\theta)\,m_{12}(\theta)\,d\theta \tag{9.89}$$

transforms the dynamics of the system into a cascade nonlinear system in strict feedback form

$$\begin{cases} \dot{z}_1 = m_{11}^{-1}(\xi_1)\,z_2 \\ \dot{z}_2 = g_1(z_1 - \gamma(\xi_1)\,, \xi_1) \\ \dot{\xi}_1 = \xi_2 \\ \dot{\xi}_2 = u \end{cases} \tag{9.90}$$

where u is the new control from collocated partial feedback linearization and $g_1(q_1, q_2) = -\partial V(q)/\partial q_1$.

Lemma 9.2. *([31], Proposition 3.9.2.) (Class-II normal form) Consider a Class-II underactuated system with two DOF. Then, the following change of coordinates obtained from the Lagrangian of the system*

$$\begin{cases} z_1 = q_1 + \gamma(q_2) \\ z_2 = m_{21}(q_2)\,p_1 + m_{22}(q_2)\,p_2 = \dfrac{\partial L}{\partial \dot{q}_2} \\ \xi_1 = q_2 \\ \xi_1 = p_2 \end{cases} \tag{9.91}$$

where

$$\gamma(q_2) = \int_0^{q_2} m_{21}^{-1}(\theta)\,m_{22}(\theta)\,d\theta \tag{9.92}$$

is defined over $U = \{q_2\,|m_{21}\,(q_2) \neq 0\}$ transforms the dynamics of the system into a cascade nonlinear system in quadratic nontriangular form where u is the new control from noncollocated partial feedback linearization.

From Definition 9.5 and MWIP system model (9.9), it is found that the MWIP system belongs to neither the Class-I nor the Class-II underactuated mechanical systems.

In general, the DSC control design requests the strict or semistrict feedback form of the nonlinear system model [18,25]. According to Lemma 9.1, the dynamics of a Class-I underactuated mechanical system may be transformed into a cascade nonlinear system in strict feedback form. However, the dynamics of a Class-II underactuated system may only be transformed into a cascade nonlinear system in a quadratic nontriangular form from Lemma 9.2. Similarly, the dynamics of an MWIP-type underactuated system may not be transformed into a cascade nonlinear system in strict feedback form. This makes it difficult to design a DSC controller for the MWIP system.

In order to solve this problem, we propose a coordinate transformation approach to transform an MWIP model into a cascade nonlinear system in semistrict feedback form.

Let us introduce the following variables:

$$\begin{cases} x_1 = \left(\hat{m}_{11} + \hat{m}_{12}\cos(x_2)\right)x_4 + \left(\hat{m}_{22} + \hat{m}_{12}\cos(x_2)\right)x_3 \\ x_2 = q_2 = \theta_b \\ x_3 = \dot{q}_2 = \dot{\theta}_b \\ x_4 = \dot{q}_1 = \dot{\theta}_w \end{cases} \tag{9.93}$$

then the MWIP system model (9.65) can be rewritten as

$$\begin{cases} \hat{m}_{11}\dot{x}_4 + \hat{m}_{12}\cos(x_2)\,\dot{x}_3 \\ = u + \hat{m}_{12}x_3^2\sin(x_2) + \tau_{d1} \\ \\ \left(\hat{m}_{11} + \hat{m}_{12}\cos(x_2)\right)\dot{x}_4 + \left(\hat{m}_{22} + \hat{m}_{12}\cos(x_2)\right)\dot{x}_3 \\ = \hat{m}_{12}x_3^2\sin(x_2) + \hat{G}_b\sin(x_2) + \tau_{d2} \end{cases} \tag{9.94}$$

Theorem 9.5. *For the MWIP system (9.94), the global change of coordinates (9.93) transforms the dynamics of the system into a cascade nonlinear system in semistrict feedback form*

$$\dot{x}_1 = -\hat{m}_{12}\sin(x_2)\,x_3x_4 + \hat{G}_b\sin(x_2) + \tau_{d2} \tag{9.95a}$$

$$\begin{cases} \dot{x}_2 = x_3 = \dfrac{x_1}{\hat{m}_{22} + \hat{m}_{12}\cos(x_2)} - \dfrac{\hat{m}_{11} + \hat{m}_{12}\cos(x_2)}{\hat{m}_{22} + \hat{m}_{12}\cos(x_2)}x_4 \\ \quad = f_1(x_1, x_2) + g_1(x_1, x_2)\,x_4 \\ \dot{x}_4 = A^{-1}\left[\hat{m}_{12}\hat{m}_{22}\sin(x_2)\left(\dfrac{x_1 - \left(\hat{m}_{11} + \hat{m}_{12}\cos(x_2)\right)x_4}{\hat{m}_{22} + \hat{m}_{12}\cos(x_2)}\right)^2 \right. \\ \\ \quad \left. - \hat{m}_{12}\hat{G}_b\sin(x_2)\cos(x_2)\right] + A^{-1}\left(\hat{m}_{22} + \hat{m}_{12}\cos(x_2)\right)u \\ \quad + A^{-1}\left[\left(\hat{m}_{22} + \hat{m}_{12}\cos(x_2)\right)\tau_{d1} - \hat{m}_{12}\cos(x_2)\,\tau_{d2}\right] \\ \quad = f_2(x_1, x_2, x_4) + g_2(x_1, x_2, x_4)\,u + \Delta(x_1, x_2, x_4, \tau_{d1}, \tau_{d2}) \end{cases} \tag{9.95b}$$

Proof of the theorem is similar to the work in [32]. After coordinate transformation the MWIP system model is represented in a semistrict feedback form as cascade of an outer (9.95b) and a core (9.95a) subsystem.

Assumption 9.1. *The uncertainties and disturbances of the system are bounded and satisfy*

$$|\tau_{d1}| \leq \bar{d}_1, |\tau_{d2}| \leq \bar{d}_2, |\tilde{\tau}_{d1}| \leq \xi_1, |\tilde{\tau}_{d2}| \leq \xi_2 \tag{9.96}$$

where $\bar{d}_1, \bar{d}_2, \xi_1$ and ξ_2 are known bounds.

Our purpose is to design a control u forcing x_2 to be stabilized around zero. Together with the proposed disturbance observer, for MWIP system (9.95) we design a new DSCNDO as follows:

$$u = u_{\mathrm{DSC}} - u_{\tau_d} \tag{9.97}$$

$$u_{\tau_d} = u_{\tau_{d1}} - u_{\tau_{d2}} \tag{9.98}$$

where

$$\begin{cases} u_{\tau_{d1}} = \hat{\tau}_{d1} \\ u_{\tau_{d2}} = \dfrac{\hat{m}_{12}\cos(x_2)}{\hat{m}_{22} + \hat{m}_{12}\cos(x_2)}\hat{\tau}_{d2} \end{cases} \tag{9.99}$$

For convenience of the mathematical derivation, we define following notations in advance:

$$\begin{cases} M_{C12} = \hat{m}_{12}\cos(x_2) = \hat{m}_{12}\cos(S_1), \\ G_{S2} = \hat{G}_b \sin(x_2) = \hat{G}_b \sin(S_1), \\ M_{C22} = \hat{m}_{22}\cos(x_2) = \hat{m}_{22}\cos(S_1), \\ S_{C2} = \dfrac{\sin(x_2)}{\cos^2(x_2)} = \dfrac{\sin(S_1)}{\cos^2(S_1)}, \\ M_{S12} = \hat{m}_{12}\sin(x_2) = \hat{m}_{12}\sin(S_1), \\ \bar{A} = \hat{m}_{11}\hat{m}_{22} - \hat{m}_{12}^2 \end{cases} \tag{9.100}$$

The pure DSC component of DSCNDO, u_{DSC}, can be obtained through the following procedure:

Step 1: Design the virtual control law $\bar{x}_4$

1. Define the first dynamic surface

$$S_1 = x_2 - x_d = x_2 - 0 = x_2. \tag{9.101}$$

Then, from the first equation of (9.95b) the derivative of S_1 can be expressed as

$$\dot{S}_1 = \dot{x}_2 = \frac{x_1}{\hat{m}_{22} + M_{C12}} - \frac{\hat{m}_{11} + M_{C12}}{\hat{m}_{22} + M_{C12}}x_4 \tag{9.102}$$

2. Select the virtual control law $\bar{x}_4$ as

$$\bar{x}_4 = \frac{\hat{m}_{22} + M_{C12}}{\hat{m}_{11} + M_{C12}}\left(\frac{1}{\hat{m}_{22} + M_{C12}}x_1 + k_1 S_1\right) \tag{9.103}$$

where $k_1 > 0$.

3. Input $\bar{x}_4$ to a first order filter, then we have

$$\tau_4 \dot{x}_{4d} + x_{4d} = \bar{x}_4, x_{4d}(0) = \bar{x}_4(0) \tag{9.104}$$

where $\tau_4 > 0$ is the filter time constant.

Step 2: Design the actual control law u_{DSC}

1. Define the second dynamic surface

$$S_2 = x_4 - x_{4d} \tag{9.105}$$

Then, from the second equation of (9.95b), (9.97)–(9.99) and (9.104), the derivative of S_2 can be expressed as

$$\begin{aligned}\dot{S}_2 &= \dot{x}_4 - \dot{x}_{4d}\\ &= A^{-1}\left(\hat{m}_{22} + M_{C12}\right) u_{\mathrm{DSC}} + A^{-1}\left[\hat{m}_{12}\hat{m}_{22}\sin(x_2)\left(\frac{x_1 - \left(\hat{m}_{11} + M_{C12}\right) x_4}{\hat{m}_{22} + M_{C12}}\right)^2 \right.\\ &\quad \left. - M_{C12} G_{S2}\right] - \frac{\bar{x}_4 - x_{4d}}{\tau_4} + A^{-1}\left[\left(\hat{m}_{22} + M_{C12}\right)\tilde{\tau}_{d1} - M_{C12}\tilde{\tau}_{d2}\right]\end{aligned} \tag{9.106}$$

2. Select the control law u_{DSC} as follows:

$$\begin{aligned}u_{\mathrm{DSC}} &= \frac{A}{\hat{m}_{22} + M_{C12}}\left(\frac{\bar{x}_4 - x_{4d}}{\tau_4} - k_2 S_2\right) - \frac{\hat{m}_{12}\hat{m}_{22}\sin(x_2)\left(x_1 - \left(\hat{m}_{11} + M_{C12}\right)x_4\right)^2}{\left(\hat{m}_{22} + M_{C12}\right)^3}\\ &\quad + \frac{M_{C12} G_{S2}}{\hat{m}_{22} + M_{C12}}\end{aligned} \tag{9.107}$$

where $k_2 > 0$.

The whole control system block diagram is shown in Figure 9.2.

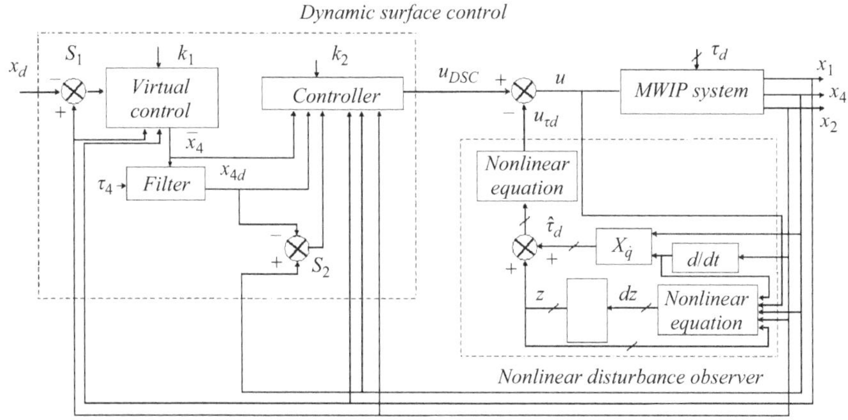

Figure 9.2 Block diagram of MWIP system with DSCNDO

To analyze the stability of system (9.95) controlled by the proposed DSCNDO, we first introduce the following lemma.

Lemma 9.3. *[33]. For a nonlinear system (9.95), suppose that a positive definite function $V(x(t)) = V(t)$ satisfies the following differential inequality:*

$$\dot{V}(t) \leq -\varsigma V + C \tag{9.108}$$

where constants ς and C satisfy $\varsigma > 0$ and $C \geq 0$. Then, for any given t_0, function $V(t)$ satisfies the following inequality:

$$0 \leq V(t) \leq [C - (C - \varsigma V(t_0)) \exp(-\varsigma(t - t_0))] / \varsigma, \forall t \geq t_0 \tag{9.109}$$

This implies that $V(t)$ is eventually bounded by C/ς, and the exponential convergence rate of $V(t)$ to the bound is ς. Thus, the state of the system (9.95) is uniformly and ultimately bounded.

The stability analysis of the whole system is concluded in the following theorem:

Theorem 9.6. *Considering a system (9.95) with modeling errors, external disturbance, unknown payloads and frictions, there exists a set of the surface gains k_1, k_2, the filter time constant τ_4 and the observer gains c_1, c_2, c_3, c_4 satisfying*

$$\gamma = \min(a_1, a_2, a_3, a_4, a_5) > 0, \exists \gamma \tag{9.110}$$

where

$$\begin{cases} a_1 = k_1 - \dfrac{\hat{m}_{11} + \hat{m}_{12}}{\hat{m}_{22}} \\ a_2 = k_2 - \dfrac{k_1}{2} - \dfrac{\hat{m}_{11} + \hat{m}_{12}}{2\hat{m}_{22}} \\ a_3 = \dfrac{1}{\tau_4} - \dfrac{3k_1}{2} - \dfrac{\hat{m}_{11} + \hat{m}_{12}}{2\hat{m}_{22}} \\ a_4 = \dfrac{c_1}{\hat{m}_{11}} - \dfrac{(\hat{m}_{11} + \hat{m}_{12})c_2}{\bar{A}} - 1 \\ a_5 = \dfrac{c_4}{\hat{m}_{22}} - \dfrac{\hat{m}_{12}c_3}{\bar{A}} - 1 \end{cases} \tag{9.111}$$

such that the NDO-based dynamic surface controller guarantees: Based on the control law (9.97)–(9.99) and (9.107), system (9.95) is semi-globally uniformly and ultimately bounded.

Proof. Choose the following Lyapunov function candidate

$$V = \frac{1}{2}S_1^2 + \frac{1}{2}S_2^2 + \frac{1}{2}e^2 + \frac{1}{2}\tilde{\tau}_d^T \tilde{\tau}_d \tag{9.112}$$

If the filter error is defined as follows:

$$e = x_{4d} - \bar{x}_4 \tag{9.113}$$

By combining the above equation with (9.103), we have

$$e = x_{4d} - \frac{\hat{m}_{22} + M_{C12}}{\hat{m}_{11} + M_{C12}} \left(\frac{x_1}{\hat{m}_{22} + M_{C12}} + k_1 S_1 \right) \tag{9.114}$$

From (9.101)–(9.103), (9.105) and (9.113), the derivative of S_1 can be written as

$$\dot{S}_1 = \frac{x_1}{\hat{m}_{22} + M_{C12}} - \frac{\hat{m}_{11} + M_{C12}}{\hat{m}_{22} + M_{C12}} x_4 = -\frac{\hat{m}_{11} + M_{C12}}{\hat{m}_{22} + M_{C12}} (S_2 + e) - k_1 S_1 \tag{9.115}$$

Substituting (9.107) to (9.106), we obtain

$$\dot{S}_2 = -k_2 S_2 + A^{-1} \left(\hat{m}_{22} + M_{C12} \right) \tilde{\tau}_{d1} - A^{-1} M_{C12} \tilde{\tau}_{d2} \tag{9.116}$$

Then, from (9.104) and (9.113) and (9.114), the derivative of e is given by

$$\begin{aligned} \dot{e} &= \dot{x}_{4d} - \frac{d}{dt} \left[\frac{\hat{m}_{22} + M_{C12}}{\hat{m}_{11} + M_{C12}} \left(\frac{x_1}{\hat{m}_{22} + M_{C12}} + k_1 S_1 \right) \right] \\ &= -\frac{e}{\tau_4} - \left[\frac{\dot{x}_1 \left(\hat{m}_{11} + M_{C12} \right) + x_1 x_3 M_{S12}}{\left(\hat{m}_{11} + M_{C12} \right)^2} \right. \\ &\quad \left. + \frac{\left(\hat{m}_{22} - \hat{m}_{11} \right) M_{S12} x_3}{\left(\hat{m}_{11} + M_{C12} \right)^2} k_1 S_1 + k_1 \dot{S}_1 \frac{\hat{m}_{22} + M_{C12}}{\hat{m}_{11} + M_{C12}} \right] \end{aligned} \tag{9.117}$$

Substituting (9.95a) and $\dot{x}_2 = x_3$ into (9.117)

$$\begin{aligned} \dot{e} = &-\frac{e}{\tau_4} + \frac{M_{S12} x_3 x_4}{\hat{m}_{11} + M_{C12}} - \frac{G_{S2} + \tau_{d2}}{\hat{m}_{11} + M_{C12}} - \frac{M_{S12} x_1 x_3}{\left(\hat{m}_{11} + M_{C12} \right)^2} \\ &- \frac{\left(\hat{m}_{22} - \hat{m}_{11} \right) M_{S12} k_1 S_1}{\left(\hat{m}_{11} + M_{C12} \right)^2} \dot{S}_1 - \frac{\hat{m}_{22} + M_{C12}}{\hat{m}_{11} + M_{C12}} k_1 \dot{S}_1 \end{aligned} \tag{9.118}$$

From (9.103), (9.105) and (9.113), we have

$$x_4 = S_2 + x_{4d} = S_2 + \bar{x}_4 + e = S_2 + \frac{x_1}{\hat{m}_{11} + M_{C12}} + \frac{\hat{m}_{22} + M_{C12}}{\hat{m}_{11} + M_{C12}} k_1 S_1 + e \tag{9.119}$$

Substituting (9.119) into (9.118)

$$\begin{aligned} \dot{e} = &-\frac{e}{\tau_4} + \frac{M_{S12} x_3}{\hat{m}_{11} + M_{C12}} \left(S_2 + \frac{x_1}{\hat{m}_{11} + M_{C12}} + \frac{\hat{m}_{22} + M_{C12}}{\hat{m}_{11} + M_{C12}} k_1 S_1 + e \right) \\ &- \frac{G_{S2} + \tau_{d2}}{\hat{m}_{11} + M_{C12}} - \frac{M_{S12} x_1 x_3}{\left(\hat{m}_{11} + M_{C12} \right)^2} - \frac{\left(\hat{m}_{22} - \hat{m}_{11} \right) M_{S12} k_1 S_1}{\left(\hat{m}_{11} + M_{C12} \right)^2} \dot{S}_1 \\ &- \frac{\hat{m}_{22} + M_{C12}}{\hat{m}_{11} + M_{C12}} k_1 \dot{S}_1 = -\frac{e}{\tau_4} + \frac{M_{S12} x_3}{\hat{m}_{11} + M_{C12}} \left(S_2 + \frac{\hat{m}_{22} + M_{C12}}{\hat{m}_{11} + M_{C12}} k_1 S_1 + e \right) \\ &- \frac{G_{S2} + \tau_{d2}}{\hat{m}_{11} + M_{C12}} - \frac{\left(\hat{m}_{22} - \hat{m}_{11} \right) M_{S12} k_1 S_1}{\left(\hat{m}_{11} + M_{C12} \right)^2} \dot{S}_1 - \frac{\hat{m}_{22} + M_{C12}}{\hat{m}_{11} + M_{C12}} k_1 \dot{S}_1 \end{aligned} \tag{9.120}$$

According to the fact that $x_2 = S_1$ and $x_3 = \dot{x}_2 = \dot{S}_1$, we have

$$\dot{e} = -\frac{e}{\tau_4} - \frac{G_{S2} + \tau_{d2}}{\hat{m}_{11} + M_{C12}} + \left[\frac{M_{S12}}{\hat{m}_{11} + M_{C12}} \left(S_2 + \frac{\hat{m}_{22} + M_{C12}}{\hat{m}_{11} + M_{C12}} k_1 S_1 + e \right) - \frac{\left(\hat{m}_{22} - \hat{m}_{11}\right) M_{S12} k_1 S_1}{\left(\hat{m}_{11} + M_{C12}\right)^2} - \frac{\hat{m}_{22} + M_{C12}}{\hat{m}_{11} + M_{C12}} k_1 \right] \dot{S}_1 \tag{9.121}$$

Substituting (9.115) into (9.121), we have

$$\begin{aligned} \dot{e} = & -\frac{e}{\tau_4} - \frac{G_{S2}}{\hat{m}_{11} + M_{C12}} - \frac{\tau_{d2}}{\hat{m}_{11} + M_{C12}} - \frac{M_{S12}(S_2 + e)^2}{\hat{m}_{22} + M_{C12}} \\ & - \frac{M_{S12} k_1 S_1 \left[2\left(S_2 + e\right) + k_1 S_1\right]}{\hat{m}_{11} + M_{C12}} + \frac{\left(\hat{m}_{22} - \hat{m}_{11}\right) M_{S12} k_1 S_1 \left(S_2 + e\right)}{\left(\hat{m}_{11} + M_{C12}\right)\left(\hat{m}_{22} + M_{C12}\right)} \\ & + k_1 \left(S_2 + e\right) + \frac{\left(\hat{m}_{22} + M_{C12}\right) k_1^2 S_1}{\hat{m}_{11} + M_{C12}} \end{aligned} \tag{9.122}$$

From (9.74), we have

$$\tilde{\tau}_d = \tau_d - \hat{\tau}_d = \hat{M}(q)\ddot{q} + \hat{N}(q, \dot{q}) - \tau - \hat{\tau}_d. \tag{9.123}$$

It follows from (9.78), (9.81)–(9.82) and (9.123) that

$$\begin{aligned} \frac{d}{dt}\left(\frac{1}{2}\tilde{\tau}_d^T \tilde{\tau}_d\right) &= \tilde{\tau}_d^T \dot{\tilde{\tau}}_d = -\tilde{\tau}_d^T \dot{\hat{\tau}}_d = -\tilde{\tau}_d^T L \tilde{\tau}_d = -\tilde{\tau}_d^T X \hat{M}^{-1} \tilde{\tau}_d \\ &= A^{-1}\left\{\left(\hat{m}_{11} + M_{C12}\right) c_2 \tilde{\tau}_{d1}^2 - \left(\hat{m}_{22} + M_{C12}\right) c_1 \tilde{\tau}_{d1}^2 \right. \\ &\quad - \left[\left(\hat{m}_{22} + M_{C12}\right) c_3 + \hat{m}_{11} c_2\right] \tilde{\tau}_{d1} \tilde{\tau}_{d2} \\ &\quad \left. + \left[\left(\hat{m}_{11} + M_{C12}\right) c_4 + M_{C12} c_1\right] \tilde{\tau}_{d1} \tilde{\tau}_{d2} + M_{C12} c_3 \tilde{\tau}_{d2}^2 - \hat{m}_{11} c_4 \tilde{\tau}_{d2}^2 \right\} \end{aligned} \tag{9.124}$$

Note that the following inequalities hold

$$\begin{aligned} \hat{m}_{11}\hat{m}_{22} \geq A \geq \bar{A} &= \left[\left(\hat{m}_b + \hat{m}_w\right)\hat{r}^2 + \hat{I}_w\right]\left(\hat{m}_b \hat{l}^2 + \hat{I}_b\right) - \hat{m}_b^2 \hat{l}^2 \hat{r}^2 \\ &= \left(\hat{m}_b + \hat{m}_w\right)\hat{m}_b \hat{l}^2 \hat{r}^2 + \left(\hat{m}_b + \hat{m}_w\right)\hat{r}^2 \hat{I}_b + \hat{m}_b \hat{l}^2 \hat{I}_w + \hat{I}_w \hat{I}_b - \hat{m}_b^2 \hat{l}^2 \hat{r}^2 \\ &= \hat{m}_w \hat{m}_b \hat{l}^2 \hat{r}^2 + \left(\hat{m}_b + \hat{m}_w\right)\hat{r}^2 \hat{I}_b + \hat{m}_b \hat{l}^2 \hat{I}_w + \hat{I}_w \hat{I}_b > 0 \end{aligned} \tag{9.125}$$

In addition, we have

$$\begin{cases} 0 \leq \dfrac{\left(\hat{m}_{11} + M_{C12}\right) c_2 \tilde{\tau}_{d1}^2}{A} \leq \dfrac{\left(\hat{m}_{11} + \hat{m}_{12}\right) c_2 \tilde{\tau}_{d1}^2}{\bar{A}}, \\ 0 \geq \dfrac{-c_1 \tilde{\tau}_{d1}^2}{\hat{m}_{11}} \geq \dfrac{-\left(\hat{m}_{22} + M_{C12}\right) c_1 \tilde{\tau}_{d1}^2}{A}, \\ 0 \leq \dfrac{M_{C12} c_3 \tilde{\tau}_{d2}^2}{A} \leq \dfrac{\hat{m}_{12} c_3 \tilde{\tau}_{d2}^2}{\bar{A}}, \\ 0 \geq \dfrac{-c_4 \tilde{\tau}_{d2}^2}{\hat{m}_{22}} \geq \dfrac{-\hat{m}_{11} c_4 \tilde{\tau}_{d2}^2}{A} \end{cases} \tag{9.126}$$

It follows from (9.124) to (9.126) that

$$\begin{aligned}\frac{d}{dt}\left(\frac{1}{2}\tilde{\tau}_d^T\tilde{\tau}_d\right) \le &\left(\frac{(\hat{m}_{11}+\hat{m}_{12})c_2}{\bar{A}} - \frac{c_1}{\hat{m}_{11}}\right)\tilde{\tau}_{d1}^2\\ &+\left(\frac{\hat{m}_{12}c_3}{\bar{A}} - \frac{c_4}{\hat{m}_{22}}\right)\tilde{\tau}_{d2}^2\\ &+A^{-1}\left[(c_1+c_4)M_{C12} + c_4\hat{m}_{11}\right.\\ &\left.-\left(\hat{m}_{11}c_2 + \hat{m}_{22}c_3 + M_{C12}c_3\right)\right]\tilde{\tau}_{d1}\tilde{\tau}_{d2}\end{aligned} \tag{9.127}$$

From (9.115)–(9.116), (9.122) and (9.127), we have

$$\begin{aligned}\dot{V} \le &-k_1S_1^2 - k_2S_2^2 - \frac{e^2}{\tau_4} + k_1eS_2 + k_1e^2 - \left(\frac{c_1}{\hat{m}_{11}} - \frac{(\hat{m}_{11}+\hat{m}_{12})c_2}{\bar{A}}\right)\tilde{\tau}_{d1}^2\\ &-\left(\frac{c_4}{\hat{m}_{22}} - \frac{\hat{m}_{12}c_3}{\bar{A}}\right)\tilde{\tau}_{d2}^2 - \frac{\hat{m}_{11}+M_{C12}}{\hat{m}_{22}+M_{C12}}(S_2+e)S_1 + A^{-1}\left(\hat{m}_{22}+M_{C12}\right)S_2\tilde{\tau}_{d1}\\ &-A^{-1}M_{C12}S_2\tilde{\tau}_{d2} - \frac{G_{S2}e}{\hat{m}_{11}+M_{C12}} - \frac{e\tau_{d2}}{\hat{m}_{11}+M_{C12}} - \frac{M_{S12}(S_2+e)^2e}{\hat{m}_{22}+M_{C12}}\\ &-\frac{M_{S12}k_1S_1\left[2(S_2+e)+k_1S_1\right]e}{\hat{m}_{11}+M_{C12}} + \frac{(\hat{m}_{22}-\hat{m}_{11})M_{S12}k_1S_1(S_2+e)e}{(\hat{m}_{11}+M_{C12})(\hat{m}_{22}+M_{C12})}\\ &+\frac{(\hat{m}_{22}+M_{C12})k_1^2S_1e}{\hat{m}_{11}+M_{C12}} + A^{-1}\left[(c_1+c_4)M_{C12} + c_4\hat{m}_{11}\right.\\ &\left.-\left(\hat{m}_{11}c_2 + \hat{m}_{22}c_3 + M_{C12}c_3\right)\right]\tilde{\tau}_{d1}\tilde{\tau}_{d2}\end{aligned} \tag{9.128}$$

According to the Young's inequality, it follows that

$$\begin{cases} k_1eS_2 \le \dfrac{k_1}{2}e^2 + \dfrac{k_1}{2}S_2^2,\\ -\dfrac{(\hat{m}_{11}+M_{C12})S_1S_2}{\hat{m}_{22}+M_{C12}} \le \dfrac{\hat{m}_{11}+\hat{m}_{12}}{2\hat{m}_{22}}\left(S_1^2+S_2^2\right),\\ -\dfrac{(\hat{m}_{11}+M_{C12})S_1e}{\hat{m}_{22}+M_{C12}} \le \dfrac{\hat{m}_{11}+\hat{m}_{12}}{2\hat{m}_{22}}\left(S_1^2+e^2\right),\\ A^{-1}\left(\hat{m}_{22}+M_{C12}\right)S_2\tilde{\tau}_{d1} \le \tilde{\tau}_{d1}^2\\ \quad + \dfrac{1}{4}A^{-2}S_2^2\left(\hat{m}_{22}+M_{C12}\right)^2,\\ -A^{-1}M_{C12}S_2\tilde{\tau}_{d2} \le \tilde{\tau}_{d2}^2 + \dfrac{1}{4}A^{-2}M_{C12}^2S_2^2,\\ -\dfrac{e\tau_{d2}}{\hat{m}_{11}+M_{C12}} \le \dfrac{e^2}{4\left(\hat{m}_{11}+M_{C12}\right)^2} + \tau_{d2}^2,\\ A^{-1}\left[(c_1+c_4)M_{C12} + c_4\hat{m}_{11}\right.\\ \quad \left.-\left(\hat{m}_{11}c_2+\hat{m}_{22}c_3+M_{C12}c_3\right)\right]\tilde{\tau}_{d1}\tilde{\tau}_{d2}\\ \quad \le A^{-2}\left[(c_1+c_4)M_{C12} + c_4\hat{m}_{11}\right.\\ \quad \left.-\left(\hat{m}_{11}c_2+\hat{m}_{22}c_3+M_{C12}c_3\right)\right]^2 + \dfrac{1}{4}\tilde{\tau}_{d1}^2\tilde{\tau}_{d2}^2 \end{cases} \tag{9.129}$$

Thus,

$$
\begin{aligned}
\dot{V} \le & -\left(k_1 - \frac{\hat{m}_{11}+\hat{m}_{12}}{\hat{m}_{22}}\right) S_1^2 - \left(k_2 - \frac{k_1}{2} - \frac{\hat{m}_{11}+\hat{m}_{12}}{2\hat{m}_{22}}\right) S_2^2 \\
& -\left(\frac{1}{\tau_4} - \frac{3k_1}{2} - \frac{\hat{m}_{11}+\hat{m}_{12}}{2\hat{m}_{22}}\right) e^2 - \left(\frac{c_1}{\hat{m}_{11}} - \frac{\left(\hat{m}_{11}+\hat{m}_{12}\right) c_2}{\bar{A}} - 1\right) \tilde{\tau}_{d1}^2 \\
& -\left(\frac{c_4}{\hat{m}_{22}} - \frac{\hat{m}_{12}c_3}{\bar{A}} - 1\right) \tilde{\tau}_{d2}^2 + \frac{1}{4} A^{-2} S_2^2 \left[\left(\hat{m}_{22}+M_{C12}\right)^2 + M_{C12}^2\right] \\
& -\frac{G_{S2}e + M_{S12}k_1S_1\left[2\left(S_2+e\right)+k_1S_1\right]e}{\hat{m}_{11}+M_{C12}} + \frac{e^2}{4\left(\hat{m}_{11}+M_{C12}\right)^2} - \frac{M_{S12}\left(S_2+e\right)^2 e}{\hat{m}_{22}+M_{C12}} \\
& +\frac{\left(\hat{m}_{22}-\hat{m}_{11}\right) M_{S12}k_1S_1\left(S_2+e\right)e}{\left(\hat{m}_{11}+M_{C12}\right)\left(\hat{m}_{22}+M_{C12}\right)} + \frac{\left(\hat{m}_{22}+M_{C12}\right) k_1^2 S_1 e}{\hat{m}_{11}+M_{C12}} \\
& + A^{-2}\left[\left(c_1+c_4\right) M_{C12} + c_4\hat{m}_{11} - \left(\hat{m}_{11}c_2 + \hat{m}_{22}c_3 + M_{C12}c_3\right)\right]^2 + \tau_{d2}^2 + \frac{1}{4}\tilde{\tau}_{d1}^2\tilde{\tau}_{d2}^2
\end{aligned}
\tag{9.130}
$$

Then, there exists a nonnegative continuous function $\phi_1(\,\cdot\,)$ satisfying

$$
\begin{aligned}
0 \le & \left| \frac{1}{4} A^{-2} S_2^2 \left[\left(\hat{m}_{22}+M_{C12}\right)^2 + M_{C12}^2\right] - \frac{G_{S2}e + M_{S12}k_1S_1\left[2\left(S_2+e\right)+k_1S_1\right]e}{\hat{m}_{11}+M_{C12}} \right. \\
& + \frac{e^2}{4\left(\hat{m}_{11}+M_{C12}\right)^2} - \frac{M_{S12}\left(S_2+e\right)^2 e}{\hat{m}_{22}+M_{C12}} + \frac{\left(\hat{m}_{22}-\hat{m}_{11}\right) M_{S12}k_1S_1\left(S_2+e\right)e}{\left(\hat{m}_{11}+M_{C12}\right)\left(\hat{m}_{22}+M_{C12}\right)} \\
& + \frac{\left(\hat{m}_{22}+M_{C12}\right) k_1^2 S_1 e}{\hat{m}_{11}+M_{C12}} + A^{-2}\left[\left(c_1+c_4\right) M_{C12} + c_4\hat{m}_{11}\right. \\
& \left. \left. - \left(\hat{m}_{11}c_2 + \hat{m}_{22}c_3 + M_{C12}c_3\right)\right]^2 \right| \\
\le & \, \varphi_1\left(k_1, S_1, S_2, c_1, c_2, c_3, c_4, e\right)
\end{aligned}
\tag{9.131}
$$

Given any $p > 0$, let us introduce a set $\Omega := \{S_1^2 + S_2^2 + e^2 < 2p\}$. Apparently set Ω is compact in R^3. Therefore, the continuous function $\phi_1(\,\cdot\,)$ has a maximum, say M on Ω. It follows that

$$
\begin{aligned}
\dot{V} \le & -\left(k_1 - \frac{\hat{m}_{11}+\hat{m}_{12}}{\hat{m}_{22}}\right) S_1^2 - \left(k_2 - \frac{k_1}{2} - \frac{\hat{m}_{11}+\hat{m}_{12}}{2\hat{m}_{22}}\right) S_2^2 \\
& -\left(\frac{1}{\tau_4} - \frac{3k_1}{2} - \frac{\hat{m}_{11}+\hat{m}_{12}}{2\hat{m}_{22}}\right) e^2 - \left(\frac{c_1}{\hat{m}_{11}} - \frac{\left(\hat{m}_{11}+\hat{m}_{12}\right) c_2}{\bar{A}} - 1\right) \tilde{\tau}_{d1}^2 \\
& -\left(\frac{c_4}{\hat{m}_{22}} - \frac{\hat{m}_{12}c_3}{\bar{A}} - 1\right) \tilde{\tau}_{d2}^2 + \bar{d}_2^2 + \frac{1}{4}\xi_1^2\xi_2^2 + M
\end{aligned}
\tag{9.132}
$$

According to Lemma 9.3, if the following inequalities are satisfied:

$$a_i > 0, i = 1, 2, 3, 4, 5 \tag{9.133}$$

then, we have

$$\dot{V} \le -2\gamma V + M_1 \tag{9.134}$$

Table 9.1 Exactly known physical parameters of the MWIP system

Mass of the wheel (kg)	Moment of inertia of the wheel (kg m^2)	Radius of the wheel (m)	Length between the wheel axle and the center of gravity of the body (m)	Viscous resistance in the driving system (N s/m)	Viscous resistance of the ground (N s/m)
$m_w = 29.0$	$I_w = 0.6$	$r = 0.254$	$l = 0.267$	$D_l = 0.1$	$D_w = 3.3$

where

$$M_1 = M + \frac{\xi_1^2 \xi_2^2}{4} + \frac{\bar{d}_2^2}{2\hat{m}_{12}} > 0 \tag{9.135}$$

$$\gamma = \min(a_1, a_2, a_3, a_4, a_5) > 0 \tag{9.136}$$

By the selections of k_1, k_2, τ_4 and c_1, c_2, c_3, c_4, we can make $\gamma > M_1/2p$. This results in $\dot{V} \leq 0$ on $V = p$. Thus, $V \leq p$ is an invariant set, i.e., if $V(0) \leq p$ then $V(t) \leq p$ for all $t \geq 0$. Therefore, $V(t)$ is bounded, so are S_1, S_2, e and $\tilde{\tau}_{d1}, \tilde{\tau}_{d2}$. According to above analysis and Lemma 9.3, the overall closed-loop control system is semiglobally uniformly and ultimately bounded. This completes the proof. □

9.5 Simulation study

9.5.1 Simulation of SMC for MWIP

Consider an MWIP system whose exactly known system parameters are listed in Table 9.1. To test the robustness of the proposed controllers, it is assumed that the actual physical parameters of the body are different from that of the nominal system. Table 9.2 shows the actual and nominal body parameters. An LQR controller designed by linearizing nominal system was found cannot stabilize the actual system.

9.5.1.1 Simulation study for the Type-I SMC controller

The function F_1 in (9.27), the parameter β_1 and η were selected as $0.1 \times \left|\hat{f}_1\right|$, 0.1 and 1, respectively. Function k was given by

$$k = -(F_1 + \beta_1 |r| + \eta)/(1 - \beta_1) - 1 \tag{9.137}$$

Assume the desired moving velocity is $x_2{}^* = \dot{\theta}_w^* = 7$ rad/s. From (9.18) we can compute the estimated equilibrium $\hat{\mathbf{x}}^*$, which is used to construct the initial sliding surface (9.17). Therefore, the final SMC controller (9.29) can be obtained from Theorem 9.2 by selecting appropriate λ_1 and λ_2.

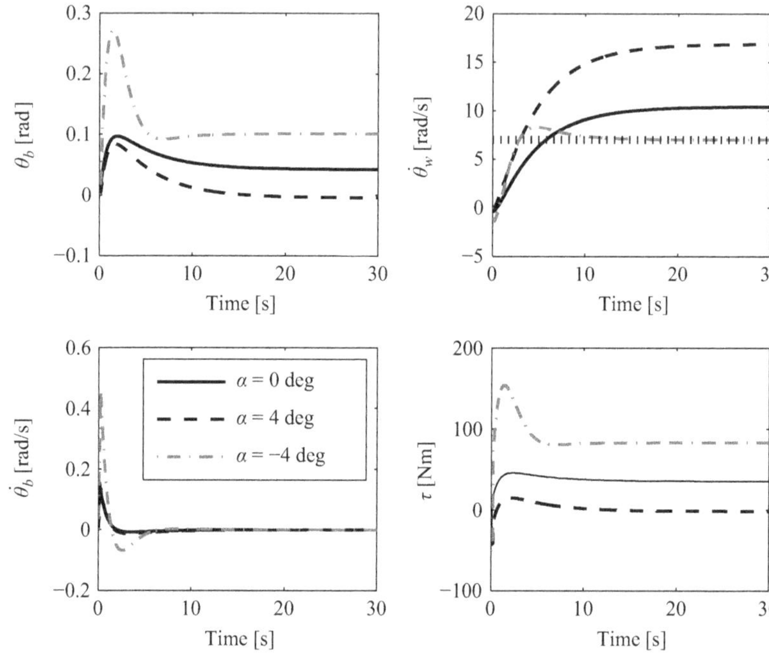

Figure 9.3 System trajectories of using the Type-I SMC controller on different slopes. (Black dot line denotes the desired velocity.)

To validate the robustness of the Type-I SMC controller, we performed the velocity control simulation of an MWIP system on different slopes. Three different cases of ground slope $\alpha = 0°$, $\alpha = 4°$ and $\alpha = -4°$ were considered in the simulation. The simulation results are shown in Figure 9.3. In all cases, the system was stabilized while the modeling uncertainties and external disturbance exist. Note that there are unavoidable steady tracking errors in all the three cases.

9.5.1.2 Simulation study for the Type-II SMC controller

The function F_1 in (9.51), the parameter β_2 and η were selected as $0.1 \times \left|\hat{f}_2\right|$, 0.1 and 1, respectively. Function $\tilde{k}$ was given by

$$\tilde{k} = -(F_2 + \beta_2 |\tilde{r}| + \eta)/(1 - \beta_2) - 1 \tag{9.138}$$

Assume the desired moving velocity is $x_2{}^* = \dot{\theta}_w^* = 7$ rad/s. According to Theorem 9.4, coefficients μ_1, μ_2, μ_3 were selected as $\mu_1 = 1.5$, $\mu_2 = 0.01$ and $\mu_3 = 0.05$. Three different cases of ground slope $\alpha = 0°$, $\alpha = 4°$ and $\alpha = -4°$ were also assumed. The final simulation results of using the Type-II SMC controller are depicted in Figure 9.4. In all the three cases, the MWIP system is asymptotically stable with very small steady tracking error, which substantiates the efficiency of the presented

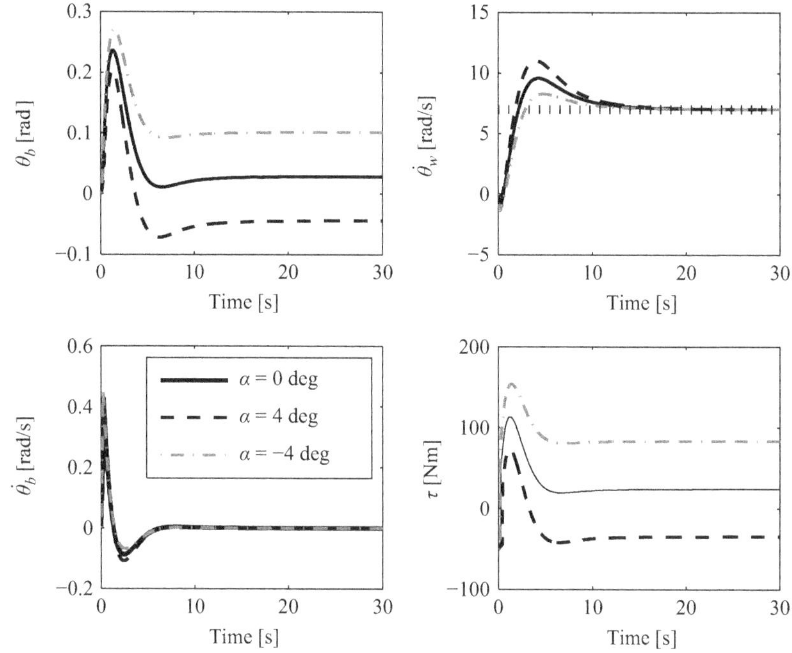

Figure 9.4 System trajectories of using the Type-II SMC controller on different slopes. (Black dot line denotes the desired velocity.)

approach. If the physical parameters of MWIP system are very close to that of the nominal system, the maximum slope we achieved in the simulation is about 45 degree. This also demonstrates the robustness of proposed approach. Compared with the Type-I SMC controller, the most significant advantage of the Type-II SMC controller is its ability to eliminate the steady tracking error.

9.5.2 Simulation of DSCNDO for MWIP

In this simulation, we use the same parameter assumptions as given in the previous subsection. The dissipation parameter D_w is assumed to be Gaussian random variable with known covariance (0.2 N s/m) and mean value (0.5 N s/m). And the external disturbances are assumed as:

$$\tau_{\text{ext1}} = 30\sin(2t + \pi/2)(\text{N m}), \tau_{\text{ext2}} = 0.2\sin(t)(\text{N m}). \tag{9.139}$$

An obvious equilibrium of the MWIP system can be easily obtained:

$$x^* = [x_2{}^*, x_3{}^*, x_4{}^*]^T = [0, 0, 0]^T \tag{9.140}$$

Based on (9.111) and Tables 9.1 and 9.2, the controller parameters are chosen as

$$k_1 = 6, \quad k_2 = 6, \quad \tau_4 = 0.1, \quad c_1 = c_4 = 100, \quad c_2 = c_3 = 0. \tag{9.141}$$

Table 9.2 Uncertain body parameters of the MWIP system

	Mass of the body (kg)	Moment of inertia of the body (kg m^2)
Actual system	$m_b = 310.6$	$I_b = 65.0$
Nominal system	$\hat{m}_b = 137.6$	$\hat{I}_b = 35.0$

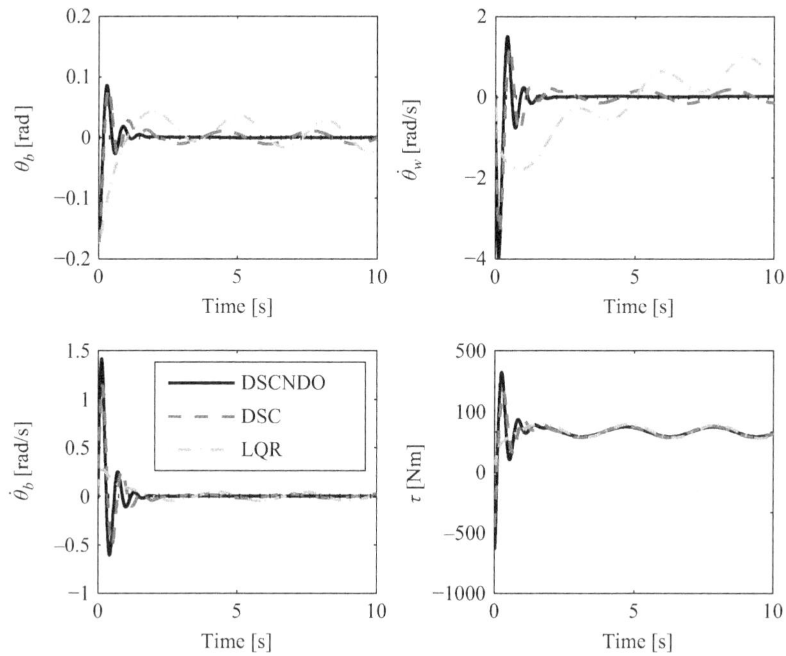

Figure 9.5 System trajectories of using DSCNDO, DSC and LQR control strategies in the balance control of MWIP system ($\theta_b(0) = -15°$)

The equilibrium control simulation results of the actual MWIP system with considering any uncertainties and disturbances using DSCNDO, DSC and LQR controller are shown in Figure 9.5. Suppose that the initial conditions of the system are $x(0) = [-0.2617, 0, 0]^T$.

As shown in Figure 9.5, the LQR controller cannot suppress the effect of system uncertainties and disturbance. The two DSC controllers have better performance than the LQR controller. Note that although the DSC controllers can make the state variables to be semiglobally uniformly and ultimately bounded, they do not guarantee that the states converge to the desired value. There are small oscillations in the response trajectories all the time due to the external disturbances. Compared with a pure DSC controller, the control performance seems better when using a DSCNDO because the amplitudes of oscillations are significantly reduced.

In a word, the simulation results show that the DSCNDO controller is more effective than a pure DSC or LQR controller when there exists model uncertainties and external disturbances.

9.6 Conclusion

Robust controller design for underactuated systems is a hard problem. In this chapter, we proposed two kinds of robust controllers for the MWIP, which is a typical underactuated system. The first kind of robust controllers is based on SMC, which is very suitable for the MWIP suffering from both parameter uncertainties and external disturbances. With a specially designed sliding surface, the proposed SMC controller not only stabilizes the MWIP but also eliminates the steady tracking error even the equilibrium is uncertain.

Considering the deficiency of chattering in SMC controllers, a novel global coordinate transformation was proposed to achieve the DSC control of an MWIP system. The proposed method partially solves the difficulty in designing a DSC controller for a non-Class-I type underactuated mechanical system. To compensate for parametric uncertainties in a real MWIP-based robot system as well as external disturbances, we combined the proposed DSC controller and an NDO. The efficiency of all proposed controllers is validated by numerical simulations.

Acknowledgment

This work was supported by the National Natural Science Foundation of China under Grant number 61473130, the Science Fund for Distinguished Young Scholars of Hubei Province (2015CFA047), the Fundamental Research Funds for the Central Universities (HUST: 2015TS028) and the Program for New Century Excellent Talents in University (NCET-12-0214).

References

[1] Pathak, K., Franch, J., Agrawal, S.K.: 'Velocity and position control of a wheeled inverted pendulum by partial feedback linearization', *IEEE Transactions on Robotics*, 2005, **21**, (3), pp. 505–513.

[2] Salerno, A., Angeles, J.: 'The control of semi-autonomous two-wheeled robots undergoing large payload-variations', *Proc. IEEE Int. Conf. on Robotics and Automation*, New Orleans, LA, 2004, pp. 1740–1745.

[3] Ha, Y.-S. and Yuta, S.: 'Trajectory tracking control for navigation of the inverse pendulum type self-contained mobile robot', *Robotics and Autonomous Systems*, 1996, **17**, pp. 65–80.

[4] Baloh, M. and Parent, M.: 'Modeling and model verification of an intelligent self-balancing two-wheeled vehicle for an autonomous urban transportation system', *Proc. Conf. Comp. Intelligence, Robot. Autonom. Syst.*, Singapore, 2003, pp. 1–7.

[5] Grasser, F., D'Arrigo, A., Colombi, S. and Rufer, A.: 'Joe: A mobile, inverted pendulum,' *IEEE Transactions on Industrial Electronics*, 2002, **49**, 1, pp. 107–114.
[6] 'nBot Balancing Robot', http://www.geology.smu.edu/~dpa-www/robo/nbot/, accessed 14 Sep 2013.
[7] 'Steve's LegWay' http://www.teamhassenplug.org/robots/legway/, accessed 17 Jan 2003.
[8] Tirmant, H., Baloh, M., Vermeiren, L., Guerra, T. M. and Parent, M.: 'B2, an alternative two wheeled vehicle for an automated urban transportation system', *Proc. IEEE Intelligent Vehicle Symposium*, Paris, France, 2002, pp. 594–603.
[9] 'SegWay' http://www.segway.com/, accessed 15 Aug 2016.
[10] Karkoub, M., Parent, M.: 'Modelling and non-linear feedback stabilization of a two-wheel vehicle', *Proceedings of the Institution of Mechanical Engineers, Part I: Journal of Systems and Control Engineering*, 2004, **218**, 8, pp. 675–686.
[11] Huang, J., Ding, F., Fukuda, T., Matsuno, T.: 'Modeling and Velocity Control for a Novel Narrow Vehicle Based on Mobile Wheeled Inverted Pendulum', *IEEE Transactions on Control Systems Technology*, 2013, **21**, 5, pp. 1607–1617.
[12] Spong, M. W.: 'The swing up control problem for the acrobat', *IEEE Control Systems Magazine*, 1995, **15**, 1, pp. 49–55.
[13] Fantoni, I., Lozano, R., and Spong, M. W.: 'Energy based control of the pendubot', *IEEE Transactions on Automatic Control*, 2000, **45**, 4, pp. 725–729.
[14] Gu, Y.-L.: 'Direct adaptive control scheme for underactuated dynamic systems', *Proc. IEEE Conference on Decision and Control*, San Antonio, Texas, 1993, pp. 1625–1627.
[15] Bloch, A. M., Leonard, N. E. and Marsden, J. E.: 'Stabilization of the pendulum on a rotor arm by the method of controlled Lagrangians', *Proc. IEEE International Conference on Robotics and Automation*, Piscataway, New Jersey, 1999, pp. 500–505.
[16] Sankaranarayanan, V. and Mahindrakar, A. D.: 'Control of a class of underactuated mechanical systems using sliding modes', *IEEE Transactions on Robotics*, 2009, **25**, 2, pp. 459–467.
[17] Park, B. S., Yoo, S. J., Park, J. B. and Choi, Y. H., 'Adaptive neural sliding mode control of non-holonomic wheeled mobile robots with model uncertainty', *IEEE Transactions on Control Systems Technology*, 2009, **17**, 1, pp. 207–214.
[18] Swaroop, D., Gerdes, J. C., Yip, P. P. and Hedrick, J. K.: 'Dynamic surface control of nonlinear systems', *Proc. the 16th American Control Conference (ACC)*, Albuquerque, USA: NM, 1997, pp. 3028–3034.
[19] Oh, S.-R., Sun, J., Li, Z.: 'Path following control of underactuated marine vessels via dynamic surface control technique', *Proc. the ASME Dynamic Systems and Control Conference*, 2008, pp. 81–88.
[20] Qaiser, N., Iqbal, N. and Hussain, A.: 'Exponential stabilization of a class of underactuated mechanical systems using dynamic surface control', *International Journal of Control, Automation and Systems*, 2007, **5**, 5, pp. 547–558.

[21] Huang, J., Guan, Z.-H., Matsuno, T., Fukuda T. and Sekiyama, K.: 'Sliding mode velocity control of mobile wheeled inverted pendulum systems', *IEEE Transactions on Robotics*, 2010, **26**, 4, pp. 750–758.

[22] Huang, J., Ri, S., Liu, L., Wang, Y., Kim, J. and Pak, G.: 'Nonlinear disturbance observer-based dynamic surface control of mobile wheeled inverted pendulum', *IEEE Transactions on Control Systems Technology*, 2015, **23**, 6, pp. 2400–2407.

[23] Mohammadi, A., Tavakoli, M., Marquez, H. J. and Hashemzadeh, F.: 'Nonlinear disturbance observer design for robotic manipulators', *Control Engineering Practice*, 2013, **21**, pp. 253–267.

[24] Chen, W. H.: 'Disturbance observer based control for nonlinear systems', *IEEE/ASME Transactions on Mechatronics*, 2004, **9**, 4, pp. 706–710.

[25] Wu, J., Huang, J., Wang, Y. and Xing, K.: 'Nonlinear disturbance observer-based dynamic surface control for trajectory tracking of pneumatic muscle system', *IEEE Transactions on Control Systems Technology*, 2014, **22**, 2, pp. 440–455.

[26] Wei, X., Zhang, H. F. and Guo, L.: 'Composite disturbance-observer-based control and terminal sliding mode control for uncertain structural systems', *International Journal of Systems Science*, 2009, **40**, 10, pp. 1009–1017.

[27] Yang, J., Li, S. and Yu, X.: 'Sliding-mode control for systems with mismatched uncertainties via a disturbance observer', *IEEE Transactions on Industrial Electronics*, 2013, **60**, 1, pp. 160–169.

[28] Formal'skii, A. M.: 'An inverted pendulum on a fixed and a moving base', *Journal of Applied Mathematics and Mechanics*, 2006, **70**, pp. 56–64.

[29] Lee, K., Coverstone-Carroll, V.: 'Control algorithms for stabilizing underactuated robots', *Journal of Intelligent and Robotic Systems*, 1998, **15**, 12, pp. 681–697.

[30] Ashrafiuon, H., Erwin, R. S.: 'Sliding control approach to underactuated multibody systems', *Proc. American Control Conference*, Boston, MA, 2004, pp. 1283–1288.

[31] Olfati-Saber, R.: '*Nonlinear Control of Underactuated Mechanical Systems with Application to Robotics and Aerospace Vehicles*', PhD Thesis, 1–307, MIT, USA, 2001.

[32] Yang, Z. J., Nagai, T., Kanae, S. and Wada, K.: 'Dynamic surface control approach to adaptive robust control of nonlinear systems in semi-strict feedbackform', *International Journal of Systems Science*, 2007, **38**, 9, pp. 709–724.

[33] Qu, Z., Dawson, D. M., Lim, S. Y. and Dorsey, J. F.: 'A new class of robust control laws for tracking of robots', *International Journal of Robotics Research*, 1994, **13**, 4, pp. 355–363.

Chapter 10

Case studies on non-linear control theory of the inverted pendulum

Xu Xu[1], *Hongwei Zhang*[1], *and Giuseppe Carbone*[1]

Abstract

This chapter deals with the control of inverted pendulums through non-linear control theory. Simulations of the proposed control method are carried out in MATLAB® environment with satisfactory results in controlling systems of single and double-link inverted pendulum installed in a cart, for a range of starting positions. The proposed control technique can be extended to the control of many other non-linear systems. Examples are proposed as referring to design and teaching applications. In particular, a single inverted pendulum system constructed using Lego bricks and controlled by Lego Mindstorms EV3 is used to prove the user-friendliness of the proposed control method as implemented by Master students at Sheffield Hallam University. Moreover, 3 degrees of freedom parallel manipulator Cassino Parallel Manipulator is investigated for replacing a cart and providing three dimensional motions to an inverted pendulum while minimising dynamics effects at design stage.

10.1 Introduction

The control of inverted pendulums and cart systems is a widely addressed topic in control literature, such as reported in [1–3]. In fact, the above-mentioned non-linear systems are highly unstable, and they cannot maintain the upright position without an appropriate control. One approach to the control problem of an inverted pendulum on a cart is to linearise the system locally around the upright position by approximating the non-linear terms to linear ones. Traditionally, inverted pendulum systems are controlled by an open-loop swing up controller forcing the pendulum to reach the top equilibrium, and then switching to a closed-loop linear controller to control the pendulum from this point to maintain the upright position [3,4]. Some more recent publications, such as in [5,6], the authors proposed non-linear techniques to extend the stability region to a bigger neighbourhood of the upright position, but these methods

[1]Department of Engineering and Mathematics, Sheffield Hallam University, UK

still involved switching between at least two control strategies and are not in the true sense of global control. For a comprehensive and thorough review of non-linear control strategies in controlling inverted pendulum systems, we refer the readers to Boubaker's survey article in 2013 [7].

In this chapter, we study an approach to non-linear control theory that is based on a generalisation of the linear quadratic optimal control theory. This approach gives approximations to non-linear optimal control systems, by solving a Riccati equation. There is a brief introduction of the theory, and detailed derivation of it can be found in [8–10]. The inverted pendulum and cart control problem can be solved by applying this non-linear control technique, so the pendulum can be controlled globally. In particular, this non-linear control technique can be easily applied to more complicated versions of the inverted pendulum and cart family: multi-link inverted pendulum and cart systems, which has not been reported in the existing literature.

In Section 10.2, a single-link pendulum is modelled and analysed. This is then extended to a double-link pendulum (the general case being similar), and Lagrange's equations for the system are derived. In Section 10.3.1, linearized equations of the pendulum (where the mass is concentrated at the top) in the form of

$$\dot{\boldsymbol{x}} = \boldsymbol{A}\boldsymbol{x} + \boldsymbol{B}\boldsymbol{u} \tag{10.1}$$

are studied. If matrices $\boldsymbol{A}$ and $\boldsymbol{B}$ form a controllable pair, then the local optimal feedback control $\boldsymbol{u}$ is given by

$$\boldsymbol{u} = -\boldsymbol{R}^{-1}\boldsymbol{B}^T\boldsymbol{P}\boldsymbol{x}, \tag{10.2}$$

where, $\boldsymbol{P}$ satisfies the algebraic Riccati equation and $\boldsymbol{R}$ is a weighting matrix. In Section 10.3.2, the global pendulum model in the form of

$$\dot{\boldsymbol{x}} = \boldsymbol{A}(\boldsymbol{x})\boldsymbol{x} + \boldsymbol{B}(\boldsymbol{x})\boldsymbol{u}, \tag{10.3}$$

and the corresponding controllability of $(\boldsymbol{A}(\boldsymbol{x}), \boldsymbol{B}(\boldsymbol{x}))$ for 'almost all' $\boldsymbol{x}$ (i.e. controllable, apart possibly from a set of measure zero [11]) are analysed. The $\boldsymbol{x}$ in $\boldsymbol{A}(\boldsymbol{x})$ and $\boldsymbol{B}(\boldsymbol{x})$ is fixed for each step; therefore, the non-linear model resembles the linear case for one step. This method also demonstrates effective robust control because of the update at each step, so any inaccuracy of $\boldsymbol{A}$ and $\boldsymbol{B}$ is adjusted automatically. Furthermore, the global control $\boldsymbol{u}$ is shown to be

$$\boldsymbol{u} = -\boldsymbol{R}^{-1}\boldsymbol{B}(\boldsymbol{x})^T\boldsymbol{P}(\boldsymbol{x})\boldsymbol{x}, \tag{10.4}$$

where $\boldsymbol{P}(\boldsymbol{x})$ satisfies the pointwise algebraic Riccati equation:

$$\boldsymbol{A}^T(\boldsymbol{x})\boldsymbol{P}(\boldsymbol{x}) + \boldsymbol{P}(\boldsymbol{x})\boldsymbol{A}(\boldsymbol{x}) + \boldsymbol{Q} - \boldsymbol{P}(\boldsymbol{x})\boldsymbol{B}(\boldsymbol{x})\boldsymbol{R}^{-1}\boldsymbol{B}(\boldsymbol{x})^T\boldsymbol{P}(\boldsymbol{x}) = \boldsymbol{0}, \tag{10.5}$$

where $\boldsymbol{Q}$ and $\boldsymbol{R}$ are weighting matrices. In Section 10.4, some simulation examples are demonstrated to verify the above non-linear control theory of the inverted pendulum. It is shown that the single pendulum installed on a cart can be controlled from a large range of initial positions, including the rest position where the pendulum hangs downwards. The double-link pendulum and car system can also be controlled from many initial positions.

Case studies have been proposed for demonstrating the proposed control architecture with single and double pendulums. In particular, in Section 10.5, a single inverted pendulum system constructed using Lego bricks and controlled by Lego Mindstorms EV3 is used to prove the user-friendliness of the proposed control method as implemented by Master students at Sheffield Hallam University. Then in Section 10.6, a case study has been proposed as referring to a 3 degrees of freedom parallel manipulator CaPaMan (Cassino Parallel Manipulator), which is designed to replace a cart and provide three dimensional motions to an inverted pendulum while minimising dynamics effects at design stage.

10.2 Modelling of the cart–pendulum systems

Often inverted pendulums are considered in combination with moving carts. The system of a single pendulum installed on a cart is drawn in Figure 10.1.

The dynamical model of the cart and the pendulum are equations of motions often obtained by applying force analysis using free body diagrams and Newton's second law $F = ma$. However, there are other methods available for achieving a system's dynamical model, for example the Lagrangian approach which calculates the difference between total kinetic energy T and the total potential energy V of the system:

$$L = T - V \tag{10.6}$$

and the Hamiltonian equation which calculates the sum of the two types of energy:

$$H = T + V. \tag{10.7}$$

It is usually easier to use the Lagrangian method than the one based on force analysis because all is required are the generalised kinetic and potential energy terms, so resolving of the forces (which is often complicated) is not needed. We adopt the

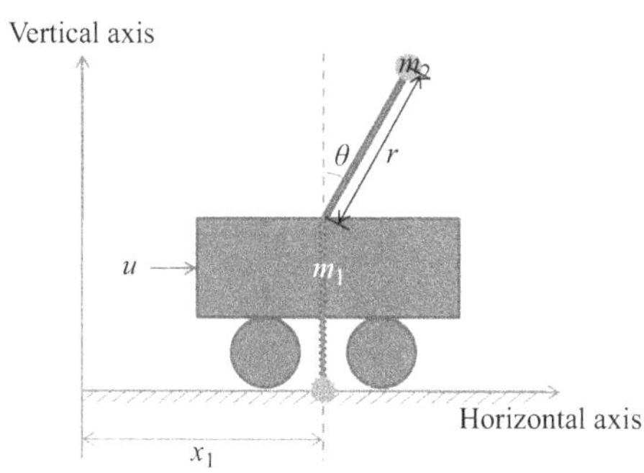

Figure 10.1 The single inverted pendulum and cart diagram

Lagrangian approach for its simplicity, where the Lagrange's equations [12,13] are given by

$$\frac{d}{dt}\frac{\partial L}{\partial \dot{x}_i} - \frac{\partial L}{\partial x_i} = f_i, \quad 1 \le i \le n, \tag{10.8}$$

where x_i represents the ith generalised coordinate and f_i the ith generalised force applied on the object. These Lagrange's equations are equivalent to Newton's laws [14].

In the case of a single pendulum–cart system, there are two x variables shown in Figure 10.1, namely the horizontal distance x_1 (m) travelled by cart from the left reference, and the angle θ (rad) between the pendulum rod and the vertical axis. $\dot{x}_1$ and $\dot{\theta}$ represent velocity of the cart along the horizontal axis and angular velocity of the rod around the rod-cart connection point, respectively. Here,

$$f_1 = u, f_2 = 0. \tag{10.9}$$

The total kinetic energy of the pendulum–cart system can be written as

$$\begin{aligned} T &= \frac{1}{2}m_1\dot{x}_1^2 + \frac{1}{2}m_2\left[\frac{d}{dt}(x_1 + r\sin\theta)\right]^2 + \frac{1}{2}m_2\left[\frac{d}{dt}(r\cos\theta)\right]^2 \\ &= \frac{1}{2}m_1\dot{x}_1^2 + \frac{1}{2}m_2\left(\dot{x}_1 + r\dot{\theta}\cos\theta\right)^2 + \frac{1}{2}m_2\left(-r\dot{\theta}\sin\theta\right)^2 \\ &= \frac{1}{2}(m_1 + m_2)\dot{x}_1^2 + m_2 r\dot{x}_1\dot{\theta}\cos\theta + \frac{1}{2}m_2 r^2\dot{\theta}^2, \end{aligned} \tag{10.10}$$

where m_1 and m_2 are the masses of cart and of pendulum, respectively, r denotes the length of the pendulum and g is acceleration due to gravity. The total potential energy of the system, using the bottom of the pendulum rest position as the vertical reference point, can be written as

$$V = m_2 g(r + r\cos\theta). \tag{10.11}$$

Therefore, the Lagrangian equation is given by

$$\begin{aligned} L &= T - V \\ &= \frac{1}{2}(m_1 + m_2)\dot{x}_1^2 + m_2 r\dot{x}_1\dot{\theta}\cos\theta + \frac{1}{2}m_2 r^2\dot{\theta}^2 - m_2 gr(1 + \cos\theta) \end{aligned} \tag{10.12}$$

Substitute (10.12) into (10.8), we obtain

$$\begin{aligned} \ddot{x}_1 &= \frac{m_2 r\dot{\theta}^2\sin\theta - m_2 g\sin\theta\cos\theta + u}{m_1 + m_2\sin^2\theta}, \\ \ddot{\theta} &= \frac{-m_2 r\dot{\theta}^2\sin\theta\cos\theta + m_2 g\sin\theta + m_1 g\sin\theta - u\cos\theta}{r\left(m_1 + m_2\sin^2\theta\right)}, \end{aligned} \tag{10.13}$$

where u represents the control as the force applied on the cart indicated in Figure 10.1. The equations in (10.13) satisfy Newtown's second law [14]. Note the two equations in (10.3) both have second derivatives on the left-hand-side and are not yet in the standard state-space model form. A state-space representation of the system can be written as

$$\begin{aligned}
\dot{x}_1 &= x_2 \\
\dot{x}_2 &= \frac{m_2 r x_4^2 \sin x_3 - m_2 g \sin x_3 \cos x_3 + u}{m_1 + m_2 \sin^2 x_3} \\
\dot{x}_3 &= x_4 \\
\dot{x}_4 &= \frac{-m_2 r x_4^2 \sin x_3 \cos x_3 + m_2 g \sin x_3 + m_1 g \sin x_3 - u \cos x_3}{r\left(m_1 + m_2 \sin^2 x_3\right)}
\end{aligned} \tag{10.14}$$

by changes of variables, such that $x_2 = \dot{x}_1$, $x_3 = \theta$ and $x_4 = \dot{x}_3$ and splitting each of the (10.13) equations into two equations. This translates into the state-space matrix form as

$$\begin{pmatrix} \dot{x}_1 \\ \dot{x}_2 \\ \dot{x}_3 \\ \dot{x}_4 \end{pmatrix} = \begin{pmatrix} 0 & 1 & 0 & 0 \\ 0 & 0 & \frac{-m_2 g \sin x_3 \cos x_3}{\left(m_1 + m_2 \sin^2 x_3\right) x_3} & \frac{m_2 r x_4 \sin x_3}{m_1 + m_2 \sin^2 x_3} \\ 0 & 0 & 0 & 1 \\ 0 & 0 & \frac{(m_1 + m_2) g \sin x_3}{r\left(m_1 + m_2 \sin^2 x_3\right) x_3} & \frac{-m_2 r x_4 \sin x_3 \cos x_3}{r\left(m_1 + m_2 \sin^2 x_3\right)} \end{pmatrix} \begin{pmatrix} x_1 \\ x_2 \\ x_3 \\ x_4 \end{pmatrix} + \begin{pmatrix} 0 \\ \frac{1}{m_1 + m_2 \sin^2 x_3} \\ 0 \\ \frac{-\cos x_3}{r\left(m_1 + m_2 \sin^2 x_3\right)} \end{pmatrix} u, \tag{10.15}$$

One particular advantage of this method is that it can be easily extended to a general multi-link pendulum case, without the need to perform complex force analysis on the new and previous pendulum objects. For example a double pendulum and cart system is illustrated in Figure 10.2, where θ_1 and θ_2 represent the angles between the first and the second pendulum rods and the vertical axis and u denotes the force (control) applied on the cart.

A similar energy analysis can be performed as before. Using the new vertical and horizontal references as indicated in Figure 10.2, the total kinetic energy and potential energy of the new system can be modified to

$$\begin{aligned}
T_2 &= T + \frac{1}{2} m_3 \left[\frac{d}{dt}(x_1 + r_1 \sin\theta_1 + r_2 \sin\theta_2) \right]^2 + \frac{1}{2} m_3 \left[\frac{d}{dt}(r_1 \cos\theta_1 + r_2 \cos\theta_2) \right]^2 \\
&= \frac{1}{2}(m_1 + m_2 + m_3)\dot{x}_1^2 + \frac{1}{2}(m_2 + m_3) r_1^2 \dot{\theta}_1^2 + \frac{1}{2} m_3 r_2^2 \dot{\theta}_2^2 + (m_2 + m_3) r_1 \dot{x}_1 \dot{\theta}_1 \cos\theta_1 \\
&\quad + m_3 r_2 \dot{x}_1 \dot{\theta}_2 \cos\theta_2 + m_3 r_1 r_2 \dot{\theta}_1 \dot{\theta}_2 \cos(\theta_1 - \theta_2),
\end{aligned} \tag{10.16}$$

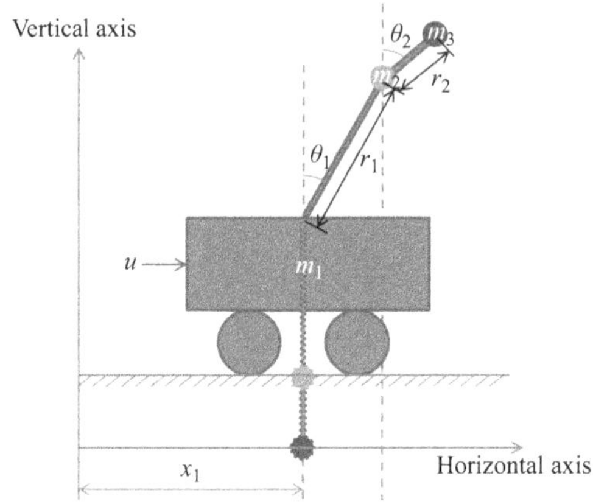

Figure 10.2 The double-link inverted pendulum and cart system diagram

and

$$
\begin{aligned}
V_2 &= V + m_3 g(r_1 + r_2 + r_1\cos\theta_1 + r_2\cos\theta_2) \\
&= (m_2 + m_3) g r_1 (1 + \cos\theta_1) + m_3 g r_2 (1 + \cos\theta_2),
\end{aligned} \tag{10.17}
$$

which lead to the new Lagrange's equation

$$
\begin{aligned}
L_2 &= T_2 - V_2 \\
&= \frac{1}{2}(m_1 + m_2 + m_3)\dot{x}_1^2 + \frac{1}{2}(m_2 + m_3) r_1^2 \dot{\theta}_1^2 + \frac{1}{2} m_3 r_2^2 \dot{\theta}_2^2 + (m_2 + m_3) r_1 \dot{x}_1 \dot{\theta}_1 \cos\theta_1 \\
&\quad + m_3 r_2 \dot{x}_1 \dot{\theta}_2 \cos\theta_2 + m_3 r_1 r_2 \dot{\theta}_1 \dot{\theta}_2 \cos(\theta_1 - \theta_2) - (m_2 + m_3) g r_1 (1 + \cos\theta_1) \\
&\quad - m_3 g r_2 (1 + \cos\theta_2),
\end{aligned} \tag{10.18}
$$

where m_3 denotes the mass of the newly added pendulum, and r_1 and r_2 are the lengths of the original rod and the new rod, respectively. The energy equations appear complicated; however, the analysis performed above is relatively straightforward in the sense that effect only comes from the new pendulum and corresponding energy terms can simply be added to the original equations.

The process of obtaining a state-space model for the double-link pendulum cart system is also similar as the one is the single pendulum case. By solving the Langrange's equation in (10.8) and splitting each differential equation with a second derivative into two equations containing only first derivatives (i.e. let $x_2 = \dot{x}_1, x_3 = \theta_1$, $x_4 = \dot{x}_3, x_5 = \theta_2$ and $x_6 = \dot{x}_5$), we obtain standard state-space model of the double-link pendulum cart system in matrix form:

$$\begin{pmatrix} \dot{x}_1 \\ \dot{x}_2 \\ \dot{x}_3 \\ \dot{x}_4 \\ \dot{x}_5 \\ \dot{x}_6 \end{pmatrix} = \begin{pmatrix} 0 & 1 & 0 & 0 & 0 & 0 \\ 0 & 0 & \frac{m_2 ga \sin x_3 \cos x_3}{-x_3 d(x_3)} & \frac{m_2 r_1 a x_4 \sin x_3}{d(x_3)} & 0 & \frac{m_2 m_3 r_2 x_6 [\sin(x_5 - 2x_3) - \sin x_5]}{-2d(x_3)} \\ 0 & 0 & 0 & 1 & 0 & 0 \\ 0 & 0 & \frac{ge(x_5) \sin x_3}{2r_1 x_3 d(x_3)} & \frac{x_4 f(x_3, x_5)}{-2d(x_3)} & \frac{m_1 m_3 g b(x_3, x_5) \sin x_5}{-r_1 x_5 d(x_3)} & \frac{r_2 x_6 h(x_3, x_5)}{-r_1 d(x_3)} \\ 0 & 0 & 0 & 0 & 0 & 1 \\ 0 & 0 & \frac{m_1 gab(x_3, x_5) \sin x_3}{-r_2 x_3 d(x_3)} & \frac{m_1 r_1 a x_4 c(x_3, x_5)}{r_2 d(x_3)} & \frac{m_1 ga \cos^2 x_3 \sin x_5}{r_2 x_5 d(x_3)} & \frac{m_1 m_3 x_6 \sin(2x_5 - 2x_3)}{-2d(x_3)} \end{pmatrix}$$

$$\times \begin{pmatrix} x_1 \\ x_2 \\ x_3 \\ x_4 \\ x_5 \\ x_6 \end{pmatrix} + \begin{pmatrix} 0 \\ \frac{m_2 + m_3 c(x_3, x_5)^2}{d(x_3)} \\ 0 \\ \frac{m_2 \cos x_3 - m_3 c(x_3, x_5) \sin x_5}{-r_1 d(x_3)} \\ 0 \\ \frac{ac(x_3, x_5) \sin x_3}{-r_2 d(x_3)} \end{pmatrix} u, \tag{10.19}$$

where constant a and functions $b(x_3, x_5), c(x_3, x_5), d(x_3), e(x_5), f(x_3, x_5)$ and $h(x_3, x_5)$ have been defined as the following:

$$a = m_2 + m_3; \tag{10.20}$$

$$b(x_3, x_5) = \cos x_3 \cos x_5; \tag{10.21}$$

$$c(x_3, x_5) = \sin(x_3 - x_5); \tag{10.22}$$

$$d(x_3) = m_2 a \sin^2 x_3 + m_1 m_3 c^2 + m_1 m_2; \tag{10.23}$$

$$e(x_5) = 2m_2^2 + m_1 m_3 + 2m_2 m_3 + 2m_1 m_2 + m_1 m_3 \cos(2x_5); \tag{10.24}$$

$$f(x_3, x_5) = m_2 a \sin(2x_3) - m_1 m_3 \sin(2x_5 - 2x_3); \tag{10.25}$$

$$h(x_3, x_5) = m_2 m_3 \sin x_3 \cos x_5 + m_1 m_3 c. \tag{10.26}$$

Having now obtained the mathematical models (state-space representations) of the single pendulum installed on a cart and the double-link pendulum and cart system, we shall discuss how to achieve linear and non-linear quadratic optimal controls of the systems in theory in Sections 10.3.1 and 10.3.2, respectively.

10.3 Control analysis

10.3.1 Linear control analysis

A general linear control system has the following form:

$$\dot{\boldsymbol{x}} = \boldsymbol{A}\boldsymbol{x} + \boldsymbol{B}\boldsymbol{u}, \tag{10.27}$$

where $\boldsymbol{x}$ is the state variable vector, $\boldsymbol{u}$ is the control vector and $(\boldsymbol{A}, \boldsymbol{B})$ is a controllable pair, or at least a stabilisable pair, to guarantee finite solutions of the Ricatti equation (see later for details). Linearised equations of the pendulum–cart system where the mass is concentrated at the top are in the form of (10.27).

The quadratic infinite-time cost function [3]:

$$J = \frac{1}{2}\int_0^{\infty} (\boldsymbol{x}^T\boldsymbol{Q}\boldsymbol{x} + \boldsymbol{u}^T\boldsymbol{R}\boldsymbol{u})dt \tag{10.28}$$

leads to the linear optimal feedback control solution:

$$\boldsymbol{u} = -\boldsymbol{R}^{-1}\boldsymbol{B}^T\boldsymbol{P}\boldsymbol{x}, \tag{10.29}$$

where $\boldsymbol{R}$ is a weighting matrix and $\boldsymbol{P}$ is a positive-definite Hermitian or real symmetric matrix. $\boldsymbol{P}$ satisfies the algebraic Riccati equation:

$$\boldsymbol{A}^T\boldsymbol{P} + \boldsymbol{P}\boldsymbol{A} + \boldsymbol{Q} - \boldsymbol{P}\boldsymbol{B}\boldsymbol{R}^{-1}\boldsymbol{B}^T\boldsymbol{P} = \boldsymbol{0}. \tag{10.30}$$

The above Riccati equation can be solved using the LQR (linear-quadratic regulator) function in MATLAB [3]. To implement the optimal control solution, substitute (10.29) into (10.27), we obtain

$$\begin{aligned}\dot{\boldsymbol{x}} &= \boldsymbol{A}\boldsymbol{x} - \boldsymbol{B}\boldsymbol{R}^{-1}\boldsymbol{B}^T\boldsymbol{P}\boldsymbol{x}\\ &= [\boldsymbol{A} - \boldsymbol{B}\boldsymbol{R}^{-1}\boldsymbol{B}^T\boldsymbol{P}]\boldsymbol{x},\end{aligned} \tag{10.31}$$

which is a stable controlled system in the linearised case [3], as long as the condition of the observability of the pair $(\boldsymbol{Q}^{\frac{1}{2}}, \boldsymbol{A})$ is satisfied.

In the cases of the two inverted pendulum and cart systems analysed in Section 10.2, the non-linear state-space models of the systems can be considered as linear representations when the inverted pendulums are kept in a small neighbourhood of the vertical upright position. Then, for the single pendulum–cart case, we can assume that x_3 and x_4 (which is $\dot{x}_3$) are small quantities, and the following approximations can be made:

$$\sin x_3 \approx x_3,\ \cos x_3 \approx 1,\ \sin^2 x_3 \approx 0 \text{ and } x_4^2 \sin x_3 = \dot{x}_3^2 \sin x_3 \approx 0. \tag{10.32}$$

The linearised state-space model can be simplified substantially from (10.14) and can then be written into the form of (10.27), where

$$\boldsymbol{A}_1 = \begin{pmatrix} 0 & 1 & 0 & 0 \\ 0 & 0 & -\frac{m_2 g}{m_1} & 0 \\ 0 & 0 & 0 & 1 \\ 0 & 0 & \frac{(m_1+m_2)g}{rm_1} & 0 \end{pmatrix} \text{ and } \boldsymbol{B}_1 = \begin{pmatrix} 0 \\ \frac{1}{m_1} \\ 0 \\ -\frac{1}{rm_1} \end{pmatrix}. \tag{10.33}$$

Similarly, for the double-pendulum–cart system, to achieve the linearised state-space model, we also assume that x_3, x_4, x_5 and x_6 are small such that

$$\sin x_3 \approx x_3, \sin x_5 \approx x_5, \cos x_3 \approx \cos x_5 \approx 1, \sin^2 x_3 \approx \sin^2 x_5 \approx 0, \cos^2 x_3 \approx \cos^2 x_5 \approx 1, \sin^2(x_3 - x_5) \approx 0 \text{ and } x_4^2 \sin x_3 \approx x_4^2 \sin x_5 \approx x_6^2 \sin x_3 \approx x_6^2 \sin x_5 \approx 0. \tag{10.34}$$

These approximations simplify the state-space model of the double-link pendulum and cart system into the linearised form of (10.27), where

$$A_2 = \begin{pmatrix} 0 & 1 & 0 & 0 & 0 & 0 \\ 0 & 0 & -\frac{(m_2+m_3)g}{m_1} & 0 & 0 & 0 \\ 0 & 0 & 0 & 1 & 0 & 0 \\ 0 & 0 & \frac{Mg}{2r_1 m_1 m_2} & 0 & -\frac{m_3 g}{r_1 m_2} & 0 \\ 0 & 0 & 0 & 0 & 0 & 1 \\ 0 & 0 & -\frac{(m_2+m_3)g}{r_2 m_2} & 0 & \frac{(m_2+m_3)g}{r_2 m_2} & 0 \end{pmatrix} \text{ and } B_2 = \begin{pmatrix} 0 \\ \frac{1}{m_1} \\ 0 \\ -\frac{1}{rm_1} \\ 0 \\ 0 \end{pmatrix}, \tag{10.35}$$

and $M = 2m_2^2 + m_1 m_3 + 2m_2 m_3 + 2m_1 m_2 + m_1 m_3$.

The linear models represent the systems adequately when the starting positions of the inverted pendulums are near the vertical upright positions. Because the linear optimal feedback controlled systems are stable at the equilibriums where $\boldsymbol{x} = 0$, the linear representations of the systems will always be valid as long as the initial positions are within a small neighbourhood of the vertical upright position. If the initial positions are not within this range, however, the linear models obtained in this section are not close approximations of the original systems and the control solutions achieved in (10.29) do not provide satisfactory control.

Traditionally, the inverted pendulum systems are controlled by an open-loop swing up controller forcing the pendulum to reach the top equilibrium, and then switching to a linear feedback controller (e.g. a linear quadratic optimal controller as the one we discussed in this section) to control the pendulum from this point. In Section 10.3.2, we shall consider and analyse a freezing technique to control non-linear dynamical systems originally introduced by Banks and Mhana in 1992 [8], using the theory of linear quadratic optimal control. The inverted pendulum systems can be controlled from a much wider range of starting positions by applying this non-linear optimal control technique.

10.3.2 Non-linear control analysis

A general non-linear control system can be represented by the following:

$$\dot{\boldsymbol{x}} = \tilde{\boldsymbol{f}}(\boldsymbol{x}, \boldsymbol{u}), \quad \boldsymbol{x}(0) = \boldsymbol{x}_0. \tag{10.36}$$

If $\tilde{\boldsymbol{f}}(\boldsymbol{x}, \boldsymbol{u})$ is a linear-analytic function, the general non-linear system equation shown in (10.36) is equivalent to

$$\dot{\boldsymbol{x}} = \boldsymbol{f}(\boldsymbol{x}) + \boldsymbol{g}(\boldsymbol{x})\boldsymbol{u}, \tag{10.37}$$

where $\boldsymbol{f}(\boldsymbol{x})$ and $\boldsymbol{g}(\boldsymbol{x})$ are non-linear, analytic functions and the controller $\boldsymbol{u}$ is affine.

If $\tilde{\boldsymbol{f}}(\boldsymbol{x}, \boldsymbol{u})$ is not a linear-analytic function, the system can always be written in the linear analytic form presented in (10.37) as long as the mild condition of the controller function $\boldsymbol{u}$ being differentiable is satisfied, although the trade-off is the increase of the dimensionality of the state space. This is done by introducing a new control function $\boldsymbol{v}$ such that

$$\dot{\boldsymbol{u}} = \boldsymbol{v}. \tag{10.38}$$

Then a new state variable y can be defined as

$$y = \begin{pmatrix} x \\ u \end{pmatrix}. \tag{10.39}$$

Now a new system representing the original one in (10.36) but in linear analytic form can be written as

$$\begin{aligned} \dot{y} &= \begin{pmatrix} \tilde{\boldsymbol{f}}(\boldsymbol{x}, \boldsymbol{u}) \\ \boldsymbol{v} \end{pmatrix} \\ &= \begin{pmatrix} \tilde{\tilde{\boldsymbol{f}}}(y) \\ \boldsymbol{0} \end{pmatrix} + \begin{pmatrix} \boldsymbol{0} \\ \boldsymbol{I} \end{pmatrix} \boldsymbol{v} \\ &= \boldsymbol{f}(y) + \boldsymbol{g}(y)\boldsymbol{v}, \end{aligned} \tag{10.40}$$

where $\boldsymbol{I}$ is the identity matrix and the new controller $\boldsymbol{v}$ is affine. It can be seen now that (10.40) is of the same form as (10.37), even though the new state space has a higher dimension.

Note that in the cases when $\tilde{\boldsymbol{f}}(\boldsymbol{x}, \boldsymbol{u})$ is not linear-analytic and the technique of rewriting the system into a linear-analytic one as explained above is adopted, the new controller function $\boldsymbol{v}$ determined might not be the optimal control for the original system, as the original system might have a non-differentiable control $\boldsymbol{u}$ which works better. However, this situation is rare, as the optimal controller would be differentiable for the vast majority of physical systems. So most physical system models obtained are either in linear-analytic form or can be easily rewritten into a higher dimensional linear-analytic function, and this process does not affect the solution of the optimal controller.

The non-linear function $\boldsymbol{f}(\boldsymbol{x})$ can be taken out of (10.37) for further analysis. The function $\boldsymbol{f}(\boldsymbol{x})$ can be written analytically into a matrix function

$$\boldsymbol{f}(x) = A(x)x, \tag{10.41}$$

provided it satisfies the condition

$$\boldsymbol{f}(x) = \boldsymbol{0} \quad at \quad x = \boldsymbol{0}. \tag{10.42}$$

The above condition essentially means that the non-linear system $\dot{\boldsymbol{x}} = \boldsymbol{f}(\boldsymbol{x})$ has an equilibrium at the origin. This condition is true for at least one of the models of any real system. For example in the single or multi-link inverted pendulum system, an unstable equilibrium exists at $\boldsymbol{x} = \boldsymbol{0}$, where the cart is not moving, and all the pendulum mass is at the top (i.e. the angles are 0).

Then by substituting (10.41) into (10.37), the non-linear control system of (10.37) can be written as

$$\dot{\boldsymbol{x}} = \boldsymbol{A}(\boldsymbol{x})\boldsymbol{x} + \boldsymbol{B}(\boldsymbol{x})\boldsymbol{u}, \tag{10.43}$$

if $\boldsymbol{B}(\boldsymbol{x}) = \boldsymbol{g}(\boldsymbol{x})$. In the case when the system function in (10.36) is not analytic, but the control $\boldsymbol{u}$ is differentiable, $\boldsymbol{g}(\boldsymbol{x})$ is simply $\begin{pmatrix} \boldsymbol{0} \\ \boldsymbol{I} \end{pmatrix}$ as shown in (10.40) and is linear.

One important point to note is that the non-linear state-space representation (10.43) of the system is not unique. If the function $\boldsymbol{f}(\boldsymbol{x})$ is analytic, then it has a convergent Taylor's series such that

$$\boldsymbol{f}(\boldsymbol{x}) = \sum_{i=0}^{\infty} \frac{1}{i!} \frac{\partial^i}{\partial \boldsymbol{x}^i} \boldsymbol{f}(\boldsymbol{0})\, \boldsymbol{x}^i = \sum_{i=0}^{\infty} \frac{\boldsymbol{f}^{(i)}(\boldsymbol{0})}{i!} \boldsymbol{x}^i, \tag{10.44}$$

where $\boldsymbol{x}^i = x_1^{i_1}, x_2^{i_2}, \ldots x_n^{i_n}$ [15]. The Taylor series starts with a constant term for any function, but if $\boldsymbol{f}(\boldsymbol{0}) = \boldsymbol{0}$, it means the constant term for the function f is 0 in this case, then every other term is a polynomial which contains at least one x (x_1 or x_2 or $x_3 \ldots$). This is true for most physical systems, as discussed earlier for the equilibrium condition in (10.42). Then take each homogeneous component, i.e. take all powers of order one (linear terms) and then all power of order two (quadratic terms) and so on. That means, in any one of those homogeneous polynomials, there are always terms with at least one x (x_1 or x_2 or x_3) in it. Then an 'x' can be pulled out to form the matrix $\boldsymbol{A}(\boldsymbol{x})$.

For any system which satisfied the origin equilibrium condition stated in (10.42), a representation in the form of (10.43) can always be found, but this representation is non-unique. A simple example to demonstrate this fact is shown below:

Example 10.1

An analytic system is governed by the following differential equations:

$$\begin{aligned} \dot{x}_1 &= x_1 x_2 \\ \dot{x}_2 &= x_2 + u_2. \end{aligned} \tag{10.45}$$

The equations can be written in matrix form as

$$\begin{pmatrix} \dot{x}_1 \\ \dot{x}_2 \end{pmatrix} = \begin{pmatrix} x_2 & 0 \\ 0 & 1 \end{pmatrix} \begin{pmatrix} x_1 \\ x_2 \end{pmatrix} + \begin{pmatrix} 0 \\ 1 \end{pmatrix} \begin{pmatrix} u_1 \\ u_2 \end{pmatrix}, \tag{10.46}$$

or

$$\begin{pmatrix} \dot{x}_1 \\ \dot{x}_2 \end{pmatrix} = \begin{pmatrix} 0 & x_1 \\ 0 & 1 \end{pmatrix} \begin{pmatrix} x_1 \\ x_2 \end{pmatrix} + \begin{pmatrix} 0 \\ 1 \end{pmatrix} \begin{pmatrix} u_1 \\ u_2 \end{pmatrix}. \tag{10.47}$$

Although the two state-space matrices are different, they represent differential equations which are equivalent. It can be seen from this example that the state-space model of a system is not unique. Depending on how the matrix is split out of the differential equations, there are often many valid mathematical representations of the same system, although some are more physical than others, or are expected to lead to simpler control calculations and implementations.

Once a state-space model of the non-linear system is constructed in the form of (10.43), the global non-linear control theory [8,9] states that by attaching the generalised quadratic infinite time optimal cost function

$$J = \frac{1}{2}\int_0^\infty (\boldsymbol{x}^T\boldsymbol{Q}(\boldsymbol{x})\boldsymbol{x} + \boldsymbol{u}^T\boldsymbol{R}(\boldsymbol{x})\boldsymbol{u})dt \tag{10.48}$$

to (10.42), it generates the non-linear optimal control solution:

$$\boldsymbol{u}^* = -\boldsymbol{R}^{-1}(\boldsymbol{x})\boldsymbol{B}^T(\boldsymbol{x})\boldsymbol{P}(\boldsymbol{x})\boldsymbol{x}, \tag{10.49}$$

where $\boldsymbol{P}(\boldsymbol{x})$ satisfies the algebraic Riccati equation:

$$\boldsymbol{A}^T(\boldsymbol{x})\boldsymbol{P}(\boldsymbol{x}) + \boldsymbol{P}(\boldsymbol{x})\boldsymbol{A}(\boldsymbol{x}) + \boldsymbol{Q}(\boldsymbol{x}) - \boldsymbol{P}(\boldsymbol{x})\boldsymbol{B}(\boldsymbol{x})\boldsymbol{R}^{-1}(\boldsymbol{x})\boldsymbol{B}^T(\boldsymbol{x})\boldsymbol{P}(\boldsymbol{x}) = \boldsymbol{0}. \tag{10.50}$$

The optimal stabilising control can be obtained by substituting (10.49) into (10.43) to produce a coupled control system:

$$\begin{aligned}\dot{\boldsymbol{x}} &= \boldsymbol{A}(\boldsymbol{x})\boldsymbol{x} - \boldsymbol{B}(\boldsymbol{x})\boldsymbol{R}^{-1}(\boldsymbol{x})\boldsymbol{B}^T(\boldsymbol{x})\boldsymbol{P}(\boldsymbol{x})\boldsymbol{x} \\ &= [\boldsymbol{A}(\boldsymbol{x}) - \boldsymbol{B}(\boldsymbol{x})\boldsymbol{R}^{-1}(\boldsymbol{x})\boldsymbol{B}^T(\boldsymbol{x})\boldsymbol{P}(\boldsymbol{x})]\boldsymbol{x}.\end{aligned} \tag{10.51}$$

To solve the equations in (10.51) numerically, an integration method for each time step can be used. For example the Euler's method or the fourth order Runge–Kutta method can be adopted for the inverted pendulum cases.

We use a freezing technique to solve for the next step of $\boldsymbol{x}$ iteratively using an integration method. Starting at $\boldsymbol{x}(0)$ (which is a known chosen initial state), $\boldsymbol{A}(\boldsymbol{x}(0))$ and $\boldsymbol{B}(\boldsymbol{x}(0))$ are then fixed matrices at this initial step. MATLAB provides a LQR solution for $\boldsymbol{P}(\boldsymbol{x}(0))$ after solving the algebraic Riccati equation because at each time step for a fixed $\boldsymbol{x}$, the system becomes essentially the same as the linear case, so linear control theory can be easily applied. Now substitute $\boldsymbol{A}(\boldsymbol{x}(0))$, $\boldsymbol{B}(\boldsymbol{x}(0))$ and $\boldsymbol{P}(\boldsymbol{x}(0))$ into (10.51) to solve for the next $\boldsymbol{x}$, say $\boldsymbol{x}(1)$. Then having obtained $\boldsymbol{x}(1)$, the same process can be repeated for the next step and so on for the system.

One point to bear in mind is that as we discussed earlier in this section, the state-space representation $\boldsymbol{A}(\boldsymbol{x})$ of any system is not unique, so the state-space solution of the controlled system $\boldsymbol{x}$ may also vary. In the case of the single inverted pendulum, for example, even if we still adopt the Lagrangian approach to obtain the equations of motions given by (10.14) in Section 10.2, it is possible to write a different state-space model in matrix by splitting the $\boldsymbol{x}$ from each equation of (10.15) differently. Therefore, the global non-linear optimal control solution (10.51) obtained by using the freezing technique is not unique, because the model of the inverted pendulum–cart system can be different to the one used for the control analysis. In Example 10.1, it is seen that, in the first representation, the system is totally uncontrollable and the second is controllable apart from on the set where $x_1 = 0$, which is of two-dimensional measure zero. Therefore, the controllability of the original system varies, depending on the state-space model written.

In the linear control case for a single inverted pendulum, the closed-loop pendulum control needs an initial position of $x_3(0) = 0$, i.e. the pendulum mass is concentrated at the top of the rod, or $x_3(0) \approx 0$, i.e. the mass is within a small neighbourhood of the top position so the approximations of $\sin x_3 \approx x_3$ and $\cos x_3 \approx 1$ are true to linearise the system state-space model. To reach this initial position, either the pendulum is manually placed, or an open-loop swinging up control is usually implemented.

In the non-linear control case, the pendulum can start from any initial position apart from two initial states of $x_3 = \pm\frac{\pi}{2}$ because $(\boldsymbol{A}(\boldsymbol{x}), \boldsymbol{B}(\boldsymbol{x}))$ is unstabilisable at these two positions [9,16]. But $(\boldsymbol{A}(\boldsymbol{x}), \boldsymbol{B}(\boldsymbol{x}))$ is controllable for all other $\boldsymbol{x}$ and in the rare situation when the pendulum does start from the two angles stated above, the pendulum will fall slightly due to gravity and hence changing the value of x_2 to other values than $\pm\frac{\pi}{2}$. Then the non-linear optimal stabilising control will kick in to control the pendulum. Therefore, we say this condition does not change the overall controllability of $(\boldsymbol{A}(\boldsymbol{x}), \boldsymbol{B}(\boldsymbol{x}))$ and the pendulum system is controllable for 'almost all' $\boldsymbol{x}$. In higher order systems, such as the multi-link inverted pendulum systems, the nature of the uncontrollable subspace will be more complicated and will require further analysis.

Another advantage of this non-linear global control method, apart from the fact that it can be controlled from nearly any position, is that it is intrinsically robust. This is because unlike the linear case, any inaccuracy in the modelling of A and B can be taken account of and corrected at each time step and would therefore not affect the eventual control result. To implement the control in real time, a powerful computer is needed to solve the Riccati equation within the required time.

The control of inverted pendulum system is a benchmark in testing the general non-linear optimal control theory discussed above. This method does not only apply to the pendulum system but also to any general non-linear control systems, for example aeronautical systems [8] or ship systems [17]. It is also worth noting that there are other techniques for solving non-linear control problems, for example a linear, time-varying approximation approach introduced in [18]. In the next section, we will demonstrate some simulation results of the single and double-link inverted pendulum and cart control systems, using the non-linear optimal control method discussed in this section.

10.4 Simulations for single and double pendulum control

We use a fourth order Runge–Kutta method (because of its high accuracy) to calculate and to apply the optimal control solutions to the inverted pendulum and cart systems numerically in MATLAB. The non-linear control theory and the simulation principles are discussed in Section 10.3.2. Simulation results for the single-pendulum and cart controlled system defined in (10.14) in Section 10.2 are plotted in Figures 10.3–10.5. Figures 10.6 and 10.7 display the simulation results for the double-pendulum and cart-controlled system presented in Section 10.2.

For the single pendulum and cart simulation, the parameters used are: mass of the cart: 2 kg, mass of the pendulum: 0.1 kg and the rod length 0.5 m. For the double pendulum and cart simulation, the parameters are: mass of the cart: 2 kg, mass of the first pendulum: 0.1 kg, mass of the second pendulum: 0.1 kg and both of the rods are 0.1 m long.

Sample MATLAB codes are attached in Appendix A for the convenience of the readers.

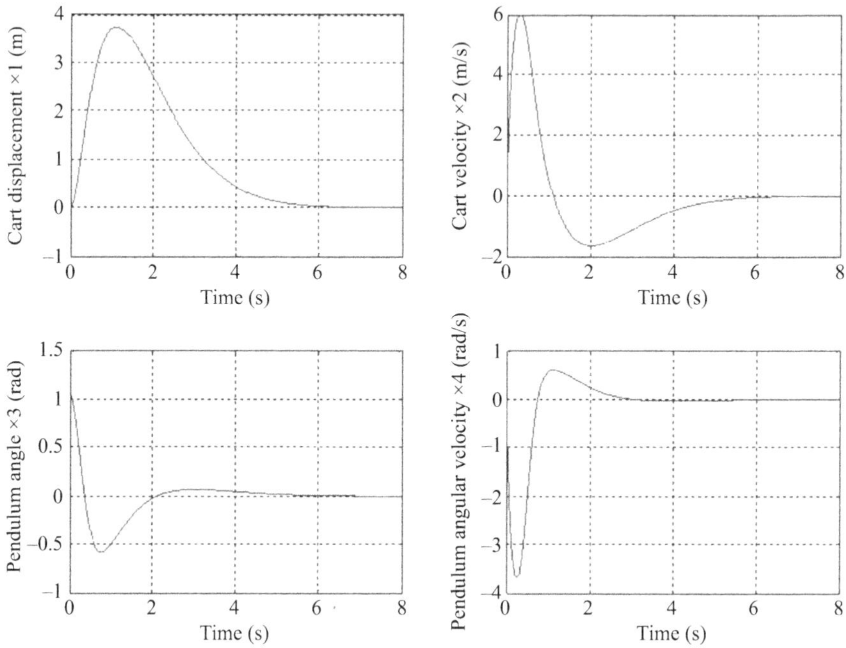

Figure 10.3 The plots of $x_1 - x_4$ for the single pendulum control system starting from $x_3 = \frac{\pi}{3}$

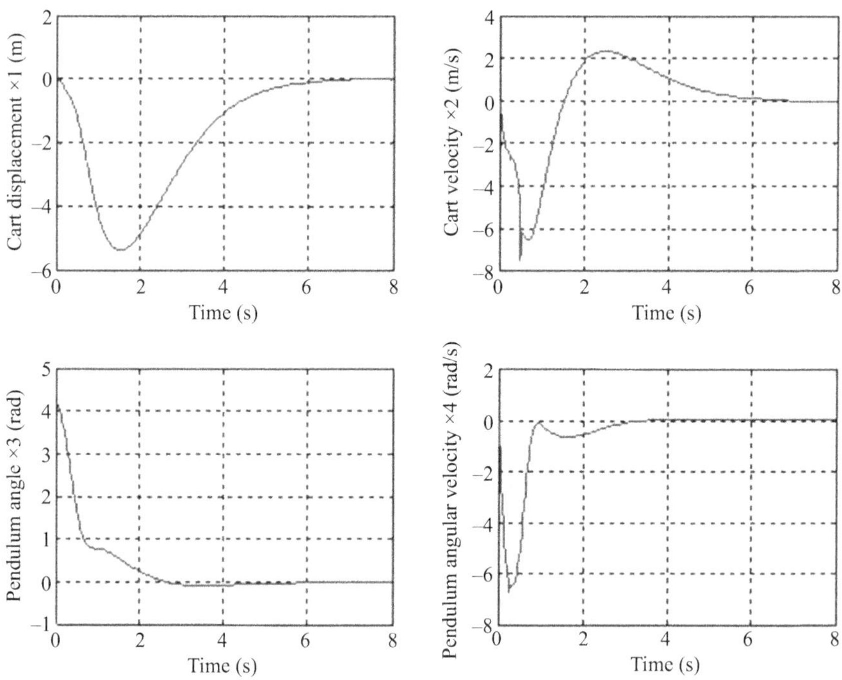

Figure 10.4 The plots of $x_1 - x_4$ for the single pendulum control system starting from $x_3 = \frac{4\pi}{3}$

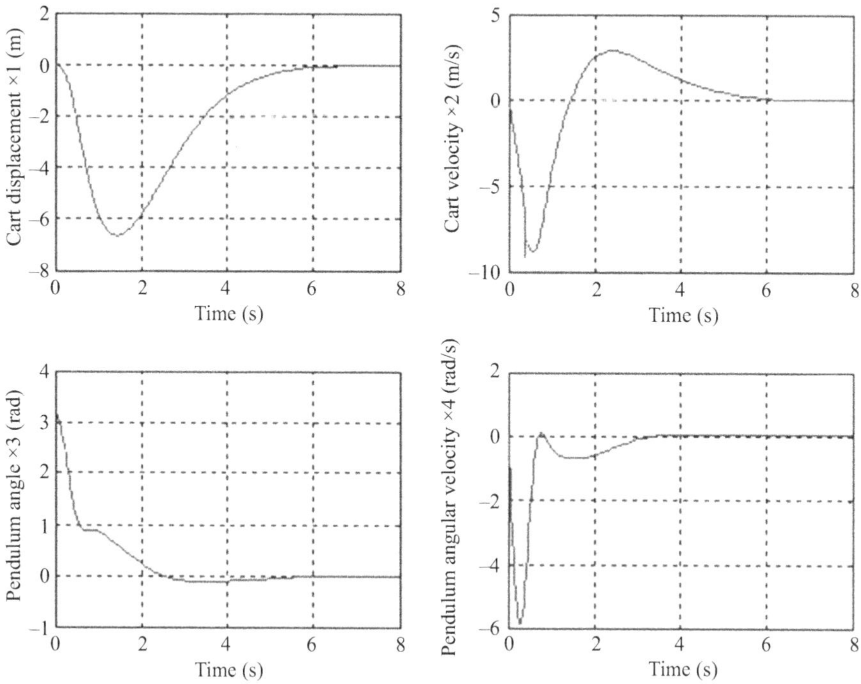

Figure 10.5 The plots of $x_1 - x_4$ for the single pendulum control system starting from $x_3 = \pi$

10.5 Case study 1 – LEGO EV3-based inverted pendulum

10.5.1 Introduction

As a classical control problem, inverted pendulum provides a good understanding about control techniques of non-linear systems and their linearisation approaches. Controlling of inverted pendulum system lays strong background for a control engineer, and the depth of the inverted pendulum study can be varying from undergraduate level to the postgraduate level. There are many inverted pendulum systems developed for educational purposes, for example the Quanser Linear Servo Cart with the Inverted Pendulum (as shown in Figure 10.8), the Quanser Rotary Inverted Pendulum (as shown in Figure 10.9) and the Feedback Instruments Digital Inverted Pendulum (as shown in Figure 10.10) are very popular among the equipment offered by engineering teaching equipment providers. Many universities and researchers around the world have also designed and built their own inverted pendulum systems [19–21] to suit different teaching needs and research investigations. In recent years, rapid advancements in microprocessor technology have led to the development of many inverted pendulums controlled by microcontrollers such as Arduino, Raspberry Pi and PIC controllers

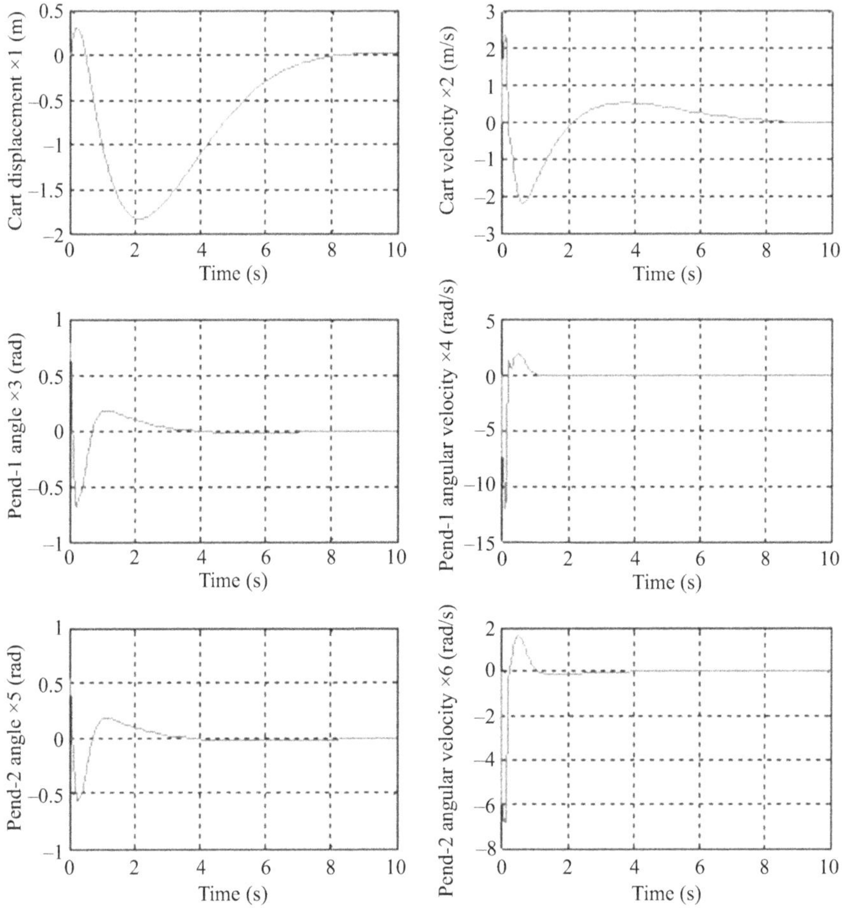

Figure 10.6 Double pendulum control system $x_1 - x_6$ plots starting from $x_3 = \frac{\pi}{4}$ and $x_5 = \frac{\pi}{6}$

which are more affordable and transportable [22–24]. Another very popular inverted pendulum development is using the LEGO Mindstorms kit [25–27].

10.5.2 LEGO Mindstorms EV3

Initially designed as a toy for school children, LEGO Mindstorms is now widely used in university education and research, attracting applications in many areas, including artificial intelligence, embedded systems, control systems, robotics and operating systems. In additional to the characteristics that they are inexpensive, easily reconfigurable, reprogrammable, versatile and robust, another unique benefit of using LEGO Mindstorms is that the complete inverted pendulum system can be constructed and controlled by LEGO Mindstorms kit including over 500 TECHNIC elements for

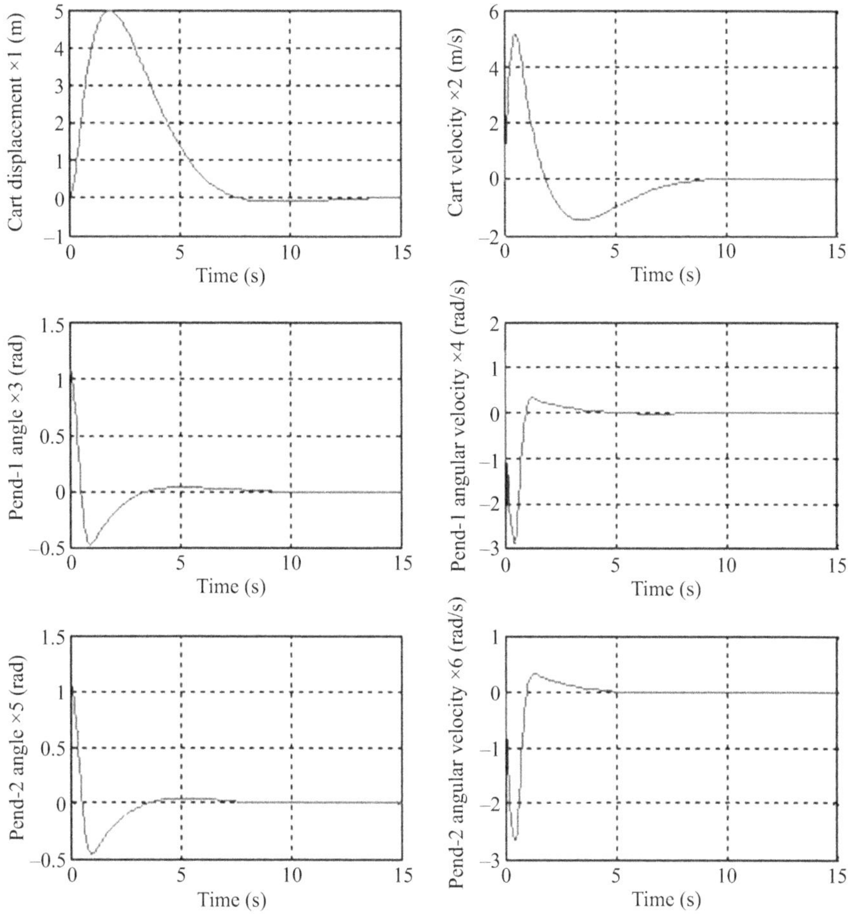

Figure 10.7 Double pendulum control system $x_1 - x_6$ plots starting from $x_3 = x_5 = \frac{\pi}{3}$

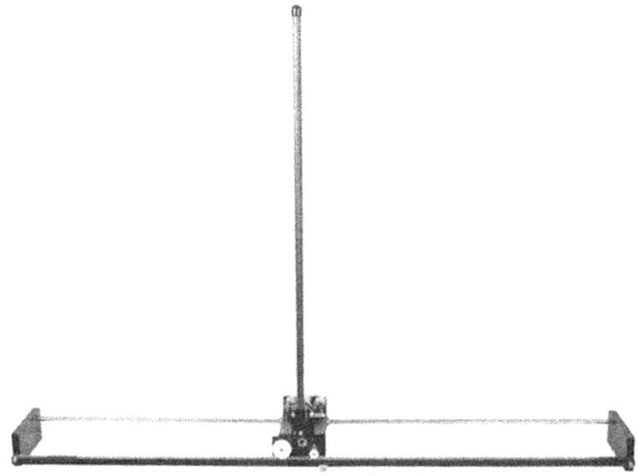

Figure 10.8 Quanser linear servo cart with the inverted pendulum [28]

Figure 10.9 Quanser rotary inverted pendulum [29]

Figure 10.10 Feedback instruments digital pendulum [30]

mechanical design and construction, 5 sensors, 3 motors and the programmable brick with powerful ARM9 CPU in its latest EV3 kit.

LEGO Mindstorms EV3 is the third generation of the LEGO robotics kit which was introduced in 2013. It has a more intelligent and powerful ARM9 CPU running Linux. A USB connector and Micro SD slot (up to 32 GB) as well as communication

Figure 10.11 LEGO Gyroboy (Segway)

capability through Wi-Fi are new features of the EV3 brick. There were many software and programming languages which can communicate with the LEGO Mindstorms. LabVIEW, MATLAB & Simulink®, RobotC for Mindstorms, C++ are the most common programming platforms and languages used in LEGO Mindstorms [31,32]. Mindstorms EV3 can use Bluetooth or wired USB connection for short-range communication and Wi-Fi facilitates for long-range networking. For the Wi-Fi communications, EV3 brick needs to connect with an external Wi-Fi dongle, and the only supported Wi-Fi module with the LEGO Mindstorms platform is NETGEAR N150 (WNA1100) Wireless Adapter [31].

LEGO Mindstorms EV3 has four output ports and four input ports which can measure the data up to 1000 samples per second. The system has 16 MB of flash memory, inbuilt 64 MB RAM and enables to extend the memory up to 32 GB by a mini HDSC memory card reader [31]. The LEGO Mindstorms EV3 has two large interactive servo motors with encoders, a medium servo motor, a colour sensor, a touch sensor (two state pressure sensor), a gyro sensor and an IR sensor [31].

10.5.3 Construction of LEGO inverted pendulum

The most popular LEGO inverted pendulum was the so-called Gyro Boy which is a self-balancing Segway on two wheels (as shown in Figure 10.11). Gyro Boy uses a combination of the large servo motors' encoders working with the Gyro sensor to balance itself upright. LEGO provided the user with complex example models, such as Gyro Boy, as an inspiration of what can be built and controlled with LEGO MINDSTORMS. The robot also responds to commands and feedback from the colour sensor and the ultrasonic sensor.

More traditional inverted pendulum applications have also been developed using either LEGO Mindstorms NXT or EV3. As shown in Figure 10.12, a single inverted

Figure 10.12 Single inverted pendulum on a cart using LEGO EV3

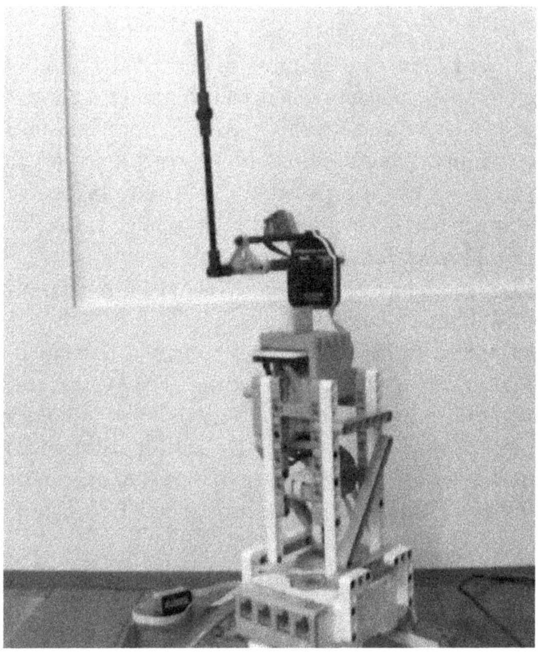

Figure 10.13 Rotary inverted pendulum using LEGO EV3

pendulum cart was developed by Professor Masakatsu Kawada and his students with LEGO EV3 [33]. In this application, a GlideWheel-AS Angle Sensor has been used to measure the angle of the pendulum and two large servo motors were used to control the cart as well as providing the cart displacement measurements. They have also developed a LEGO EV3 rotary inverted pendulum system as shown in Figure 10.13 [34–36].

Figure 10.14 Single inverted pendulum on a cart based on the Lego Mindstorms EV3

Figure 10.14 has shown the structure of the single inverted pendulum that was built using simple LEGO bricks, and the pendulum was attached onto the cart which was also made from LEGO bricks. With two heavy wheels on top, the centre of gravity of the pendulum was assumed at the top of pendulum. Two side bars were used for the protection of the brick screen and the pendulum. Pendulum was attached to the rotatable wheel on cart, and it could slow down the pendulum by using gears if required. Two large servo motors were used to move the cart, and the position of the cart was measured with the feedback from the motors' encoders. The Gyro sensor was used for measuring the angle and the angular velocity of the pendulum. The physical parameters of the LEGO-based single inverted pendulum system are shown in Table 10.1.

10.5.4 Controller design and Simulink model

Controller was designed in MATLAB and Simulink while Wi-Fi network connection was used to communicate the measurement and control information with the EV3 brick. The LEGO Mindstorms Simulink support package was used to develop the Simulink model as shown in Figure 10.15. The pendulum angle was obtained from taking an integration to the output from the Gyro sensor (in degrees/s). The cart position was measured by the other two large motors' encoders. The encoder

Table 10.1 Physical parameters of single inverted pendulum

Measurement	Value (units)
Mass of the cart	718 g
Mass of the pendulum	120 g
Length of the pendulum	345 mm
Diameter of a cart wheel	56 mm
Circumference of a cart wheel	17.6 cm
Length of the cart (among two wheels)	20 cm
Width of the cart (among two wheels)	10 cm
Height of the cart (from ground to pivot point of pendulum)	15 cm

measurements were given in degrees, and it was converted to radians by adding a gain element after the encoder blocks. The cart position was measured in lengths but the obtained measurement was from the angle for the wheels. Therefore, angle measurements were converted into length. Low-pass filter was used before the position signals were differentiated. The Simulink controller used a MATLAB-function, as shown in Figure 10.16, which was using the four-state variables described in (10.14) as inputs and calculate the control signal using the non-linear optimal control method introduced in Section 10.3.2.

10.5.5 Experimental results

Experimental results of the controlled single inverted pendulum and cart system, using the non-linear optimal control method and the Runge–Kutta integration method, are presented in Figure 10.17, where pendulum started at initial position of 10° ($\pi/18$). Sampling time was set at 0.1 s. The speed of the two large motors was controlled using calculated controller output between a range of -100 to 100 as a percentage of the full speed (at 160 rpm) where the negative sign indicates reverse rotation. The outputs of the Gyro sensor were all integers in degree/s, thus the staircase shape of the graphs. The experiment results show that the inverted pendulum could stabilise fairly and quickly, which demonstrated the effectiveness of the non-linear optimal control method.

10.6 Case study 2 – a parallel manipulator as the cart

A parallel kinematic manipulator is a mechanical system that uses several serial chains to support a single movable platform, or end-effector. They are investigated and implemented in several application tasks ranging from industrial manipulations to service robotics, as described for example in [37]. There is a wide range of design solutions for PKM according to their specific application(s) also. A successful PKM laboratory prototype is CaPaMan. This robot has been built in several different design versions, and it has been investigated for various applications including, for example, robotic surgery [38] or earthquake shaking tests [39].

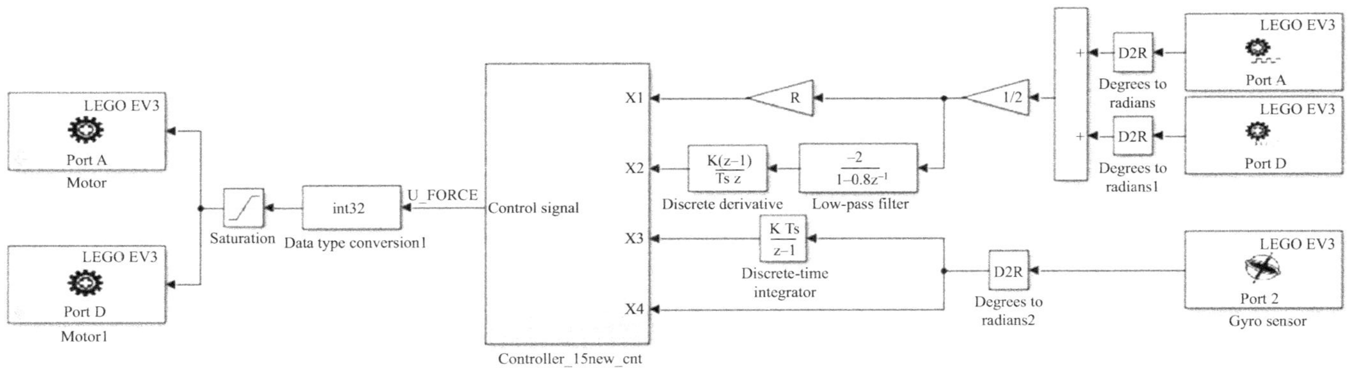

Figure 10.15 Simulink model for single inverted pendulum on a cart

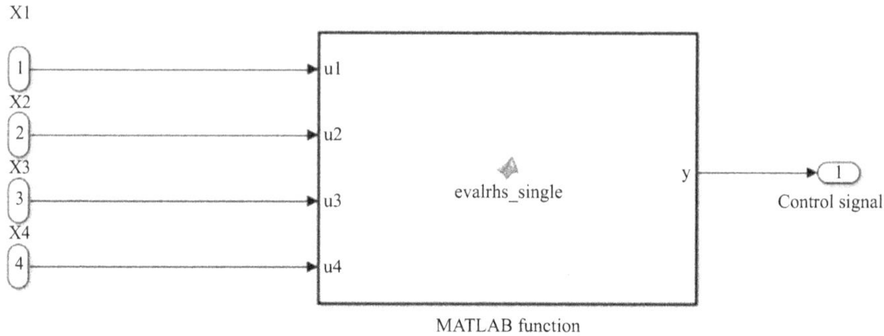

Figure 10.16 Simulink model for control subsystem

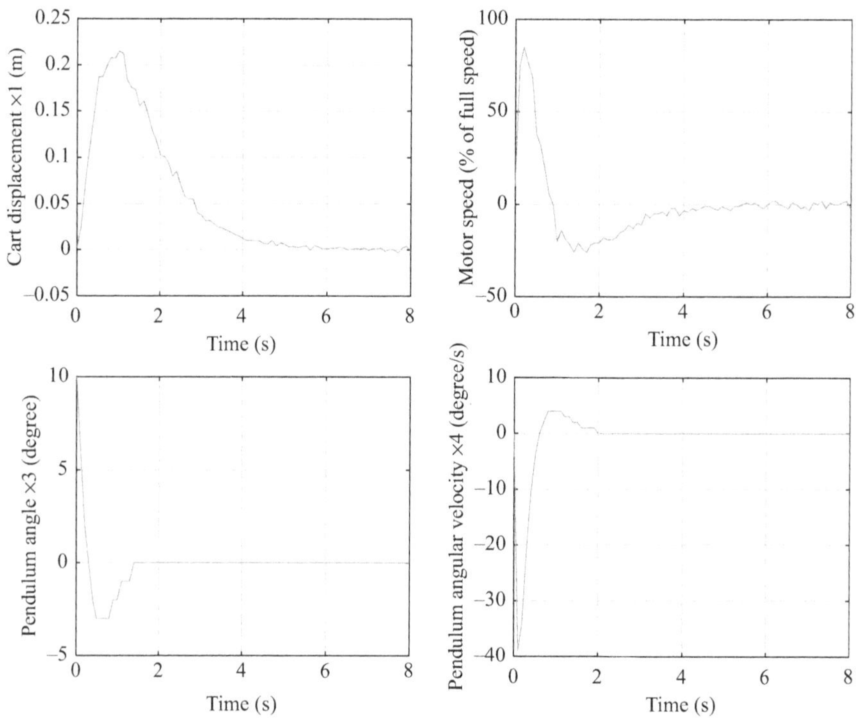

Figure 10.17 Simulation results – single inverted pendulum starting at $\pi/18$

Figure 10.18(a) and (b) shows, respectively, a kinematic model of CaPaMan and a built prototype, which has a two links mechanism on its movable plate. CaPaMan is composed of a movable plate MP that is connected to a fixed plate FP by means of three identical articulated parallelograms AP, whose coupler carries a prismatic joint

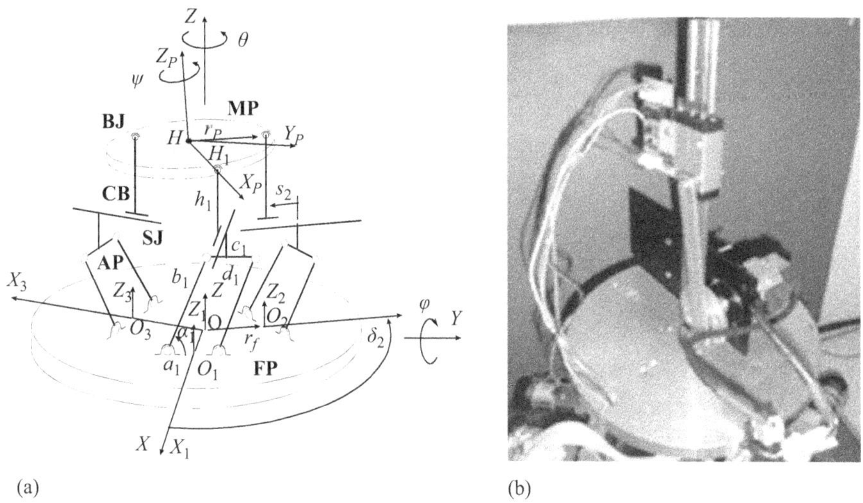

Figure 10.18 CaPaMan (Cassino Parallel Manipulator): (a) a kinematic scheme; (b) a built prototype with a pendulum on the movable plate [41,42]

Table 10.2 Sizes and motion parameters of the built prototype of CaPaMan, as referring to (Figure 10.18(a))

$a_k = c_k$ (mm)	$b_k = d_k$ (mm)	H_k (mm)	$r_P = r_f$ (mm)	α_κ (deg)	s_k (mm)
200	80	100	109.5	45; 135	−50; 50

SJ and a spherical joint BJ, as shown in the scheme of Figure 10.18(a). This specific kinematic architecture allows to keep the motors on the fixed base to reduce inertial forces. Additionally, the number of active degrees of freedom is limited to 3 while 6 degrees of mobility can be achieved with coupled motions of the movable plate. For example a synchronised motion of the three motors will allow a vertical motion along Z-axis coupled with a rotation about Z-axis.

Table 10.2 reports the values of main sizes of CaPaMan as referring to the model in Figure 10.18. A detailed modelling of CaPaMan, including close loop direct kinematics formulation, has been described, for example, in [40]. In particular, direct kinematics can provide the following expression for the centre point H of MP as

$$\begin{aligned} x &= \frac{y_3 - y_2}{\sqrt{3}} - \frac{r_P}{2}(1 - \sin\varphi)\cos(\psi - \theta) \\ y &= y_1 - r_P(\sin\psi\cos\theta + \cos\psi\sin\varphi\sin\theta) \\ z &= \frac{z_1 + z_2 + z_3}{3} \end{aligned} \tag{10.52}$$

and the orientation Euler angles of MP as

$$\begin{aligned}
\psi &= \tan^{-1}\left[\sqrt{3}\frac{z_3 - z_2}{2z_1 - z_2 - z_3}\right] \\
\theta &= \sin^{-1}\left[2\frac{y_1 + y_2 + y_3}{3r_P(1 + \sin\varphi)}\right] - \psi \\
\varphi &= \cos^{-1}\left[\pm\frac{2}{3r_P}\sqrt{z_1^2 + z_2^2 + z_3^2 - z_1 z_2 - z_2 z_3 - z_1 z_3}\right] \\
&(z \geq z_1 \Rightarrow +; z < z_1 \Rightarrow -)
\end{aligned} \tag{10.53}$$

The sign ambiguity in (10.53) for ψ, θ and φ is solved when the values of z_1 and z give the up and down orientation of the mobile plate and, therefore, the sign $-$ or $+$ is chosen, respectively.

In this chapter, CaPaMan is considered as a cart of an inverted pendulum, such as shown in Figure 10.18(b). Accordingly, (10.52) and (10.53) have been expressed as a function of H_k coordinates y_k, z_k ($k = 1, 2, 3$), which can be given by the geometry of Figure 10.18(a) from the input variable α_k

$$\begin{aligned}
y_k &= b_k \cos\alpha_k \\
z_k &= b_k \sin\alpha_k + h_k
\end{aligned} \tag{10.54}$$

so that they can be considered the input coordinates for the platform motion.

The symmetry characteristics of CaPaMan architecture have been useful to formulate analytical dynamic equations to compute the input torques as reported in [40]. Assumptions have been made in order to simplify the equations such as the effects of link elasticity and viscous damping of the joints have been neglected; links are assumed to be rigid bodies and the joints are frictionless and have no clearance. In addition, only the inertial effects of the movable platform have been considered since the legs of parallel architectures are lighter than the movable plate. Based on the above-mentioned assumption, the only forces that are applied to the rods CB by the mobile platform are those, which are contained in the plane of the articulated parallelogram, i.e. F_{ky} and F_{kz} as shown in Figure 10.19.

The F_{kx} component determines the sliding of the prismatic joint along s_k coordinate. Thus, the components of the resultant force $\mathbf{F}$ and torque $\mathbf{N}$ can be computed as

$$\begin{Bmatrix} F_x \\ F_y \\ F_z \\ N_x \\ N_y \\ N_z \end{Bmatrix} = \begin{Bmatrix} -\frac{\sqrt{3}}{2}F_{2y} + \frac{\sqrt{3}}{2}F_{3y} \\ F_{1y} - \frac{1}{2}F_{2y} - \frac{1}{2}F_{3y} \\ F_{1z} + F_{2z} + F_{3z} \\ -u_{1z}F_{1y} + \frac{1}{2}u_{2z}F_{2y} + \frac{1}{2}u_{3z}F_{3y} + u_{1y}F_{1z} + u_{2y}F_{2z} + u_{3y}F_{3z} \\ -\frac{\sqrt{3}}{2}u_{2z}F_{2y} + \frac{\sqrt{3}}{2}u_{3z}F_{3y} - u_{1x}F_{1z} - u_{2x}F_{2z} - u_{3x}F_{3z} \\ \frac{1}{2}\left(\sqrt{3}u_{2y} - u_{2x}\right)F_{2y} - \frac{1}{2}\left(\sqrt{3}u_{3y} + u_{3x}\right)F_{3y} + u_{1x}F_{1y} \end{Bmatrix} \tag{10.55}$$

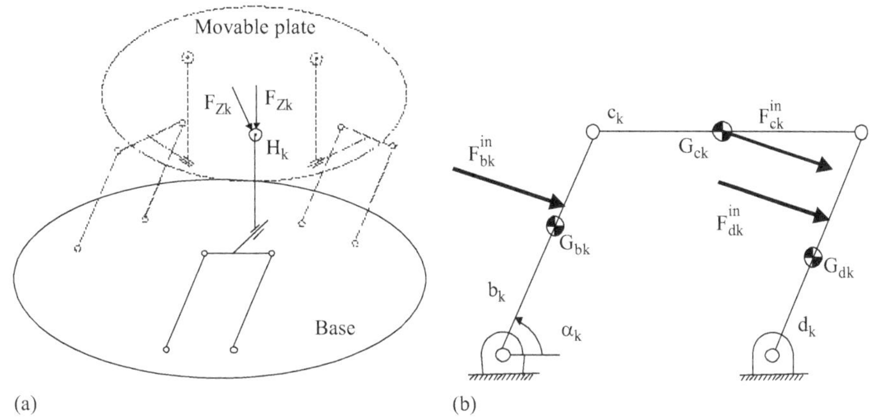

Figure 10.19 Details for free body modelling of CaPaMan: (a) a detail of the reaction forces acting on one of the spherical joints; (b) location of the centre of mass and inertial forces

with

$$\begin{Bmatrix} u_{kx} \\ u_{ky} \\ u_{kz} \end{Bmatrix} = r_p R \begin{Bmatrix} \cos\delta_k \\ \sin\delta_k \\ 0 \end{Bmatrix}; \quad (k = 1, 2, 3) \tag{10.56}$$

when the values $\delta_1 = 0$; $\delta_2 = 2\pi/3$; $\delta_3 = 4\pi/3$ are considered.

Equations (10.55) and (10.56) can be solved in a closed form to obtain an explicit expression for forces F_{ky} and F_{kz}. Referring to Figure 10.19 and once the reaction forces in the spherical joints H_k are computed, the torque $\boldsymbol{\tau}_{Pk}$ $(k = 1, 2, 3)$ on the input crank shaft of each articulated parallelogram can be obtained by considering only the inertial effects of the movable platform in the form

$$\tau_{Pk} = \frac{F_{kz} b_k}{2} \frac{\sin 2\alpha_k}{\sin\alpha_k} - F_{ky} b_k \left(\frac{h_k}{c_k \tan\alpha_k} + 1 \right) \left(1 - \frac{h_k}{h_k \cos\alpha_k + c_k \sin\alpha_k} \right) \sin\alpha_k \tag{10.57}$$

where b_k, c_k and h_k are the geometrical dimensions shown in Figure 10.19; F_{ky} and F_{kz} are the reaction forces in the spherical joints H_k. The contribution of the legs to the inverse dynamics of the CaPaMan can be determined by including a kinetostatic analysis of the articulated parallelograms. The reaction forces, centres of mass of the links and inertial forces can be identified by using free body diagrams as proposed in the schemes of Figure 10.19. Further details on the dynamic modelling and path planning of CaPaMan can be found in [40].

Simulations have been carried out by using the formulation in (10.12)–(10.14), (10.27), (10.29), (10.36), (10.37) and (10.40) in order to control an inverted pendulum

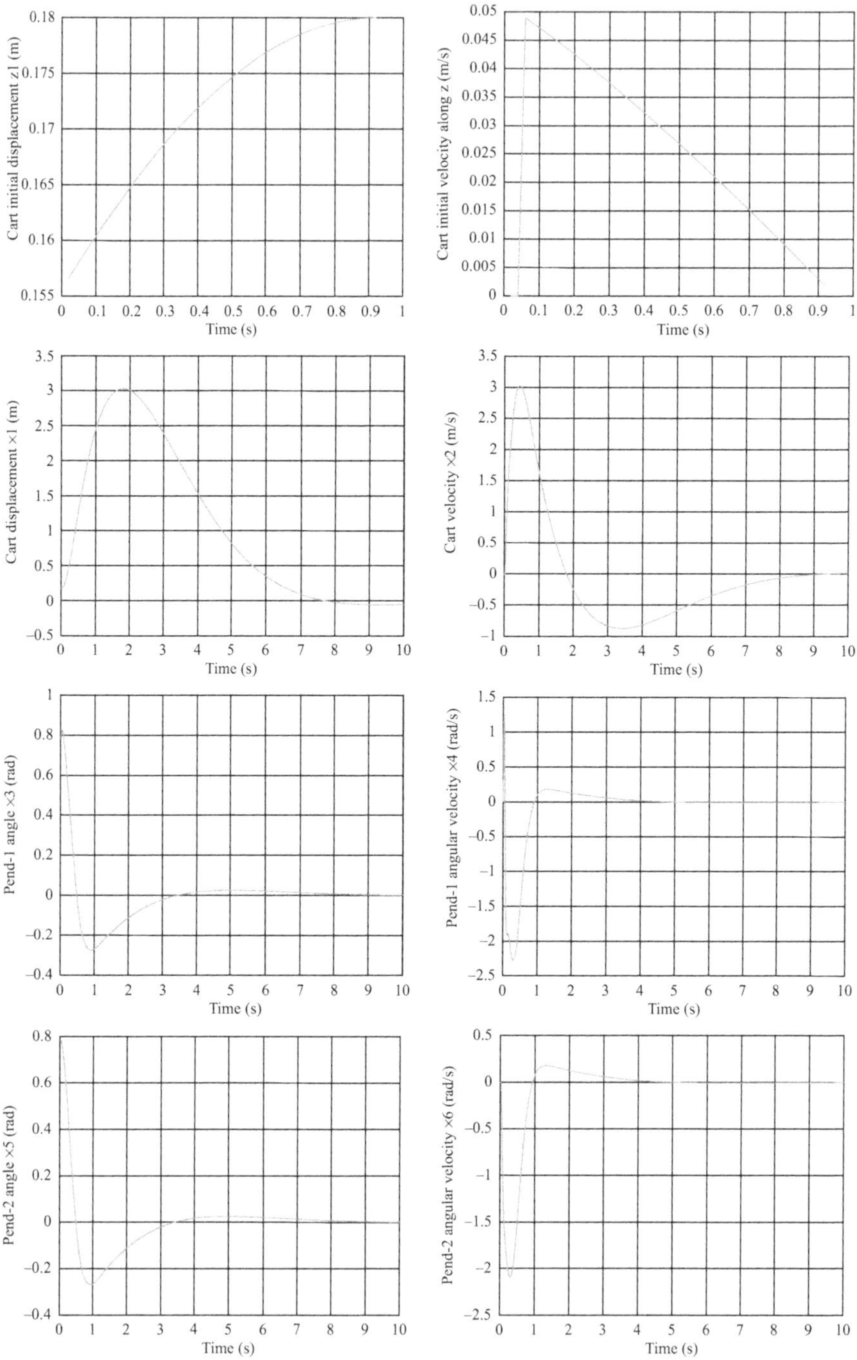

Figure 10.20 Simulation results for a double inverted pendulum mounted on CaPaMan cart with links starting both at $\pi/4$

accommodated on the movable plate of CaPaMan. Calculations through (10.41)–(10.43) have been made to compute the initial position and velocity of the movable plate. Results of the controlled inverted pendulum and cart system, using the non-linear optimal control method and the Runge–Kutta integration method, are presented in Figure 10.20, where pendulum started at initial positions $x_3 = \frac{\pi}{4}$ and $x_5 = \frac{\pi}{4}$. It shows that the inverted pendulum was able to stabilise fairly quickly.

It is to note that the presence of disturbances due to the initial cart dynamics can be minimised by statically/dynamically balancing the CaPaMan cart system. The dynamic balancing can be achieved using several methods such as using

- counter-rotary counter-masses;
- separate counter-rotations;
- idler loops;
- duplicated mechanisms/movements;
- counterweights on the movable links;
- harmonic balancing by two counter-rotating masses.

A case study as referring to CaPaMan 2bis (Cassino Parallel Manipulator version 2bis) has been carried out in [42]. Given the architecture of CaPaMan, one can use a total of six Counter-Rotary Counterweights and three Counterweights. A constant linear momentum can be obtained if the total centre of mass is stationary. For a proper balancing, one can impose a stationary centre of mass and search for the design parameters that can reduce to zero (or at least minimise) the moments of inertia, shaking forces and shaking moments, as reported in [42].

10.7 Conclusions

In this chapter, we have studied a freezing technique to control non-linear systems globally in general. We have shown two specific examples by using the non-linear optimal control technique: a single inverted pendulum and cart system and a two-link inverted pendulum and cart system. We have demonstrated good control results of the two pendulum systems in simulations. Two case studies further demonstrating the effectiveness of the non-linear control method have also been illustrated in this chapter. The theory presented in this chapter extends to multi-link pendulum control systems and indeed many other non-linear systems, e.g. aeronautical systems [8] or ship systems [17]. Note that we have only considered the control problem in this chapter, but not the state estimation problem. It is assumed that there is an effective state estimator for the optimal feedback control. It is also worth noting that there are other techniques for solving non-linear control problems, for example, a linear, time-varying approximation approach introduced in [18]. Future work will be focused on studying the non-uniqueness problem, and the identification of the uncontrollable subspace and how to avoid it.

Appendix A Solving and implementing control solutions using MATLAB

Appendix A.1 Non-linear control of a single pendulum and cart system

Script file: single_pendulum_RK.m

```
clear all;
close all;

x1(1)=0; %set initial cart displacement to be 0 (m)
x2(1)=0; %set initial cart velocity to be 0 (m/s)
x3(1)=pi; %set initial pendulum angle to be pi (rad)
x4(1)=0; %set initial pendulum angular velocity to be 0 (rad/s)

h=0.01; %time step length
t=800; %set number of iterations

for i=1:t
        x=[x1(i); x2(i); x3(i); x4(i)];
        %define and update the x vector

  k1=evalrhs_single(x);
  xtemp=x+k1*h/2;
  k2=evalrhs_single(xtemp);
  xtemp=x+k2*h/2;
  k3=evalrhs_single(xtemp);
  xtemp=x+k3*h;
  k4=evalrhs_single(xtemp);
  %call the 'evalrhs_single' function to obtain k1-k4

  x=x+h*(k1+2*k2+2*k3+k4)/6;
    %calculate the new 'x' vector using a 4th order Runge-Kutta
    %integration method

  x1(i+1)=x(1);
  x2(i+1)=x(2);
  x3(i+1)=x(3);
  x4(i+1)=x(4);
    %Reset the x1, x2, x3 & x4 variables to new values and get ready
    %for the next iteration.
end
```

```
subplot(2,2,1);
plot([0:h:(t*h)],x1); grid;
xlabel('Time (sec)'); ylabel('Cart displacement x1 (m)');

subplot(2,2,2);
plot([0:h:(t*h)],x2); grid;
xlabel('Time (sec)'); ylabel('Cart velocity x2 (m/s)');

subplot(2,2,3);
plot([0:h:(t*h)],x3); grid;
xlabel('Time (sec)'); ylabel('Pendulum angle x3 (rad)');

subplot(2,2,4);
plot([0:h:(t*h)],x4); grid;
xlabel('Time (sec)'); ylabel('Pendulum angular velocity x4 (rad/s)');
%plot x1, x2, x3 & x4 as 4 subplots in one figure window
```

Function file: evalrhs_single.m

```
function [fx] = evalrhs_single(x)

  m1=2; %Mass of the cart (kg)
  m2=0.1; %mass of the pendulum (kg)
  r=0.5; %the rod length (m)
  g=9.8; %acceleration due to gravity (m^2/s)

  x1=x(1);
  x2=x(2);
  x3=x(3);
  x4=x(4);
  %initialise x1-x4 using the input 'x' vector

  A=[0, 1, 0, 0;
    0, 0,...
    -m2*g*sin(x3)*cos(x3)/((m1+m2*(sin(x3))^2)*x3),...
    m2*r*x4*sin(x3)/(m1+m2*(sin(x3)^2));
    0, 0, 0, 1;
    0, 0,...
    (m1+m2)*g*sin(x3)/(r*(m1+m2*(sin(x3))^2)*x3),...
    -m2*r*x4*sin(x3)*cos(x3)/(r*(m1+m2*(sin(x3))^2))];
  B=[0;
    1/(m1+m2*(sin(x3))^2);
    0;
    -cos(x3)/(r*(m1+m2*(sin(x3))^2))];
```

```
        %define A and B using a non-linear state-space single pendulum
        %and cart model

    Q=[1, 0, 0, 0; 0, 1, 0, 0; 0, 0, 10, 0; 0, 0, 0, 1];
    R=0.1;
    %set Q and R matrices

    [K,P,E]=lqr(A,B,Q,R);
    %use the MATLAB 'lqr' function to solve Riccati equation and
    %work out P

    fx=(A-B*inv(R)*B'*P)*x;
    %calculate the function output 'fx' based on values of A, B, P
    %and x.
end
```

Non-linear control of a double-link pendulum and cart system:

Script file: two_link_pendulum_RK.m

```
clear all;
close all;

x1(1)=0; %set initial cart displacement to be 0 (m)
x2(1)=0; %set initial cart velocity to be 0 (m/s)
x3(1)=pi/3; %set initial pendulum 1 angle to be pi/3 (rad)
x4(1)=0; %set initial pendulum 1 angular velocity to be 0 (rad/s)
x5(1)=pi/3; %set initial pendulum 2 angle to be pi/6 (rad)
x6(1)=0; %set initial pendulum 2 angular velocity to be 0 (rad/s)

s=0.001; %time step length
t=10000; %set number of iterations

for i=1:t
    x=[x1(i); x2(i); x3(i); x4(i); x5(i); x6(i)];
    %define and update the x vector

    k1=evalrhs_double(x);
    xtemp=x+k1*s/2;
    k2=evalrhs_double(xtemp);
    xtemp=x+k2*s/2;
    k3=evalrhs_double(xtemp);
    xtemp=x+k3*s;
    k4=evalrhs_double(xtemp);
    %call the 'evalrhs_single' function to obtain k1-k4
```

```
    x=x+s*(k1+2*k2+2*k3+k4)/6;
      %calculate the new 'x' vector using a 4th order Runge-Kutta
      %integration method

    x1(i+1)=x(1);
    x2(i+1)=x(2);
    x3(i+1)=x(3);
    x4(i+1)=x(4);
    x5(i+1)=x(5);
    x6(i+1)=x(6);
      %Reset the x1, x2, x3, x4, x5 & x6 variables to new values and
      %get ready for the next iteration.
end

subplot(3,2,1);
plot([0:s:(t*s)],x1); grid;
xlabel('Time (sec)'); ylabel('Cart displacement x1 (m)');

subplot(3,2,2);
plot([0:s:(t*s)],x2); grid;
xlabel('Time (sec)'); ylabel('Cart velocity x2 (m/s)');

subplot(3,2,3);
plot([0:s:(t*s)],x3); grid;
xlabel('Time (sec)'); ylabel('Pend-1 angle x3 (rad)');

subplot(3,2,4);
plot([0:s:(t*s)],x4); grid;
xlabel('Time (sec)'); ylabel('Pend-1 angular velocity x4 (rad/s)');

subplot(3,2,5);
plot([0:s:(t*s)],x5); grid;
xlabel('Time (sec)'); ylabel('Pend-2 angle x5 (rad)');

subplot(3,2,6);
plot([0:s:(t*s)],x6); grid;
xlabel('Time (sec)'); ylabel('Pend-2 angular velocity x6 (rad/s)');
%plot x1, x2, x3, x4, x5 & x6 as 6 subplots in one figure window
```

Function file: evalrhs_double.m

```
function [fx] = evalrhs_double( x )

    m1=2; %Mass of the cart (kg)
    m2=0.1; %mass of the pendulum 1 (kg)
```

```
m3=0.1; %mass of the pendulum 2 (kg)
r1=0.1; %length of rod 1 (m)
r2=0.1; %length of rod 2 (m)
g=9.8; %acceleration due to gravity (m^2/s)

x1=x(1);
x2=x(2);
x3=x(3);
x4=x(4);
x5=x(5);
x6=x(6);
  %initialise x1-x6 using the input 'x' vector

a=m2+m3;
b=cos(x3)*cos(x5);
c=sin(x3-x5);
d=m2*a*(sin(x3))^2 + m1*m3*c^2 +m1*m2;
e=2*m2^2+m1*m3+2*m2*m3+2*m1*m2+m1*m3*cos(2*x5);
f=m2*a*sin(2*x3)-m1*m3*sin(2*x5-2*x3);
h=m2*m3*sin(x3)*cos(x5)+ m1*m3*c;
  %define a-h to reduce the coding complexity of A and B matrices

A=[0, 1, 0, 0, 0, 0;
   0, 0,...
   m2*a*g*sin(x3)*cos(x3)/(-x3*d),...
   m2*a*r1*x4*sin(x3)/d, 0, ...
   m2*m3*r2*x6*(sin(x5-2*x3)-sin(x5))/(-2*d);
   0, 0, 0, 1, 0, 0;
   0, 0,...
   g*sin(x3)*e/(2*r1*d*x3), x4*f/(-2*d), ...
   m1*m3*g*sin(x5)*b/(-r1*d*x5), ...
   r2*x6*h/(-r1*d);
   0, 0, 0, 0, 0, 1;
   0, 0,...
   m1*a*g*sin(x3)*b/(-r2*d*x3), ...
   m1*a*r1*c*x4/(r2*d), ...
   m1*a*g*(cos(x3))^2*sin(x5)/(r2*d*x5), ...
   m1*m3*sin(2*x5-2*x3)*x6/(-2*d)];

B=[0;
   (m2+m3*c^2)/d;
   0;
   (m2*cos(x3)-m3*sin(x5)*c)/(-r1*d);
   0;
   a*sin(x3)*c/(-r2*d)];
```

```
        %define A and B using a non-linear state-space 2-link pendulum
        %and cart model

    Q=[1, 0, 0, 0, 0, 0; 0, 1, 0, 0, 0, 0; 0, 0, 1, 0, 0, 0; ...
       0, 0, 0, 1, 0, 0; 0, 0, 0, 0, 1, 0; 0, 0, 0, 0, 0, 1];
    R=1;
    %set Q and R matrices

    [K,P,E]=lqr(A,B,Q,R);
       %use the MATLAB 'lqr' function to solve Riccati equation and
       %work out P

    fx=(A-B*inv(R)*B'*P)*x;
       %calculate the function output 'fx' based on values of A, B, P
       %and x.
end
```

Acknowledgement

The first author Xu Xu would like to thank emeritus professor Stephen Paul Banks for the discussions and clarifications he generously provided regarding the optimal non-linear control technique presented in this chapter.

References

[1] Dutton, K., Thompson, S., & Barraclough, B. *The art of control engineering.* Harlow, England: Pearson Education Limited; 1997 Jun 1.

[2] d'Azzo, J. J., & Houpis, C. D. *Linear control system analysis and design: conventional and modern.* New York City, USA: McGraw-Hill Higher Education; 1995 Jan 1.

[3] Ogata, K., & Yang, Y. *Modern control engineering.* 1970.

[4] Schaefer, J. F., & Cannon, R. H. 'On the control of unstable mechanical systems'. In *Automatic and Remote Control III: Proceedings of the Third Congress of the International Federation of Automatic Control, Paper C*, 1966 June, Vol. 6, p. 1966.

[5] Mazenc, F., & Praly, L. 'Adding integrations, saturated controls, and stabilization for feedforward systems'. *IEEE Transactions on Automatic Control.* 1996; 41(11):1559–78.

[6] Zhao, J. & Spong, M.W. 'Hybrid control for global stabilization of the cart–pendulum system'. *Automatica*, 2001, *37*(12), pp. 1941–1951.

[7] Boubaker, O. 'The inverted pendulum benchmark in nonlinear control theory: a survey'. *International Journal of Advanced Robotic Systems.* 2013; 10(5):233.

[8] Banks, S. P., & Mhana, K. J. 'Optimal control and stabilization for nonlinear systems'. *IMA Journal of Mathematical Control and Information*. 1992; 9(Jan (2)):179–96.
[9] McCaffrey, D., & Banks, S. P. 'Lagrangian manifolds and asymptotically optimal stabilizing feedback control'. *Systems & Control Letters*. 2001; 43(Jul (3)):219–24.
[10] Mracek, C. P., & Cloutier, J. R. 'Control designs for the nonlinear benchmark problem via the state-dependent Riccati equation method'. *International Journal of Robust and Nonlinear Control*. 1998;8(Apr (4–5)):401–33.
[11] Halmos, P. R. *Measure theory*. New York, NY, USA: Springer-Verlag; 2013 Dec 19.
[12] Lanczos, C. The Variational Principles of Mechanics, 4th ed. New York: Dover, pp. 53 and 61, 1986.
[13] Goldstein, H. *Classical mechanics*, 2nd ed. Reading, MA: Addison-Wesley, p. 44, 1980.
[14] Hand, L. N., & Finch, J. D. *Analytical mechanics*, 2nd ed. Cambridge, UK: Cambridge University Press, p. 23, 2008.
[15] Perko, L. *Differential equations and dynamical systems*. New York, NY, USA: Springer-Verlag; 2013 Nov 21.
[16] Harrison, R. F. 'Asymptotically optimal stabilising quadratic control of an inverted pendulum'. *IEE Proceedings-Control Theory and Applications*. 2003; 150(1):7–16.
[17] ÇImen, T., & Banks, S. P. 'Nonlinear optimal tracking control with application to super-tankers for autopilot design'. *Automatica*. 2004; 40(11):1845–63
[18] Tomás-Rodriguez, M., & Banks, S. P. *Linear, time-varying approximations to nonlinear dynamical systems: with applications in control and optimization*. New York, NY, USA: Springer-Verlag; 2010 Feb 4.
[19] Bröckelt, R., & Li, H. 'A light weight rotary double pendulum: maximizing the domain of attraction'. In *Proceedings of the IEEE Conference on Decision and Control*, 2003, Vol. 4, pp. 3299–304.
[20] Solis, J., & Takanishi, A. 'Development of a wheeled inverted pendulum robot and a pilot experiment with master students'. In *Mechatronics and Its Applications (ISMA). 2010 7th International Symposium on*, 1–6.
[21] Turner, M., & Cooley, T. R. 'A low-cost and flexible open-source inverted pendulum for feedback control laboratory courses'. In *2015 American Society for Engineering Education Annual Conference and Exposition*, Seattle, Washington, pp. 26.63.1–13.
[22] Grămescu, B. S., Niţu, C. I., Phuc, N., & Borzea, C. 'PID control for two-wheeled inverted pendulum (WIP) system'. *Romanian Review Precision Mechanics, Optics and Mechatronics*. 2015; (48):246–51.
[23] Reck, R., & Sreenivas, R. 'Developing an affordable and portable control systems laboratory kit with a Raspberry Pi'. *Electronics*. 2016; 5(3):36.
[24] Kang, M. T., Vo, H. D., Kim, H. K., & Kim, S. B. 'Control system design for a mobile inverted pendulum via sliding mode technique'. In *Mechatronics, ICM2007 4th IEEE International Conference on*, 2007, pp. 1–6.

[25] Kapitonov, A. A., Bobtsov, A. A., Kapitanyuk, Y. A. *et al.* 'Course of lab activities on control theory based on the LEGO NXT'. *IFAC Proceedings Volumes.* 2014; 47(3):9063–8.

[26] Canale, M., & Casale-Brunet, S. 'A multidisciplinary approach for Model Predictive Control Education: a LEGO Mindstorms NXT-based framework'. *International Journal of Control, Automation and Systems.* 2014; 12(5): 1030–9.

[27] Mathew, S. T. & Mija, S. J. 'Design of H2 controller for stabilization of two-wheeled inverted pendulum'. In *International Conference on Advanced Communication Control and Computing Technologies (ICACCCT)*, 2014, pp. 174–9.

[28] Quanser. *'Linear Servo Base Unit with Inverted Pendulum'.* Viewed September 2016, http://www.quanser.com/products/linear_servo_IP02.

[29] Quanser. *'Rotary Inverted Pendulum'.* Viewed September 2016, http://www.quanser.com/products/rotary_pendulum.

[30] Feedback Instruments Ltd. *'Digital Pendulum 33-005-PCI'.* Viewed September 2016 http://www.feedback-instruments.com/products/education/control_instrumentation/digital_pendulum.

[31] LEGO. *'LEGO Mindstorms EV3 Support'.* Viewed September 2016. https://www.lego.com/en-gb/mindstorms/support.

[32] STARBLUE. *'LEGO Answers Online'.* Last updated 2012 November. http://bricks.stackexchange.com/questions/679/is-there-any-other-way-to-program-the-lego-mindstorms-nxt-other-than-the-provide.

[33] Masakatsukawata. *'Inverted Pendulum on a Cart (LEGO MINDSTORMS EV3)'.* 2014. Available at: https://www.youtube.com/watch?v=vQ607X6Rid4 (accessed: 10/11/2016).

[34] Masakatsukawata. *'Rotary Inverted Pendulum (LEGO MINDSTORMS EV3)'.* 2014. Available at: https://www.youtube.com/watch?v=xrPAGlFUDlQ (accessed: 10/11/2016).

[35] Kawada, M. 'Development of control engineering educational contents using LEGO MINDSTORMS and Simulink Support Package'. In *Proceedings of Automatic Control Allied Lecture Papers*, 2014, Vol. 57, pp. 598–605.

[36] Masakatsu, M. 'Development of rotary inverted pendulum using LEGO MINDSTORMS'. *Measurement and control.* 2015; 54(3):192–5.

[37] Ceccarelli, M. *Fundamentals of mechanics of robotic manipulation.* Springer, Dordrecht; 2004.

[38] Carbone, G., & Ceccarelli, M. 'A serial-parallel robotic architecture for surgical tasks'. *Robotica: An International Journal.* 2005; 23:345–54.

[39] Selvi, Ö., & Ceccarelli, M. 'Interpretation of earthquake effects on mechanism operation: an experimental approach'. *Journal of Naval Science and Engineering.* 2012; 8(2):31–45.

[40] Carbone, G., Ceccarelli, M., Oliveira, P. J., Saramago, S. F. P., & Carvalho, J. C. M. 'An optimum path planning of CaPaMan (Cassino Parallel Manipulator) by using inverse dynamics'. *Robotica: An International Journal.* 2008; 26(2):229–39.

[41] LARM Webpage. Last accessed on November 2016. http://webuser.unicas.it/weblarm/larmindex.htm.

[42] Cafolla, D., Carbone, G., & Ceccarelli, M. 'Chapter 8 in dynamic balancing of mechanisms and synthesizing of parallel robots'. *Balancing of a 3-DOFs Parallel Manipulator*. Springer, Dordrecht; 2015, pp. 173–91. (DOI) 10.1007/978-3-319-17683-3_8.

Chapter 11

Bipedal-double-pendulum walking robot control using recurrent hybrid neural network

Şahin Yildirim[1]

Abstract

This chapter presents neural control scheme of a planar-like double-pendulum-bipedal robot. For simplicity, only a five-link planar system is considered. The system effectively acts as two dynamically interacting planar robot arms. The scheme employs a single neural controller for the whole biped. Recurrent networks have feedback connections and thus an inherent memory for dynamics which makes them suitable for dynamic system modeling. A feature of the networks adopted is their hybrid hidden layer which includes both linear and nonlinear neurons. The standard proportional derivative (PD) controller is also employed for comparison. The results presented show the superior ability of the proposed neural control scheme at adapting to changes in the dynamics parameters of the bipedal robot.

11.1 Introduction

A bipedal-like double-pendulum robot can be seen as an extension of a twin-arm Selectively Compliant Assembly Robot Arm (SCARA) system if the third dimension is ignored. The main problem with a bipedal robot, however, is that of balance control to prevent the robot from falling over and enable it to walk.

Several researchers have proposed dynamic balance control systems for bipedal robots. A five-degree-of-freedom biped model moving in the sagittal plane has been investigated [1]. Three criteria have been introduced to evaluate the walk-'stability', 'maximum speed', and 'energy consumption'. In their investigation, relations between these three criteria and the parameters (gait, speed, period, stride, length of the leg, joint angles, etc.) are formulated accordingly to dynamics of robot [2]. A systematic formulation was provided that yields itself to an increase in dimensionality of the model, nonlinear studies and simulation and analysis of single and double support. The dynamics of a three dimensional model of the human body has been derived by Huston and Passerello [3]. Their model was subjected to several specific motions and the ground reaction force and its distance from the center of gravity

[1]Mechatronic Engineering Department, University of Erciyes, Turkey

of the body are calculated. State feedback laws have been suggested [3], since they are a natural control method for multivariable systems. Control methods using inverted pendulum type models to generate simple control laws have also been proposed [4]. A method of deriving the linear Quadratic Regulator for an n-legged robot that could be adapted for bipeds has been presented by Channon *et al.* [5].

A speed control problem of heavy duty vehicle trough angular position of throttle valve is researched by Yadav *et al.* [6]. Modified internal model control with fuzzy supervisor is used to control the speed of heavy duty vehicle. The time domain performance values were taken into consideration for performance analysis of the heavy duty vehicle for different uncertains.

An intelligent evolutionary least learning machine tool investigated to forecast the vehicle speed sequence [7]. Mozaffari *et al.* were used the driving data collected on the San Francisco urban roads by a private Honda insight vehicle. As results, proposed method by Mozaffari *et al.* was a powerful tool for predicting the vehicle speeds. Tagne *et al.* [8] researched to lateral displacement control of autonomous vehicle with respect to a given reference path. The proposed control law was validated a driving simulation engine according to several real driving scenarios. Simulations were also performed using experimental data acquired by a Peugeot 308.

Alam *et al.* [9] investigated the problem of finding a safety criterion between neighboring heavy duty vehicle platooning and real-life experimental results were presented in an attempt to validate the theoretical results in practice. Adaptive intelligent cruise control of heavy duty vehicle was investigated by Alam *et al.* [10]. In their study, a linear quadratic control framework used for the controller design. The proposed controller performance was evaluated through numerical and experimental studies. The experimental and simulation results showed that heavy duty vehicle platooning could be conducted at close spacing with standardized sensors and control units. Rödönyi *et al.* [11] researched novel numerical methods for analyzing robust peak-to-peak performance of heterogeneous platoons. The proposed method was indicated on three platoon controllers. A new speed controller for internal combustion engine designed by Tibola *et al.* [12]. The proposed controller was used for two approaches for the internal combustion engine speed control. The stability analysis for the developed controller was presented.

In this chapter, a neural control approach is used to control hip, foot trajectory and stability of a bipedal robot. The neural network scheme is employed a single neural controller for the whole bipedal robot.

11.2 Bipedal-like double-pendulum robot

Figures 11.1 and 11.2 illustrate a five-link planar bipedal robot. As can be seen from the figures, the torso has mass m_t and inertia J_t measured about its center of mass which is located at distance l_t above the hip joints.

The upper and lower leg elements are identical, with length $2l_e$, mass m_e and inertia J_e measured about their centers of mass. The feet have mass m_f located at the ankle joint and are assumed to have negligible inertia. With both feet on the ground, the robot acts almost like the closed-loop linkage of two cooperating planar

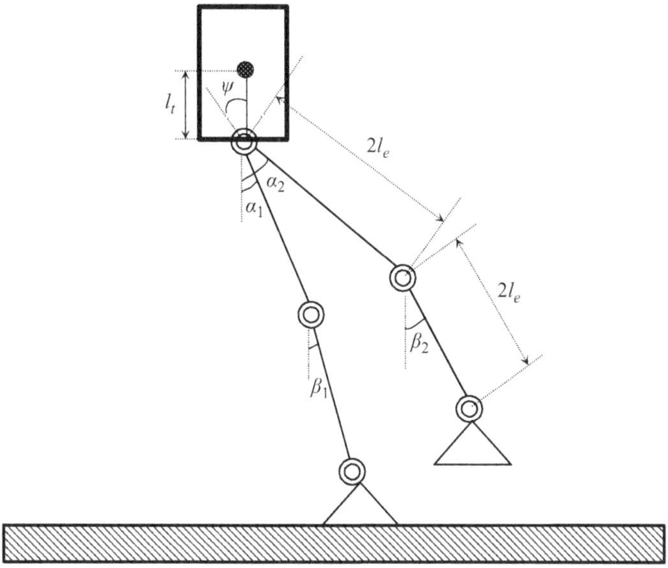

Figure 11.1 A five-link planar bipedal robot

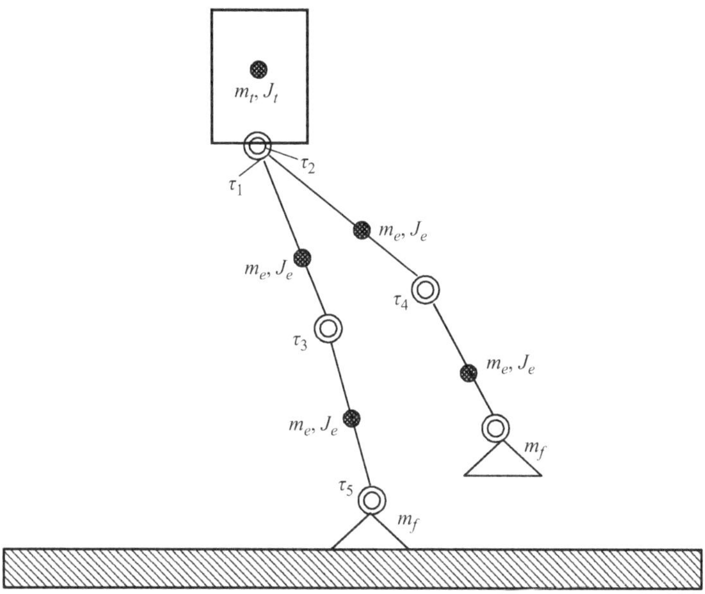

Figure 11.2 Schematic configuration of the bipedal robot

SCARA robots. However, the goal of this chapter is to propose a control law for the walking is the control of the walking movements of the robot where only one foot remains on the ground.

11.2.1 Walking

Walking commences at time $t = 0$ with one foot, called the swing foot, leaving the ground as depicted in Figure 11.3(a). Then, over a period T, the swing foot moves forward a distance $2L$ and the hip a distance L. Walking ends at time $t = T^-$, the instant just prior to the swing foot making contact with the ground. Throughout this period, the other foot, called the stance foot, remains fixed on the ground, thus providing a single support for the biped.

At a given instant during the single-support phase, the position of the bipedal robot relative to the stance foot can be specified by the following coordinate vector $\mathbf{\Theta}$:

$$\mathbf{\Theta} = [\psi \quad \alpha_1 \quad \alpha_2 \quad \beta_1 \quad \beta_2]^T \tag{11.1}$$

where ψ represents the angle of the torso with respect to the vertical, α_1 and α_2 are the angles of the upper leg elements relative to the vertical and β_1 and β_2 are the angles of the lower leg elements relative to the vertical, as shown in Figure 11.1.

Motors located at the robot's hip, knees and stance foot generate torques τ_1 to τ_5 as shown in Figure 11.2. These can be grouped into vector $\boldsymbol{\tau}$ where

$$\boldsymbol{\tau} = [\tau_1 \quad \tau_2 \quad \tau_3 \quad \tau_4 \quad \tau_5] \tag{11.2}$$

Note that the torque arising from the swing foot ankle joint motor must be zero due to the assumption of negligible foot inertia. Hence, it is not included in (11.2).

The joint accelerations $\ddot{\mathbf{\Theta}}$ arising from the application of torques $\boldsymbol{\tau}$ can be derived from the following [13]:

$$\mathbf{M}(\mathbf{\Theta})\ddot{\mathbf{\Theta}} + \mathbf{V}(\mathbf{\Theta}, \dot{\mathbf{\Theta}}) + \mathbf{G}(\mathbf{\Theta}) = \mathbf{C}_1\mathbf{F}_1 + \mathbf{C}_2\mathbf{F}_2 + \mathbf{D}\boldsymbol{\tau} \tag{11.3}$$

where $\mathbf{M}(\mathbf{\Theta})$ is the mass matrix, $\mathbf{V}(\mathbf{\Theta}, \dot{\mathbf{\Theta}})$ is a vector containing centripetal and Coriolis terms, $\mathbf{G}(\mathbf{\Theta})$ is a vector of torques arising from gravitational loading, $\mathbf{F}_1$ and $\mathbf{F}_2$ are the foot forces, $\mathbf{C}_1$ and $\mathbf{C}_2$ are the foot force transformation matrices and $\mathbf{D}$ is the joint transformation matrix. General equation of biped motion is described in Appendix.

11.2.2 Impact

The second stage of a step results in the swing foot hitting the ground and the stance foot breaking contact with the ground. This stage, termed impact, is assumed to be instantaneous, commencing at time $t = T^-$ and finishing at some infinitesimally short time later $t = T^+$ (Figure 11.3(b)). Consequently, the position of the biped will not appreciably change during impact.

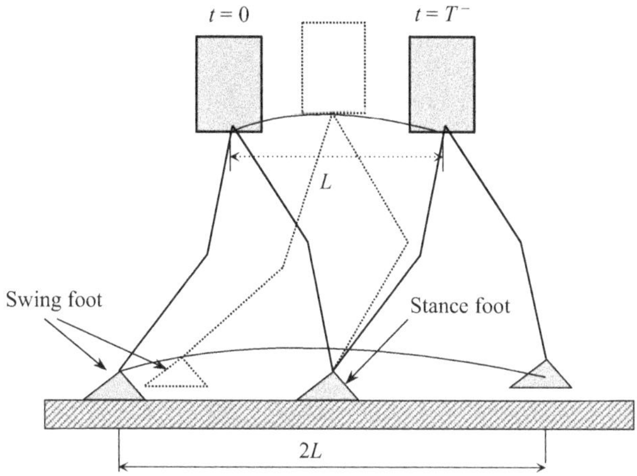

Figure 11.3(a) *Walking with swing foot while stance foot provides support*

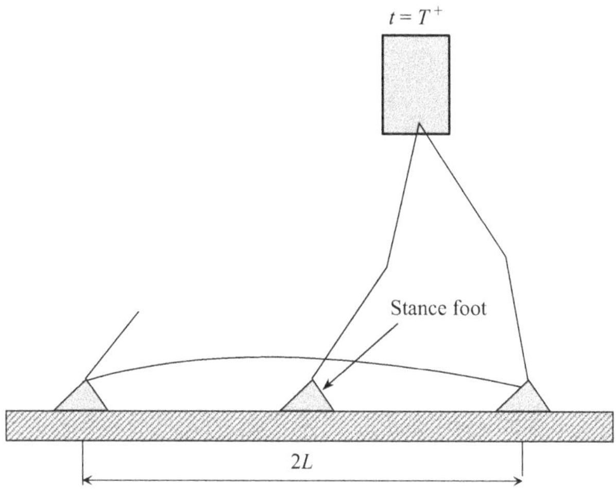

Figure 11.3(b) *Impact of swing foot with the ground*

11.3 Proposed neural network

Neural network can be represented in a general diagrammatic form as illustrated in Figure 11.4. This diagram depicts the hybrid hidden layer as comprising a linear part and a nonlinear part and shows that, in addition to the usual feedforward connections, the networks also have feedback connections from the output layer to the hidden layer and self-feedback connections in the hidden layer [14].

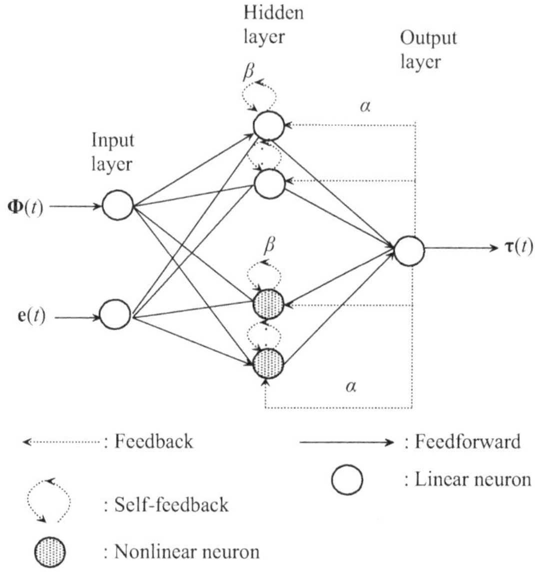

Figure 11.4 Neural network controller

At a given discrete time t, let $\mathbf{\Phi}(t)$ be the input to a recurrent hybrid network, $\boldsymbol{\tau}(t)$ the output of the network, $\mathbf{x}_1(t)$ the output of the linear part of the hidden layer and $\mathbf{x}_2(t)$ the output of the nonlinear part of the hidden layer.

The operation of the proposed network is summarized by the following equations:

$$\mathbf{x}_1(t+1) = \mathbf{W}^{\mathrm{I1}}\mathbf{\Phi}(t+1) + \beta\mathbf{x}_1(t) + \alpha\mathbf{J}_1\boldsymbol{\tau}(t) \tag{11.4}$$

$$\mathbf{x}_2(t+1) = \mathbf{F}\{\mathbf{W}^{\mathrm{I2}}\mathbf{\Phi}(t+1) + \beta\mathbf{x}_2(t) + \alpha\mathbf{J}_2\boldsymbol{\tau}(t)\} \tag{11.5}$$

$$\boldsymbol{\tau}(t+1) = \mathbf{W}^{\mathrm{H1}}\mathbf{x}_1(t+1) + \mathbf{W}^{\mathrm{H2}}\mathbf{x}_2(t+1) \tag{11.6}$$

where $\mathbf{W}^{\mathrm{I1}}$ is the matrix of weights of connections between the input layer and the linear hidden layer, $\mathbf{W}^{\mathrm{I2}}$ is the matrix of weights of connections between the input layer and the nonlinear hidden layer, $\mathbf{W}^{\mathrm{H1}}$ is the matrix of weights of connections between the linear hidden layer and the output layer, $\mathbf{W}^{\mathrm{H2}}$ is the matrix of weights of connections between the nonlinear hidden layer and the output layer, $\mathbf{F}\{\}$ is the activation function of neurons in the nonlinear hidden layer and α and β are the weights of the self-feedback and output feedback connections. $\mathbf{J}_1$ and $\mathbf{J}_2$ are respectively $n_{\mathrm{H1}} \times n_{\mathrm{O}}$ and $n_{\mathrm{H2}} \times n_{\mathrm{O}}$ matrices with all elements equal to 1, where n_{H1} and n_{H2} are the numbers of linear and nonlinear hidden neurons and n_{O} is the number of output neurons.

If only linear activation is adopted for the hidden neurons, the above equations simplify to:

$$\boldsymbol{\tau}(t+1) = \mathbf{W}^{\mathrm{H1}}\mathbf{x}(t+1) \tag{11.7}$$

$$\mathbf{x}(t+1) = \mathbf{W}^{\mathrm{I1}}\boldsymbol{\Phi}(t+1) + \beta\mathbf{x}(t) + \alpha\mathbf{J}_1\boldsymbol{\tau}(t) \tag{11.8}$$

Replacing $\boldsymbol{\tau}(t+1)$ by $\mathbf{W}^{\mathrm{H1}}\ \mathbf{x}(t+1)$ in (11.8) gives

$$\mathbf{x}(t+1) = (\beta\mathbf{I} + \alpha\mathbf{J}_1\mathbf{W}^{\mathrm{H1}})\mathbf{x}(t) + \mathbf{W}^{\mathrm{I1}}\boldsymbol{\Phi}(t+1) \tag{11.9}$$

where $\mathbf{I}$ is a $n_{\mathrm{H1}} \times n_{\mathrm{H1}}$ identity matrix.

Equation (11.9) is of the form

$$\mathbf{x}(t+1) = \mathbf{A}\mathbf{x}(t) + \mathbf{B}\boldsymbol{\Phi}(t+1) \tag{11.10}$$

where $\mathbf{A} = \beta\mathbf{I} + \alpha\mathbf{J}\mathbf{W}^{\mathrm{H1}}$ and $\mathbf{B} = \mathbf{W}^{\mathrm{I1}}$. Equation (11.10) represents the state equation of a linear system of which $\mathbf{x}$ is the state vector. The elements of $\mathbf{A}$ and $\mathbf{B}$ can be adjusted through training so that any arbitrary linear system of order n_{H1} can be modeled by the given network. When nonlinear neurons are adopted, this gives the network the ability to perform nonlinear dynamics mapping and thus model nonlinear dynamic systems. The existence in the recurrent hybrid network of a hidden layer with both linear and nonlinear neurons facilitates the modeling of practical nonlinear systems comprising linear and nonlinear parts. The inverse dynamics equation of the robot includes both linear terms and nonlinear terms. Thus, in theory, the inverse dynamics model of the robot and therefore an inverse model controller for it can be implemented using the proposed recurrent hybrid neural network structure.

In this work, the values of the weights of the recurrent connections, α and β, were fixed. This means only the weights of the feedforward connections, $\mathbf{W}^{\mathrm{I}}$ and $\mathbf{W}^{\mathrm{H}}$, needed to be adjusted and thus, it was possible to employ the standard backpropagation algorithm to train the neural network.

11.4 Control system

Figure 11.5 shows the control system adopted [14]. The system comprises a PD controller and a Neural Network (NN) controller, which is a recurrent hybrid network used to model the inverse dynamics of the biped. A trajectory generator (not shown) produces the required sequence of rotations for each joint to yield the desired walking movements of the biped. The NN is trained online during control to give the system the ability to adapt to change. Relative positions $\boldsymbol{\Phi}$ of the robot joint angles are used as reference signals to the controller:

$$\boldsymbol{\Phi} = [\phi_1 \quad \phi_2 \quad \phi_3 \quad \phi_4 \quad \phi_5]^T \tag{11.11}$$

$$\phi_1 = \phi_1 - \psi, \ \phi_2 = \alpha_2 - \psi, \ \phi_3 = \beta_1 - \alpha_1, \ \phi_4 = \beta_2 - \alpha_2, \ \phi_5 = \gamma_1 - \beta_1 \tag{11.12}$$

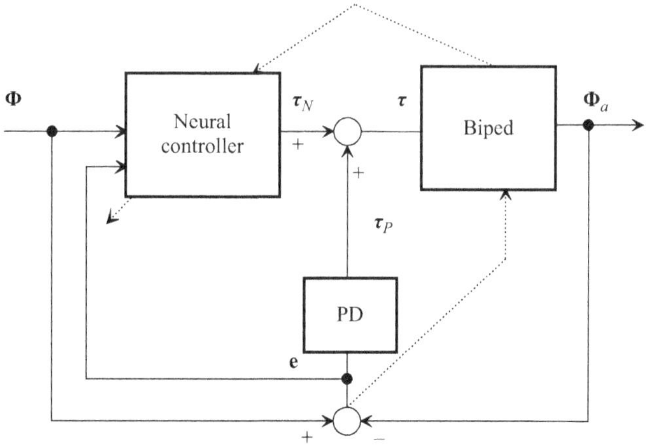

Figure 11.5 Bipedal robot control system 1

From Figure 11.5, the vector of input torques to the bipedal robot is given by

$$\boldsymbol{\tau} = \boldsymbol{\tau}_N + \boldsymbol{\tau}_P \tag{11.13}$$

where

$$\boldsymbol{\tau}_N = f(\boldsymbol{\Phi}, \mathbf{e}) \tag{11.14}$$

$$\boldsymbol{\tau}_P = \mathbf{K}_P\mathbf{e} + \mathbf{K}_D\dot{\mathbf{e}} \tag{11.15}$$

$\boldsymbol{\tau}_N$ is the output of the NN controller (Figure 11.4), which is seen as performing a nonlinear mapping f on the set point $\boldsymbol{\Phi}$ and the control error $\mathbf{e}$. $\boldsymbol{\tau}_P$ is the output of the PD controller. $\mathbf{K}_P$ and $\mathbf{K}_D$ are the gains of the controller. In this work, the following empirically determined values were adopted for $\mathbf{K}_P$ and $\mathbf{K}_D$: $\mathbf{K}_P = \text{diag}\,[4000\ \ 4000\ \ 4000\ \ 4000\ \ 4000]$ and $\mathbf{K}_D = \text{diag}\,[20\ \ 20\ \ 20\ \ 20\ \ 20]$.

11.4.1 Neural network controller

The NN controllers for biped control system 1 had three layers, as shown in Figure 11.4. The first layer received reference inputs from the trajectory generator and also the control errors. Again, the second layer consisted of two parts, a linear part and a nonlinear part. Finally, the third layer of the network produced outputs to drive the robot joints.

The controller was trained by backpropagating the error **e** through the biped in a way similar to that shown in Figure 11.5, with a minor modification as shown below.

Let E_k be the joint angle error for joint k. Using the chain rule, it follows that the gradient of E_k in weight space is

$$\frac{\partial \mathrm{E}_k}{\partial \mathrm{w}_{ij}} = \frac{\partial \mathrm{E}_k}{\partial \Phi_{ak}} \frac{\partial \Phi_{ak}}{\partial \boldsymbol{\tau}_k} \frac{\partial \boldsymbol{\tau}_k}{\partial \boldsymbol{\tau}_{Nk}} \frac{\partial \boldsymbol{\tau}_{Nk}}{\partial \mathrm{w}_{ij}} \tag{11.16}$$

Table 11.1 Parameters of the bipedal robot

Kinematics parameters	$l_t = 0.25\,\text{m}$, $l_e = 0.175\,\text{m}$
Mass parameters	$m_t = 20\text{–}200\,\text{kg}$, $m_e = 5\,\text{kg}$, $m_f = 2\,\text{kg}$
Inertia parameters	$I_t = 1\,\text{kg}\,\text{m}^2$, $I_e = 0.05\,\text{kg}\,\text{m}^2$

Table 11.2 Structural and training parameters

NN	η	μ	α	β	n	N	**AF**
RHN	0.0001	0.01	0.8	0.8	12 + 12	35,000	HT

In (11.16), $\dfrac{\partial \boldsymbol{\tau}_{Nk}}{\partial \mathrm{w}_{ij}}$ can be readily obtained through the Back Propogation (BP) algorithm but $\dfrac{\partial \Phi_{ak}}{\partial \boldsymbol{\tau}_k}$ is difficult to determine exactly because the robot dynamics are unknown. To overcome this problem, $\dfrac{\partial \Phi_{ak}}{\partial \boldsymbol{\tau}_k}$ is computed as follows:

$$\Delta \Phi_{ak} = \frac{\partial \Phi_{ak}}{\partial \boldsymbol{\tau}_k} \Delta \boldsymbol{\tau}_k + \sum_m \frac{\partial \Phi_{ak}}{\partial \mathrm{F}_m} \Delta \mathrm{F}_m \tag{11.17a}$$

$$\Delta \Phi_{ak} = \frac{\partial \Phi_{ak}}{\partial \boldsymbol{\tau}_k} \Delta \boldsymbol{\tau}_k + \sum_m \frac{\partial \Phi_{ak}}{\partial \boldsymbol{\tau}_k} \frac{\partial \boldsymbol{\tau}_k}{\partial F_m} \Delta F_m \tag{11.17b}$$

$$\Delta \Phi_{ak} = \frac{\partial \Phi_{ak}}{\partial \boldsymbol{\tau}_k} \Delta \boldsymbol{\tau}_k + \frac{\partial \Phi_{ak}}{\partial \boldsymbol{\tau}_k} \sum_m \frac{\partial \boldsymbol{\tau}_k}{\partial F_m} \Delta F_m \tag{11.17c}$$

$$\frac{\partial \Phi_{ak}}{\partial \boldsymbol{\tau}_k} = \frac{1}{\left(\dfrac{\Delta \boldsymbol{\tau}_k}{\Delta \Phi_{ak}} + \displaystyle\sum_m \frac{\partial \boldsymbol{\tau}_k}{\partial F_m} \frac{\Delta F_m}{\Delta \Phi_{ak}} \right)} \tag{11.17d}$$

11.5 Results

The control system 1 architecture illustrated in Figure 11.5 was implemented on a Pentium 200-MHz personal computer using the MATLAB® software [15]. The kinematic and inertia parameter values used for the bipedal robot are given in Table 11.1. The structural and training parameters adopted for the neural controller are given in Table 11.2.

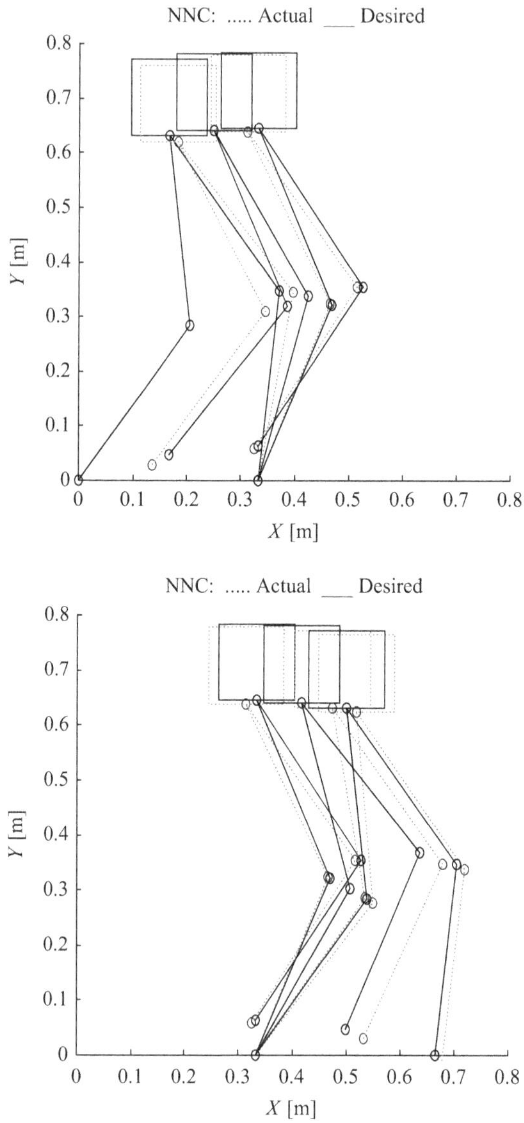

Figure 11.6 Foot and hip trajectories of the biped using control system 1 ($m_t = 200$ kg)

The simulation results are presented below. They illustrate the effects of different torso masses, stances and step lengths on the performance of the control system 1.

Using the control system of Figure 11.5, the biped was trained for 35,000 iterations to carry a torso mass of 20 kg to 1,000 points randomly located in the vertical

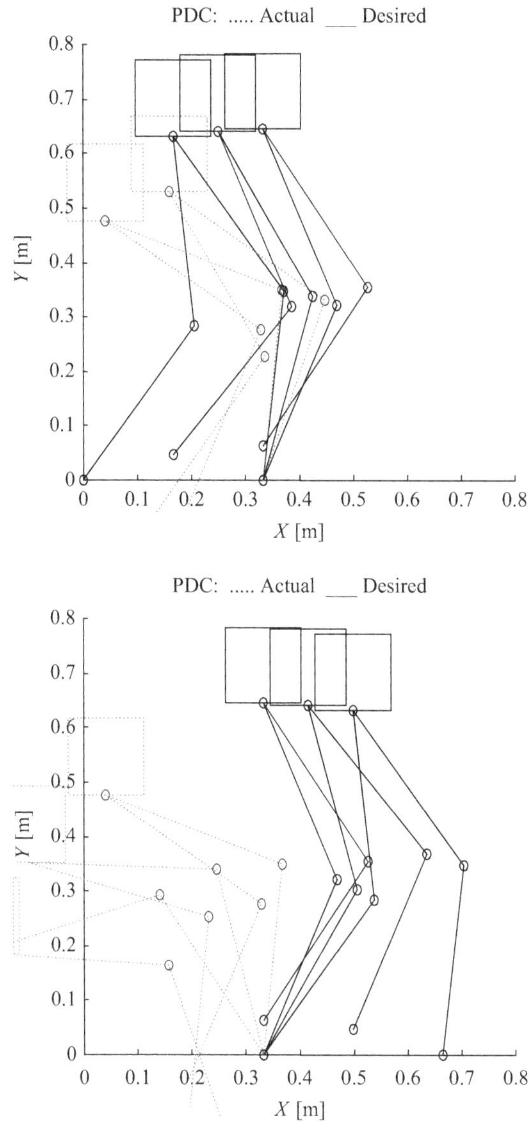

Figure 11.7 Foot and hip trajectories of the biped using PD controller ($m_t = 200\,kg$)

(X–Y) plane. Following training, the biped was prescribed regular foot and hip trajectories. After just one attempt, it was able to trace the desired trajectories with only very small errors.

The biped was then made to carry a torso mass of 200 kg. The foot and hip trajectories of the biped on the second attempt, after the torso mass had increased to

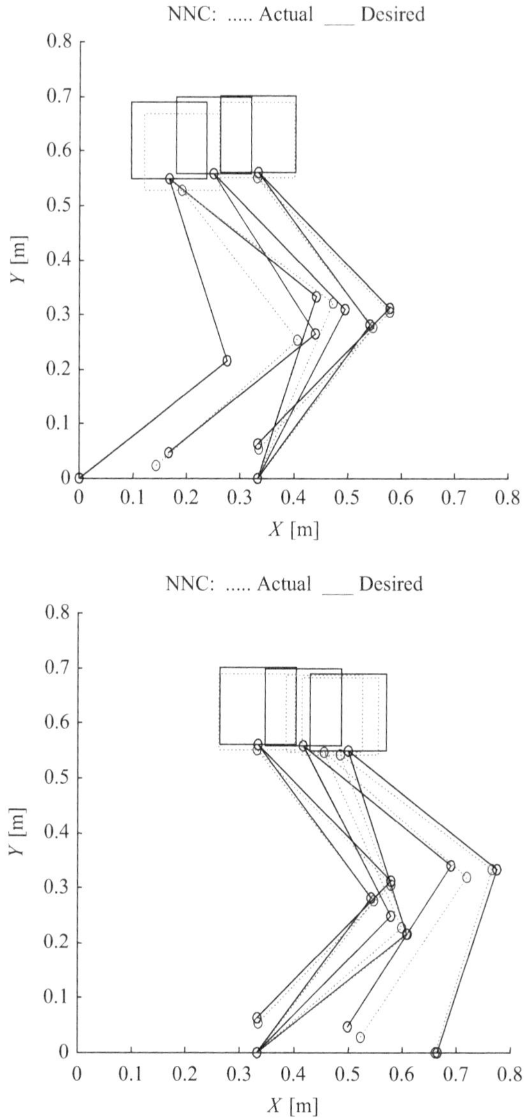

Figure 11.8 Foot and hip trajectories of the biped using control system 1 for a different stance ($m_t = 200\,kg$)

200 kg, are shown in Figure 11.6. For comparison, foot and hip trajectories obtained using the PD controller alone are depicted in Figure 11.7 for the cases with the increased torso mass.

The biped was next tested on different stances. The result with the increased torso mass is shown in Figure 11.8. For comparison, the performance of the PD controller is illustrated in Figure 11.9.

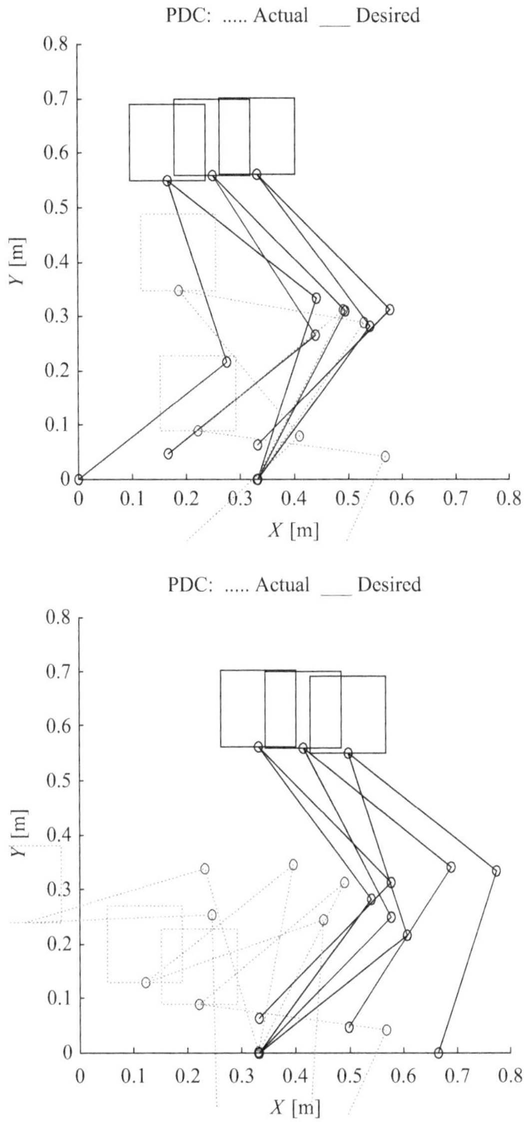

Figure 11.9 Foot and hip trajectories of the biped using PD controller for a different stance ($m_t = 200$ kg)

Tables 11.3–11.5 summarize all the results obtained for the different controllers by giving the root mean square errors (RMSEs) in each case.

It can be noted from the figures and tables presented that of the proposed neural control system performed better than the PD controller alone.

Table 11.3 RMSEs for various step lengths

Step length (m)	PD controller (RMSE)	Biped control system 1 (RMSE)
0.62	0.2067	0.0643
0.57	0.2421	0.0731
0.51	0.1787	0.0481
0.43	0.2982	0.0792

Table 11.4 RMSEs errors for various stances

Stance H (m)	PD controller (RMSE)	Biped control system 1 (RMSE)
0.65	0.2173	0.0621
0.61	0.2362	0.0639
0.58	0.2376	0.0648
0.55	0.2409	0.0656

Table 11.5 RMSEs errors for different torso masses

m_t (kg)	PD controller (RMSE)	Biped control system 1 (RMSE)
20	0.2638	0.0711
40	0.4011	0.1281
60	0.5821	0.1804
100	35.8588	3.2285
200	63.0643	6.8149

11.6 Conclusion

This paper has described the neural control systems for a whole bipedal robot. The main contribution of the paper consists in the solution of biped stability by means of neural network controllers. The system controls all of the robot joints simultaneously. Simulations have shown that proposed controllers are better than simple standard PD controllers at rejecting large load disturbances and accommodating different step lengths and walking stances.

Appendix A

Defining foot force vectors $\mathbf{F}_1$ and $\mathbf{F}_2$ which contain foot force components for foot 1 and 2, respectively:

$$\mathbf{F}_1 = \begin{bmatrix} \mathrm{R}_{x1} \\ \mathrm{R}_{y1} \end{bmatrix} \tag{A.1}$$

$$\mathbf{F}_2 = \begin{bmatrix} \mathrm{R}_{x2} \\ \mathrm{R}_{y2} \end{bmatrix} \tag{A.2}$$

where R_{x1} and R_{y1} are the horizontal and vertical components of the force acting on foot 1; R_{x2} and R_{y2} are similar components for foot 2.

The position and orientation of the biped in space may be completely specified by the general coordinate vector:

$$\mathbf{\Theta} = [x\ y\ \psi\ \alpha_1\ \alpha_2\ \beta_1\ \beta_2] \tag{A.3}$$

This vector consists of the joint coordinates $\mathbf{\Theta}$ with the addition of x and y, which are the Cartesian coordinates of the hip relative to some fixed frame of reference. The equations of motion corresponding to this general coordinate system may be derived using the method of Lagrange [6], thus yielding:

$$\mathbf{M}(\mathbf{\Theta})\ddot{\mathbf{\Theta}} + \mathbf{V}(\mathbf{\Theta}, \dot{\mathbf{\Theta}}) + \mathbf{G}(\mathbf{\Theta}) = \mathbf{C}_1\mathbf{F}_1 + \mathbf{C}_2\mathbf{F}_2 + \mathbf{D}\boldsymbol{\tau} \tag{A.4}$$

where the mass matrix $\mathbf{M}(\mathbf{\Theta})$, the centrifugal and Coriolis component vector $\mathbf{V}(\mathbf{\Theta}, \dot{\mathbf{\Theta}})$, the gravity vector $\mathbf{G}(\mathbf{\Theta})$, the foot force matrices $\mathbf{C}_1$ and $\mathbf{C}_2$ and the torque matrix $\mathbf{D}$ are given by

$$\mathbf{M}(\mathbf{\Theta}) = \begin{bmatrix} k_1 & 0 & -k_2C\psi & k_3C\alpha_1 & k_3C\alpha_2 & k_4C\beta_1 & k_2C\beta_2 \\ 0 & k_1 & -k_2S\psi & k_3S\alpha_1 & k_3S\alpha_2 & k_4S\beta_1 & k_4S\beta_2 \\ -k_2C\psi & -k_2S\psi & k_5 & 0 & 0 & 0 & 0 \\ k_3C\alpha_1 & k_3S\alpha_1 & 0 & k_6 & 0 & k_8C(\alpha_1-\beta_1) & 0 \\ k_3C\alpha_2 & k_3S\alpha_2 & 0 & 0 & k_6 & 0 & k_8C(\alpha_2-\beta_2) \\ k_4C\beta_1 & k_4S\beta_1 & 0 & k_8C(\alpha_1-\beta_1) & 0 & k_7 & 0 \\ k_4C\beta_2 & k_4S\beta_2 & 0 & 0 & k_8C(\alpha_2-\beta_2) & 0 & k_7 \end{bmatrix} \tag{A.5}$$

$$\mathbf{V}(\mathbf{\Theta}, \dot{\mathbf{\Theta}}) = \begin{bmatrix} k_2S\psi\dot{\psi}^2 - k_3(S\alpha_1\dot{\alpha}_1^2 + S\alpha_2\dot{\alpha}_2^2) - k_4(S\beta\alpha_1\dot{\beta}_1^2 + S\beta\alpha_2\dot{\beta}_2^2) \\ -k_2C\psi\dot{\psi}^2 - k_3(C\alpha_1\dot{\alpha}_1^2 + C\alpha_2\dot{\alpha}_2^2) - k_4(C\beta\alpha_1\dot{\beta}_1^2 + C\beta\alpha_2\dot{\beta}_2^2) \\ 0 \\ k_8S(\alpha_1-\beta_1)\dot{\beta}_1^2 \\ k_8S(\alpha_2-\beta_2)\dot{\beta}_2^2 \\ -k_8S(\alpha_1-\beta_1)\dot{\alpha}_1^2 \\ -k_8S(\alpha_2-\beta_2)\dot{\alpha}_2^2 \end{bmatrix} \tag{A.6}$$

$$\mathbf{G}(\boldsymbol{\Theta}) = \begin{bmatrix} 0 \\ k_1 g \\ -k_2 g S\psi \\ k_3 g S\alpha_1 \\ k_3 g S\alpha_2 \\ k_3 g S\beta_1 \\ k_3 g S\beta_2 \end{bmatrix} \tag{A.7}$$

$$\mathbf{D} = \begin{bmatrix} 0 & 0 & 0 & 0 & 0 \\ 0 & 0 & 0 & 0 & 0 \\ -1 & -1 & 0 & 0 & 0 \\ 1 & 0 & -1 & 0 & 0 \\ 0 & 1 & 0 & -1 & 0 \\ 0 & 0 & 1 & 0 & -1 \\ 0 & 0 & 0 & 1 & 0 \end{bmatrix} \tag{A.8}$$

$$\mathbf{C}_1 = \begin{bmatrix} 1 & 0 \\ 0 & 1 \\ 0 & 0 \\ 2l_e C\alpha_1 & 2l_e S\alpha_1 \\ 0 & 0 \\ 2l_e C\beta_1 & 2l_e S\beta_1 \\ 0 & 0 \end{bmatrix} \tag{A.9}$$

$$\mathbf{C}_2 = \begin{bmatrix} 1 & 0 \\ 0 & 1 \\ 0 & 0 \\ 0 & 0 \\ 2l_e C\alpha_2 & 2l_e S\alpha_2 \\ 0 & 0 \\ 2l_e C\beta_2 & 2l_e S\beta_2 \end{bmatrix} \tag{A.10}$$

where C and S represent the cosine and sine functions, respectively, and

$$k_1 = m_t + 4m_e + 2m_f \tag{A.11a}$$

$$k_2 = m_t l_t \tag{A.11b}$$

$$k_3 = 3m_e l_e + 2m_f l_e \tag{A.11c}$$

$$k_4 = m_e l_e + 2m_f l_e \tag{A.11d}$$

$$k_5 = J_t + m_t l_t^2 \tag{A.11e}$$

$$k_6 = J_e + 5m_e l_e^2 + 4m_f l_e^2 \tag{A.11f}$$

$$k_7 = J_e + m_e l_e^2 + 4m_f l_e^2 \tag{A.11g}$$

$$k_8 = 2m_e l_e^2 + 4m_f l_e^2 \tag{A.11h}$$

References

[1] Hemami H., Farnsworth R.L. 'Postural and gait stability of a planar five link biped by simulation', *IEEE Trans. Autom. Control.* 1977; **22**: 452–458.

[2] Kimua H., Shimoyama I., Miura H. 'Criteria for dynamic walk of the quadruped', *Proceedings of Int. Symposium and Exposition on Robots*; Sydney, Australia, 1988. pp. 595–600.

[3] Huston R.L., Passerella C.E. 'On the dynamics of human body model', *J. Biomech.* 1971; **4**: 369–379.

[4] Goddard R.E., Hemami H., Weimer F.C. 'Biped side step in the frontal plane', *IEEE Trans. Autom. Control.* 1983; **28**: 179–186.

[5] Channon P.H., Hopkins S.H., Pham D.T. 'Optimal control of an n-legged robot', *IMechE.* 1996; **210**: 51–63.

[6] Yadav A.K., Gaur P. 'Intelligent modified internal model control for speed control of nonlinear uncertain heavy duty vehicles', *ISA Trans.* 2015; **56**: 288–298.

[7] Mozaffari L., Mozaffari A., Azad N.L. 'Vehicle speed prediction via a sliding-window time series analysis and an evolutionary least learning machine: A case study on San Francisco urban roads', *Eng. Sci. Technol. Int. J.* 2015; **18**(2): 150–162.

[8] Tagne G., Talj R., Charara, A. 'Design and validation of a robust immersion and invariance controller for the lateral dynamics of intelligent vehicles', *Control Eng. Pract.* 2015; **40**: 81–92.

[9] Alam A., Gattami A., Johansson K.H. Tomlin C.J. 'Guaranteeing safety for heavy duty vehicle platooning: Safe set computations and experimental evaluations', *Control Eng. Pract.* 2014; **24**: 33–41.

[10] Alam A., Martensson J., Johansson K.H. 'Experimental evaluation of decentralized cooperative cruise control for heavy-duty vehicle platooning', *Control Eng. Pract.* 2015; **38**: 11–25.

[11] Rödönyi G., Gaspar P., Bokor J., Palkovics L. 'Experimental verification of robustness in a semi-autonomous heavy vehicle platoon', *Control Eng. Pract.* 2014; **28**: 13–25.

[12] Tibola J.R., Lanzanova T.D.M., Martins M.E.S., Gründling H.A., Pinheiro H. 'Modeling and speed control design of an ethanol engine for variable speed gensets', *Control Eng. Pract.* 2015; **35**: 54–66.

[13] Craig J.J. *Introduction to Robotics: Mechanics and Control*. Reading, MA; Addison-Wesley, 1989.

[14] Pham D.T., Yıldırım Ş. 'Control of the trajectory of a planar robot using recurrent hybrid networks', *Int. J. Mach. Tools Manuf.: Des., Res. Appl.* 1999; **39**: 415–429.

[15] Matlab Release 2014, Simulink Toolbox, Mathworks Inc. Ltd., 2014.

Part IV

Robust controllers-based observers via Takagi–Sugeno or linear approaches

Chapter 12
A survey on the polytopic Takagi–Sugeno approach: application to the inverted pendulum

Souad Bezzaoucha[1], Holger Voos[1], and Mohamed Darouach[2]

Abstract

This chapter gives a general scope, states the main results obtained and methods used for the polytopic Takagi–Sugeno approach with a detailed application to the inverted pendulum. Modeling, observer and controller design will be considered.

12.1 Introduction

The modeling of complex dynamic systems such as in biochemical processes or mobile robotic applications often results in nonlinear models. In order to enhance their performance, this nonlinear behavior has to be taken into account in the control and/or observer design. The complexity of the resulting models to be dealt with then also requires the use of heavy mathematical tools.

In order to cope with the complexity introduced by various kind of nonlinearities (nonlinear behavior, time-varying parameters, saturation, hysteresis, sin and cos functions, ...) and since a considerable amount of results have been established in the linear systems framework, the idea behind the so-called Takagi–Sugeno (T–S) polytopic model is to extend these existing results for nonlinear models.

However, the direct transition from linear models to the nonlinear one remains delicate. Hence, new modeling techniques have been proposed in order to make "easier" and "elegant" this transition [1]. In fact, the polytopic T–S or multiple models (MMs) decomposition provides an alternative and attractive path to deal with complex nonlinear systems and to obtain an equivalent representation by bounding the parameters and using the well-known sector-nonlinearity transformation (SNT). This

[1]Interdisciplinary Centre for Security, Reliability and Trust (SnT), Automatic Control Research Group, University of Luxembourg, Luxembourg
[2]Research Center for Automatic Control of Nancy (CRAN), Université de Lorraine, France

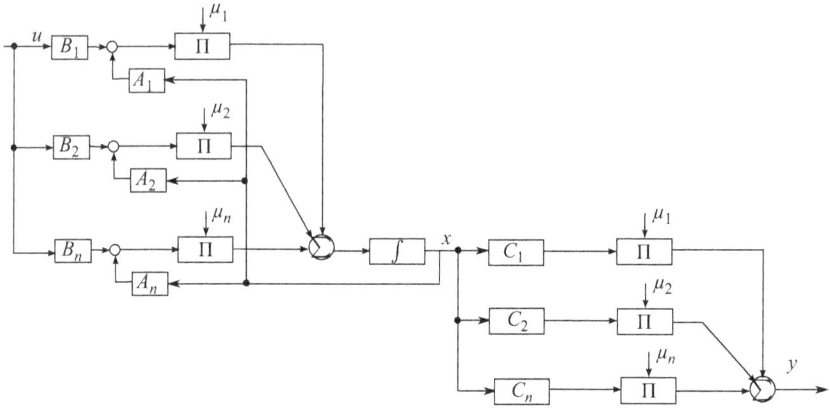

Figure 12.1 Coupled T–S models

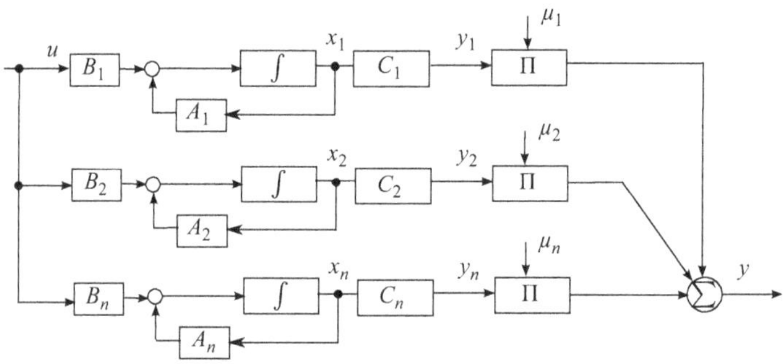

Figure 12.2 Decoupled T–S models

structure offers an efficient representation of nonlinear behaviors while remaining relatively simple compared to general nonlinear models.

The strength of these techniques consists in the fact that the properties of the nonlinear systems are expressed by a collection of linear subsystems, i.e., in a polytopic paradigm, where the overall model of the system is obtained by merging the local subsystems through a time-varying blending called weighting functions. Thanks to the convexity of the weighting functions and to the linearity of the subsystems defining the vertices, some results can be established for nonlinear systems [2–5] using tools borrowed from the linear theory [6,7].

As pointed in [1,8], different MMs can be obtained using different aggregation structures of the submodels. Basically, two major structures can be distinguished. As explained in the following table and illustrated by Figures 12.1 and 12.2, in the first structure, the submodels share the same state vector (the so-called polytopic T–S model); in the second one, known as decoupled MM, the submodels are decoupled, and their state vectors are different.

Coupled T–S models	**Decoupled T–S models**
The global state vector x is common to all submodels	Each submodel has its own independant state vector x_i
$\begin{cases} \dot{x}(t) = \sum_{i=1}^{n} \mu_i(\xi(t))(A_i x(t) + B_i u(t)) \\ y(t) = \sum_{i=1}^{n} \mu_i(\xi(t)) C_i x(t) \end{cases}$	$\begin{cases} \dot{x}_i(t) = A_i x_i(t) + B_i u(t) \\ y_i(t) = C_i x_i(t) \\ y(t) = \sum_{i=1}^{n} \mu_i(\xi(t)) y_i(t) \end{cases}$

In the following, only the coupled class will be addressed since its model may be obtained in a systematic way from the analytic expression of the nonlinear model without any loss of information by the SNT (presented in the next section). The decoupled representation is more interesting for the identification where the different dimension of the submodels may avoid an extra parametrization in certain working set [9]. However, since the work presented in this chapter does not concern the identification and for the estimation and control, the decoupled model may be easily transposed into a coupled one (by concatenating the local states x_i), the presented study will be only about the coupled polytopic T–S models.

In the present chapter, a survey about the T–S polytopic approach and its application to an inverted pendulum is presented; two main problems will be addressed: the first one is the modeling and the second one is the control design (observer based).

The modeling process through the well-known SNT is introduced. This procedure is used as a systematic and analytical method to rewrite and thus transform a nonlinear system into a polytopic T–S form without any loss of information [10,11]. The observer design is then presented with a Lyapunov stability analysis used to derive sufficient existence conditions of the observer. These conditions are expressed in terms of $\mathcal{LMI}$s, which can easily be solved by dedicated softwares (LMI toolbox in MATLAB®, YALMIP [12] or the SeDuMi package for solving optimization problems with linear, quadratic and semidefiniteness constraints [13].

The inverted pendulum example is considered to illustrate the effectiveness of the proposed approach. From the nonlinear equations of the inverted pendulum, using the SNT transformation, a T–S polytopic model of the system is deduced and used.

The overview of this chapter is as follows. A short introduction to T–S systems and SNT transformation is presented in Section 12.2, and some useful results on the observer and controller design are detailed in Sections 12.3 and 12.4. A numerical example of an inverted pendulum to illustrate each step is given for all sections.

12.2 Systematic transformation of a nonlinear system into a polytopic T–S form

An efficient way to deal with nonlinear system estimation or fault diagnosis consists in rewriting the original nonlinear system in a simpler form, like the polytopic T–S model. Originally introduced by [14], the T–S representation allows to exactly describe nonlinear systems, under the condition that the nonlinearities are bounded. This is reasonable since state variables as well as parameters of physical systems are bounded ([10,15] and the references therein). The polytopic T–S model is based on a nonlinear interpolation between certain linear submodels with adequate weighting functions.

The T–S systems, as proposed by Takagi and Sugeno [14], are described by local models merged together using fuzzy IF-THEN rules, as follows:

IF $z_1(t)$ is in M_{i1} and …and $z_p(t)$ is in M_{ip},

$$\text{THEN} \begin{cases} \dot{x}(t) = A_i x(t) + B_i u(t) \\ y(t) = C_i x(t) \end{cases} \quad i = 1, 2, \ldots, r \tag{12.1}$$

where M_{ij} is the fuzzy set, and r is the number of model rules; $x(t) \in \mathbb{R}^{n_x}$ is the system state, $u(t) \in \mathbb{R}^m$ is the input and $y(t) \in \mathbb{R}^p$ is the system output. $z_i(t)$ are known premise variables and each linear consequent equation represented by $A_i x(t) + B_i u(t)$ is called subsystem.

Note that for the "fuzzy" model, the fuzzification process consists in transforming crisp values into fuzzy linguistic variables. The membership function is used to associate a grade to each linguistic variable. Selection of the number of membership functions and their initial values is based on the process knowledge and intuition.

In order to obtain a polytopic T–S representation from a nonlinear system, in a systematic and more structured way, the SNT is applied. The SNT is a systematic procedure that guarantees an exact model construction for nonlinear systems with bounded nonlinearities. Starting with a general form of nonlinear systems, a quasi-linear parameter-varying (LPV) state representation is realized. The polytopic T–S form is then obtained by using the convex polytopic transformation where each vertex defines a linear submodel, and nonlinear parts are rejected into the weighting functions $\mu_i(\xi(t))$.

On the opposite of a classical fuzzy model, the weighting functions premise variables of the polytopic T–S form may be either measurable or unmeasurable. For simplicity reasons, it is easier to consider a system with measurable premise variables (most of the available works in the literature have considered models with measurable premise variables). However, in most practical cases (and realistic models), the premise variables depend on the state variables of the plant (and even sometimes unknown input dependent), which brings some additional difficulties but leads to more challenging problems in the observer/controller design. In this chapter, both cases of measurable and unmeasurable premise variables are considered.

Consider a dynamic nonlinear system described by the following state equations:

$$\begin{cases} \dot{x}(t) = f(t, x, u) \\ y(t) = g(t, x, u) \end{cases} \tag{12.2}$$

where $f(.)$ and $g(.)$ are vector functions of dimension $\mathbb{R}^{n_x}$ and $\mathbb{R}^p$, respectively.

The SNT allows to express nonlinear dynamic systems into a convex combination of linear submodels:

$$\begin{cases} \dot{x}(t) = \sum_{i=1}^{n} \mu_i(t)(A_i x(t) + B_i u(t)) \\ y(t) = \sum_{i=1}^{n} \mu_i(t)(C_i x(t) + D_i u(t)) \end{cases} \tag{12.3}$$

with $X_i(.)$, for $X \in \{A, B, C, D\}$, are matrices functions of appropriate dimensions. The functions $\mu_i(t)$ represent the weights of the submodels (A_i, B_i, C_i, D_i) in the global model and satisfy the convex sum property:

$$\sum_{i=1}^{n} \mu_i(t) = 1, \ \forall t \tag{12.4}$$

In order to obtain the polytopic T–S form, first, the system (12.2) is transformed in a linear time-varying form:

$$\begin{cases} \dot{x}(t) = A(\xi(t))x(t) + B(\xi(t))u(t) \\ y(t) = C(\xi(t))x(t) + D(\xi(t))u(t) \end{cases} \tag{12.5}$$

Note that several choices for the system (12.5) are possible (details on how to obtain (12.5) from (12.2) are detailed in [10]). The nonlinear entries of the matrices $X(\xi(t))$, with $X \in \{A, B, C, D\}$, are called premise variables and denoted as $\xi_j(x, u)$, $j = 1, \ldots, q$.

From (12.5), a convex polytopic transformation is performed for each premise variable as follows:

$$\xi_j(x, u) = F_{j,1}(\xi_j(x, u))\xi_{j,1} + F_{j,2}(\xi_j(x, u))\xi_{j,2} \tag{12.6}$$

where

$$\xi_{j,1} = \max_{x,u}\{\xi_j(x, u)\}, \quad \xi_{j,2} = \min_{x,u}\{\xi_j(x, u)\} \tag{12.7}$$

$$\begin{cases} F_{j,1}(\xi_j(x, u)) = \dfrac{\xi_j(x, u) - \xi_{j,2}}{\xi_{j,1} - \xi_{j,2}} \\ F_{j,2}(\xi_j(x, u)) = \dfrac{\xi_{j,1} - \xi_j(x, u)}{\xi_{j,1} - \xi_{j,2}} \end{cases} \tag{12.8}$$

For q premise variables, $n = 2^q$ submodels are obtained. The weighting functions $\mu_i(t)$ are defined by

$$\mu_i(t) = \prod_{j=1}^{n} F_{j,\sigma_i^j}(\xi_j(x, u)) \tag{12.9}$$

where the indexes σ_i^j ($i = 1, \ldots, 2^q$ and $j = 1, \ldots, q$) are equal to 1 or 2 and indicate which partition of the jth premise variable ($F_{j,1}$ or $F_{j,2}$) is involved in the ith submodel.

The relation between the submodel number i and indexes σ_i^j is given by

$$i = 2^{n-1}\sigma_i^1 + 2^{n-2}\sigma_i^2 + \cdots + 2^0\sigma_i^n - (2^1 + 2^2 + \cdots + 2^{n-1}) \tag{12.10}$$

The constant matrices X_i are obtained by replacing the variables ξ_j in the matrices A, B, C and D with the scalars defined in (12.7):

$$X_i = X(\xi_{1,\sigma_i^1}, \ldots, \xi_{q,\sigma_i^q}), \quad X \in \{A, B, C, D\} \tag{12.11}$$

To sum up, the SNT decomposition enables to represent the nonlinear system (12.2) in a compact polytopic T–S model (12.3) without any approximation or loss of information. For a better understanding of the previous procedure, let us consider the following equation of motion for the inverted pendulum given by the nonlinear model (12.12):

$$\begin{cases} \dot{x}_1(t) = x_2(t) \\ \dot{x}_2(t) = \frac{mlx_4^2(t)\sin(x_3(t)) - mg\sin(x_3(t))\cos(x_3(t)) + u}{M + m\sin^2(x_3(t))} \\ \dot{x}_3(t) = x_4(t) \\ \dot{x}_4(t) = \frac{-mlx_4^2(t)\sin(x_3(t))\cos(x_3(t)) + g(m+M)\sin(x_3(t)) - u\cos(x_3(t))}{l(M + m\sin^2(x_3(t)))} \end{cases} \tag{12.12}$$

where $x_1(t)$ denotes the cart position (m) and $x_3(t)$ the angle (radians) displacement of the pendulum from the vertical configuration. The velocities are denoted by $x_2(t)$ and $x_4(t)$.

g is the gravity constant ($g = 9.81$ m/s^2), m and M are, respectively, the point mass attached at the end of the pendulum and the cart mass, l is the pendulum length and $u(t)$ is the force applied to the cart (newtons).

According to the nonlinear model, four premise variables are defined

$$\begin{cases} \xi_1(t) = \frac{1}{M + m\sin^2(x_3(t))} \\ \xi_2(t) = \frac{\sin(x_3(t))}{x_3(t)} \\ \xi_3(t) = \cos(x_3(t)) \\ \xi_4(t) = x_4(t)\sin(x_3(t)) \end{cases} \tag{12.13}$$

Note that $x_3(t) \in (-\pi/2, \pi/2)$ and $x_4(t) \in [-\alpha, \alpha]$. The system is uncontrollable when $x_3(t) = \pm\pi/2$. To maintain the controllability of the system model, it is assumed that $x_3(t) \in [-88°, 88°]$.

The following quasi-LPV representation is chosen as follows:

$$\dot{x}(t) = A(\xi(t))x(t) + B(\xi(t))u(t) \tag{12.14}$$

where $A(\xi(t))$ and $B(\xi(t))$ are defined as

$$A(\xi(t)) = \begin{pmatrix} 0 & 1 & 0 & 0 \\ 0 & 0 & -mg\xi_1(t)\xi_2(t)\xi_3(t) & ml\xi_1(t)\xi_4(t) \\ 0 & 0 & 0 & 1 \\ 0 & 0 & \frac{g(m+M)}{l}\xi_1(t)\xi_2(t) & -m\xi_1(t)\xi_3(t)\xi_4(t) \end{pmatrix},$$

$$B(\xi(t)) = \begin{pmatrix} 0 \\ \xi_1(t) \\ 0 \\ -\frac{1}{l}\xi_1(t)\xi_3(t) \end{pmatrix} \tag{12.15}$$

The SNT is now applied for each premise variable. We have then

$$\xi_1(t) = F_{1,1}(\xi_1(t))\xi_{1,1} + F_{1,2}(\xi_1(t))\xi_{1,2} \tag{12.16}$$

where

$$\xi_{1,1}(t) = \max\{\xi_1(t)\} = \frac{1}{M}, \quad \xi_{1,2} = \min\{\xi_1(t)\} = \frac{1}{M + m\sin^2(88\pi/180)} \tag{12.17}$$

and

$$\begin{cases} F_{1,1}(\xi_1(t)) = \dfrac{\xi_1(t) - \xi_{1,2}}{\xi_{1,1} - \xi_{1,2}} \\ F_{1,2}(\xi_1(t)) = \dfrac{\xi_{1,1} - \xi_1(t)}{\xi_{1,1} - \xi_{1,2}} \end{cases} \tag{12.18}$$

The same decomposition is applied for the second premise variable $\xi_2(t)$, with $\xi_{2,1} = 1$, $\xi_{2,2} = 2/\pi$ and

$$\begin{cases} F_{2,1}(\xi_2(t)) = \dfrac{\xi_2(t) - \xi_{2,2}}{\xi_{2,1} - \xi_{2,2}} \\ F_{2,2}(\xi_2(t)) = \dfrac{\xi_{2,1} - \xi_2(t)}{\xi_{2,1} - \xi_{2,2}} \end{cases} \tag{12.19}$$

For numerical reasons, this premise variable $\xi_2(t)$ is defined for an angle $x_3(t) \neq 0$, if $x_3(t) = 0$, then $\xi_2(t)$ is set to be equal to 1.

Following the same procedure for $\xi_3(t) = \cos(x_3(t))$, we have

$$\xi_3(t) = F_{3,1}(\xi_3(t))\xi_{3,1} + F_{3,2}(\xi_3(t))\xi_{3,2} \tag{12.20}$$

where

$$\xi_{3,1}(t) = \max\{\xi_3(t)\} = 1, \quad \xi_{3,2} = \min\{\xi_3(t)\} = \cos(88\pi/180) \tag{12.21}$$

and

$$\begin{cases} F_{3,1}(\xi_3(t)) = \dfrac{\xi_3(t) - \xi_{3,2}}{\xi_{3,1} - \xi_{3,2}} \\ F_{3,2}(\xi_3(t)) = \dfrac{\xi_{3,1} - \xi_3(t)}{\xi_{3,1} - \xi_{3,2}} \end{cases} \tag{12.22}$$

Finally, for $\xi_4(t) = x_4(t)\sin(x_3(t))$, we have

$$\xi_4(t) = F_{4,1}(\xi_4(t))\xi_{4,1} + F_{4,2}(\xi_4(t))\xi_{4,2} \tag{12.23}$$

where

$$\xi_{4,1}(t) = \max\{\xi_4(t)\} = \alpha, \quad \xi_{4,2} = \min\{\xi_4(t)\} = -\alpha \tag{12.24}$$

and

$$\begin{cases} F_{4,1}(\xi_4(t)) = \dfrac{\xi_4(t) - \xi_{4,2}}{\xi_{4,1} - \xi_{4,2}} \\ F_{4,2}(\xi_4(t)) = \dfrac{\xi_{4,1} - \xi_4(t)}{\xi_{4,1} - \xi_{4,2}} \end{cases} \tag{12.25}$$

For $q = 4$, premise variables, and replacing all of them by their polytopic decomposition into the system model (12.12), we obtain the following polytopic system (with $2^4 = 16$ submodels):

$$\begin{aligned}\dot{x}(t) = \Bigg(& \underbrace{F_{11}(\xi_1)F_{21}(\xi_2)F_{31}(\xi_3)F_{41}(\xi_4)}_{\mu_1(\xi(t))} A_1 + \underbrace{F_{11}(\xi_1)F_{12}(\xi_2)F_{31}(\xi_3)F_{42}(\xi_4)}_{\mu_2(\xi(t))} A_2 \\ & + \underbrace{F_{11}(\xi_1)F_{11}(\xi_2)F_{32}(\xi_3)F_{41}(\xi_4)}_{\mu_3(\xi(t))} A_3 + \cdots + \underbrace{F_{12}(\xi_1)F_{22}(\xi_2)F_{32}(\xi_3)F_{41}(\xi_4)}_{\mu_{15}(\xi(t))} A_{15} \\ & + \underbrace{F_{12}(\xi_1)F_{22}(\xi_2)F_{32}(\xi_3)F_{44}(\xi_4)}_{\mu_{16}(\xi(t))} A_{16} \Bigg) x(t) + (\mu_1(\xi(t))B_1 + \cdots + \mu_{16}(\xi(t))B_{16})\, u(t)\end{aligned} \tag{12.26}$$

where $x = [x_1, x_2, x_3, x_4]^T$ is the system state and $F(t)$ the system input. The system matrices defined as

$$A_i = A(\xi_{1,\sigma_i^1}, \ldots, \xi_{4,\sigma_i^4}), \quad B_i = B(\xi_{1,\sigma_i^1}, \xi_{3,\sigma_i^3}) \tag{12.27}$$

where the relation between the indexes σ_i^j ($i = 1, \ldots, 16$ and $j = 1, \ldots, 4$) and the submodel number i is given by (12.10).

Then, the nonlinear model of the inverted pendulum may be written in a polytopic T–S form as follows:

$$\dot{x}(t) = \sum_{i=1}^{16} \mu_i(\xi(t))(A_i x(t) + B_i u(t)) \tag{12.28}$$

12.3 Stability analysis of polytopic T–S systems

Let us consider the following time-varying system:

$$\dot{x}(t) = A(t)x(t) \tag{12.29}$$

this system is asymptotically stable (quadratic stability), if there exists a Lyapunov matrix P positive symmetric $P = P^T > 0$ such that the Lyapunov function

$$V(x(t)) = x^T(t)Px(t) > 0 \tag{12.30}$$

is positive, null for $x = 0$ and decreasing in time. The same reasoning is applied for the nonlinear system represented by the autonomous T–S model:

$$\dot{x}(t) = \sum_{i=1}^{n} \mu_i(\xi(t))A_i x(t) \tag{12.31}$$

with the weighting functions $\mu_i(\xi(t))$ satisfying the convex sum property (12.4).

Lemma 12.1. *The system (12.31) is globally asymptotically stable if there exists a matrix P positive symmetric* $P = P^T > 0 \in \mathbb{R}^{n_x \times n_x}$ *such that the following conditions are verified for* $i = 1, \ldots, n$ *[10]:*

$$A_i^T P + PA_i < 0 \tag{12.32}$$

Proof. The proof is based on the choice of a Lyapunov function $V(x(t)) = x^T(t)Px(t)$ where $P = P^T > 0 \in \mathbb{R}^{n_x \times n_x}$. The sum convex property of the weighting functions allows to obtain n matrices inequalities ($\mathcal{LMI}s$) to solve simultaneously. □

Later research has focused in conceiving more efficient $\mathcal{LMI}s$ conditions, such as the ones in [16–18]. In fact, a key point for the T–S system study is the search for an adequate Lyapunov function. A usual approach is to call for a single quadratic Lyapunov function but a significant improvement can be obtained by considering time-varying or polytopic Lyapunov functions that incorporate the system dynamics [19–21].

12.4 Observer design for polytopic T–S models

The observability concept for a system consists on establishing the conditions under which the state may be reconstructed from the inputs and the measurable outputs.

For the nonlinear systems, the observability problem has been widely studied in the literature (see [22–24] for example).

For the T–S approach, a necessary condition for the observability may be given by the observability constraint of the local systems, i.e., for each pair (A_i, C_i) of the system submodels, or at least their detectability [25,26].

Let us consider the following polytopic T–S model given by the following equation:

$$\begin{cases} \dot{x}(t) = \sum_{i=1}^{n} \mu_i(\xi(t))(A_i x(t) + B_i u(t)) \\ y(t) = Cx(t) \end{cases} \tag{12.33}$$

The most common observer for these systems is a polytopic Luenberger observer [27]. It basically consists in extending the design of the popular Luenberger (proportional) observer used in the linear system framework. It is defined by

$$\begin{cases} \dot{\hat{x}}(t) = \sum_{i=1}^{n} \mu_i(\hat{\xi}(t))(A_i\hat{x}(t) + B_iu(t) + L_i(y(t) - \hat{y}(t))) \\ \hat{y}(t) = C\hat{x}(t) \end{cases} \tag{12.34}$$

with $\hat{x}(t)$ and $\hat{\xi}(t)$ represent, respectively, the state $x(t)$ and premise variables $\xi(t)$ estimate.

The observer gains L_i are designed such that the state estimate $\hat{x}(t)$ asymptotically converges toward the state $x(t)$. Then, the error dynamics stability $e(t) = x(t) - \hat{x}(t)$ has to be ensured.

The weighting functions $\mu_i(\xi(t))$ verify the following convex sum property:

$$\sum_{i=1}^{r} \mu_i(\xi(t)) = 1, \quad 0 \leq \mu_i(\xi(t)) \leq 1, \quad i = 1, \ldots, M, \quad \forall t \tag{12.35}$$

$\xi(t)$ may depend on measurable premises variables (a part of the input $u(t)$ or the output $y(t)$) or unmeasurable premises variables (as the system states, UIs). The state estimation for nonlinear systems using the T–S polytopic model approach has been addressed with different methods, but most of the available works in the literature have considered models with measurable premise variables.

For simplicity reasons, it is easier to consider a system with measurable premise variables. However, in most practical cases (and realistic models), the premise variables depend on the state variables of the plant (and even sometimes unknown input dependent). In this chapter, both cases of measurable premise variables and the general case of unmeasurable premise variables are addressed; the second case brings some additional difficulties but leads to more challenging problems in the observer design [28,29].

12.4.1 Measurable premise variables

In the measurable premise variables case, the observer design is based on the availability of the premise variables. Then, the system and the polytopic observer share the same weighting (also called activation) functions. The main advantage is the possibility to factorize these functions during the estimation error dynamic study $e(t)$ defined as $x(t)-\hat{x}(t)$. We then have

$$\dot{e}(t) = \sum_{i=1}^{n} \mu_i(\xi(t))(A_i - L_iC)e(t) \tag{12.36}$$

A simple stability analysis of the system (12.36) allows us to find the desired gains L_i and that may be given for example by the following theorem:

Lemma 12.2. *[30] The state estimation error converges asymptotically to zero if there exists a positive symmetric matrix $P = P^T > 0 \in \mathbb{R}^{n_x \times n_x}$ and matrices $K_i \in \mathbb{R}^{n_x \times m}$ such that the following conditions are satisfied for $i = 1, \ldots, n$:*

$$PA_i + A_i^T P - K_i C - C^T K_i^T < 0 \tag{12.37}$$

The observer gains are then given by

$$L_i = P^{-1} K_i \tag{12.38}$$

The same study may be carried out for the case of a polytopic output ($y(t) = \sum_{i=1}^{n} \mu_i(\xi(t)) C_i x(t)$) where the conservatism introduced by a system defined by a double sum may be reduced (see [16,31] for more details).

12.4.2 Unmeasurable premise variables

For the unmeasurable premise variables case, the estimation error dynamics cannot be easily computed directly, since in the considered case, the weighting functions depend on the unmeasurable variables $\xi(t)$, whereas those of the observer depend on their estimates $\hat{\xi}(t)$. In order to overcome this difficulty, taking benefits of the convex sum property, the state derivative is rewritten as an uncertain T–S system where the uncertainties are bounded. The observer gains are then determined in a way to ensure the stability of the state estimation error dynamics as well as an $\mathscr{L}_2$ attenuation of the transfer from the uncertainties to the state estimation error.

The polytopic T–S model (12.33) is then rewritten as

$$\dot{x}(t) = \sum_{i=1}^{n} \Big(\mu_i(\hat{\xi}(t))(A_i x(t) + B_i u(t)) + (\mu_i(\xi(t)) - \mu_i(\hat{\xi}(t)))(A_i x(t) + B_i u(t)) \Big) \tag{12.39}$$

From (12.34) and (12.39), the state estimation error dynamics is then given by

$$\dot{e}(t) = \sum_{i=1}^{n} \mu_i(\hat{\xi}(t))(A_i - L_i C)e(t) + \Delta A(t)x(t) + \Delta B(t)u(t) \tag{12.40}$$

where $\Delta A(t)$ and $\Delta B(t)$ are considered uncertainties and defined as follows:

$$\begin{cases} \Delta A(t) = \displaystyle\sum_{i=1}^{n} \Big(\mu_i(\xi(t)) - \mu_i(\hat{\xi}(t)) \Big) A_i \\ \qquad\;\; = \mathscr{A} \Sigma_A(t) E_A \\ \Delta B(t) = \displaystyle\sum_{i=1}^{n} \Big(\mu_i(\xi(t)) - \mu_i(\hat{\xi}(t)) \Big) B_i \\ \qquad\;\; = \mathscr{B} \Sigma_B(t) E_B \end{cases} \tag{12.41}$$

with

$$\mathscr{A} = \begin{bmatrix} A_1 & \dots & A_n \end{bmatrix},\ \Sigma_A(t) = \operatorname{diag}(\delta_1(t)I_{n_x}, \dots, \delta_n(t)I_{n_x}),$$
$$\mathscr{B} = \begin{bmatrix} B_1 & \dots & B_n \end{bmatrix},\ \Sigma_B(t) = \operatorname{diag}(\delta_1(t)I_{n_u}, \dots, \delta_n(t)I_{n_u}), \tag{12.42}$$
$$E_A = \begin{bmatrix} I_{n_x} & \dots & I_{n_x} \end{bmatrix}^T,\ E_B = \begin{bmatrix} I_{n_u} & \dots & I_{n_u} \end{bmatrix}^T,\ \delta_i(t) = \mu_i(\xi(t)) - \mu_i(\hat{\xi}(t))$$

The observer gains are then given by the following theorem:

Theorem 12.1. *[32] The system (12.40) is globally asymptotically stable and the gain $\mathscr{L}_2$ of the transfer of the pair $u(t), x(t)$ to the estimation error is bounded by γ, if there exists two positive symmetric matrices $P_1 \in \mathbb{R}^{n_x \times n_x}$ and $P_2 \in \mathbb{R}^{n_x \times n_x}$, matrices $K_i \in \mathbb{R}^{n_x \times m}$ and positive scalars λ_1, λ_2 et $\overline{\gamma}$ such that the following $\mathscr{LMI}$s are satisfied for all $i, j = 1, \dots, n$:*

$$\begin{pmatrix} \mathbb{S}(P_1A_i - K_iC) + I_{n_x} & 0 & 0 & P_1\mathscr{A} & P_1\mathscr{B} \\ * & \mathbb{S}(P_2A_j) + \lambda_1 E_A^T E_A & P_2B_j & 0 & 0 \\ * & * & -\overline{\gamma} I_{n_u} + \lambda_2 E_B^T E_B & 0 & 0 \\ * & * & * & -\lambda_1 I & 0 \\ * & * & * & 0 & -\lambda_1 I \end{pmatrix} < 0 \tag{12.43}$$

where $\mathbb{S}(M) := M + M^T$.

The observer gains and the attenuation level γ of the transfer from the uncertainties to the state estimation error are given by

$$L_i = P_1^{-1}K_i \tag{12.44}$$

$$\gamma = \sqrt{\overline{\gamma}} \tag{12.45}$$

12.5 Polytopic T–S models stabilization

The most common strategy for the nonlinear polytopic T–S models stabilization is the so-called parallel distributed compensation (PDC). Introduced in [33,34], this command law depends on the same weighting functions used for the polytopic T–S model. These controllers may be state feedback, observer based or output feedback (static or dynamic).

In this section, the controller design for both state and output feedback is presented. Pros and cons for each method will be given and illustrated with an application to the inverted pendulum system (if applicable).

12.5.1 State feedback stabilization

12.5.1.1 Classical state feedback stabilization

Let us consider the following polytopic T–S model:

$$\dot{x}(t) = \sum_{i=1}^{n} \mu_i(\xi(t))(A_i x(t) + B_i u(t)) \tag{12.46}$$

The state feedback PDC control law is given by the following equation:

$$u(t) = -\sum_{j=1}^{n} \mu_j(\xi(t))K_j x(t) \tag{12.47}$$

The controlled system is then written as

$$\dot{x}(t) = \sum_{i=1}^{n}\sum_{j=1}^{n} \mu_i(\xi(t))\mu_j(\xi(t))(A_i - B_i K_j)\, x(t) \tag{12.48}$$

Let us consider the Lyapunov function $V(x(t)) = x^T(t)Px(t)$, the system (12.48) asymptotic stability is ensured if there exists a positive symmetric matrix $P_1 \in \mathbb{R}^{n_x \times n_x}$ and matrices $R_i \in \mathbb{R}^{n_x \times m}$ such that the following $\mathscr{LMI}s$ are verified for $i,j = 1,\ldots,n$ ([10]):

$$P_1 A_i^T + A_i P_1 - R_j^T B_i^T - B_i R_j < 0 \quad i,j = 1,\ldots,n \tag{12.49}$$

The controller gains K_j are then given by

$$K_j = R_j P_1^{-1} \quad \text{with} \quad P_1 = P^{-1} \tag{12.50}$$

The control objective of this subsection is to balance the inverted pendulum around the origin ($x_3 \simeq 0$). To illustrate this application, a PDC control is applied. In order to minimize the design effort and complexity, we will use as few rules as possible. So, we consider our model around small angle, which implies the following simplification:

$$\cos(x_3) \simeq 1, \qquad \sin(x_3) \simeq x_3 \tag{12.51}$$

The premise variables are then resumed to

$$\xi_1(t) = \frac{1}{M + m x_3^2(t)}, \qquad \xi_4(t) = x_4(t).x_3(t) \tag{12.52}$$

The corresponding polytopic T–S model will be then represented by four submodels. The rank of the controllability matrix is checked.

From (12.49) and (12.50), the controller gains are calculated and given by

$$\begin{cases} K_1 = \begin{pmatrix} -0.6359 & -1.9782 & -29.1095 & -7.3509 \end{pmatrix} \\ K_2 = \begin{pmatrix} -0.6359 & -1.9782 & -29.1095 & -7.3509 \end{pmatrix} \\ K_3 = \begin{pmatrix} -0.4563 & -1.4258 & -23.2823 & -5.6007 \end{pmatrix} \\ K_4 = \begin{pmatrix} -1.0079 & -3.1316 & -41.5012 & -11.2301 \end{pmatrix} \end{cases} \tag{12.53}$$

The nonlinear system is simulated with the polytopic T–S controller, the obtained results are depicted in Figure 12.3.

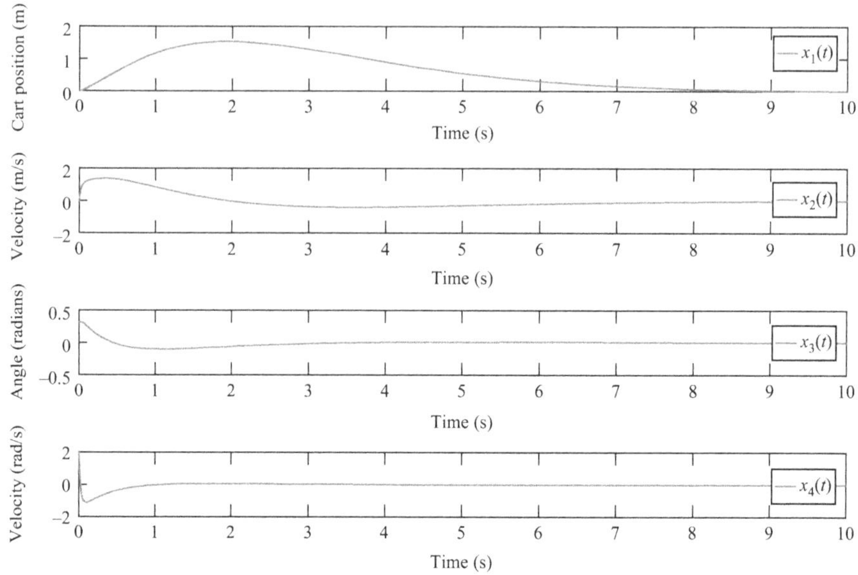

Figure 12.3 State feedback controller

We can note that the state feedback controller is based on the fact that all the states are available. If it is not the case, either the observer-based state feedback stabilization or the output feedback stabilization can be used.

12.5.2 Observer-based state feedback stabilization

For the case where the system state is not fully available, a solution may be an observer-based controller. As it was done for the observer design, both cases of measurable and unmeasurable premise variables are presented.

12.5.2.1 Measurable premise variables

The observer-based feedback controller for polytopic T–S models with measurable premise variables is given by

$$u(t) = -\sum_{j=1}^{n} \mu_j(\xi(t))K_j\hat{x}(t) \tag{12.54}$$

where the observer dynamics is given by

$$\begin{cases} \dot{\hat{x}}(t) = \sum_{i=1}^{n} \mu_i(\xi(t))(A_i\hat{x}(t) + B_iu(t) + L_i(y(t) - \hat{y}(t))) \\ \hat{y}(t) = C\hat{x}(t) \end{cases} \tag{12.55}$$

Combining both the controller law (12.75) and the observer (12.76) with an estimation error defined as $e(t) = x(t) - \hat{x}(t)$, the following equations are obtained:

$$\begin{cases} \dot{x}(t) = \sum_{i=1}^{n}\sum_{j=1}^{n} \mu_i(\xi(t))\mu_j(\xi(t)) \big((A_i - B_iK_j)x(t) + B_iK_je(t)\big) \\ \dot{e}(t) = \sum_{i=1}^{n}\sum_{j=1}^{n} \mu_i(\xi(t))\mu_j(\xi(t)) \big(A_i - L_iC_j\big)e(t) \end{cases} \quad (12.56)$$

The following augmented state system is then deduced:

$$\dot{x}_a(t) = \sum_{i=1}^{n}\sum_{j=1}^{n} \mu_i(\xi(t))\mu_j(\xi(t))G_{ij}x_a(t) \quad (12.57)$$

with

$$x_a(t) = \begin{pmatrix} x(t) \\ e(t) \end{pmatrix}, G_{ij} = \begin{pmatrix} A_i - B_iK_j & B_iK_j \\ 0 & A_i - L_iC_j \end{pmatrix} \quad (12.58)$$

Let us consider the following Lyapunov function:

$$V(x_a(t)) = x_a^T(t)Px_a(t), \quad P = P^T > 0 \quad (12.59)$$

The augmented system stability (12.57) is then ensured if the following conditions are satisfied for $i,j = 1, \dots, n$:

$$G_{ij}^TP + PG_{ij} < 0 \quad (12.60)$$

Let us choose a bloc diagonal Lyapunov function, s.t.

$$P = \begin{pmatrix} P_1 & 0 \\ 0 & P_2 \end{pmatrix} \quad (12.61)$$

Then, condition (12.60) becomes

$$\begin{pmatrix} P_1A_i + A_i^TP_1 - P_1B_iK_j - K_j^TB_i^TP_1 & P_1B_iK_j \\ * & P_2A_i + A_i^TP_2 - P_2L_iC_j - C_j^TL_i^TP_2 \end{pmatrix} \quad (12.62)$$

where $*$ refers to the transpose of $P_1B_iK_j$.

Condition (12.62) is nonlinear, a solution may be to choose $P_1 = \alpha I$ where α is a positive scalar and proceed to a change of variable.

To summarizer, the augmented system is globally asymptotically stable if there exists a positive symmetric matrix P_2, a positive scalar α and matrices $\overline{K}_j$ and $\overline{L}_i$ such that the following $\mathscr{LMI}s$ admit a solution for $i,j = 1, \dots, n$:

$$\begin{pmatrix} \alpha A_i + \alpha A_i^T - B_i\overline{K}_j - \overline{K}_j^TB_i^T & B_i\overline{K}_j \\ (B_i\overline{K}_j)^T & P_2A_i + A_i^TP_2 - \overline{L}_iC_j - C_j^T\overline{L}_i^T \end{pmatrix} \quad (12.63)$$

the observer and controller gains are then given by

$$L_i = P_2^{-1}\overline{L}_i, \quad K_j = \alpha^{-1}\overline{K}_j \quad (12.64)$$

12.5.2.2 Unmeasurable premise variables

For nonlinear systems represented by unmeasurable premise variable polytopic T–S models, the state feedback PDC controller as well as the observer dynamics are given by the following equations:

$$\begin{cases} u(t) = -\sum_{j=1}^{n} \mu_j(\hat{\xi}(t))K_j\hat{x}(t) \\ \dot{\hat{x}}(t) = \sum_{i=1}^{n} \mu_i(\hat{\xi}(t))(A_i\hat{x}(t) + B_iu(t) + L_i(y(t) - \hat{y}(t))) \\ \hat{y}(t) = C\hat{x}(t) \end{cases} \tag{12.65}$$

The augmented system $x_a(t) = \begin{pmatrix} \hat{x}(t) \\ e(t) \end{pmatrix}$ with $e(t) = x(t) - \hat{x}(t)$ is given by [35]:

$$\dot{x}_a(t) = \sum_{i=1}^{n}\sum_{j=1}^{n}\sum_{k=1}^{n} \mu_i(\xi(t))\mu_j(\hat{\xi}(t))\mu_k(\hat{\xi}(t))G_{ijk}x_a(t) \tag{12.66}$$

with

$$G_{ijk} = \begin{pmatrix} R_{ik} & B_iK_k \\ S_{ijk} & \theta_{jk} + \Delta B_{ij}K_k \end{pmatrix} \tag{12.67}$$

$$R_{ik} = A_i - B_iK_k,\ S_{ijk} = \Delta A_{ij} - \Delta B_{ij}K_k + L_jC,\ \theta_{jk} = A_i - L_iC_j \tag{12.68}$$

$$\Delta A_{ij} = A_i - A_j,\ \Delta B_{ij} = B_i - B_j \tag{12.69}$$

The augmented state (12.66) is rewritten as

$$\begin{aligned} \dot{x}_a(t) &= \sum_{i=1}^{n}\sum_{j=1}^{n} \mu_i(\xi(t))\mu_j(\hat{\xi}(t))\mu_j(\hat{\xi}(t))G_{ijj}x_a(t) \\ &\quad + 2\sum_{i=1}^{n}\sum_{j<k} \mu_i(\xi(t))\mu_j(\hat{\xi}(t))\mu_k(\hat{\xi}(t))\frac{G_{ijk} + G_{ikj}}{2}x_a(t) \end{aligned} \tag{12.70}$$

Let us consider the Lyapunov function $V(x_a(t)) = x_a^T(t)Px_a(t)$. The augmented system asymptotic stability (12.70) is ensured if the following conditions are satisfied for $i,j,k = 1,\ldots,n$ [34,35]:

$$\begin{aligned} G_{ijj}^TP + PG_{ijj} &< 0 \\ \left(\frac{G_{ijk} + G_{ikj}}{2}\right)^T P + P\left(\frac{G_{ijk} + G_{ikj}}{2}\right) &\le 0, \quad i,j < k\ s.t.\ \mu_i \cap \mu_j \cap \mu_k = \emptyset \end{aligned} \tag{12.71}$$

The notation $\mu_i \cap \mu_j \cap \mu_k = \emptyset$ means that for all $i,j < k$ excepting $\mu_i\mu_j\mu_k = 0, \forall\xi$.

Conditions (12.71) are not convex (due to $\mu_i(\xi) \neq \mu_i(\hat{\xi})$), it is difficult to transform them into $\mathscr{LMI}$s in P, K_i and L_i. To solve those issues, the following technique was proposed in [35]:

Let us consider the following bloc diagonal Lyapunov function:

$$P = \begin{pmatrix} P_1 & 0 \\ 0 & P_2 \end{pmatrix} \tag{12.72}$$

Replacing (12.72) into (12.71), it becomes

$$\begin{pmatrix} R_{ij}^T P_1 + P_1 R_{ij} & P_1 B_i K_j + S_{ijj}^T P_2 \\ * & \Sigma_{ijk} \end{pmatrix} < 0$$

$$\begin{pmatrix} (R_{ik} + R_{ij})^T P_1 + P_1 (R_{ik} + R_{ij}) & P_2 (S_{ijk} + S_{ikj}) + (K_k + K_j)^T B_i^T P_1 \\ * & \Sigma_{ijk} + \Sigma_{ikj} \end{pmatrix} < 0 \tag{12.73}$$

with $\Sigma_{ijk} = (\theta_{jk} + \Delta B_{ij} K_k)^T P_2 + P_2 (\theta_{jk} + \Delta B_{ij} K_k)$.

Since (12.73) is still in a $\mathscr{BMI}$ form, one way to solve the problem is the following:

Since the $\mathscr{BMI}$ (12.73) imply that

$$\begin{aligned} R_{ij}^T P_1 + P_1 R_{ij} &< 0 \\ (R_{ik} + R_{ij})^T P_1 + P_1 (R_{ik} + R_{ij}) &< 0 \end{aligned} \tag{12.74}$$

These last inequalities are easy to transform into $\mathscr{LMI}$. Once P_1 and K_i are obtained, we substitute them into (12.73) and obtain $\mathscr{LMI}$ conditions in P_2 and L_i that can be solved by a convex optimization technique such as the interior point method.

In order to illustrate the observer-based feedback controller, let us consider the previous example with four submodels.

For an output matrix $C = \begin{pmatrix} 1 & 0 & 0 & 0 \\ 0 & 0 & 1 & 0 \end{pmatrix}$, the obtained observer gains are given by

$$\begin{cases} L_1 = \begin{pmatrix} 5.2254 & -7.8695 \\ 55.7035 & -225.1685 \\ -5.3685 & 35.1068 \\ -149.9143 & 927.3724 \end{pmatrix} \\[2ex] L_2 = \begin{pmatrix} 5.2254 & -7.8695 \\ 55.7036 & -225.1684 \\ -5.3685 & 35.1068 \\ -149.9146 & 927.3723 \end{pmatrix} \\[2ex] L_3 = \begin{pmatrix} 5.2173 & -7.9696 \\ 55.4060 & -228.3607 \\ -5.3184 & 35.6670 \\ -148.6749 & 940.5266 \end{pmatrix} \\[2ex] L_4 = \begin{pmatrix} 5.2236 & -7.9102 \\ 55.7293 & -227.0121 \\ -5.3914 & 35.4560 \\ -150.2795 & 935.7448 \end{pmatrix} \end{cases} \tag{12.75}$$

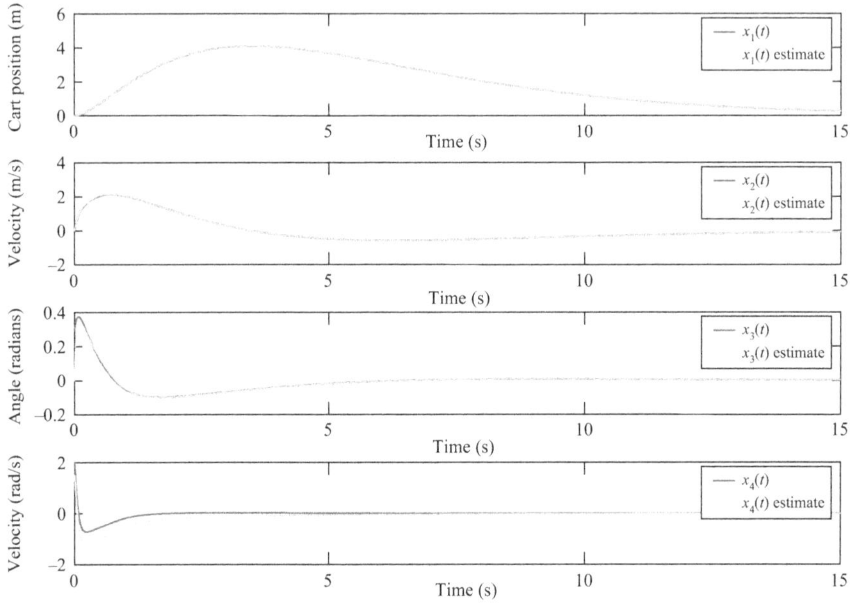

Figure 12.4 Observer-based state feedback controller

To simplify the computation (less $\mathscr{LMI}$s to solve), the controller gains were set to be equal, they are given by

$$K = \begin{pmatrix} -0.2146 & -0.8185 & -16.8005 & -3.3704 \end{pmatrix} \tag{12.76}$$

The nonlinear system is simulated with the polytopic T–S controller, the obtained results are depicted in Figure 12.4.

From the depicted figure, it is shown that the efficiency of the proposed observer-based feedback controller. Note that the control law $u(t)$ is in a polytopic T–S form, and it was applied to the nonlinear model.

Remark 12.1. *Another way would be to use the descriptor approach [36,37]. In fact, this approach is well known to avoid the coupling terms between the feedback gains and the Lyapunov matrices and thus facilitates the $\mathscr{LMI}$s resolution. As a consequence, the number of $\mathscr{LMI}$s to solve decreases, and relaxed conditions are obtained [37].*

The design procedure is then summarized in the following steps:

- *Deduce the augmented state dynamics $x_a(t) = \begin{pmatrix} x^T(t) & e^T(t) & u^T(t) \end{pmatrix}^T$ from the system, observer and controller (12.46) and (12.65):*

$$E\dot{x}_a(t) = \sum_{i=1}^{n} \mu_i(\xi(t))\mathscr{A}_i x_a(t) \tag{12.77}$$

with

$$E = \begin{pmatrix} I_{n_x} & 0 & 0 \\ 0 & I_{n_x} & 0 \\ 0 & 0 & 0_{n_u} \end{pmatrix}, \quad \mathscr{A}_i = \begin{pmatrix} A_i + \Delta A(t) & 0 & B_i + \Delta B(t) \\ \Delta A(t) & A_i - L_i C & \Delta B(t) \\ K_i & K_i & I \end{pmatrix} \tag{12.78}$$

where $\Delta A(t)$ and $\Delta B(t)$ are considered uncertainties and defined as in (12.41) and (12.42).

- *For positive symmetric matrices P_1 and P_2 and a matrix P_3, the condition to satisfy for the asymptotic stability of the augmented system (12.77) is given by*

$$\begin{pmatrix} \mathbb{S}(P_1 A_i + P_1 \Delta A(t)) & \Delta A^T(t) P_2 & K_i^T P_3 + P_1 B_i + P_1 \Delta B(t) \\ * & \mathbb{S}(P_2 A_i - P_2 L_i C) & P_2 \Delta B(t) + K_i^T P_3 \\ * & * & P_3 + P_3^T \end{pmatrix} < 0 \tag{12.79}$$

where $*$ refers to the transpose term and $\mathbb{S}(M) := M + M^T$.

- *Condition (12.79) is then decomposed into the sum of a constant term and a time-varying one, i.e.,*

$$Q + \mathscr{Q}(t) + \mathscr{Q}^T(t) < 0 \tag{12.80}$$

with

$$Q = \begin{pmatrix} \mathbb{S}(P_1 A_i) & \Delta A^T(t) P_2 & K_i^T P_3 + P_1 B_i \\ * & \mathbb{S}(P_2 A_i - P_2 L_i C) & K_i^T P_3 \\ * & * & P_3 + P_3^T \end{pmatrix} \tag{12.81}$$

$$\mathscr{Q}(t) = \begin{pmatrix} \mathbb{S}(P_1 \Delta A(t)) & \Delta A^T(t) P_2 & P_1 \Delta B(t) \\ * & 0 & P_2 \Delta B(t) \\ * & * & 0 \end{pmatrix} \tag{12.82}$$

- *Using the definition of $\Delta A(t)$ and $\Delta B(t)$ (as it was done in Section 12.4.2), and the following lemma:*

Lemma 12.3. *[38] Let us consider matrices X and Y of appropriate dimensions, a time-varying matrix $\Delta(t)$ and a positive scalar ε. The following property is true:*

$$X^T \Delta^T(t) Y + Y^T \Delta(t) X \leq \varepsilon X^T X + \varepsilon^{-1} Y^T Y \tag{12.83}$$

if $\Delta^T(t)\Delta(t) \leq I$.

The inequality (12.79) is linearized, and the observer and controller gains deduced (by a simple change of variables).

Another interesting study case would be the uncertain polytopic T–S systems. In fact, if the model presents some uncertainties or modeling error, they have to be taken into account in the design process. A short overview of this case is given in the next subsection.

12.5.2.3 State feedback stabilization for uncertain polytopic T–S systems

In practice, there exists several uncertainties sources that affect the physical systems; these uncertainties must be taken into account during the modeling.

They may be due to an imperfect knowledge of the system parameters numerical values, to some modeling approximation, unknown inputs or disturbances. All these phenomena introduce a difference between the real system behavior and the one obtained by its model simulation.

In order to avoid this difference, or at least to minimize their effect, several robust control techniques have been proposed [39–41]. In the following, the state feedback stabilization for uncertain polytopic T–S systems is presented.

Let us consider the following model:

$$\dot{x}(t) = \sum_{i=1}^{n} \mu_i(\xi(t))((A_i + \Delta A(t))x(t) + (B_i + \Delta B(t))u(t)) \tag{12.84}$$

where the uncertainties $\Delta_A(t)$, $\Delta_B(t)$ that affect the state and control matrices are common to all submodels and are given in the form

$$\Delta A(t) = A\Sigma_A(t)E_A \tag{12.85}$$

$$\Delta B(t) = B\Sigma_B(t)E_B \tag{12.86}$$

The time-varying matrices $\Sigma_A(t)$ and $\Sigma_B(t)$ are unknown but bounded as follows:

$$\Sigma_A^T(t)\Sigma_A(t) \leq I, \ \forall t \tag{12.87}$$

$$\Sigma_B^T(t)\Sigma_B(t) \leq I, \ \forall t \tag{12.88}$$

Let us consider the following control law for the system (12.84) where the premise variables are supposed to be measurable:

$$u(t) = -\sum_{j=1}^{n} \mu_j(\xi(t))K_j x(t) \tag{12.89}$$

where the controller gains $K_j \in \mathbb{R}^{n_u \times n_x}$ are to be determined.

Replacing (12.89) into (12.84), we obtain

$$\dot{x}(t) = \sum_{i=1}^{n}\sum_{j=1}^{n} \mu_i(\xi(t))\mu_j(\xi(t))(A_i - B_iK_j + \Delta A(t) - \Delta B(t)K_j)x(t) \tag{12.90}$$

In order to analyze the system state stability, let us consider the following Lyapunov function:

$$V(x(t)) = x^T(t)Px(t) \tag{12.91}$$

where $P \in \mathbb{R}^n$ is a positive symmetric matrix.

In order to ensure (12.90) global asymptotic stability, the following conditions must be verified:

$$\begin{pmatrix} \mathbb{S}(A_iP_1 - B_iR_j) + \lambda_A AA^T + \lambda_B BB^T & P_1E_A^T & R_j^TE_B^T \\ * & -\lambda_A I & 0 \\ * & * & -\lambda_B I \end{pmatrix} < 0, \quad i,j = 1,\ldots,n \quad (12.92)$$

with $P_1 = P^{-1}$, λ_A, λ_B are positive scalars and $\mathbb{S}(A_iP_1 - B_iR_j) := A_iP_1 - B_iR_j + P_1A_i^T - R_j^TB_i^T$. In this case, the controller gains K_j are given by

$$K_j = R_jP_1^{-1} \quad (12.93)$$

for $i, j = 1, \ldots, n$.

The presented results are given for the measurable premise variable. They may be extended to the general case of unmeasurable premise variable as well as to the observer-based control by applying the same idea and based on the results presented in the previous subsection.

As it was shown, the state feedback or observer-based feedback approach is quite interesting for the system stabilization. In other alternative, we have to achieve a stability purpose in the case where the system state is not fully available and/or to avoid the additional computations due to an observer design would be the output feedback control.

12.5.3 Static output feedback stabilization

In this section, the static output feedback control is considered. Contrarily to the observer-based or output dynamic controllers [42–44], static controllers are particularly interesting to reduce real-time computational cost when implementing practical applications since it does not need any ODE solving [2,45]. However, it is important to highlight that the drawback of output feedback controllers is the coupling introduced that makes the $\mathcal{LMI}s$ to be solved quite conservative and restrictive for some cases [46].

In the following section, the descriptor approach applied to the polytopic T–S models is also introduced. The descriptor approach consists on introducing virtual dynamics in the outputs equations, then, a descriptor formulation can be obtained considering an extended state vector for the closed-loop dynamics. Rewriting the closed-loop system by the use of descriptor form allows to avoid appearance of crossing terms between the state-space matrices and the controller's ones. Therefore, the benefit of the descriptor formulation is to make easier the $\mathcal{LMI}s$ formulation of nonquadratic stability conditions [36,37,47,48]. Herein, the stabilization condition for polytopic T–S models with static output feedback is based on the descriptor approach. For simplicity reasons, only the case of measurable premise variables is considered.

Let us consider the following polytopic T–S model:

$$\begin{cases} \dot{x}(t) = \sum_{i=1}^{n} \mu_j(\xi(t))(A_ix(t) + B_iu(t)) \\ y(t) = \sum_{i=1}^{n} \mu_j(\xi(t))(C_ix(t) + D_iu(t)) \end{cases} \quad (12.94)$$

and the static output feedback control defined by

$$u(t) = \sum_{j=1}^{n} \mu_j(\xi(t))K_j y(t) \tag{12.95}$$

Based on the descriptor approach and introducing a virtual dynamics in the input and outputs equations, we can write

$$\begin{cases} 0.\dot{u}(t) = \sum_{i=1}^{n} \mu_i(\xi(t))K_i y(t) - u(t) \\ 0.\dot{y}(t) = \sum_{i=1}^{n} \mu_i(\xi(t))(C_i x(t) + D_i u(t)) - y(t) \end{cases} \tag{12.96}$$

From (12.94) and (12.96), the augmented state vector dynamics $x_a(t) = \left(x^T(t)\ u^T(t)\ y^T(t)\right)^T$ is given by

$$E\dot{x}_a(t) = \sum_{i=1}^{n} \mu_i(\xi(t))\mathscr{A}_i^s x_a(t) \tag{12.97}$$

with

$$E = \begin{pmatrix} I_{nx} & 0 & 0 \\ 0 & 0_{n_u} & 0 \\ 0 & 0 & 0_m \end{pmatrix}, \mathscr{A}_i^s = \begin{pmatrix} A_i & B_i & 0 \\ 0 & -I_{n_u} & K_i \\ C_i & D_i & -I_m \end{pmatrix} \tag{12.98}$$

Let us consider the Lyapunov function:

$$V(t) = x_a^T(t)E^T P x_a(t) \tag{12.99}$$

such that

$$E^T P = P^T E \geq 0 \tag{12.100}$$

A possible choice of the matrix P ensuring the condition (12.100) is a block diagonal form $P = \mathrm{diag}(P_1, P_2, P_3)$ with $P_1 = P_1^T > 0$ and $P_2, P_3 > 0$. In this case, $V(t)$ is a quadratic function of the state system, in fact, $V(t) = x^T(t)P_1x(t)$. The time derivative of the Lyapunov function (12.99) is then given by

$$\dot{V}(t) = \sum_{i=1}^{n} h_i(\xi(t))x_a^T(t)((\mathscr{A}_i^s)^T P + P^T \mathscr{A}_i^s)x_a(t) \tag{12.101}$$

From (12.98) and (12.101), in order to ensure the system (12.97) stability, i.e., $\dot{V}(t) < 0$ the conditions to solve regarding P_1, P_2, P_3 and R_j, $i = 1, \ldots, n$ are

$$\begin{pmatrix} A_i^T P_1 + P_1 A_i & P_1 B_i & C_i^T P_3 \check{z} \\ * & -P_2 - P_2^T & R_i + D_i^T P_3 \\ * & * & -P_3 - P_3^T \end{pmatrix} < 0 \tag{12.102}$$

The controller gains are given by

$$K_i = (P_2^T)^{-1} R_i \tag{12.103}$$

From the presented results, it is important to highlight that the polytopic T–S approach may be really efficient (when it is applicable). In fact, this approach does not fit to all systems, in many cases, we will have some structural constraints that make it simply impossible to be applied like for the proposed static output feedback controller and the inverted pendulum.

It is clear that the proposed controller (12.95) exists only if the (12.102) conditions are satisfied, which is not the case for the considered application. In fact, the $\mathscr{LMI}s$ (12.102) admit a solution only if each local system (or subsystem) is stable, i.e., real part of the eigenvalues of A_i is negative, which is not the case for the inverted pendulum.

A solution in this case would be to search for a relaxed conditions by introducing a polytopic Lyapunov function or a dynamic state feedback controller for example, such that introducing extra dynamics may be required to increase the number of degrees of freedom in the design and improve the closed-loop transient response.

Some more results regarding the structural constraints for polytopic T–S models may be found in the following articles [49–51].

12.6 Conclusion

In this chapter, a review on the polytopic T–S approach with an application to the inverted pendulum was presented. A general scope of the approach with the most common results was presented and applied to a nonlinear model of the inverted pendulum. It was shown in one hand that the polytopic T–S approach may be an effective tool with several advantages and satisfying results; however, in the other hand, that approach is subject to some structural constraints that do not make it applicable for any system.

References

[1] Orjuela R, Marx B, Maquin D, Ragot J. Design of robust H_∞ observers for nonlinear systems using a multiple model. In: *17th World Congress The International Federation of Automatic Control*. Seoul, Korea; 2008.

[2] Chadli M, Maquin D, Ragot J. Static output feedback for Takagi–Sugeno systems: An LMI approach. In: *10th Mediterranean Conference on Control and Automation*. Lisbon, Portugal; 2002.

[3] Guerra TM, Kruszewski A, Vermeiren L, Tirmant H. Conditions of output stabilization for nonlinear models in the Takagi–Sugeno's form. *Fuzzy Sets and Systems*. 2006;157(9):1248–1259.

[4] Kruszewski A, Wang R, Guerra TM. Nonquadratic stabilization conditions for a class of uncertain nonlinear discrete time TS fuzzy models: A new approach. *IEEE Transactions on Automatic Control*. 2008;53:606–611.

[5] Bezzaoucha S, Voos H, Darouach M. A new polytopic approach for the unknown input functional observer design. *International Journal of Control*. 2017 Published online February; https://dx.doi.org/10.1080/00207179.2017.1288299.

[6] Luenberger DG. An introduction to observers. *IEEE Transactions on Automatic Control*. 1971;16(6):596–602.

[7] Darouach M, Zasadzinski M, Xu S. Full-order observers for linear systems with unknown inputs. *IEEE Transactions on Automatic Control*. 1994;39(3): 606–609.

[8] Filev D. Fuzzy modeling of complex systems. *International Journal of Approximate Reasoning*. 1991;5(3):281–290.

[9] Orjuela R, Marx B, Ragot J, Maquin D. Nonlinear system identification using heterogeneous multiple models. *Applied Mathematics and Computer Science*. 2013;23(1):103–115.

[10] Tanaka K, Wang HO. *Fuzzy Control Systems Design and Analysis: A Linear Matrix Inequality Approach*. John Wiley & Sons, Inc.; 2001.

[11] Kawamoto S, Tada K, Ishigame A, Taniguchi T. An approach to stability analysis of second order fuzzy systems. In: *IEEE International Conference on Fuzzy Systems*. San Diego, California, USA; 1992.

[12] Lofberg J. YALMIP: A toolbox for modeling and optimization in MATLAB. In: *In Computer Aided Control Systems Design Conference*. Taipei, Taiwan; 2004.

[13] Labit Y, Peaucelle D, Henrion D. SEDUMI Interface 1.02: a tool for solving LMI problems with SEDUMI. In: *IEEE International Symposium on Computer Aided Control System Design*; 2002.

[14] Takagi T, Sugeno M. Fuzzy identification of systems and its applications to modeling and control. *IEEE Transactions on Systems, Man and Cybernetics*. 1985;15(1):116–132.

[15] Bezzaoucha S, Marx B, Maquin D, Ragot J. Nonlinear joint state and parameter estimation. Application to a wastewater treatment plant. *Control Engineering Practice*. Accepted paper in May 2013. 2013;21(10):1377–1385.

[16] Sala A, Arino C. Asymptotically necessary and sufficient conditions for stability and performance in fuzzy control: Applications of Polya's theorem. *Fuzzy Sets and Systems*. 2007;158(24):2671–2686.

[17] Fang CH, Liu YS, Kau SW, Hong L, Lee ITFS C H. A new LMI-based approach to relaxed quadratic stabilization of T–S fuzzy control systems. *IEEE Transaction on Fuzzy Systems*. 2006;14:286–397.

[18] Teixeira M, Assuncao E, Avellar RITFS. On relaxed LMI-based designs for fuzzy regulators and fuzzy observers. *IEEE Transaction on Fuzzy Systems*. 2003;11(5):613–623.

[19] Tanaka K, Hori T, Wang HO. A multiple Lyapunov function approach to stabilization of fuzzy control systems. *IEEE Transactions on Fuzzy Systems*. 2003;11:582–589.

[20] Kruszewski A. *Lois de commande pour une classe de modèles non linéaires sous la forme Takagi–Sugeno: Mise sous forme LMI*. Université de Valenciennes et du Hainaut-Cambresis. France; 2006.

[21] Sename O, Gaspar P, Bokor J. *Lecture Notes in Robust Control and Linear Parameter Varying approaches: Application to Vehicle Dynamics*. Springer, editor; 2013.

[22] Besançon G. *Nonlinear Observers and Applications*. Springer, editor. Springer; 2007.

[23] Fossard A, Normand-Cyrot D. *Systèmes non linéaires: Modélisation – Estimation (Tome I)*. Masson, editor. Masson; 1993.

[24] Diop S, Fliess M. On nonlinear observability. In: *First European Control Conference*. Grenoble, France; 1991.

[25] Kamidi R. *Controllability and observability of polytopic linear models*. Eindhoven University of Technology. The Netherlands; 2000.

[26] Angelis GZ. *System analysis, modeling and control with Polytopic Linear Models*. Technische Universiteit Eindhoven. Netherland; 2001.

[27] Luenberger DG. An introduction to observers. *IEEE Transactions on Automatic Control*. 1971;16(6):596–602.

[28] Yoneyama Y. H_∞ filtering for fuzzy systems with immeasurable premise variables: An uncertain system approach. *Fuzzy Sets and Systems*. 2008;39(10):1033–1040.

[29] Bergsten P, Palm R. Thau-Luenberger observers for TS fuzzy systems. In: *Ninth IEEE International Conference on Fuzzy Systems*. San Antonio, Texas, USA; 2000.

[30] Patton R, Chen J, Lopez-Toribio C. Fuzzy observers for nonlinear dynamic systems fault diagnosis. In: *37th IEEE Conference on Decision and Control*. Tampa, Florida, USA; 1998.

[31] Tuan HD, Apkarian P, Narikiyo T, Yamamoto Y. Parameterized Linear Matrix Inequality Techniques in Fuzzy Control System Design. *IEEE Transactions on Fuzzy Systems*. 2001;9(2):324–332.

[32] Bezzaoucha S. *Fault tolerant control for Takagi–Sugeno nonlinear systems*. Lorraine University. France; 2013.

[33] Wang HO, Tanaka K, Griffin M. An approach to fuzzy control of non linear systems: Stability and design issues. *IEEE Transactions on Fuzzy Systems*. 1996;4(1):14–23.

[34] Tanaka K, Ikeda T, Wang H. Fuzzy regulators and fuzzy observers: Relaxed stability conditions and LMI-based design. *IEEE Transactions on Fuzzy Systems*. 1998;6(2):250–265.

[35] Chadli M, Maquin D, Ragot J. Observer-based controller for Takagi–Sugeno models. In: *IEEE International Conference on Systems, Man and Cybernetics*. vol. 5; 2002.

[36] Tanaka K, Ohtake H, Wang HO. A descriptor system approach to fuzzy control system design via fuzzy Lyapunov functions. *IEEE Transactions on Fuzzy Systems*. 2007;15(3):333–341.

[37] Guelton K, Bouarar T, Manamanni N. Robust dynamic output feedback fuzzy Lyapunov stabilization of Takagi–Sugeno systems – A descriptor redundancy approach. *Fuzzy Sets and Systems*. 2009;160(19): 2796–2811.

[38] Zhou K, Khargonekar PP. Robust stabilization of linear systems with norm-bounded time-varying uncertainty. *Systems and Control Letters*. 1988;10(1):17–20.

[39] Henrion D. *Stabilité des systèmes linéaires incertains à commande contrainte*. Institut National des Sciences Appliquées de Toulouse. Toulouse, France; 1999.
[40] Daafouz J. *Robustesse en performance des systèmes linéaires incertains. Placement de pôles et coût garanti*. Institut National des Sciences Appliquées. Toulouse, France; 1997.
[41] Zhou K, Doyle JC, Glover K. *Robust and Optimal Control*. Englenwood Cliffs, New Jersey, USA: Prentice-Hall; 1996.
[42] Li J, Wang HO, Newmann D, Tanaka K. Dynamic parallel distributed compensation For Takagi–Sugeno systems: An LMI approach. *Information Sciences*. 2000;123(4):201–221.
[43] Yoneyama J, Nishikawa M, Katayama H, Ichikawa A. Design of output feedback controllers for Takagi–Sugeno fuzzy systems. *Fuzzy Sets and Systems*. 2001;121(1):127–148.
[44] Yoneyama Y. Output feedback stabilization of fuzzy systems with unobservable premise variables. In: *International Conference on System of Systems Engineering*. Los Angeles, CA, USA; 2006.
[45] Huang D, Nguang SK. Robust H_∞ static output feedback control of fuzzy systems: An ILMI approach. *IEEE Transactions on Systems, Man and Cybernetics*. 2006;36(1):216–222.
[46] Bouarar T. *Contribution à la synthèse de lois de commande pour les descripteurs de type Takagi–Sugeno incertains et perturbés*. Université de Reims Champagne Ardenne. France; 2009.
[47] Lin C, Wang QG, Lee TH. H_∞-control for Takagi–Sugeno fuzzy descriptor systems. In: *International Conference on Systems, Man and Cybernetics*. Tokyo, Japan; 1999.
[48] Taniguchi T, Tanaka K, Othake H, Wang HO. Model construction, rule reduction, and robust compensation for generalized form of Takagi–Sugeno fuzzy systems. *Transactions on Fuzzy Systems*. 2001;9(4):525–537.
[49] Bezzaoucha S, Voos H, Darouach M, Dentler J. On the unknown input functional observers design: a polytopic Takagi–Sugeno approach. In: *24th Mediterranean Conference on Control and Automation*. Athens, Greece; 2016.
[50] Bezzaoucha S, Voos H, Darouach M, Dentler J. A polytopic approach for the nonlinear unknown input functional observers design: application to a quadrotor aerial robots landing. In: *European Control Conference (ECC'16)*. Aalborg, Danemark; 2016.
[51] Bezzaoucha S, Marx B, Maquin D, Ragot J. State constrained tracking control for nonlinear systems. *Journal of the Franklin Institute*. 2015;352(7): 2866–2886.

Chapter 13

Robust fault-tolerant control of nonlinear inverted pendulum and cart system with simultaneous actuator and sensor faults sliding-mode observer

Slim Dhahri[1], Ali Ben Brahim[1], Fayçal Ben Hmida[1] and Anis Sellami[1]

In this chapter, we propose a robust active fault-tolerant control (AFTC) scheme for a class of uncertain nonlinear systems with simultaneous actuator and sensor faults described via Takagi–Sugeno (T–S) multiple models. First, by transforming the sensor fault into pseudoactuator fault, a novel T–S sliding-mode observer (TS-SMO) with two discontinuous terms is developed to provide separate estimates of the actuator and sensor faults for the purpose of fault compensation. The robustness of the proposed observer against uncertainties has been taken into account via H_∞ norm minimization. Second, we use obtained on-line fault estimation information to design dynamic output feedback controller (DOFC) for robustly compensating the effects of actuator and sensor faults from the system inputs and outputs and guarantee the stability of the overall closed-loop system. The stability proof with H_∞ performances and D-stability constraints is formulated as a linear matrix inequalities (LMI) optimization problem. The effectiveness of the proposed robust AFTC approach to treat simultaneous actuator and sensor faults is illustrated using a nonlinear inverted pendulum with cart system.

13.1 Introduction

Fault disturbs the normal system operations, thus causing an unacceptable deterioration of these performances or leading to wrong dangerous situations. This fault can occur in any part of the system such as actuators and sensors. Sensor fault degrades the feedback system performances and so touches the control stability. Similarly, actuator fault may lead to minimize the performances of design controllers, or even the overall

[1]Research Unit on Control, Monitoring and Safety of Systems, Ecole Nationale Supérieure d'Ingénieurs de Tunis (ENSIT), University of Tunis, Tunisia

system execution. In order to preserve system performances and human security, fault diagnosis (FD) and fault-tolerant control (FTC) technology have caused wide attention over the past few years. A very powerful approach for FD is fault estimation (FE) in the sense that it not only detects and isolates the fault, but determines the size, location and dynamics behavior of the fault along a system trajectories. Therefore, FE is considered a major problem in the modern control theory that received a considerable amount of attention during the past few years. Fruitful results can be found in several excellent books [1–3] and survey papers [4–6].

In practical applications, most of the systems are nonlinear, so it is significant to study FD and FTC of nonlinear systems. For the case of Lipschitz nonlinear systems, LMI-based FTC designs to compensate actuator fault are proposed in [7–9]. The authors in [10,11] addresses the FTC design for Lipschitz nonlinear systems subject to matched uncertainty along with sensor faults using a sliding-mode observer (SMO). Since its excellent ability of nonlinear system description, very interesting approaches have represented these systems under Takagi–Sugeno (T–S) fuzzy form. T–S models have been introduced in [12]. The feature is to understand the overall system behavior by a set of local linear models. Each local model represents the system's operation in a particular area. The local models are then aggregated using an interpolation mechanism by premise variables satisfying the convex sum property. Taking the T–S representation, several attempts have been oriented to the FE and FTC of nonlinear systems, see for instance [13–20].

Thanks to its robustness to uncertainties, special attentions have already been made in the application of SMO design to robust FE and FTC schemes for T–S systems subject to actuator and/or sensor faults [21–27]. In the context of actuator FE, constructing diagnosis model in order to estimate faults is not possible if sensor faults occur simultaneously. The same difficulty is present when trying to estimate sensor faults. Several design methods have been developed in a precise and effective way when actuator and sensor faults estimation is divided on two steps:

Step 1: Actuator FE is proven without considering sensor faults [28–40].
Step 2: Sensor FE is solved without considering actuator faults [41–48].

Nevertheless, in practical systems, it is often the case when actuator and sensor faults occur simultaneously. In this framework, simultaneous FE is highly important. So far, in [49,50], the authors propose simultaneously estimating actuator and sensor faults study for Lipschitz nonlinear systems. Two subsystems are constructed where each one contains a particular class of faults. In [51,52], the problem of simultaneous actuator and sensor FE is considered for linear parameter varying systems. The idea of this work is based on the transformation of sensor fault in order to obtain an augmented actuator fault vector.

The information issued from actuator and sensor FE scheme are used by the FTC module to reconfigure the control law in order to compensate the faults and ensure acceptable system performances. FTC for various forms of T–S model has received considerably increasing attention both in modern research theory and application domains even in faulty situations. Wonderful results can be obtained by [18,53–57].

These studies guarantee the stability of systems with various faults. In addition, the problem of fault-tolerant tracking control is discussed in [58].

Based on the above works, the objective of this chapter is to develop robust AFTC based on FE of simultaneous actuator and sensor faults for T–S fuzzy systems. The main contributions of this chapter are two aspects:

- First, robust TS-SMO with two discontinuous terms is designed for simultaneous actuator and sensor FE. The stability conditions of the proposed observer is studied with Lyapunov theories and LMI convex optimization.
- Second, using the obtained FE, a dynamic output feedback controller (DOFC)-based FTC is designed to guarantee the system stability in the presence of actuator and sensor faults.

The remainder of this chapter is organized as follows: In Section 13.2, we introduce nonlinear and T–S model structures of an inverted pendulum from the literature [59]. In Section 13.3, we describe the LMI-based design of the proposed robust T–S fuzzy SMO for both the actuator and sensor fault cases. Section 13.4 presents the design of T–S DOFC scheme to compensate the effect of simultaneous actuator and sensor faults. Simulation example is given in Section 13.5 illustrating the effectiveness of the proposed methods. Finally, Section 13.6 presents some concluding remarks.

13.2 Problem formulation

13.2.1 Motivation

Simultaneous actuator and sensor faults represent a common challenge for AFTC methods. This motivates the proposal of a strategy to ensure system robustness against this challenging and practically probable scenario of simultaneous faults via an estimation and compensation approach. Hence, the objective of this chapter is to develop an AFTC strategy for regulator problems based on robust FE and compensation of simultaneous actuator faults and sensor faults. The compensation process requires two steps. The first is to design a robust TS-SMO with two discontinuous terms able to provide separate estimates of the actuator and/or sensor faults despite the presence of uncertainties. Second, based on the faults estimation, a T–S DOFC is used in order to maintain the performance and stability of the nominal control system during both faulty and fault-free cases. The schematic diagram of this approach is shown in Figure 13.1.

13.2.2 Inverted-pendulum modeling

13.2.2.1 Mathematical model

The inverted-pendulum benchmark, in particular, the cart version illustrated in Figure 13.2, has been considered in many references to solve the problem of nonlinear control [59]. The cart with inverted pendulum consists of a moveable carriage with one degree of freedom on which a pendulum is mounted and freely rotatable in driving

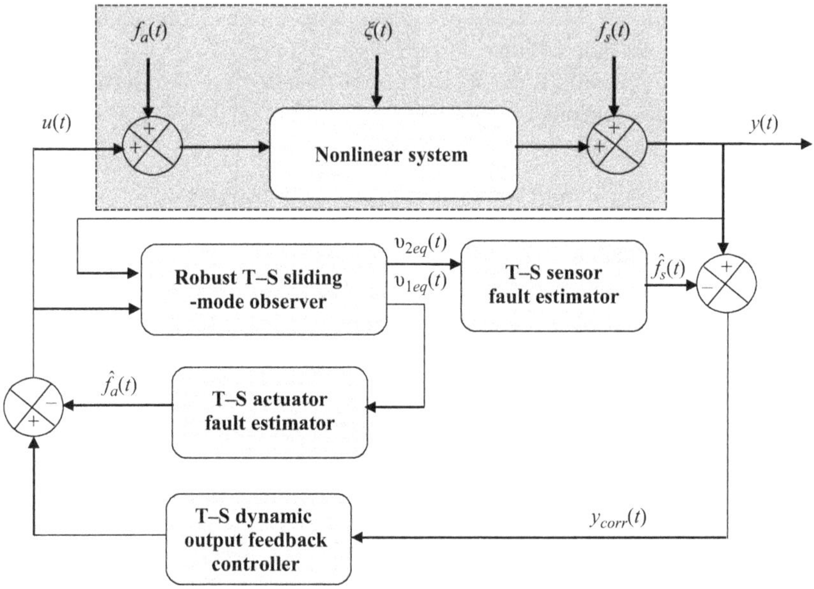

Figure 13.1 Proposed FTC architecture

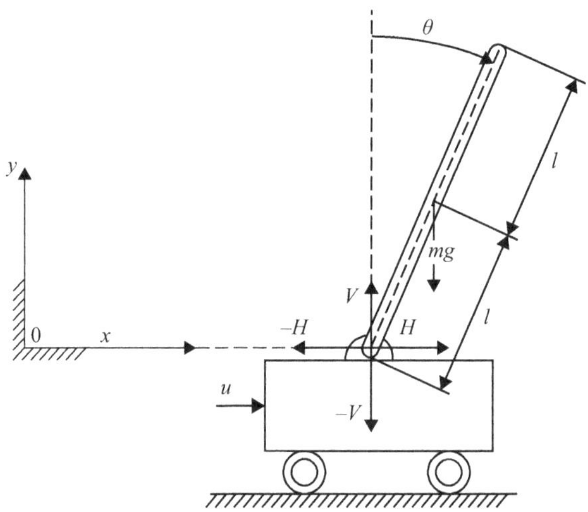

Figure 13.2 Nonlinear inverted pendulum and cart system [59]

Table 13.1 Numerical values of parameters of cart with inverted pendulum

Symbol	Description	Value	Unit
m	Point mass of the pendulum	0.2	kg
M	Mass of the cart	0.8	kg
l	Distance from the joint to the mass point m	0.5	m
b	Viscous friction of the joint	0.06	N s/rad
ρ	Cart friction coefficient	0.05	N s/m
g	Gravitational constant	9.81	ms^{-2}
L	Total length of rail	2	m

direction. The carriage is driven by a motor that exerts a force F through a belt-drive transmission.

The process can be described by the following state-space model

$$\begin{cases} \dot{x}_1(t) = x_2(t) \\ \dot{x}_2(t) = \dfrac{g\sin(x_1(t)) - m\,l\,a\,x_2^2(t)\,\frac{\sin(2x_1(t))}{2} - ba\cos(x_1(t))\,x_4(t) - a\cos(x_1(t))(F - f_c)}{\frac{4l}{3} - m\,l\,a\,\cos(x_1(t))^2} \\ \dot{x}_3(t) = x_4(t) \\ \dot{x}_4(t) = \dfrac{-m\,g\,a\,\frac{\sin(2x_1(t))}{2} + \frac{4\,m\,l\,a}{3}x_2^2(t)\,\sin(x_1(t)) - b\,a\,x_4(t) + \frac{4a}{3}(F - f_c)}{\frac{4}{3} - m\,a\,\cos(x_1(t))^2} \end{cases} \tag{13.1}$$

where $x_1(t) = \theta(t)$, $x_2(t) = \dot{\theta}(t)$, $x_3(t) = x(t)$ and $x_4(t) = \dot{x}(t)$ represent angular position of the pendulum, angular velocity of the pendulum, cart position and the cart velocity, respectively, with $a = 1/(m + M)$, $f_c = \rho\ \text{sign}\,(x_4(t))$ denotes the friction force between the cart and the track, and g denotes the gravitational acceleration. The external force F is the input $u := F$ and is limited to $|F| \leq F_{\max}$. The parameters and variables of the inverted pendulum on the cart are recorded in Table 13.1.

13.2.2.2 T–S model

In order to prepare for the SMO design, a T–S model is derived using the local approximation approach [60]. First, we rewrite (13.1) in the form

$$\underbrace{\begin{bmatrix} \dot{x}_1(t) \\ \dot{x}_2(t) \\ \dot{x}_3(t) \\ \dot{x}_4(t) \end{bmatrix}}_{\dot{x}(t)} = \underbrace{\begin{bmatrix} x_2(t) \\ \dfrac{g\sin(x_1(t)) + ba\cos(x_1(t))\,x_4(t)}{\frac{4l}{3} - m\,l\,a\,\cos(x_1(t))^2} \\ x_4(t) \\ \dfrac{-m\,g\,a\,\frac{\sin(2x_1(t))}{2} - b\,a\,x_4(t)}{\frac{4}{3} - m\,a\,\cos(x_1(t))^2} \end{bmatrix}}_{A(x(t))} + \underbrace{\begin{bmatrix} 0 \\ \dfrac{-a\cos(x_1(t))}{\frac{4l}{3} - m\,l\,a\,\cos(x_1(t))^2} \\ 0 \\ \dfrac{\frac{4a}{3}}{\frac{4}{3} - m\,a\,\cos(x_1(t))^2} \end{bmatrix}}_{B(x(t))}$$

$$\times \left(u(t) \underbrace{- f_c + m\,l\,x_2^2(t)\,\sin(x_1(t))}_{\xi(t)} \right) \tag{13.2}$$

The system is linearized around $x_1(t) = 0$ and $x_1(t) = \pm\pi/4$.

When $x_1(t)$ is near zero, the nonlinear equations can be simplified as

$$\begin{bmatrix} \dot{x}_1(t) \\ \dot{x}_2(t) \\ \dot{x}_3(t) \\ \dot{x}_4(t) \end{bmatrix} = \begin{bmatrix} x_2(t) \\ \frac{g\,x_1(t)+b\,a\,x_4(t)}{\frac{4l}{3}-m\,l\,a} \\ x_4(t) \\ \frac{-m\,g\,a\,x_1(t)-b\,a\,x_4(t)}{\frac{4}{3}-m\,a} \end{bmatrix} + \begin{bmatrix} 0 \\ \frac{-a}{\frac{4l}{3}-m\,l\,a} \\ 0 \\ \frac{\frac{4a}{3}}{\frac{4}{3}-m\,a} \end{bmatrix} (u(t)+\xi(t)) \tag{13.3}$$

When $x_1(t)$ is near $\pm\pi/4$, the nonlinear equations can be simplified as

$$\begin{bmatrix} \dot{x}_1(t) \\ \dot{x}_2(t) \\ \dot{x}_3(t) \\ \dot{x}_4(t) \end{bmatrix} = \begin{bmatrix} x_2(t) \\ \frac{g\frac{2\sqrt{2}}{\pi}}{\frac{4l}{3}-\frac{m\,l\,a}{2}}x_1(t) + \frac{b\,a\frac{\sqrt{2}}{2}}{\frac{4l}{3}-\frac{m\,l\,a}{2}}x_4(t) \\ x_4(t) \\ \frac{-m\,g\,a\frac{2}{\pi}}{\frac{4}{3}-\frac{m\,a}{2}}x_1(t) - \frac{b\,a}{\frac{4}{3}-\frac{m\,a}{2}}x_4(t) \end{bmatrix} + \begin{bmatrix} 0 \\ \frac{-a\frac{\sqrt{2}}{2}}{\frac{4l}{3}-\frac{m\,l\,a}{2}} \\ 0 \\ \frac{\frac{4a}{3}}{\frac{4}{3}-\frac{m\,a}{2}} \end{bmatrix} (u(t)+\xi(t)) \tag{13.4}$$

Note that (13.3) and (13.4) are now linear systems. We arrive at the following fuzzy model based on the linear subsystems:

Rule1: IF $x_1(t)$ *is about* 0

THEN $\dot{x}(t) = A_1x(t) + B_1(u(t)+\xi(t))$

Rule2: IF $x_1(t)$ *is about* $\pm\frac{\Pi}{4}$

THEN $\dot{x}(t) = A_2x(t) + B_2(u(t)+\xi(t))$

The matrices A_i and B_i of the local model $i = 1, 2$ corresponding to rules are

$$A_1 = \begin{bmatrix} 0 & 1 & 0 & 0 \\ \frac{g}{\frac{4l}{3}-m\,l\,a} & 0 & 0 & \frac{b\,a}{\frac{4l}{3}-m\,l\,a} \\ 0 & 0 & 0 & 1 \\ \frac{-m\,g\,a}{\frac{4}{3}-m\,a} & 0 & 0 & \frac{-b\,a}{\frac{4}{3}-m\,a} \end{bmatrix}, \quad B_1 = \begin{bmatrix} 0 \\ \frac{-a}{\frac{4l}{3}-m\,l\,a} \\ 0 \\ \frac{\frac{4a}{3}}{\frac{4}{3}-m\,a} \end{bmatrix} \tag{13.5}$$

$$A_2 = \begin{bmatrix} 0 & 1 & 0 & 0 \\ \frac{g\frac{2\sqrt{2}}{\pi}}{\frac{4l}{3}-\frac{m\,l\,a}{2}} & 0 & 0 & \frac{b\,a\frac{\sqrt{2}}{2}}{\frac{4l}{3}-\frac{m\,l\,a}{2}} \\ 0 & 0 & 0 & 1 \\ \frac{-m\,g\,a\frac{2}{\pi}}{\frac{4}{3}-\frac{m\,a}{2}} & 0 & 0 & -\frac{b\,a}{\frac{4}{3}-\frac{m\,a}{2}} \end{bmatrix}, \quad B_2 = \begin{bmatrix} 0 \\ \frac{-a\frac{\sqrt{2}}{2}}{\frac{4l}{3}-\frac{m\,l\,a}{2}} \\ 0 \\ \frac{\frac{4a}{3}}{\frac{4}{3}-\frac{m\,a}{2}} \end{bmatrix} \tag{13.6}$$

Membership functions for rules 1 and 2 can be simply defined as

$$\mu_1(\zeta(t)) = \frac{1 - \frac{1}{1+\exp(-14(x_1-\frac{\pi}{8}))}}{1+\exp(-14(x_1+\frac{\pi}{8}))} \tag{13.7}$$

$$\mu_1(\zeta(t)) = 1 - \mu_1(\zeta(t)) \tag{13.8}$$

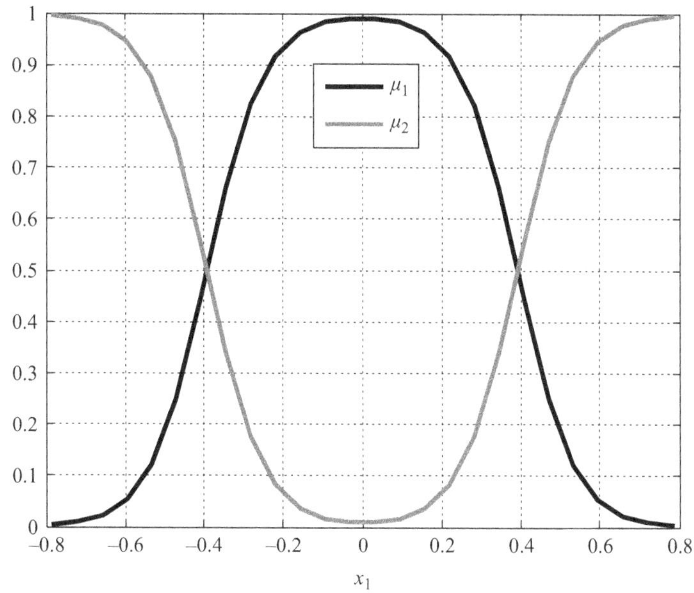

Figure 13.3 Membership functions of two-rule model

and shown in Figure 13.3. The state-space model (13.1) can thus be transformed into the T–S form

$$\begin{cases} \dot{x}(t) = \sum_{i=1}^{2} \mu_i(\zeta(t)) \left[A_i x(t) + B_i(u(t) + \xi(t))\right] \\ y(t) = Cx(t) \end{cases} \tag{13.9}$$

The system (13.9) was driven by the control input $u = 10\sin(3t)$. Figure 13.4 shows the response of the real system and its fuzzy approximation.

13.3 Fault reconstruction by sliding-mode observation

13.3.1 System description

We consider the general case where the uncertain T–S model structure is affected by additive actuator faults $f_a(t)$ and sensor faults $f_s(t)$

$$\begin{cases} \dot{x}(t) = \sum_{i=1}^{k} \mu_i(\zeta(t)) \left[A_i x(t) + B_i u(t) + M_i f_a(t) + D_i \xi(t)\right] \\ y(t) = Cx(t) + N f_s(t) \end{cases} \tag{13.10}$$

where $x(t) \in \mathbb{R}^n$, $u(t) \in \mathbb{R}^m$ and $y(t) \in \mathbb{R}^p$ denote, respectively, the state vector, the control inputs and the output vector. $A_i \in \mathbb{R}^{n\times n}$, $B_i \in \mathbb{R}^{n\times m}$, $M_i \in \mathbb{R}^{n\times q}$, $D_i \in \mathbb{R}^{n\times l}$,

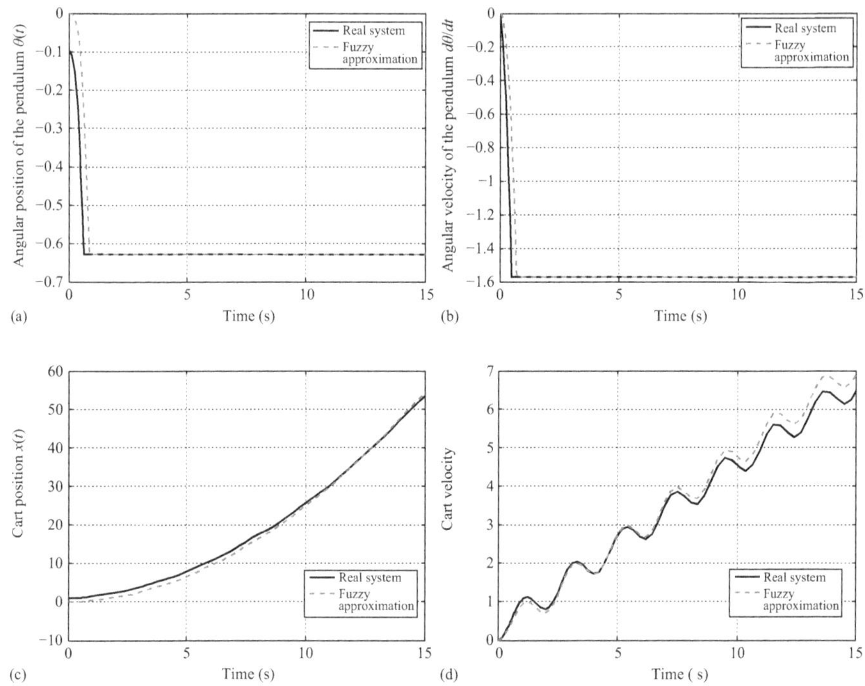

Figure 13.4 Comparison between the real (solid) and the fuzzy (dashed) system: (a) trajectory of x_1, (b) trajectory of x_2, (c) trajectory of x_3 and (d) trajectory of x_4

$C \in \mathbb{R}^{p\times n}$ and $N \in \mathbb{R}^{p\times h}$ are the matrices describing the so-called ith submodel. It is supposed that $rank(C) = p$, $rank(N) = h$ where $n > p \geq \tilde{q}$ ($\tilde{q} = q + h$), the pairs (A_i, B_i) are controllable, and the pairs (A_i, C_i) are observable. $f_a(t) \in \mathbb{R}^q$ represents the vector of actuator faults, and $f_s(t) \in \mathbb{R}^h$ represents the vector of sensor faults. The integer k represents the number of submodels. The weighing functions μ_i are nonlinear and depend on the premise variable $\zeta(t)$ is supposed to be measurable.

Remark 13.1. *The T–S model (13.9) of inverted pendulum with cart is a special case of (13.10), while assuming that the uncertainties and actuator fault affect the system in the same direction as the control input; therefore, $D_i = B_i$ and $M_i = B_i$.*

Remark 13.2. *The common C and N in (13.10) are only a small restriction, since many applications [17,19,20] comprise outputs that are linear in the system states and sensor faults.*

The weighting functions satisfy the following convex sum property:

$$\mu_i(\zeta(t)) \geq 0, i = 1, \ldots, k, \quad \sum_{i=1}^{k} \mu_i(\zeta(t)) = 1 \tag{13.11}$$

for any $\zeta(t)$. Before starting the main results of this chapter, the following assumptions are made:

Assumption 13.1. *The actuator fault distribution matrix M_i in (13.10) satisfies:*

$$rank(CM_i) = rank(M_i) = q \tag{13.12}$$

Assumption 13.2. *All invariant zeros of (A_i, M_i, C) must lie in the open left half of the complex plane, that is*

$$rank \begin{bmatrix} sI_n - A_i & M_i \\ C & 0 \end{bmatrix} = n + q \tag{13.13}$$

holds for all complex number s with $Re(s) \geq 0$.

Assumption 13.3. *The actuator fault $f_a(t)$ and sensor fault $f_s(t)$, and the uncertainty vector $\xi(t)$ satisfy the following norm bounded constraints:*

$$\|f_a(t)\| \leq \rho_a, \|f_s(t)\| \leq \rho_s, \|\xi(t)\| \leq \xi_0 \tag{13.14}$$

where ρ_a, ρ_s and ξ_0 are three known positive constants.

In order to transform the sensor faults into a fictitious actuator faults, we assume that there exists an orthogonal matrix $T_R \in \mathbb{R}^{p\times p}$, obtained by the QR transformation of sensor faults matrix N, such that

$$T_R y(t) := \begin{cases} y_1(t) = C_1 x(t) \\ y_2(t) = C_2 x(t) + N_1 f_s(t) \end{cases} \tag{13.15}$$

where $y_2(t) \in \mathbb{R}^h$ and $N_1 \in \mathbb{R}^{h\times h}$ are invertible.

To avoid the direct multiplication of the sensor and/or noise by the observer gain, an augmented system state with output filter states is constructed. The filtered output is given as follows:

$$\begin{aligned} \dot{w}(t) &= -A_f w(t) + A_f y_2(t) \\ &= -A_f w(t) + A_f C_2 x(t) + A_f N_1 f_s(t) \end{aligned} \tag{13.16}$$

where $-A_f \in \mathbb{R}^{h\times h}$ is any design matrix with stable eigenvalues.

Based on (13.10) and (13.16), an augmented T–S fuzzy system can be formulated of order $(n + h)$ as

$$\begin{cases} \dot{\bar{x}}(t) = \sum_{i=1}^{k} \mu_i(\zeta(t)) \left[\bar{A}_i \bar{x}(t) + \bar{B}_i u(t) + \bar{M}_i f_a(t) + \bar{D}_i \xi(t)\right] + \bar{N} f_s(t) \\ \bar{y}(t) = \bar{C} x(t) \end{cases} \tag{13.17}$$

where

$$\begin{aligned} &\bar{x}(t) = \begin{bmatrix} x(t) \\ w(t) \end{bmatrix}, \quad \bar{y}(t) = \begin{bmatrix} y_1(t) \\ w(t) \end{bmatrix}, \quad \bar{A}_i = \begin{bmatrix} A_i & 0 \\ A_f C_2 & -A_f \end{bmatrix}, \\ &\bar{B}_i = \begin{bmatrix} B_i \\ 0 \end{bmatrix}, \quad \bar{M}_i = \begin{bmatrix} M_i \\ 0 \end{bmatrix}, \quad \bar{N} = \begin{bmatrix} 0 \\ A_f N_1 \end{bmatrix}, \\ &\bar{D}_i = \begin{bmatrix} D_i \\ 0 \end{bmatrix}, \quad \bar{C} = \begin{bmatrix} C_1 & 0 \\ 0 & I_h \end{bmatrix} \end{aligned} \tag{13.18}$$

In order to simultaneously estimate the actuator and sensor faults, we construct the following TS-SMO:

$$\begin{cases} \dot{\hat{\bar{x}}}(t) = \sum_{i=1}^{k} \mu_i(\zeta(t)) \left[\bar{A}_i \hat{\bar{x}}(t) + \bar{B}_i u(t) + \bar{G}_{l,i} \bar{e}_y(t) + \bar{G}_{n,i} \upsilon_1(t) \right] + \bar{N} \upsilon_2(t) \\ \hat{\bar{y}}(t) = \bar{C} \hat{\bar{x}}(t) \end{cases} \tag{13.19}$$

where $\hat{\bar{x}}(t)$ is the observer state, $\hat{\bar{y}}(t)$ is the observer output and $\bar{e}_y(t) = \bar{y}(t) - \hat{\bar{y}}(t)$ is the output estimation error. $\upsilon_1(t)$ and $\upsilon_2(t)$ are discontinuous switching term necessary to maintain a sliding motion. $\bar{G}_{l,i}$ and $\bar{G}_{n,i}$ are appropriate gain matrices.
Defining $\bar{A}_{0,i} = \bar{A}_i - \bar{G}_{l,i}\bar{C}$ and the state estimation error as $\bar{e}(t) = \bar{x}(t) - \hat{\bar{x}}(t)$. From (13.10) and (13.19), the following error system can be obtained:

$$\begin{aligned} \dot{\bar{e}}(t) = & \sum_{i=1}^{k} \mu_i(\zeta(t)) \left[\bar{A}_{0,i} e(t) + \bar{M}_i f_a(t) - \bar{G}_{n,i} \upsilon_1(t) + \bar{D}_i \xi(x, u, t) \right] \\ & + \bar{N} \left(f_s(t) - \upsilon_2(t) \right) \end{aligned} \tag{13.20}$$

13.3.2 Existence conditions

For a stable TS-SMO to exist, the following existence conditions in Lemma 13.1 have to be fulfilled.

Lemma 13.1.

1. *The observer matching condition for augmented system (13.17)*

$$rank \left(\bar{C} \left[\bar{M}_i \ \bar{N} \right] \right) = rank \left(\left[\bar{M}_i \ \bar{N} \right] \right) \tag{13.21}$$

   holds if and only if Assumptions 13.1 holds.
2. *Each local model for the system (13.17) is minimum phase, i.e.,*

$$rank \begin{bmatrix} sI_{n+h} - \bar{A}_i & \bar{M}_i & \bar{N} \\ \bar{C} & 0 & 0 \end{bmatrix} = n + \tilde{q} \tag{13.22}$$

   holds for all complex number s with Re(s) ≥ 0 if and only if Assumption 13.2 holds.

Proof.

1. From the matrices partition of (13.18), it is easy to see that $\forall i \in [1, \ldots, k]$

$$\bar{C} \left[\bar{M}_i \ \bar{N} \right] = \begin{bmatrix} C_1 M_i & 0 \\ 0 & A_f N_1 \end{bmatrix} \tag{13.23}$$

   Premultiply $\bar{C} \left[\bar{M}_i \ \bar{N} \right]$ in (13.23) with a nonsingular matrix $\begin{bmatrix} I_{p-h} & 0 \\ 0 & A_f^{-1} \end{bmatrix}$, we obtain

$$\begin{bmatrix} C_1 M_i & 0 \\ 0 & N_1 \end{bmatrix} \tag{13.24}$$

It follows that

$$rank\left(\bar{C}\left[\begin{array}{cc}\bar{M}_i & \bar{N}\end{array}\right]\right) = rank(C_1 M_i) + rank(N_1) \tag{13.25}$$

Since $N_1 \in \mathbb{R}^{h\times h}$ is full rank, the condition (13.25) will be satisfied if and only if

$$rank\,[C_1 M_i] = rank\,[CM_i] = q \tag{13.26}$$

2. Expressing (13.22) in terms of the partitioned matrices of (13.18), it follows that $\forall\, i \in [1,\ldots,k]$

$$rank\begin{bmatrix} sI_n - A_i & M_i & 0 \\ -A_f C_2 & 0 & A_f N_1 \\ C_1 & 0 & 0 \end{bmatrix} = n + q + h \tag{13.27}$$

Premultiplying the matrix in (13.27) by the following nonsingular matrix

$$\begin{bmatrix} I_n & 0 \\ 0 & T_R^{-1} \end{bmatrix}\begin{bmatrix} I_n & 0 & 0 \\ 0 & 0 & I_{p-h} \\ 0 & A_f^{-1} & 0 \end{bmatrix}, \tag{13.28}$$

we have

$$rank\begin{bmatrix} sI_n - A_i & M_i & 0 \\ C & 0 & N \end{bmatrix} = n + q + h \tag{13.29}$$

Since N is full rank, we obtain

$$rank\begin{bmatrix} sI_n - A_i & M_i \\ C & 0 \end{bmatrix} = n + rank(M_i) \tag{13.30}$$

which must be verified for the original T–S fuzzy system (13.10), $\forall s \in \mathbb{C}$ such that $Re(s) \geq 0$. □

13.3.3 T–S model in canonical form

The design of a TS-SMO for simultaneous actuator and sensor faults reconstruction for the class of T–S systems (13.17) is carried out in a special canonical form. Consider a nonlinear change of coordinates obtained by combining linear coordinate transformations for each linear system of the T–S fuzzy model (13.17).

$$\begin{bmatrix} \bar{x}_1(t) \\ \bar{x}_2(t) \end{bmatrix} = T(\zeta)\bar{x}(t) \text{ where } T(\zeta) = \sum_{i=1}^{k} \mu_i(\zeta) T_i \tag{13.31}$$

The linear coordinate transformations are obtained from a series of transformations $T_i = T_{1,i} T_0$. Under conditions (13.21) and (13.22), there always exists a T_0 and $T_{1,i}$ such that the transformed system matrices in (13.18) have the following block structures:

$$\begin{gathered} T_i \bar{A}_i T_i^{-1} = \left[\begin{array}{c|c} \bar{A}_{11,i} & \bar{A}_{12,i} \\ \hline \bar{A}_{211,i} & \multirow{2}{*}{$\bar{A}_{22,i}$} \\ \bar{A}_{221,i} & \end{array}\right], \quad T_i \bar{B}_i = \begin{bmatrix} \bar{B}_{1,i} \\ \bar{B}_{2,i} \end{bmatrix} \\ T_i\left[\begin{array}{cc}\bar{M}_i & \bar{N}\end{array}\right] = \begin{bmatrix} 0 & 0 \\ \bar{M}_{2,i} & \bar{N}_2 \end{bmatrix}, \quad T_i \bar{D}_i = \begin{bmatrix} \bar{D}_{1,i} \\ \bar{D}_{2,i} \end{bmatrix} \end{gathered} \tag{13.32}$$

where $\bar{A}_{11,i} \in \mathbb{R}^{(n+h-p)\times(n+h-p)}$, $\bar{D}_{1,i} \in \mathbb{R}^{(n+h-p)\times l}$, $\bar{M}_{2,i} \in \mathbb{R}^{p\times q}$ and $\bar{N}_2 \in \mathbb{R}^{p\times h}$.

Define $\bar{A}_{221,i}$ as the top $p-(q+h)$ rows of $\bar{A}_{21}$, the subblocks $\bar{M}_{2,i}$ and $\bar{N}_2$ when partitioned have the following structures:

$$\bar{M}_{2,i} = \begin{bmatrix} 0 \\ \bar{M}_{22,i} \end{bmatrix}, \quad \bar{N}_2 = \begin{bmatrix} 0 \\ \bar{N}_{22} \end{bmatrix} \tag{13.33}$$

where $\bar{M}_{22,i} \in \mathbb{R}^{(q+h)\times q}$ and $\bar{N}_2 \in \mathbb{R}^{(q+h)\times h}$.

The output distribution matrix, in the new coordinates, has the following structure:

$$\bar{C} = CT_i^{-1} = \begin{bmatrix} 0 & \bar{C}_2 \end{bmatrix} \tag{13.34}$$

where $\bar{C}_2 \in \mathbb{R}^{p\times p}$ is nonsingular.

Lemma 13.2. *[28] The pair $(\bar{A}_{11,i}, \bar{A}_{211,i})$ is detectable if and only if the invariant zeros of $(\bar{A}_i, \begin{bmatrix} \bar{M}_i & \bar{N} \end{bmatrix}, \bar{C})$ are stable.*

We define the Lyapunov matrix $\bar{P}$ as

$$\bar{P} = \begin{bmatrix} \bar{P}_1 & \bar{P}_2 \\ \bar{P}_2^T & \bar{P}_3 \end{bmatrix} > 0 \text{ where } \bar{P}_2 := \begin{bmatrix} \bar{P}_{21} & 0 \end{bmatrix} \tag{13.35}$$

with $\bar{P}_1 \in \mathbb{R}^{(n+h-p)\times(n+h-p)}$ and $\bar{P}_{21} \in \mathbb{R}^{(n+h-p)\times(p-q-h)}$.

Then, the error system (13.20) can be partitioned as

$$\begin{aligned} \dot{\bar{e}}_1(t) &= \sum_{i=1}^{k} \mu_i(\zeta(t)) \big[\bar{A}_{11,i}\bar{e}_1(t) + (\bar{A}_{12,i} - \bar{G}_{l1,i})\bar{e}_2(t) + \bar{G}_{n1,i}\upsilon_1(t) \\ &\quad + \bar{D}_{1,i}\xi(t)\big] \end{aligned} \tag{13.36}$$

$$\begin{aligned} \dot{\bar{e}}_2(t) &= \sum_{i=1}^{k} \mu_i(\zeta(t)) \big[\bar{A}_{21,i}\bar{e}_1(t) + (\bar{A}_{22,i} - \bar{G}_{l2,i})\bar{e}_2(t) + \bar{M}_{2,i}f_a(t) + \bar{G}_{n2,i}\upsilon_1(t) \\ &\quad + \bar{D}_{2,i}\xi(t)\big] + \bar{N}_2(f_s(t) - \upsilon_2(t)) \end{aligned} \tag{13.37}$$

where

$$\begin{bmatrix} \bar{e}_1(t) \\ \bar{e}_2(t) \end{bmatrix} = T(\zeta)\bar{e}(t), \quad \begin{bmatrix} \bar{G}_{l1,i} \\ \bar{G}_{l2,i} \end{bmatrix} = T(\zeta)\bar{G}_{l,i} \text{ and } \begin{bmatrix} \bar{G}_{n1,i} \\ \bar{G}_{n2,i} \end{bmatrix} = T(\zeta)\bar{G}_{n,i} \tag{13.38}$$

In the coordinates of (13.32), let

$$\begin{bmatrix} \bar{G}_{n1,i} \\ \bar{G}_{n2,i} \end{bmatrix} = \begin{bmatrix} -\bar{L}_i\bar{C}_2^{-1} \\ \bar{C}_2^{-1} \end{bmatrix} \text{ and } \bar{L}_i = \begin{bmatrix} \bar{L}_{1,i} & 0 \end{bmatrix} \tag{13.39}$$

where $\bar{C}_2$ is obtained in (13.34), and appropriate $\bar{L}_{1,i}$ can be obtained using LMIs as discussed later in this section.

In order to identify the sliding motion, it is performing to apply a further change of coordinates according to

$$T_{L,i} = \begin{bmatrix} I_{n+h-p} & \begin{bmatrix} \bar{L}_{1,i} & 0_{(n-p)\times\tilde{q}} \end{bmatrix} \\ 0_{p\times(n-p)} & \bar{C}_2 \end{bmatrix} \tag{13.40}$$

Then, in the new coordinate system, it is straightforward to see that the transformed system matrices have the form

$$\begin{aligned} \tilde{A}_i &= \begin{bmatrix} \tilde{A}_{11,i} & \tilde{A}_{12,i} \\ \tilde{A}_{21,i} & \tilde{A}_{22,i} \end{bmatrix}, \quad \begin{bmatrix} \tilde{M}_i & \tilde{N} \end{bmatrix} = \begin{bmatrix} 0 & 0 \\ \tilde{M}_{2,i} & \tilde{N}_2 \end{bmatrix} \\ \tilde{D}_i &= \begin{bmatrix} \tilde{D}_{1,i} \\ \tilde{D}_{2,i} \end{bmatrix}, \quad \tilde{C} = \begin{bmatrix} 0 & I_p \end{bmatrix} \end{aligned} \tag{13.41}$$

where $\tilde{A}_{11,i} = \bar{A}_{11,i} + \bar{L}_{1,i}\bar{A}_{211,i}$, $\tilde{M}_{2,i} = \bar{C}_2\bar{M}_{2,i}$, $\tilde{N}_2 = \bar{C}_2\bar{N}_2$ and $\tilde{D}_{2,i} = \bar{C}_2\bar{D}_{2,i}$. The nonlinear gain matrix will have the form

$$\begin{bmatrix} \tilde{G}_{n1,i} \\ \tilde{G}_{n2,i} \end{bmatrix} = T_L(\zeta) \begin{bmatrix} \bar{G}_{n1,i} \\ \bar{G}_{n2,i} \end{bmatrix} = \begin{bmatrix} 0 \\ I_p \end{bmatrix} \tag{13.42}$$

and the Lyapunov matrix will be

$$\tilde{P} = (T_{L,i}^T)^{-1}\bar{P}T_{L,i}^{-1} = \begin{bmatrix} \tilde{P}_1 & 0 \\ 0 & \tilde{P}_0 \end{bmatrix} \tag{13.43}$$

where $\tilde{P}_0 = -\bar{P}_2^T\bar{P}_1^{-T}\bar{P}_2 + \bar{P}_3$.

The error system from (13.36) and (13.37) in the new coordinate system is

$$\dot{\tilde{e}}_1(t) = \sum_{i=1}^{k} \mu_i(\zeta(t)) \left[\tilde{A}_{11,i}e_1(t) + (\tilde{A}_{12,i} - \tilde{G}_{l1,i})e_y(t) + \tilde{D}_{1,i}\xi(t) \right] \tag{13.44}$$

$$\begin{aligned} \dot{e}_y(t) = {} & \sum_{i=1}^{k} \mu_i(\zeta(t)) \left[\tilde{A}_{21,i}e_1(t) + (\tilde{A}_{22,i} - \tilde{G}_{l2,i})e_y(t) + \tilde{D}_{2,i}\xi(t) \right. \\ & \left. + \tilde{M}_{2,i}f_a(t) - \upsilon_1(t) \right] + \tilde{N}_2(f_s(t) - \upsilon_2(t)) \end{aligned} \tag{13.45}$$

where

$$\begin{bmatrix} \tilde{e}_1(t) \\ e_y(t) \end{bmatrix} = T_L(\zeta) \begin{bmatrix} \bar{e}_1(t) \\ \bar{e}_2(t) \end{bmatrix} \tag{13.46}$$

and

$$\begin{bmatrix} \tilde{G}_{l1,i} \\ \tilde{G}_{l2,i} \end{bmatrix} = T_L(\zeta) \begin{bmatrix} \bar{G}_{l1,i} \\ \bar{G}_{l2,i} \end{bmatrix} \tag{13.47}$$

A choice of observer gain $\tilde{G}_{l,i}$ from (13.47) is then

$$\tilde{G}_{l,i} = \begin{bmatrix} \tilde{G}_{l1,i} \\ \tilde{G}_{l2,i} \end{bmatrix} = \begin{bmatrix} \tilde{A}_{12,i} \\ \tilde{A}_{22,i} - \tilde{A}_{22}^s \end{bmatrix} \tag{13.48}$$

where $\tilde{A}_{22}^s$ is stable design matrix. Substituting (13.48) into (13.44) and (13.45) yields

$$\dot{e}_1(t) = \sum_{i=1}^{k} \mu_i(\zeta(t)) \left[\tilde{A}_{11,i}e_1(t) + \tilde{D}_{1,i}\xi(t)\right] \tag{13.49}$$

$$\begin{aligned} \dot{e}_y(t) &= \sum_{i=1}^{k} \mu_i(\zeta(t)) \left[\tilde{A}_{21,i}e_1(t) + \tilde{A}_{22}^s e_y(t) + \tilde{D}_{2,i}\xi(t) + \tilde{M}_{2,i}f_a(t) - \upsilon_1(t)\right] \\ &\quad + \tilde{N}_2(f_s(t) - \upsilon_2(t)) \end{aligned} \tag{13.50}$$

The discontinuous output error injection terms $\upsilon_1(t)$ and $\upsilon_2(t)$ are defined by

$$\upsilon_1(t) := \begin{cases} \eta_a \frac{\tilde{P}_0 e_y(t)}{\|\tilde{P}_0 e_y(t)\|} & \text{if } e_y(t) = 0 \\ 0 & \text{otherwise} \end{cases} \tag{13.51}$$

$$\upsilon_2(t) := \begin{cases} \eta_s \frac{\tilde{P}_0 e_y(t)}{\|\tilde{P}_0 e_y(t)\|} & \text{if } \bar{e}_y(t) = 0 \\ 0 & \text{otherwise} \end{cases} \tag{13.52}$$

where η_a and η_s are positive scalars which will be determined formally later, but loosely speaking, they must be respectively an upper bound on the magnitude of the actuator and sensor fault signal.

The objective now is to present robust SMO with two discontinuous terms for reconstruct both actuator and sensor faults, as well as the T–S fuzzy system states. It will be shown that the sufficient conditions for the stability with H_∞ performances of the observer error (13.49) and (13.50) are established by using Lyapunov stability and LMI techniques.

For simplicity, we introduce the following notations

$$\mu_i = \mu_i(\zeta(t)), \ \ \tilde{A}_{11}(\mu) = \sum_{i=1}^{k} \mu_i(\zeta(t))\tilde{A}_{11,i}, \ \ \tilde{A}_{21}(\mu) = \sum_{i=1}^{k} \mu_i(\zeta(t))\tilde{A}_{21,i}, \tag{13.53}$$

$$\begin{aligned} \tilde{M}_2(\mu) &= \sum_{i=1}^{k} \mu_i(\zeta(t))\tilde{M}_{2,i}, \ \ \tilde{D}_1(\mu) = \sum_{i=1}^{k} \mu_i(\zeta(t))\tilde{D}_{1,i}, \\ \tilde{D}_2(\mu) &= \sum_{i=1}^{k} \mu_i(\zeta(t))\tilde{D}_{2,i} \end{aligned} \tag{13.54}$$

Then, the error system from (13.49) and (13.50) can be rewritten as

$$\dot{e}_1(t) = \tilde{A}_{11}(\mu)e_1(t) + \tilde{D}_1(\mu)\xi(t) \tag{13.55}$$

$$\begin{aligned} \dot{e}_y(t) &= \tilde{A}_{21}(\mu)e_1(t) + \tilde{A}_{22}^s e_y(t) + \tilde{D}_2(\mu)\xi(t) + \tilde{M}_2(\mu)f_a(t) - \upsilon_1(t) \\ &\quad + \tilde{N}_2(f_s(t) - \upsilon_2(t)) \end{aligned} \tag{13.56}$$

Define the controlled estimation error r as

$$r(t) = H\tilde{e}(t) = H\begin{bmatrix} \tilde{e}_1(t) \\ e_y(t) \end{bmatrix} \tag{13.57}$$

where H is a prespecified weight matrix and assumed to have full rank

$$H := \begin{bmatrix} H_1 & 0 \\ 0 & H_2 \end{bmatrix} \tag{13.58}$$

Consider the following worst case performance measure:

$$\|H\|_\infty := \sup_{\|\xi\|_2 \neq 0} \frac{\|r\|_2^2}{\|\xi\|_2^2} \tag{13.59}$$

We now present Theorem 13.1 which establishes the sufficient conditions for the existence of the proposed TS-SMO with a prescribed $\mathscr{H}_\infty$ performance $\|H\|_\infty \leq \mu$, where μ is a small positive constant. In other words, the $\|H\|_\infty$ gain of the transfer function from the system uncertainty $\xi(t)$ to the estimation error r is bounded by $\sqrt{\mu}$.

13.3.4 Sliding motion stability

Theorem 13.1. *Under Assumptions 13.1–13.3, the estimation error dynamics are asymptotically stable with the prescribed $\mathscr{H}_\infty$ tracking performance $\|r\|_2^2 \leq \sqrt{\mu}\,\|\xi\|_2^2$, if there exist matrices $\tilde{P}_1 = \tilde{P}_1^T > 0$, $\tilde{P}_0 = \tilde{P}_0^T > 0$, and positive scalar μ such that*

$$\begin{bmatrix} \Pi_1 + H_1^T H_1 & \bar{A}_{21,i}^T \bar{C}_2^T \tilde{P}_0 & \tilde{P}_1 \bar{D}_{1,i} & \bar{W}_i \bar{D}_{2,i} \\ * & \Pi_2 + H_2^T H_2 & 0 & \tilde{P}_0 \bar{C}_2 \bar{D}_{2,i} \\ * & 0 & -\mu I & 0 \\ * & * & 0 & -\mu I \end{bmatrix} < 0 \tag{13.60}$$

where

$$\Pi_1 = \tilde{P}_1 \bar{A}_{11,i} + \bar{A}_{11,i}^T \tilde{P}_1 + \bar{W}_i \bar{A}_{21,i} + \bar{A}_{21,i}^T \bar{W}_i^T \tag{13.61}$$

$$\Pi_2 = \tilde{A}_{22}^{sT} \tilde{P}_0 + \tilde{P}_0 \tilde{A}_{22}^s \tag{13.62}$$

Once the LMI (13.60) is solved, $\bar{L}_i$ can be obtained from $\bar{L}_i = \tilde{P}_1^{-1} \bar{W}_i$.

Proof. Consider the Lyapunov function as

$$V(t) = \tilde{e}^T(t) \tilde{P} \tilde{e}(t) \tag{13.63}$$

It is worth noting that $\tilde{P}$ in the new coordinate (13.41) has the quadratic form (13.43). Therefore, the Lyapunov function can be rewritten as

$$\begin{aligned} V(t) &= V_1(t) + V_2(t) \\ &= \tilde{e}_1^T(t) \tilde{P}_1 \tilde{e}_1(t) + e_y^T(t) \tilde{P}_0 e_y(t) \end{aligned} \tag{13.64}$$

The time derivative of $V_1(t)$ along the trajectories of system (13.55) can be shown to be equal to

$$\begin{aligned} \dot{V}_1(t) &= \dot{\tilde{e}}_1^T(t)\tilde{P}_1\tilde{e}_1(t) + \tilde{e}_1^T(t)\tilde{P}_1\dot{\tilde{e}}_1(t) \\ &= \tilde{e}_1^T(t)(\tilde{A}_{11}^T(\mu)\tilde{P}_1 + \tilde{P}_1\tilde{A}_{11}(\mu))\tilde{e}_1(t) + 2\tilde{e}_1^T\tilde{P}_1\tilde{D}_1\xi(t) \end{aligned} \tag{13.65}$$

Similarly, we can obtain the time derivative of $V_2(t)$ as

$$\begin{aligned} \dot{V}_2(t) &= e_y^T(t)(\tilde{A}_{22}^{sT}\tilde{P}_0 + \tilde{P}_0\tilde{A}_{22}^s)e_y(t) + 2e_y^T(t)\tilde{P}_0\tilde{A}_{21}(\mu)\tilde{e}_1(t) \\ &\quad + 2e_y^T(t)\tilde{P}_0\tilde{D}_2(\mu)\xi(t) + 2e_y^T(t)\tilde{P}_0\tilde{M}_2(\mu)f_a(t) - 2e_y^T(t)\tilde{P}_0\upsilon_1(t) \\ &\quad + 2e_y^T(t)\tilde{P}_0\tilde{N}_2(f_s(t) - \upsilon_2(t)) \end{aligned} \tag{13.66}$$

From the definition of $\upsilon_1(t)$ in (13.51) and the bound on $f_a(t)$, we can obtain

$$\begin{aligned} &2e_y^T(t)\tilde{P}_0\tilde{M}_2(\mu)f_a(t) - 2e_y^T(t)\tilde{P}_0\upsilon_1(t) \\ &\quad = \sum_{i=1}^{k}\mu_i(\zeta(t))\left(2e_y^T(t)\tilde{P}_0\tilde{M}_{2,i}f_a(t) - 2e_y^T(t)\tilde{P}_0\upsilon_1(t)\right) \\ &\quad \le \sum_{i=1}^{k}\mu_i(\zeta(t))\left(2\left\|\tilde{P}_0e_y(t)\right\|\kappa_i - 2\eta_a\right) \end{aligned} \tag{13.67}$$

with $\kappa_i := \left\|\tilde{M}_{2,i}\right\|\rho_a$. Due to the convexity condition of the T–S membership functions $\left(\sum_{i=1}^{k}\mu_i(\zeta(t)) = 1\right)$, it holds that $\sum_{i=1}^{k}\mu_i(\zeta(t))\kappa_i \le \kappa_{\max} := \left\|\tilde{M}_2\right\|_{\max}\rho_a$. It thus follows from (13.67) that

$$\begin{aligned} 2e_y^T(t)\tilde{P}_0\tilde{M}_2(\mu)f_a(t) - 2e_y^T(t)\tilde{P}_0\upsilon_1(t) &\le 2\left\|\tilde{P}_0e_y(t)\right\|(\kappa_{\max} - \eta_a) \\ &= -2\eta_1\left\|\tilde{P}_0e_y(t)\right\| < 0 \end{aligned} \tag{13.68}$$

where the positive scalar η_1 satisfies $\eta_1 < \eta_a - \kappa_{\max}$. Similarly, from the definition of $\upsilon_2(t)$ in (13.52) and the bound on $f_s(t)$, we can obtain

$$\begin{aligned} 2e_y^T(t)\tilde{P}_0\tilde{N}_2(f_s(t) - \upsilon_2(t)) &\le 2\left\|\tilde{P}_0e_y(t)\right\|\left\|\tilde{N}_2\right\|(\rho_s - \eta_s) \\ &\le -2\eta_2\left\|\tilde{P}_0e_y(t)\right\|\left\|\tilde{N}_2\right\| < 0 \end{aligned} \tag{13.69}$$

where the positive scalar η_2 satisfies $\eta_s > \rho_s + \eta_2$.

From (13.68) and (13.69), we can obtain $\dot{V}(t)$ as

$$\begin{aligned}\dot{V}(t) &= \dot{V}_1(t) + \dot{V}_2(t) \\ &\leq \tilde{e}_1^T(t)(\tilde{A}_{11}^T(\mu)\tilde{P}_1 + \tilde{P}_1\tilde{A}_{11}(\mu))\tilde{e}_1(t) + e_y^T(t)(\tilde{A}_{22}^{sT}\tilde{P}_0 + \tilde{P}_0\tilde{A}_{22}^s)e_y(t) \\ &\quad + 2e_y^T(t)\tilde{P}_0\tilde{A}_{21}(\mu)\tilde{e}_1(t) + 2\tilde{e}_1^T(t)\tilde{P}_1\tilde{D}_1(\mu)\xi(x,u,t) \\ &\quad + 2e_y^T(t)\tilde{P}_0\tilde{D}_2(\mu)\xi(t) \\ &= \tilde{e}_1^T(t)\Pi_1\tilde{e}_1(t) + e_y^T(t)\Pi_2 e_y(t) + 2e_y^T(t)\tilde{P}_0\tilde{A}_{21}(\mu)\tilde{e}_1(t) \\ &\quad + 2\tilde{e}_1^T(t)\tilde{P}_1\tilde{D}_1(\mu)\xi(t) + 2e_y^T(t)\tilde{P}_0\tilde{D}_2(\mu)\xi(t) \end{aligned} \tag{13.70}$$

when $\xi(x, u, t) = 0$, it follows that

$$\dot{V}(t) \leq \begin{bmatrix} \tilde{e}_1(t) \\ e_y(t) \end{bmatrix}^T \underbrace{\begin{bmatrix} \Pi_1 & \tilde{A}_{21}^T(\mu)\tilde{P}_0 \\ \tilde{P}_0\tilde{A}_{21}(\mu) & \Pi_2 \end{bmatrix}}_{\Lambda} \begin{bmatrix} \tilde{e}_1(t) \\ e_y(t) \end{bmatrix} \tag{13.71}$$

If there exists a feasible solution to (13.60), then we can conclude that $\Lambda < 0$, thus $\dot{V}(t) < 0$. This implies that $\tilde{e}(t) \to 0$ as $t \to \infty$. Therefore, the error dynamics is asymptotically stable when $\xi(x, u, t) = 0$.

When $\xi(x, u, t) \neq 0$, to attain the robustness of the proposed TS-SMO to the uncertainties $\xi(x, u, t)$ in L_2 sense, we define

$$V_0(t) = \dot{V}(t) + r^T(t)r(t) - \mu\xi^T(t)\xi(t) \tag{13.72}$$

If condition (13.60) is satisfied, it follows that

$$V_0(t) \leq \begin{bmatrix} \tilde{e}_1(t) \\ e_y(t) \\ \xi(t) \end{bmatrix} \Omega \begin{bmatrix} \tilde{e}_1(t) \\ e_y(t) \\ \xi(t) \end{bmatrix} < 0 \tag{13.73}$$

Then, under zero initial conditions, we obtain the following:

$$\begin{aligned}\int_0^\infty (\|r\|^2 - \mu\|\xi\|^2)dt &= \int_0^\infty (\|r\|^2 - \mu\|\xi\|^2 + \dot{V})dt - \int_0^\infty \dot{V}dt \\ &= \int_0^\infty (\|r\|^2 - \mu\|\xi\|^2 + \dot{V})dt - V(\infty) + V(0) \\ &\leq \int_0^\infty V(0)dt < 0 \end{aligned} \tag{13.74}$$

which implies that

$$\int_0^T (r^T r)dt \leq \mu \int_0^T (\xi^T\xi)dt \tag{13.75}$$

namely

$$\|r\|_2^2 \leq \sqrt{\mu}\,\|\xi\|_2^2 \tag{13.76}$$

This completes the proof. □

The state estimation error $\tilde{e}(t) := col(\tilde{e}_1(t), e_y(t))$ is stable on the sliding mode surface (13.78), then $\tilde{e}_1(t)$ is bounded a long T–S fuzzy system trajectories (13.17). Thus, it has written, as a positive scalar ϖ, $\sup_{0\le t\le\infty} \|\tilde{e}_1(t)\| \le \varpi$.

Remark 13.3. *The effect of the system uncertainties $\xi(t)$ on the estimation errors (13.55) and (13.56) is decided by the value of μ. The accuracy of FE increases with smaller value of μ. The minimization of μ can be found by solving the following LMI optimization problem:*

$$\text{minimize}(\mu) \text{ subject to } \tilde{P}_1 = \tilde{P}_1^T > 0,\ \tilde{P}_0 = \tilde{P}_0^T > 0 \text{ and } (13.60) \tag{13.77}$$

The objective now is to determine the constant gain η_a in (13.51) and η_s in (13.52) such that the error systems can be driven to the sliding surface $\mathscr{S}$ which is defined as

$$\mathscr{S} = \left\{ \left(\tilde{e}_1(t), e_y(t)\right) \left| \tilde{e}_y(t) = 0 \right. \right\} \tag{13.78}$$

in finite time and maintain on it thereafter.

13.3.5 Sliding motion reachability

Theorem 13.2. *Given T–S system (13.17) with existence conditions (13.21) and (13.22) and the proposed TS-SMO (13.19). Then, the error dynamics (13.55) and (13.56) can be driven to the sliding surface given by (13.78) in finite time and remain on it if the LMI optimization problem formulated in (13.60) is solvable and the gains η_a and η_s satisfy*

$$\eta_a + \eta_s \left\| \tilde{N}_2 \right\| \ge \chi_{\max} + \eta_3 \tag{13.79}$$

where $\chi_{\max} := \left\| \tilde{A}_{21,i} \right\|_{\max} \varpi + \left\| \tilde{D}_{2,i} \right\|_{\max} \xi_0 + \left\| \tilde{M}_{2,i} \right\|_{\max} \rho_a + \left\| \tilde{N}_2 \right\| \rho_s$ *and η_3 is a positive scalar.*

Proof. Consider the Lyapunov function $V_s(t) = \tilde{e}_y^T(t)\tilde{P}_0 e_y(t)$. The time derivative of $V_s(t)$ can be obtained as

$$\begin{aligned}\dot{V}_s(t) &= e_y^T(t)(\tilde{A}_{22}^{sT}\tilde{P}_0 + P_2\tilde{A}_{22}^s)e_y(t) + 2e_y^T(t)\tilde{P}_0\tilde{A}_{21}(\mu)\tilde{e}_1(t) \\ &\quad + 2e_y^T(t)\tilde{P}_0\tilde{D}_2(\mu)\xi(t) + 2e_y^T(t)\tilde{P}_0\tilde{M}_2(\mu)f_a(t) - 2e_y^T(t)\tilde{P}_0\upsilon_1(t) \\ &\quad + 2e_y^T(t)\tilde{P}_0\tilde{N}_2(f_s(t) - \upsilon_2(t))\end{aligned} \tag{13.80}$$

Since $\tilde{A}_{22}^s$ is a stable matrix by design, therefore, $\tilde{A}_{22}^{sT}P_2 + P_2\tilde{A}_{22}^s < 0$. Then, (13.80) becomes

$$\begin{aligned}\dot{V}_s(t) &\le 2e_y^T(t)\tilde{P}_0\tilde{A}_{21}(\mu)\tilde{e}_1(t) + 2e_y^T(t)\tilde{P}_0\tilde{D}_2(\mu)\xi(t) + 2e_y^T(t)\tilde{P}_0\tilde{M}_2(\mu)f_a(t) \\ &\quad - 2e_y^T(t)\tilde{P}_0\upsilon_1(t) + 2e_y^T(t)\tilde{P}_0\tilde{N}_2\,(f_s(t) - \upsilon_2(t)) \\ &= \sum_{i=1}^{k} \mu_i(\zeta(t)) \left(2e_y^T(t)\tilde{P}_0\left[\tilde{A}_{21,i}\tilde{e}_1(t) + \tilde{D}_{2,i}\xi(t) + \tilde{M}_{2,i}f_a(t) - \upsilon_1(t)\right]\right) \\ &\quad + 2e_y^T(t)\tilde{P}_0\tilde{N}_2\,(f_s(t) - \upsilon_2(t))\end{aligned} \tag{13.81}$$

Using Assumptions (13.1)–(13.3) for the T–S fuzzy system (13.10) and the argument above $\|\tilde{e}_1(t)\| \leq \varpi$, the following relations are readily obtained:

$$2e_y^T(t)\tilde{P}_0\tilde{A}_{21,i}\tilde{e}_1(t) \leq 2\left\|\tilde{A}_{21,i}\right\|\left\|\tilde{P}_0e_y(t)\right\|\varpi \tag{13.82}$$

$$2e_y^T(t)\tilde{P}_0\tilde{D}_{2,i}\xi(t) \leq 2\left\|\tilde{D}_{2,i}\right\|\left\|\tilde{P}_0e_y(t)\right\|\xi_0 \tag{13.83}$$

$$2e_y^T(t)\tilde{P}_0\tilde{M}_{2,i}f_a(t) \leq 2\left\|\tilde{M}_{2,i}\right\|\left\|\tilde{P}_0e_y(t)\right\|\rho_a \tag{13.84}$$

$$2e_y^T(t)\tilde{P}_0\tilde{N}_2f_s(t) \leq 2\left\|\tilde{N}_2\right\|\left\|Pe_y(t)\right\|\rho_s \tag{13.85}$$

Applying relations (13.82)–(13.85), the following estimate for $\dot{V}_s(t)$ holds:

$$\dot{V}_s(t) \leq \sum_{i=1}^{k}\mu_i(\zeta(t))\left[2\left\|\tilde{P}_0e_y(t)\right\|\chi_i - 2e_y^T(t)\tilde{P}_0\upsilon_1(t) - 2e_y^T(t)\tilde{P}_0\tilde{N}_2\upsilon_2(t)\right] \tag{13.86}$$

with $\chi_i := \left\|\tilde{A}_{21,i}\right\|\varpi + \left\|\tilde{D}_{2,i}\right\|\xi_0 + \left\|\tilde{M}_{2,i}\right\|\rho_a + \left\|\tilde{N}_2\right\|\rho_s$. Due to the convexity condition of the T–S membership functions $\left(\sum_{i=1}^{k}\mu_i(\zeta(t)) = 1\right)$, it holds that $\sum_{i=1}^{k}\mu_i(\zeta(t))\chi_i \leq \chi_{\max} := \left\|\tilde{A}_{21,i}\right\|_{\max}\varpi + \left\|\tilde{D}_{2,i}\right\|_{\max}\xi_0 + \left\|\tilde{M}_{2,i}\right\|_{\max}\rho_a + \left\|\tilde{N}_2\right\|\rho_s$. It thus follows from (13.86) that

$$\dot{V}_s(t) \leq 2\left\|\tilde{P}_0e_y(t)\right\|\chi_{\max} - 2\eta_a\left\|\tilde{P}_0e_y(t)\right\| - 2\eta_s\left\|\tilde{N}_2\right\|\left\|\tilde{P}_0e_y(t)\right\| \tag{13.87}$$

If the condition (13.79) holds, then

$$\dot{V}_s(t) \leq -2\eta_3\left\|\tilde{P}_0e_y(t)\right\| \leq -2\eta_3\sqrt{\lambda_{\min}(\tilde{P}_0)}V_s^{1/2} \tag{13.88}$$

where $\lambda_{\min}(\tilde{P}_0)$ is the smallest eigenvalue of $\tilde{P}_0$. This shows that the reachability condition [61] is satisfied. As a consequence, an ideal sliding motion is achieved and maintained after some finite time.

This completes the proof. □

Finally, the observer gains in the original coordinates (13.19) are given

$$\bar{G}_{l,i} = \left(T_{L,i}T_i\right)^{-1}\begin{bmatrix}\tilde{A}_{12,i}\\ \tilde{A}_{22,i} - \tilde{A}_{22}^s\end{bmatrix} \tag{13.89}$$

$$\bar{G}_{n,i} = \left(T_{L,i}T_i\right)^{-1}\begin{bmatrix}0_{(n+h-p)\times p}\\ I_p\end{bmatrix} \tag{13.90}$$

13.3.6 Simultaneous actuator and sensor faults estimation

Assume the robust TS-SMO in the form of (13.19) has been designed, the objective now is to simultaneously estimate actuator and sensor faults for T–S fuzzy system (13.17). From Theorem 13.2, we know that an ideal sliding motion (13.78) will take place after some finite time if the condition (13.79) is satisfied. During the sliding motion, (13.56) becomes

$$\begin{aligned} 0 &= \tilde{A}_{21}(\mu)\tilde{e_1}(t) + \tilde{D}_2(\mu)\xi(t) + \tilde{M}_2(\mu)f_a(t) - \upsilon_{1eq}(t) + \tilde{N}_2 f_s(t) - \upsilon_{2eq}(t) \\ &= \sum_{i=1}^{k} \mu_i(\zeta(t)) \left(\tilde{A}_{21,i}\tilde{e_1}(t) + \tilde{D}_{2,i}\xi(t) + \tilde{M}_{2,i}f_a(t) - \upsilon_{1eq}(t) \right) \\ &\quad + \tilde{N}_2 f_s(t) - \upsilon_{2eq}(t) \end{aligned} \tag{13.91}$$

where $\upsilon_{1eq}(t)$ and $\upsilon_{2eq}(t)$ denote the equivalent output error injection signals which are required to maintain a sliding motion. It can be approximated to any degree of accuracy by replacing (13.51) and (13.52) with

$$\upsilon_{1eq} = \eta_a \frac{\tilde{P}_0 e_y(t)}{\left\| \tilde{P}_0 e_y(t) \right\| + \delta_1} \tag{13.92}$$

and

$$\upsilon_{2eq} = \eta_s \frac{\tilde{P}_0 e_y(t)}{\left\| \tilde{P}_0 e_y(t) \right\| + \delta_2} \tag{13.93}$$

where δ_1 and δ_2 are two small positive scalars to reduce the chattering effect.

Let

$$\phi(\tilde{e}_1, x, u) = \sum_{i=1}^{k} \mu_i(\zeta(t)) \left(\tilde{A}_{21,i}\tilde{e}_1(t) + \tilde{D}_{2,i}\xi(t) \right) \tag{13.94}$$

From (13.94), it follows that

$$\|\phi(\tilde{e}_1, x, u)\|_2 \le \sum_{i=1}^{k} \mu_i(\zeta(t))\varsigma_i \tag{13.95}$$

with $\varsigma_i := \left\| \tilde{A}_{21,i} \right\|_2 \varpi + \left\| \tilde{D}_{2,i} \right\|_2 \xi_0$.

Due to the convexity condition of the T–S membership functions $\left(\sum_{i=1}^{k} \mu_i(\zeta(t)) = 1 \right)$, it holds that $\sum_{i=1}^{k} \mu_i(\zeta(t))\varsigma_i \le \varsigma_{\max} := \left\| \tilde{A}_{21} \right\|_{\max} \varpi + \left\| \tilde{D}_2 \right\|_{\max} \xi_0$ where $\|.\|_{\max}$ denotes the maximum of each matrix norm for some i. It thus follows from (13.95) that

$$\|\phi(\tilde{e}_1, x, u)\| \le \varsigma_{\max} \tag{13.96}$$

Thus, approximately, for a small $\varsigma_{\max}$,

$$\left(\tilde{M}_2(\mu)f_a(t) - \upsilon_{1eq}(t) \right) + \tilde{N}_2 \left(f_s(t) - \upsilon_{2eq}(t) \right) \to 0, (t \to \infty) \tag{13.97}$$

where $\tilde{M}_2(\mu) := \sum_{i=1}^{k} \mu_i(\zeta)\tilde{M}_{2,i}$.

Note that $\begin{bmatrix} \tilde{M}_{2,i} & \tilde{N}_2 \end{bmatrix}$ is of full column rank, and hence, it can be concluded that [62]

$$\tilde{M}_{2,i}f_a(t) - \upsilon_{1eq}(t) \rightarrow 0, (t \rightarrow \infty) \tag{13.98}$$

$$f_s(t) - \upsilon_{2eq}(t) \rightarrow 0, (t \rightarrow \infty) \tag{13.99}$$

The analysis above shows that the actuator faults can be estimated from

$$\hat{f}_a(t) \approx \left(\sum_{i=1}^{k} \mu_i(\zeta(t))\tilde{M}_{2,i} \right)^{+} \upsilon_{1eq}(t) \tag{13.100}$$

The sensor faults can be estimated as

$$\hat{f}_s(t) \approx \tilde{N}_2^{+} \upsilon_{2eq}(t) \tag{13.101}$$

where $()^{+}$ denotes a left pseudoinverse.

13.4 Fault-tolerant controller design for T–S fuzzy systems

The goal of this section is to design an AFTC strategy based on robust FE of simultaneous actuator and sensor faults. The proposed strategy is able to guarantee stability and ensure the robustness of the overall closed-loop system. A sufficient condition is formulated via LMI to govern the performance of controller by assigning their eigenvalues in specific LMI region of the complex plan. The proposed strategy uses the T–S DOFC scheme [63,64] to compensate the effects of actuator and sensor faults from the system inputs and outputs and to ensure the robustness against uncertainties. Our controller is delivered by the corrected output which obtained by subtracting the reconstructed sensor faults from the (faulty) outputs (see Figure 13.1).

$$\begin{aligned} y_{\text{corr}}(t) &= Cx(t) + Nf_s(t) - N\hat{f}_s(t) \\ &= Cx(t) + N\left(f_s(t) - \hat{f}_s(t)\right) \end{aligned} \tag{13.102}$$

where the quality of the fault compensation is determined by the quality of the FE, i.e., whether the difference $f_s(t) - \hat{f}_s(t)$ is close to zero. The T–S fuzzy model given in (13.10) becomes

$$\begin{cases} \dot{x}(t) = A(\mu)x(t) + B(\mu)u(t) + M(\mu)f_a(t) + D(\mu)\xi(t) \\ y_{\text{corr}}(t) = Cx(t) + Ne_{fs}(t) \end{cases} \tag{13.103}$$

where $e_{f_s}(t) = f_s(t) - \hat{f}_s(t)$ is the sensor FE error and the system matrices $A(\mu) = \sum_{i=1}^{k} \mu_i A_i$, $B(\mu) = \sum_{i=1}^{k} \mu_i B_i$, $M(\mu) = \sum_{i=1}^{k} \mu_i M_i$ and $D(\mu) = \sum_{i=1}^{k} \mu_i D_i$.

Based on on-line FE, the T–S DOFC for T–S fuzzy models (13.10) is constructed as follows:

$$\begin{cases} \dot{x}_K(t) = A_K(\mu,\mu)x_K(t) + B_K(\mu)y_{\text{corr}}(t) \\ u(t) = C_K(\mu)x_K(t) + D_K(\mu)y_{\text{corr}}(t) - K(\mu)\hat{f}_a(t) \end{cases} \tag{13.104}$$

where $x_K(t) \in \mathbb{R}^{n\times n}$ is the controller state and $\hat{f}_a(t)$ is defined in (13.100). $A_K(\mu,\mu) \in \mathbb{R}^{n\times n}$, $B_K(\mu) \in \mathbb{R}^{n\times p}$, $C_K(\mu) \in \mathbb{R}^{m\times n}$ and $D_K(\mu) \in \mathbb{R}^{m\times p}$ are the T-SDOFTC matrices to be designed, and

$$A_K(\mu,\mu) = \sum_{i=1}^{k}\sum_{j=1}^{k}\mu_i\mu_j A_{K,ij},\; B_K(\mu) = \sum_{i=1}^{k}\mu_i B_{K,i} \tag{13.105}$$

$$C_K(\mu) = \sum_{i=1}^{k}\mu_i C_{K,i},\; D_K(\mu) = \sum_{i=1}^{k}\mu_i D_{K,i} \tag{13.106}$$

Substituting (13.103) into (13.104), we obtain the dynamic of the closed-loop system

$$\begin{cases} \dot{x}_K(t) = A_K(\mu,\mu)x_K(t) + B_K(\mu)Cx(t) + B_K(\mu)Ne_{fs}(t) \\ u(t) = C_K(\mu)x_K(t) + D_K(\mu)Cx(t) + D_K(\mu)Ne_{fs}(t) - K(\mu)\hat{f}_a(t) \end{cases} \tag{13.107}$$

Combining $u(t)$ in (13.107) with (13.103), we obtain

$$\begin{aligned} \dot{x}(t) = {} & A(\mu)x(t) + B(\mu)C_K(\mu)x_K(t) + B(\mu)D_K(\mu)Cx(t) \\ & + B(\mu)D_K(\mu)Ne_{fs}(t) + B(\mu)K(\mu)\hat{f}_a(t) - M(\mu)f_a(t) \\ & + D(\mu)\xi(t) \end{aligned} \tag{13.108}$$

The gain $K(\mu)$ is designed so that the effect of $\hat{f}_a(t)$ is either completely decoupled from the closed-loop system. In [65], the authors have shown that if $Im\,(M(\mu)) \subseteq Im\,(B(\mu))$, then the decoupling is achieved by selecting $K(\mu) = B^{+}(\mu)M(\mu)$ where $B^{+}(\mu)$ is the pseudoinverse of $B(\mu)$. Hence, $K(\mu)$ is considered a known gain in the derivation of the LMI-based FTC design.

We further obtain

$$\begin{aligned} \dot{x}(t) = {} & A(\mu)x(t) + B(\mu)C_K(\mu)x_K(t) + B(\mu)D_K(\mu)Cx(t) \\ & + B(\mu)D_K(\mu)Ne_{fs}(t) + M(\mu)e_{fa}(t) + D(\mu)\xi(t) \\ = {} & (A(\mu) + B(\mu)D_K(\mu)C)\,x(t) + B(\mu)C_K(\mu)x_K(t) \\ & + B(\mu)D_K(\mu)Ne_{fs}(t) + M(\mu)e_{fa}(t) + D(\mu)\xi(t) \end{aligned} \tag{13.109}$$

where $e_{f_a}(t) = f_a(t) - \hat{f}_a(t)$ is the actuator FE error. From (13.103) and (13.107), the following form is obtained

$$\begin{cases} x_a(t) = A_a(\mu,\mu)x_a(t) + E_a(\mu,\mu)\omega_a(t) \\ y_{\text{corr}}(t) = C_a x_a(t) + D_a\omega_a(t) \end{cases} \tag{13.110}$$

where

$$A_a(\mu,\mu) = \begin{bmatrix} A(\mu) + B(\mu)D_K(\mu)C & B(\mu)C_K(\mu) \\ B_K(\mu)C & A_K(\mu,\mu) \end{bmatrix} \tag{13.111}$$

$$E_a(\mu,\mu) = \begin{bmatrix} M(\mu) & B(\mu)D_K(\mu)N & D(\mu) \\ 0 & B_K(\mu)N & 0 \end{bmatrix} \tag{13.112}$$

$$C_a = \begin{bmatrix} C & 0 \end{bmatrix},\ D_a = \begin{bmatrix} 0 & N & 0 \end{bmatrix},\ x_a(t) = \begin{bmatrix} x(t) \\ x_K(t) \end{bmatrix},\ \omega_a(\mathrm{t}) = \begin{bmatrix} e_{fa}(t) \\ e_{fs}(t) \\ \xi(t) \end{bmatrix} \tag{13.113}$$

Here, we give an FTC design method in which the performances of the controller are specified by a regional pole assignment and an H_∞ performance index. The fuzzy H_∞ performance index is formulated as follows: given a scalar γ_K, determine an output feedback fuzzy controller in the form of (13.104) such that

- The closed-loop system described by (13.110) is asymptotically stable when $\omega_a(t) = 0$.
- Under zero-initial condition, the controlled output $y_{\text{corr}}(t)$ satisfies $\|y_{\text{corr}}(t)\|_2 < \gamma_K\|\omega_a(t)\|_2, \forall\omega_a(t) \in L_2[0,\infty), \omega_a(t) \neq 0$.

Theorem 13.3. *The eigenvalues of $A_a(\mu,\mu)$ are located in the disc region of the negative complex plane characterized by radius(α), center(β) and the closed-loop system (13.110) is stable with guaranteed H_∞ performance and with an attenuation level γ_K, if there exist symmetric positive definite matrices $X,\ Y \in \mathbb{R}^{n\times n}$ and matrices $\hat{A}_{Kij} \in \mathbb{R}^{n\times n}$, $\hat{B}_{Ki} \in \mathbb{R}^{n\times p}$, $\hat{C}_{Ki} \in \mathbb{R}^{m\times n}$, $\hat{D}_{Ki} \in \mathbb{R}^{m\times p}$ $(i,j = 1,\ldots,k)$ such that the following inequalities hold*

minimize(γ_K) subject to

$$\begin{cases} \Xi_{ii} < 0, & i = 1,\ldots,k, \\ \Xi_{ij} + \Xi_{ji} < 0, & 1 \le i < j \le k, \\ \Phi_{ii} < 0, & i = 1,\ldots,k, \\ \Phi_{ij} + \Phi_{ji} < 0, & 1 \le i < j \le k \end{cases} \tag{13.114}$$

with

$$\Xi_{ij} = \begin{bmatrix} -\alpha X & -\alpha I & A_iX + B_i\hat{C}_{Kj} + \beta X & A_i + B_i\hat{D}_{Kj}C + \beta I \\ -\alpha I & -\alpha Y & \hat{A}_{Kij} + \beta I & YA_i + \hat{B}_{Ki}C + \beta Y \\ * & * & -\alpha X & -\alpha I \\ * & * & -\alpha I & -\alpha Y \end{bmatrix} < 0 \tag{13.115}$$

and

$$\Phi_{ij} = \begin{bmatrix} \Pi_{11} & \Pi_{12} & M_i & B_i\hat{D}_{Ki}N & D_i & XC^T \\ * & \Pi_{22} & YM_i & \hat{B}_{Ki}N & YD_i & C^T \\ * & * & -\gamma_K I_q & 0 & 0 & 0 \\ * & * & * & -\gamma_K I_h & 0 & N^T \\ * & * & * & * & -\gamma_K I_l & 0 \\ * & * & * & * & * & -\gamma_K I_p \end{bmatrix} < 0 \tag{13.116}$$

$$\Pi_{11} = A_iX + XA_i^T + B_i\hat{C}_{Kj} + \hat{C}_{Kj}^T B_i^T \tag{13.117}$$

$$\Pi_{12} = \hat{A}_{Kij}^T + A_i + B_i\hat{D}_{Kij}C \tag{13.118}$$

$$\Pi_{22} = YA_i + A_i^T Y^T + \hat{B}_{Ki}C + C^T\hat{B}_{Ki}^T \tag{13.119}$$

The controller gain matrices are calculated by

$$D_{Ki} = \hat{D}_{Ki} \tag{13.120}$$

$$C_{Ki} = (\hat{C}_{Ki} - D_{Ki}CX)Q^{-T} \tag{13.121}$$

$$B_{Ki} = S^{-1}(\hat{B}_i - YB_iD_{Ki}) \tag{13.122}$$

$$A_{Kij} = S^{-1}(\hat{A}_{Kij} - Y(A_i + B_i\hat{D}_{Ki}C)X)Q^{-T} - S^{-1}YB_iC_{Kj} - B_{Ki}CXQ^{-T} \tag{13.123}$$

where $Q, S \in \mathbb{R}^{n\times n}$ *satisfy* $QS^T = I - XY$.

Proof. Consider the following candidate Lyapunov function:

$$V_3(t) = e_a^T(t)P_a e_a(t) \tag{13.124}$$

where P_a is a symmetric positive definite matrix. Its derivative with respect to time is

$$\dot{V}_3(t) = \tilde{x}_a^T(t)\left(A_a^T(\mu,\mu)P_a + P_aA_a(\mu,\mu)\right) + 2e_a^T(t)P_aE_a(\mu,\mu)\omega_a(t) \tag{13.125}$$

The robustness of the controller against the augmented input $\omega_a(t)$ can then be represented by minimizing the mathematical objective given below:

$$J = \frac{\|y\|_2}{\|\omega_a\|_2} \leq \gamma_K = \int_0^\infty \left[\frac{1}{\gamma_K}y_{\text{corr}}^T(t)y_{\text{corr}}(t) - \gamma_K\omega_a^T(t)\omega_a(t)\right]dt \leq 0 \tag{13.126}$$

It can be shown that

$$J \leq \int_0^\infty \left[\dot{V}_3(t) + \frac{1}{\gamma_K}y_{\text{corr}}^T(t)y_{\text{corr}}(t) - \gamma_K\omega_a^T(t)\omega_a(t)\right]dt \tag{13.127}$$

Subsequently, a sufficient condition for (13.126) is

$$\dot{V}_3(t) + \frac{1}{\gamma_K}y_{\text{corr}}^T(t)y_{\text{corr}}(t) - \gamma_K\omega_a^T(t)\omega_a(t) < 0 \tag{13.128}$$

Substituting (13.125) into (13.128), one obtains

$$\begin{aligned}
&\dot{V}_3(t) + \frac{1}{\gamma_K} y_{\text{corr}}^T(t) y_{\text{corr}}(t) - \gamma_K \omega_a^T(t)\omega_a(t) \\
&\quad = x_a^T(t)\left(A_a^T(\mu,\mu)P_a + P_a A_a(\mu,\mu)\right)x_a(t) + 2x_a^T(t)P_a E_a(\mu,\mu)\omega_a(t) \\
&\qquad + \frac{1}{\gamma_K} y^T(t)y(t) - \gamma_K \omega_a^T(t)\omega_a(t) \\
&\quad = x_a^T(t)\left(A_a^T(\mu,\mu)P_a + P_a A_a(\mu,\mu)\right)x_a(t) + 2x_a^T(t)P_a E_a(\mu,\mu)\omega_a(t) \\
&\qquad + \frac{1}{\gamma_K} x_a^T(t)C_a^T C_a x_a(t) + \frac{2}{\gamma_K} x_a^T(t)C_a^T D_a(\mu)\omega_a(t) \\
&\qquad + \frac{1}{\gamma_K}\omega_a^T(t)D_a^T(\mu)D_a(\mu)\omega_a(t) - \gamma_K \omega_a^T(t)\omega_a(t) \\
&\quad = \varsigma^T(t)\Theta\varsigma^T < 0
\end{aligned} \tag{13.129}$$

where

$$\varsigma(t) = \begin{bmatrix} x_a(t) \\ \omega_a(t) \end{bmatrix} \quad \text{and} \quad \Theta = \begin{bmatrix} \Theta_{11} & \Theta_{12} \\ * & \Theta_{22} \end{bmatrix} \tag{13.130}$$

with

$$\Theta_{11} = P_a A_a(\mu,\mu) + A_a^T(\mu,\mu)P_a + \frac{1}{\gamma_K} C_a^T C_a \tag{13.131}$$

$$\Theta_{12} = P_a E_a(\mu,\mu) + \frac{1}{\gamma_K} C_a^T D_a(\mu) \tag{13.132}$$

$$\Theta_{22} = \frac{1}{\gamma_K} D_a^T(\mu)D_a(\mu) - \gamma_K I \tag{13.133}$$

Using the Schur complement, $\Theta < 0$ is equivalent to

$$\begin{bmatrix} P_a A_a(\mu,\mu) + A_a^T(\mu,\mu)P_a & P_a E_a(\mu,\mu) & C_a^T \\ * & -\gamma_K I & D_a^T(\mu) \\ * & * & -\gamma_K I \end{bmatrix} < 0 \tag{13.134}$$

Let us define the symmetric positive definite matrix P_a and its inverse matrix P_a^{-1}

$$P_a = \begin{bmatrix} Y & S \\ S^T & W \end{bmatrix} \quad \text{and} \quad P_a^{-1} = \begin{bmatrix} X & Q \\ Q^T & Z \end{bmatrix} \tag{13.135}$$

Due to $P_a P_a^{-1} = I_{2n}$, one gets

$$P_a \begin{bmatrix} X \\ Q^T \end{bmatrix} = \begin{bmatrix} I_n \\ 0 \end{bmatrix} \quad \text{and} \quad P_a \begin{bmatrix} X & I_n \\ Q^T & 0 \end{bmatrix} = \begin{bmatrix} I_n & Y \\ 0 & S^T \end{bmatrix} \tag{13.136}$$

We will also define the matrices

$$\Pi_1 = \begin{bmatrix} X & I \\ Q^T & 0 \end{bmatrix}, \quad \Pi_2 = P_a \Pi_1 = \begin{bmatrix} I & Y \\ 0 & S^T \end{bmatrix} \tag{13.137}$$

Pre- and postmultiplying (13.134) by diag $\left(\Pi_1^T, I, I\right)$ and its transpose in the above inequality, we get

$$\begin{bmatrix} \Pi_1^T\left(P_aA_a(\mu,\mu)+A_a^T(\mu,\mu)P_a^T\right)\Pi_1 + & \Pi_1^TP_aE_a(\mu,\mu) & \Pi_1^TC_a^T \\ * & -\gamma_K I & D_aT \\ * & * & -\gamma_K I \end{bmatrix} < 0 \tag{13.138}$$

we have

$$\Pi_1^TP_aA_a(\mu,\mu)\Pi_1 = \begin{bmatrix} A(\mu)X + B(\mu)\left(D_K(\mu)CX + C_K(\mu)Q^T\right) & A(\mu)+B(\mu)D_K(\mu)C \\ Y\left(A(\mu)+B(\mu)D_K(\mu)C\right)X + YB(\mu)C_K(\mu)Q^T + SA_K(\mu,\mu)Q^T & YA(\mu)+\left(YB(\mu)D_K(\mu)+SB_K(\mu)\right)C \end{bmatrix} \tag{13.139}$$

$$\Pi_1^TP_aE_a(\mu,\mu) = \begin{bmatrix} M(\mu) & B(\mu)D_K(\mu)N & D(\mu) \\ YM(\mu) & \left(YB(\mu)D_K(\mu)+SB_K(\mu)\right)N & YD(\mu) \end{bmatrix} \tag{13.140}$$

$$\Pi_1^TC_a^T = \begin{bmatrix} XC^T \\ C^T \end{bmatrix} \tag{13.141}$$

Let the variable change

$$\begin{aligned} \hat{A}_K(\mu,\mu) &= Y\left(A(\mu)+B(\mu)D_K(\mu)C\right)X + YB(\mu)C_K(\mu)Q^T \\ &\quad + SB_K(\mu)CX + SA_K(\mu,\mu)Q^T \end{aligned} \tag{13.142}$$

$$\hat{B}_K(\mu) = YB(\mu)D_K(\mu)+SB_K(\mu) \tag{13.143}$$

$$\hat{C}_K(\mu) = C_K(\mu)Q^T + D_K(\mu)CX \tag{13.144}$$

$$\hat{D}_K(\mu) = D_K(\mu) \tag{13.145}$$

Substituting them into (13.138) yields

$$\Phi = \begin{bmatrix} \Pi_{11} & \Pi_{12} & M(\mu) & B(\mu)\hat{D}_K(\mu)N & D(\mu) & XC^T \\ * & \Pi_{22} & YM(\mu) & \hat{B}_K(\mu)N & YD(\mu) & C^T \\ * & * & -\gamma_K I_q & 0 & 0 & 0 \\ * & * & * & -\gamma_K I_h & 0 & N^T \\ * & * & * & * & -\gamma_K I_l & 0 \\ * & * & * & * & * & -\gamma_K I_p \end{bmatrix} < 0 \tag{13.146}$$

where

$$\Pi_{11} = A(\mu)X + XA^T(\mu) + B(\mu)\hat{C}_K(\mu) + \hat{C}_K^T(\mu)B^T(\mu) \tag{13.147}$$

$$\Pi_{12} = \hat{A}_K^T(\mu,\mu) + A(\mu) + B(\mu)\hat{D}_K(\mu)C \tag{13.148}$$

$$\Pi_{22} = YA(\mu) + A^T(\mu)Y^T + \hat{B}_K(\mu)C + C^T\hat{B}_K^T(\mu) \tag{13.149}$$

Therefore, $\Phi < 0$ can be rewritten as

$$\Phi = \sum_{i=1}^{k}\sum_{j=1}^{k}\mu_i(\zeta)\mu_j(\zeta)\Phi_{ij}$$

$$= \sum_{i=1}^{k}\mu_i^2(\zeta)\Phi_{ii} + \sum_{i=1}^{k}\sum_{i<j}^{k}\mu_i(\zeta)\mu_j(\zeta)\left(\Phi_{ij} + \Phi_{ji}\right) < 0 \tag{13.150}$$

Then, (13.116) can be obtained easily.

To prove LMI (13.115), we first need the following lemma from [66]. □

Lemma 13.3. *The matrix A is D-stable if and only if there exists a symmetric matrix X such that*

$$\begin{bmatrix} -\alpha X & \beta X + XA \\ \beta X + (XA)^T & -\alpha X \end{bmatrix} < 0, X > 0 \tag{13.151}$$

By using (13.151) of Lemma 13.3 and the variable change $A = A_K(\mu, \mu)$, $X = P_a$, one has

$$\begin{bmatrix} -\alpha P_a & \beta P_a + P_a A_a(\mu, \mu) \\ \beta P_a + (P_a A_a(\mu, \mu))^T & -\alpha P_a \end{bmatrix} < 0 \tag{13.152}$$

Then, by pre- and postmultiplying by diagonal diag(Π_1^T, Π_1^T) and its transpose, respectively, and then using the definition $\hat{A}_K(\mu, \mu)$, $\hat{B}_K(\mu)$, $\hat{C}_K(\mu)$ and $\hat{D}_K(\mu)$, one gets

$$\Xi := \begin{bmatrix} -\alpha X & -\alpha I & \phi_{11} & \phi_{12} \\ -\alpha I & -\alpha Y & \phi_{13} & \phi_{14} \\ * & * & -\alpha X & -\alpha I \\ * & * & -\alpha I & -\alpha Y \end{bmatrix} < 0 \tag{13.153}$$

where

$$\phi_{11} = A(\mu)X + B(\mu)\hat{C}_K(\mu) + \beta X \tag{13.154}$$

$$\phi_{12} = A(\mu) + B(\mu)\hat{D}_K(\mu)C + \beta I \tag{13.155}$$

$$\phi_{13} = \hat{A}_K(\mu, \mu) + \beta I \tag{13.156}$$

$$\phi_{14} = YA(\mu) + \hat{B}_K(\mu)C + \beta Y \tag{13.157}$$

which can be rewritten as

$$\Xi = \sum_{i=1}^{k}\sum_{j=1}^{k}\mu_i(\zeta)\mu_j(\zeta)\Xi_{ij}$$

$$= \sum_{i=1}^{k}\mu_i^2(\zeta)\Xi_{ii} + \sum_{i=1}^{k}\sum_{i<j}^{k}\mu_i(\zeta)\mu_j(\zeta)\left(\Xi_{ij} + \Xi_{ji}\right) < 0 \tag{13.158}$$

So if (13.115) and (13.116) hold, then the DOFC (13.107) makes the T–S fuzzy system (13.103) asymptotically stable with H_∞ performance bound γ_K. Thus, the proof is complete.

13.5 Simulation results

To illustrate the proposed AFTC strategy encompassing the possibility of simultaneous actuator and sensor faults, consider the inverted pendulum and cart system presented in Section 13.2.2. The submodel matrices A_i, $B_i (i = 1, 2)$ are given by

$$A_1 = \begin{bmatrix} 0 & 1 & 0 & 0 \\ 17.3118 & 0 & 0 & 0.0882 \\ 0 & 0 & 0 & 1 \\ -1.7312 & 0 & 0 & -0.0441 \end{bmatrix}, \quad B_1 = \begin{bmatrix} 0 \\ -1.7647 \\ 0 \\ 1.1765 \end{bmatrix} \tag{13.159}$$

$$A_2 = \begin{bmatrix} 0 & 1 & 0 & 0 \\ 14.3223 & 0 & 0 & 0.0573 \\ 0 & 0 & 0 & 1 \\ -1.0127 & 0 & 0 & -0.0405 \end{bmatrix}, \quad B_2 = \begin{bmatrix} 0 \\ -1.1467 \\ 0 \\ 1.0811 \end{bmatrix} \tag{13.160}$$

$$C = \begin{bmatrix} 1 & 0 & 0 & 0 \\ 0 & 1 & 0 & 0 \\ 0 & 0 & 1 & 1 \end{bmatrix} \tag{13.161}$$

In order to point up the proposed approach, additional actuator and sensor faults are used and are injected to the T–S model as

$$\begin{cases} \dot{x}(t) = \sum_{i=1}^{2} \mu_i(\zeta(t)) \left(A_i x(t) + B_i(u(t) + \xi(t)) + M_i f_a(t)\right) \\ y(t) = Cx(t) + Nf_s(t) \end{cases} \tag{13.162}$$

It is assumed here that the actuator fault affects the system in the same direction as the control input; therefore, $M_i = B_i$. Also, an additive cart position sensor fault has been introduced and

$$N = \begin{bmatrix} 0 \\ 1 \\ 0 \end{bmatrix} \tag{13.163}$$

It can be easily shown that $\forall\, i \in [1, 2]$, (A_i, M_i, C) is minimum phase and $rank(CM_i) = rank(M_i)$. It deserves to note that $n > p \geq \tilde{q}$, so the TS-SMO method presented in Section 13.3.4 can be used for this system to achieve simultaneous actuator and sensor faults estimation.

13.5.1 TS-SMO design

Suitable choice of the matrix T_R from (13.15) can be shown to be

$$T_R = \begin{bmatrix} 0 & 0 & 1 \\ -1 & 0 & 0 \\ 0 & -1 & 0 \end{bmatrix} \quad (13.164)$$

where we have

$$N_1 = -1$$

$$C_1 = \begin{bmatrix} 0 & 0 & 1 & 1 \\ -1 & 0 & 0 & 0 \end{bmatrix} \quad (13.165)$$

$$C_2 = \begin{bmatrix} 0 & 0 & -1 & 0 \end{bmatrix}$$

The design parameters were chosen as $A_f = 1$, $\tilde{A}^s_{22} = diag(-3, -5, -7)$, $H_1 = I_{2\times 2}$ and $H_2 = I_{3\times 3}$. By solving the LMI conditions given in Theorem 13.1, we find that

$$\mu = 0.8619, \quad L_1 = \begin{bmatrix} 1.2064 & 0 & 0 \\ 4.2188 & 0 & 0 \end{bmatrix}, \quad L_2 = \begin{bmatrix} 0.4117 & 0 & 0 \\ 0.6566 & 0 & 0 \end{bmatrix} \quad (13.166)$$

The T–S fuzzy SMO gains are given by

$$G_{l,1} = 10^3 \times \begin{bmatrix} -0.0015 & -0.7086 & 0 \\ -0.0044 & -3.5684 & 0 \\ 0.0010 & 1.5126 & 0 \\ 0.0030 & -1.8724 & 0 \\ 0 & -0.3782 & 0.0060 \end{bmatrix} \quad (13.167)$$

$$G_{n,1} = \begin{bmatrix} 0 & -1 & 0 \\ -1.5000 & -703.5529 & 0 \\ 0 & 378.1567 & 0 \\ 1 & -378.1567 & 0 \\ 0 & 0 & 1 \end{bmatrix} \quad (13.168)$$

$$G_{l,2} = \begin{bmatrix} -1.0607 & -5.0000 & 0 \\ -3.1246 & -14.3223 & 0 \\ 1.0000 & 0 & 0 \\ 2.9595 & 1.0127 & 0 \\ 0 & 0 & 6 \end{bmatrix}, \quad G_{n,2} = \begin{bmatrix} 0 & -1 & 0 \\ -1.0607 & 0 & 0 \\ 0 & 0 & 0 \\ 1 & 0 & 0 \\ 0 & 0 & 1 \end{bmatrix} \quad (13.169)$$

First, it is assumed that actuator and sensor faults are created, respectively, as

$$f_a(t) = \begin{cases} 0 & t < 2 \\ 0.5\sin(\pi(t-2)) & t \geq 2 \end{cases} \quad (13.170)$$

$$f_s(t) = \begin{cases} 0 & t < 5 \\ 1 & t \geq 5 \end{cases} \quad (13.171)$$

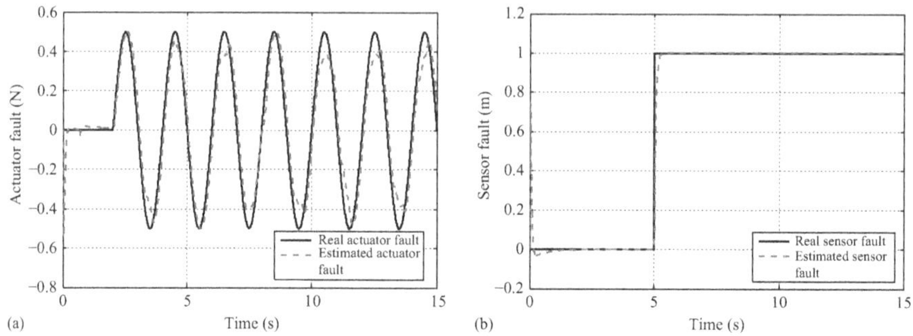

Figure 13.5 Simultaneously actuator and sensor faults estimation: (a) actuator fault estimation and (b) sensor fault estimation

Simulations are performed with

- initial conditions $x_{10} = \frac{\pi}{20}$, $x_{20} = 0$, $x_{30} = 1$ and $x_{40} = 0$,
- uncertainties are assumed to be $\xi(x,u,t) = mlx_2^2 \sin(x_1) - \mu_c sign(\dot{x}_4)$ where μ_c is the cart friction coefficient.
- $\eta_a = 10$, $\eta_s = 7$, $\delta_a = 0.1$, $\delta_s = 0.01$.

13.5.2 Simultaneous faults estimation design

Figure 13.5(a) is concerned with the actuator FE. It shows that the proposed TS-SMO can reject the effects of system uncertainties successfully and accurately estimate the actuator fault. Additive cart position sensor fault has been introduced to show the ability of the proposed strategy to handle simultaneous faults. The sensor FE is shown in Figure 13.5(b). It can be seen that despite the presence of uncertainties, the proposed method could still track the sensor fault $f_s(t)$.

13.5.3 TS-DOFC design

Solving the LMI conditions in Theorem 13.3 with the circle region $D(3,3)$ gives

$$X = \begin{bmatrix} 6.4317 & -20.1433 & 1.2865 & -5.6488 \\ -20.1433 & 64.0667 & -3.3509 & 15.6849 \\ 1.2865 & -3.3509 & 4.9831 & -5.4158 \\ -5.6488 & 15.6849 & -5.4158 & 11.5707 \end{bmatrix}, \tag{13.172}$$

$$Y = 10^5 \times \begin{bmatrix} 0.0461 & -0.0072 & -0.0101 & -0.0098 \\ -0.0072 & 0.0013 & 0.0018 & 0.0017 \\ -0.0101 & 0.0018 & 584.0238 & 0.0024 \\ -0.0098 & 0.0017 & 0.0024 & 0.0022 \end{bmatrix}, \tag{13.173}$$

$$Q = 10^6 \times \begin{bmatrix} -75.1259 & 0.0334 & 0 & 0 \\ 195.6623 & -0.1127 & 0 & 0 \\ -291.0231 & -0.0562 & 0 & 0 \\ 316.2868 & 0.0260 & 0 & 0 \end{bmatrix}, \tag{13.174}$$

$$S = 10^{-2} \times \begin{bmatrix} 0.0216 & -96.4039 & 26.5362 & 1.4535 \\ -0.0035 & 15.9460 & 62.1326 & -76.7155 \\ 100 & 0.0224 & 0 & 0 \\ -0.0047 & 21.2605 & 73.7250 & 64.1297 \end{bmatrix} \tag{13.175}$$

The minimum attenuation value is $\gamma_K = 3.1005$. The method of Theorem 13.3 provides the following fuzzy DOFC:

$$\dot{x}_K(t) = \sum_{i=1}^{2}\sum_{j=1}^{2} \mu_i(\zeta)\mu_j(\zeta)A_{Kij}x_K(t) + \sum_{i=1}^{2} \mu_i(\zeta)B_{Ki}y(t) \tag{13.176}$$

$$u(t) = \sum_{i=1}^{2} \mu_i(\zeta)\left(C_{Ki}x_K(t) + D_{Ki}y(t) - K_i\hat{f}_a(t)\right) \tag{13.177}$$

where

$$A_{K11} = A_{K12} = \begin{bmatrix} -0.9997 & -1.3840 & -5.4200 & -2.1335 \\ 0.0014 & -7.2931 & -11.8617 & -25.9868 \\ 0 & 0.0214 & -4.1859 & 0.3172 \\ 0 & 0 & -0.0004 & -4.6543 \end{bmatrix} \tag{13.178}$$

$$A_{K21} = A_{K22} = \begin{bmatrix} -0.9999 & -0.6025 & -1.9098 & -88.6122 \\ 0.0006 & -4.0112 & 0.9529 & -266.5920 \\ 0 & 0.0559 & -4.1308 & -27.4326 \\ 0 & 0 & 0.0003 & -4.5991 \end{bmatrix} \tag{13.179}$$

$$B_{K1} = 10^5 \times \begin{bmatrix} 0.0570 & 0.0006 & -584.0401 \\ 0.2641 & 0.0030 & -0.2144 \\ -0.0017 & 0 & 0 \\ 0 & 0 & 0 \end{bmatrix} \tag{13.180}$$

$$B_{K2} = 10^5 \times \begin{bmatrix} 0 & -0.0006 & -584.0336 \\ 0.0245 & -0.0026 & -0.1873 \\ -0.0042 & 0 & 0.0003 \\ 0 & 0 & 0 \end{bmatrix} \tag{13.181}$$

$$\begin{aligned} C_{K1} &= C_{K2} = 10^{-3} \times \begin{bmatrix} 0.0022 & -9.6078 & -37.5980 & -121.7464 \end{bmatrix}, \\ D_{K1} &= D_{K2} = \begin{bmatrix} 74.1897 & 0.4200 & -7.7610 \end{bmatrix} \end{aligned} \tag{13.182}$$

Figure 13.6(a)–(c) illustrates a comparison between the outputs of the fault-free model (i.e., without any fault), the outputs of the faulty system without FTC and finally the

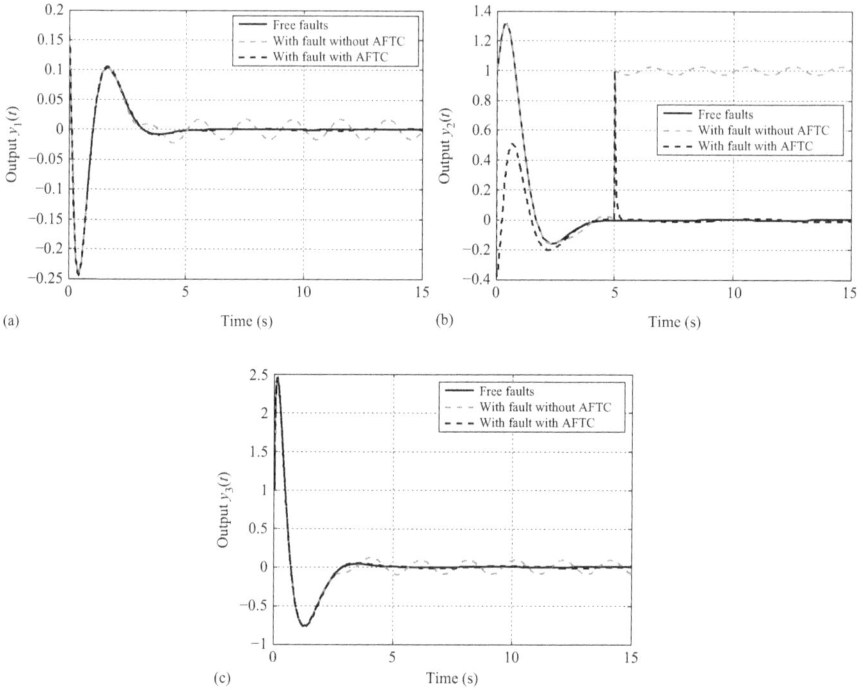

Figure 13.6 Outputs of the closed-loop system with the proposed FT control: nominal output, output without AFTC and output with AFTC: (a) output $y_1(t)$, (b) output $y_2(t)$ and (c) output $y_3(t)$

outputs with our proposed FTC. It can be noticed that the outputs without FTC do not converge to the fault-free dynamic: it underlines that the system is perturbed by both actuator and sensor faults, and it is not robust against such faults. The proposed T–S fuzzy SMO under the FTC (13.107) makes the system robust against actuator and sensor faults, since the outputs' trajectories of the system with FTC reach the outputs of nominal model.

13.6 Conclusion

This chapter proposes a FTC scheme for a class of uncertain nonlinear systems described by Takagi–Sugeno (TS) fuzzy representation. The aim of this FTC focuses on the compensation of simultaneous actuator and sensor faults effects by using a robust SMO technique. We present robust H_∞ faults reconstruction scheme for T–S fuzzy system subject both to actuator, sensor faults and disturbances. An augmented system has been constructed by assuming the sensor faults as an auxiliary actuator faults. So that, the proposed TS-SMO with two discontinuous terms has been performed by the search of suitable Lyapunov matrices in order to decouple the simultaneous faults. The LMIs conditions have been formulated by using convex optimization technique for minimizing the disturbance attenuation level. Finally, simulation results are presented to verify the effectiveness of the proposed method in this design framework. The extension of our work to FTC of T–S systems is under study. The effectiveness of the proposed FTC approach to treat simultaneous faults is illustrated by considering a parametric change position sensor and linear time-varying fault frequencies on the cart actuator.

References

[1] Isermann R., *Fault-Diagnosis Applications: Model-Based Condition Monitoring: Actuators, Drives, Machinery, Plants, Sensors, and Fault-Tolerant Systems*, Springer Science and Business Media, 2011.

[2] Chen J., Patton R.J., *Robust Model-Based Fault Diagnosis for Dynamic Systems*, Springer Science and Business Media, 2012.

[3] Alwi H., Edwards C., Tan C.P., *Fault Detection and Fault-Tolerant Control Using Sliding Modes*, Springer Science and Business Media, 2011.

[4] Gao Z., Cecati C., Ding S.X., A survey of fault diagnosis and fault-tolerant techniques – Part I: fault diagnosis with model-based and signal-based approaches, *IEEE Transactions on Industrial Electronics*, 2015, 62, (6), pp. 3757–3767.

[5] Ding S.X., Integrated design of feedback controllers and fault detectors, *Annual Reviews in Control*, 2009, 33, pp. 124–135.

[6] Wang J., H_∞ fault-tolerant controller design for networked control systems with time-varying actuator faults, *International Journal of Innovative Computing, Information and Control*, 2015, 11, pp. 1471–1481.

[7] Khosrowjerdi M.J., Barzegary S., Fault tolerant control using virtual actuator for continuous-time Lipschitz nonlinear systems, *International Journal of Robust and Nonlinear Control*, 2014, 24, (16), pp. 2597–2607.

[8] Yang J., Zhu F., Wang X., Bu X., Robust sliding-mode observer-based sensor fault estimation, actuator fault detection and isolation for uncertain nonlinear systems, *International Journal of Control, Automation and Systems*, 2015, 13, (5), pp. 1037–1046.

[9] Assidi M., Ben Brahim A., Dhahri S., Ben Hmida F., Simultaneous Actuator and Sensor faults tolerant control for uncertain linear Systems, *IEEE Conference on Control, Automation and Diagnosis*, 2017.

[10] Zhang J., Swain A.K., Nguang S.K., Robust sensor fault estimation and fault-tolerant control for uncertain Lipschitz nonlinear systems, *Proceedings of the American Control Conference*, 2014, pp. 5515–5520.

[11] Assidi M., Ben Brahim A., Dhahri S., Ben Hmida F., Sensor fault tolerant control for uncertain linear systems, *IEEE International Conference on Sciences and Techniques of Automatic Control Engineering*, 2016.

[12] Takagi T., Sugeno M., Fuzzy identification of systems and its applications to modeling and control, *IEEE Transactions on Systems, Man and Cybernetics*, 1985, 15, (1), pp. 116–132.

[13] Aouaouda S., Chadli M., Cocquempot V., Tarek Khadir M., Multi-objective H_-/H_∞ fault detection observer design for Takagi–Sugeno fuzzy systems with unmeasurable premise variables: descriptor approach, *International Journal of Adaptive Control and Signal Processing*, 2013, 27, (12), pp. 1031–1047.

[14] Wu H.-N., Zhang H.-Y., Reliable mixed L_2/H_∞ fuzzy static output feedback control for nonlinear systems with sensor faults, *Automatica*, 2005, 41, (11), pp. 1925–1932.

[15] Yang G.-H., Wang H., Fault detection and isolation for a class of uncertain state-feedback fuzzy control systems, *IEEE Transactions on Fuzzy Systems*, 2015, 23, (1), pp. 139–151.

[16] Wang H., Ye D., Yang G.-H., Actuator fault diagnosis for uncertain T–S fuzzy systems with local nonlinear models, *Nonlinear Dynamics*, 2014, 76, (4), pp. 1977–1988.

[17] Ichalal D., Marx B., Ragot J., Maquin D., Fault detection, isolation and estimation for Takagi–Sugeno nonlinear systems, *Journal of the Franklin Institute*, 2014, 351, (7), pp. 3651–3676.

[18] Zhang K., Jiang B., Dynamic output feedback fault tolerant controller design for Takagi–Sugeno fuzzy systems with actuator faults, *IEEE Transactions on Fuzzy Systems*, 2010, 18, (1), pp. 194–201.

[19] Gao Z.F., Jiang B., Shi P., Xu Y.F., Fault accommodation for near space vehicle attitude dynamics via T-S fuzzy models, *International Journal of Innovative Computing, Information and Control*, 2010, 6, (11), pp. 4843–4856.

[20] Zhao Y., Lam J., Gao H., Fault detection for fuzzy systems with intermittent measurements, *Fuzzy Systems IEEE Transactions*, 2009, 17, (2), pp. 398–410.

[21] Akhenak A., Chadli M., Ragot J., Design of a sliding mode fuzzy observer for uncertain Takagi–Sugeno fuzzy model: application to automatic steering of vehicles, *International Journal Vehicle Autonomous Systems*, 2007, (5), pp. 288–305.

[22] Akhenak A., Chadli M., Ragot J., Maquin D., Fault detection and isolation using sliding mode observer for uncertain Takagi–Sugeno fuzzy model, *IEEE Mediterranean Conference on Control and Automation*, Ajaccio, 2008, pp. 286–291.

[23] Xu D., Jiang B., Shi P., Nonlinear actuator fault estimation observer: an inverse system aproach via a T–S fuzzy model, *International Journal of Applied Mathematics and Computational Science*, 2012, 22, (1), pp. 183–196.

[24] Ichalal D., Marx B., Ragot J., Maquin D., Simultaneous state and unknown inputs estimation with PI and PMI observers for Takagi Sugeno model with unmeasurable premise variables, *IEEE Mediterranean Conference on Control and Automation*, 2009, pp. 353–358.

[25] Ichalal D., Marx B., Ragot J., Maquin D., State estimation of nonlinear systems using multiple model approach, *American Control Conference, ACC 2009*, Jun 2009, St Louis, Missouri, United States, pp. 4636–4641.

[26] Hamdi H., Rodrigues M., Mechmeche C., Belhadjbraiek N., Robust fault detection and estimation for descriptor systems based on multi-models concept, *International Journal of Control, Automation, and Systems*, 2012, 10, (6), pp. 1260–1266.

[27] Ghorbel H., Souissi M., Chaabane M., Hajjaji A., Observer design for fault diagnosis for the Takagi–Sugeno model with unmeasurable premise variables, *IEEE Mediterranean Conference on Control and Automation*, 2012, pp. 303–308.

[28] Edwards C., Spurgeon S. K., Patton R. J., Sliding mode observers for fault detection and isolation, *Automatica*, 2000, 36, pp. 541–553.

[29] Ng K.Y., Tan C.P., Edwards C., Kuang Y.C., New results in robust actuator fault reconstruction for linear uncertain systems using sliding mode observers, *International Journal of Robust and Nonlinear Control*, 2007, 17, pp. 1294–1319.

[30] Dhahri S., Sellami A., Benhmida F., Fault detection and reconstruction via both full-order and reduced-order sliding mode observers for uncertain nonlinear systems, *IEEE Conference on Modelling, Identification and Control*, 2015, DOI: 10.1109/ICMIC.2015.7409425.

[31] Dhahri S., Sellami A., Benhmida F., Robust H_∞ sliding mode observer design for fault estimation in a class of uncertain nonlinear systems with LMI optimization approach, *International Journal of Control, Automation and Systems*, 2012, 10, (5), pp. 1032–1041.

[32] Dhahri S., Sellami A., Benhmida F., Robust H_∞ sliding mode observer design for fault estimation in a class of uncertain nonlinear systems with LMI optimization approach, *IEEE Mediterranean Electrotechnical Conference*, 2012, pp. 450–454.

[33] Dhahri S., Sellami A., Benhmida F., Robust sliding mode observer-based fault reconstruction for uncertain linear systems, *American Journal of Applied Sciences*, 2011, 8, (10), pp. 1032–1040.

[34] Raoufi R., Marquez H.J., Zinober A.S.I., H_∞ sliding mode observers for uncertain nonlinear Lipschitz systems with fault estimation synthesis, *International Journal of Robust and Nonlinear Control*, 2010, 20, pp. 1785–1801.

[35] Xing G.Y., Edwards C., Nonlinear robust fault reconstruction and estimation using a sliding mode observer, *Automatica*, 2007, 40, pp. 1605–1614.

[36] Junqi Y., State and unknown information estimation for non-linear systems with both input uncertainty and output disturbance, *IMA Journal of Mathematical Control and Information*, 2016, 33, (2), pp. 427–439.

[37] Junqi Y., Fanglai Z., Xiaojie S., State estimation and simultaneous unknown input and measurement noise reconstruction based on associated observers, *International Journal of Adaptive Control and Signal Processing*, 2013, 27, pp. 846–858.

[38] Dong J.L., Youngjin P., Youn-sik P., Robust sliding mode descriptor observer for fault and output disturbance estimation of uncertain systems, *IEEE Transactions on Automatic and Control,* 2012, 57, (11), pp. 2928–2934.

[39] Veluvolu K.C., Soh Y.C., Fault reconstruction and state estimation with sliding mode observers for Lipschitz non-linear systems, *IET Control Theory and Applications*, 2010, 5, (11), pp. 1255–1263.

[40] Corless M., Tu J., State and input estimation for a class of uncertain systems, *Automatica*, 1998, 34, pp. 757–764.

[41] Tan C.P., Edwards C., Sliding mode observers for detection and reconstruction of sensor faults, *Automatica*, 2002, 38, pp. 1815–1821.

[42] Alwi H., Edwards C., Tan C.P., Sliding mode estimation schemes for incipient sensor faults, *Automatica*, 2009, 45, pp. 1679–1685.

[43] Xing G.Y., Edwards C., Sensor fault detection and isolation for nonlinear systems based on a sliding mode observer, *International Journal of Adaptive Control and Signal Processing*, 2007, 21, pp. 657–673.

[44] Zhang J., Swain A.K., Nguang S.K., Detection and isolation of incipient sensor faults for a class of uncertain non-linear systems, *IET Control Theory and Applications*, 2012, 6, (12), pp. 1870–1880.

[45] Zhang J., Akshya K.S., Nguang S.K., Robust sensor fault estimation scheme for satellite attitude control systems, *Journal of the Franklin Institute*, 2013, 350, (9), pp. 2581–2604.

[46] Zhang K., Jiang B., Peng S., *Observer-Based Fault Estimation and Accommodation for Dynamic Systems*, Springer Heidelberg, New York, Dordrecht, London, 2013.

[47] Khosrowjerdi M.J., Robust sensor fault reconstruction for Lipschitz nonlinear systems, *Mathematical Problems in Engineering*, 2011, 2011, doi:10.1155/2011/146038.

[48] Michael D., Kalyana C.V., Jagat J.R., Mohamed D., Adaptive sensor and actuator fault estimation for a classof uncertain Lipschitz nonlinear systems, *International Journal of Adaptive Control and Signal Processing*, 2016, 30, pp. 271–283.

[49] Zhang J., Akshya K.S., Nguang S.K., Simultaneous robust actuator and sensor fault estimation for uncertain non-linear Lipschitz systems, *IET Control Theory and Applications*, 2014, 8, (14), pp. 1364–1374.

[50] Ben Brahim A., Dhahri S., Ben Hmida F., Sellami A., Simultaneous actuator and sensor faults reconstruction for Lipschitz nonlinear systems, *IEEE Conference on Systems and Control*, 2015, pp. 97–103. DOI: 10.1109/ICoSC.2015.7153276.

[51] Li X., Zhu F., Simultaneous time-varying actuator and sensor fault reconstruction based on PI observer for LPV systems, *International Journal of Adaptive Control and Signal Processing*, 2014, 29, (9), pp. 1086–1098.

[52] Ben Brahim A., Dhahri S., Ben Hmida F., Sellami A., Simultaneous actuator and sensor faults reconstruction using polytopic sliding mode observer for linear parameter varying systems, *IEEE Conference on Sciences and Techniques*

of Automatic Control and Computer Engineering, 2013, pp. 480–486. DOI: 10.1109/STA.2013.6783175.

[53] Zhang K., Jiang B., Shi P., A new approach to observer-based fault-tolerant controller design for Takagi–Sugeno fuzzy systems with state delay, *Circuits, Systems and Signal Processing*, 2009, 28, (5), pp. 679–697.

[54] Han J., Zhang H., Wang Y., Liu Y., Disturbance observer based fault estimation and dynamic output feedback fault tolerant control for fuzzy systems with local nonlinear models, *ISA Transactions*, 2009, 59, pp. 114–124.

[55] Dong J., Yang G.-H., Reliable state feedback control of T–S fuzzy systems with sensor faults, *IEEE Transactions on Fuzzy Systems*, 2015, 23 (2), pp. 421–433.

[56] Ben Brahim A., Dhahri S., Ben Hmida F., Sellami A., Simultaneous actuator and sensor faults reconstruction based on robust sliding mode observer for a class of nonlinear systems, *Asian Journal of Control*, 2017, 19, (1), pp. 1–10.

[57] Ben Brahim A., Dhahri S., Ben Hmida F., Sellami A., H_∞ sliding mode observer for Takagi–Sugeno nonlinear systems with simultaneous actuator and sensor faults, *International Journal of Applied Mathematics and Computer Sciences*, 2015, 25, (3), pp. 547–559.

[58] Aouaouda S., Chadli M., Khadir M.T., Bouarar T., Robust fault tolerant tracking controller design for unknown inputs T–S models with unmeasurable premise variables, *Journal of Process Control*, 2012, 22, (5), pp. 861–872.

[59] Ogata K., *Modern Control Engineering*, 3rd ed. Englewood Cliffs, NJ: Prentice-Hall, 1997.

[60] Teixeira M.C.M., Zak S.H., Stabilizing controller design for uncertain nonlinear systems using fuzzy models, *IEEE Transaction on Fuzzy Systems*, 7 (2), pp. 133–142, 1999.

[61] Utkin V.I., *Sliding Modes in Control Optimization*, 1992, Springer, Berlin.

[62] Chen W., Saif, M., Observer-based strategies for actuator fault detection, isolation and estimation for certain class of uncertain nonlinear systems, *IET Control Theory & Applications*, 2007, 1(6), pp. 1672–1680.

[63] Dong J., Yang G.H., Dynamic output feedback H control synthesis for discrete-time T–S fuzzy systems via switching fuzzy controllers, *Fuzzy Sets and Systems*, 2009, 160(4), pp. 482–499.

[64] Li J., Wang H.O., Niemann D., Tanaka K., Dynamic parallel distributed compensation for Takagi–Sugeno fuzzy systems: an LMI approach, *Information Sciences*, 2000, 123(3–4), pp. 201–221.

[65] Gao Z., Antsaklis P. J., Reconfigurable control system design via perfect model following, *International Journal of Control*, 1992, 56, pp. 783–798.

[66] Chilali M., Gahinet P., H_∞ design with pole placement constraints: an LMI approach, *IEEE Transactions Automatic Control*, 1996, 41, (3), pp. 358–367.

Chapter 14
LMI-based control design for balancing and attitude stabilization of inverted pendulums

Magdi S. Mahmoud[1]

Abstract

This chapter explores the dynamic modeling and feedback stabilization of different types of inverted pendulums (IPs). It contains a theoretical analysis of the system dynamics and control methods, as well as a summary of MATLAB®/Simulink® simulation results. There are two primary objectives of this chapter:

1. To provide technical results pertaining to robust control-design approaches using convex optimization methods over linear-matrix inequalities. The design approaches include linear quadratic regulator (LQR), linear quadratic regulator with integral gain (LQRI), model-predictive control, H_2 control, H_∞ control and multiobjective state feedback.
2. To demonstrate the application of these approaches to the balancing and attitude stabilization of IP systems including pendulums on cart, double and triple link systems on pivot, rotational double IP and IP-type assistant robot.

14.1 Introduction

Over the past decades, inverted pendulum (IP) has been the subject of numerous studies in control systems. An IP is a pendulum which has its links rotating above its pivot point. The IP is a typical representative of a class of high-order nonlinear and nonminimum phase systems. It is often implemented either

- with the pivot point connected with a base arm that can rotate horizontally, as described in [1] or
- mounted on a cart that can move in a fixed horizontal line as introduced in [2].

The links of the pendulum are usually limited to one degree of freedom by affixing the links to an axis of rotation. It is obvious that an IP is inherently unstable and must

[1]Systems Engineering Department, King Fahd University of Petroleum and Minerals, Saudi Arabia

be actively balanced in order to remain upright, while a normal pendulum is stable when hanging downwards. This can be done by applying a torque at the pivot point for a rotational IP as considered in this thesis or moving the pivot point horizontally for the case of an IP on a cart. A simple demonstration of moving the pivot point to control the pendulum is achieved by balancing an upturned broomstick on the end of one's finger.

The uniqueness and wide application of technology derived from this unstable system have drawn interest of many researchers [3–7]. In recent years, researchers have applied the idea of a mobile IP model to various applications including the design of walking gaits for humanoid robots, robotic wheelchairs and personal transport systems.

Controlling the IP is a classic problem in dynamics and system theory and is used to verify the performance and demonstrating the effectiveness of control algorithms. An IP without attached springs is essentially unstable, and it is hence obvious that feedback of the state of the pendulum is needed to stabilize the pendulum. The IP problem resembles the control systems that exist in robotic arms. The dynamics of IP simulates the dynamics of robotic arm in the condition when the center of pressure lies below the center of gravity (CoG) for the arm so that the system is also unstable. Robotic arm behaves very much like IP under this condition.

The most familiar types of IP are the rotational single-arm pendulum [8], the cart IP [9] and the double inverted pendulum (DIP) [10]. The less common versions are the rotational two-link pendulum [11], the parallel-type dual IP [12], the triple inverted pendulum (TIP) [13], the quadruple IP [14] and the 3D or spherical pendulum [15]. An excellent survey is presented in [16].

In [17], a linearized model of two-wheeled mobile robot for balancing control design was presented. A nonlinear model of same system was studied in [18]. By developing a nonlinear observer to handle the nonlinear disturbance, recent results were reported in [19]. Through an experimental verification, both linearized and nonlinear models were treated in [20]. Employing an model-predictive control (MPC) controller in [21], a design scheme was considered to guarantee balancing the robot.

The uncertainty plays a crucial role in robotic systems. Sources of uncertainty include

- Parametric due to changes in system elements such as friction, inertia or mass,
- Environmental as a result of recording the robot output variable using sensor,
- External due to variations in working loads.

The major task is how to handle these uncertainties without degrading the stability and performance of the robot. This can be approached by generating proper and resilient controllers.

A recently identified technique for overcoming the complications inherent in wheeled inverted pendulum (WIP) systems is conveniently presented in [22,23]. Control problems involving different configurations of IPs have been considered by many researchers. A comprehensive literature review of the topic is given in [24–32]. In applications, two-wheeled robots (TWR) have two coaxial wheels mounted on either side of an intermediate body, with a center of mass above the wheel axles and therefore must actively stabilize themselves to prevent toppling. TWR have a number of advantages over other mobile robots. Although they are more difficult to control than statically stable wheeled robots, TWRs are still much easier to control than legged robots. This wheel configuration makes them highly maneuverable, because of their ability to turn on the spot, similar to differential drive robots [33–44].

On a parallel development, the rotational double inverted pendulum (RDIP) takes the classic rotational single pendulum problem to the next level of complexity. The RDIP is composed of a rotary arm that attaches to a servo system which provides a torque to the base arm to control the whole system, a short bottom rod connected to the arm and a top long rod. It is an underactuated in the sense that it has fewer inputs that degrees of freedom and extremely nonlinear unstable system due to the gravitational forces and the coupling arising from the Coriolis and centripetal forces. Since the RDIP presents considerable control-design challenges, it is an attractive tool utilized for developing different control techniques and testing their performances. Related applications include stabilizing the takeoff of a multistage rocket, as well as modeling the human posture system [45–50].

Nearly, all works on pendulum control concentrate on two problems: stabilization of the IPs and pendulums swing-up control design. The first topic is concerned with the controller design to maintain the pendulum in the upright position. In the RDIP case, controllers are designed to balance two vertical rods by manipulating the angle of the base arm. The second one refers to an adequate algorithm to swing up the pendulum from its stable equilibrium [51], the downward position to the upright position.

In this chapter, we develop and establish complete control design results based on several distinct approaches. We investigate issues regarding robust stability and robust performance of control design for IPs. The controller design is based on an approximate continuous-time linear model of the pendulum system subject to external disturbance. The contributions of our work are summarized below:

1. The analytical results are derived in systematic and unique way, thereby leading to improved techniques over the techniques in the literature.
2. All the design procedures are conveniently cast into the format of feasibility problem over linear-matrix inequalities (LMI). By this way, effective computational methods are established yielding guaranteed quality solution.
3. Simulation studies using Matlab/Simulink environment are performed for all the approaches yielding good performance results.

The chapter is organized into two parts: The first part concerns dynamic model development and occupies Sections 14.2–14.6. The second part composed of Sections 14.7 and 14.8 and focuses on the control design methods. Section 14.2 discusses basics of dynamic modeling to characterize the behavior of the multilink system. In Section 14.3, modeling issues pertaining to single and double pendulums on cart are presented. DIP on pivot and TIP on pivot are provided in Section 14.4. The case of RDIP is examined in Section 14.5. Finally, Inverted PENdulum Type Assistant Robot (I-PENTAR) as a humanoid-type robot is the subject of Section 14.6. Turning to the control side, linear control design methods are formulated in Section 14.7 as convex optimization problems over LMIs. In Section 14.8, multiobjective state feedback is treated. The chapter is concluded in Section 14.9 which summarizes the simulation studies.

Notations: In the sequel, the Euclidean norm is used for vectors. We use W^t and W^{-1} to denote, respectively, the transpose and the inverse of any square matrix W. We use $W < 0$ (≤ 0) to denote a symmetric negative definite (negative semidefinite) matrix W. O_{pq}, I_{pq} denote the $p \times q$ null and identity matrices, respectively. Let $Tr[.]$ denote the trace operator. Matrices, if their dimensions are not explicitly stated, are assumed to be compatible for algebraic operations. In symmetric block matrices or complex matrix expressions, we use the symbol • to represent a term that is induced by symmetry.

14.2 Dynamic modeling

In this chapter, we visualize IP as a system of n rods that are isotropic, homogeneous and conveniently connected. Such arrangement can be classified into two groups:

(a) A nonlinear mechanical system of n IPs on a cart can be obtained by mounting the IP on a stable and moving base. The $n + 1$ outputs are the cart position (m) and pendulum angles (rad), and the system input is the force acting upon the cart.

(b) Employing ball bearings, we obtain a nonlinear mechanical system of n IPs where the n outputs are the pendulum angles (rad), and the system inputs are the input voltages at the motors at each joint.

The purpose of dynamic modeling is to characterize the behavior of the multilink system. In the literature, there are two basic approaches to dynamic modeling:

1. Using Euler–Lagrangian energy relations and
2. Applying Newton's second law and third law on every part of the system.

In this chapter, we focus on the first approach which suits control and systems applications. It has been pointed out that using the Lagrangian approach eventually leads to effective formulation of multilink system dynamics. The mathematical models for different types of IPs [25] are frequently derived via Lagrange equations:

$$\frac{dq}{dt}\frac{\partial L}{\partial \dot{q}_j} - \frac{\partial L}{\partial q_j} + \frac{\partial D(t)}{\partial q_j} = Q(t) \tag{14.1}$$

where

- q_j stands for a particular generalized coordinates,
- T and V represent the system kinetic and potential energy, respectively,
- $L \triangleq T - V$ is the Lagrange function,
- $Q(t)$ is the vector of generalized external forces acting upon the system and
- $D(t)$ (Rayleigh, dissipation function) expresses the viscous (friction) forces.

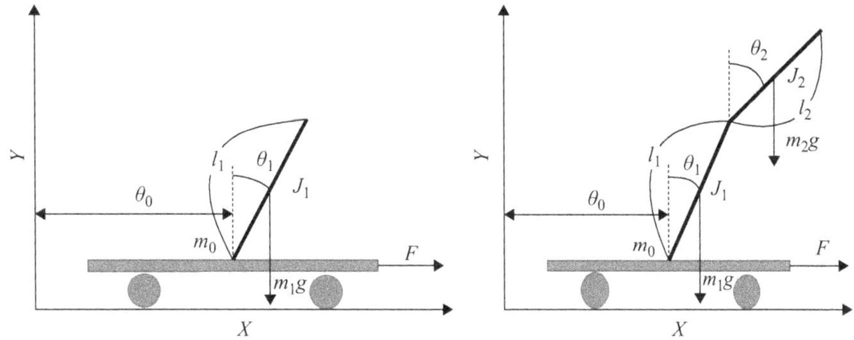

Figure 14.1 Inverted pendulum schematic: (left) single and (right) double

The use of Lagrange mechanics therefore transforms the process of deriving the motion equations of the IPs system into the determination of kinetic, potential and dissipation energies related to the cart and all pendulums. Using well-known formulas, general relations that describe the energetic balances of the cart and *j*th pendulum in an IP systems were derived and can be found in [31].

In what follows, we deal with IPs without friction forces, that is, $D(t) \equiv 0$.

14.3 Pendulums on cart

In this section, we begin looking at the dynamic models that belong to the first group.

14.3.1 Single pendulum on cart

By setting $n = 1$, we obtain the single inverted pendulum (SIP) system (Figure 14.1, left), which encompasses two subsystems:

S1: The cart subsystem:

$$(m_0 + m_1)\ddot{\theta}_0(t) + \delta_0\dot{\theta}_0(t) + \frac{1}{2}m_1\ell_1(\ddot{\theta}_1(t)\cos\theta_1(t) - \dot{\theta}_1^2(t)\sin\theta_1(t)) = F(t) \tag{14.2}$$

S2: The pendulum subsystem:

$$J_1\ddot{\theta}_1(t) + \delta_1\dot{\theta}_1(t) + \frac{1}{2}m_1\ell_1(\ddot{\theta}_0(t)\cos\theta_1(t) - \frac{1}{2}m_1 g\ell_1\sin\theta_1(t)) = 0 \tag{14.3}$$

where m_0 is the cart mass, m_1 is the pendulum mass, ℓ_1 is the pendulum length, δ_0 is the friction coefficient of the cart against the surface, δ_1 is the damping constant of the pendulum, $F(t)$ is the applied force and $J_1 = \frac{1}{3}m_1\ell_1^2$ is the single pendulum's moment of inertia with respect to the pivot.

14.3.2 Double pendulums on cart

The case of $n = 2$ yields the mathematical model of the DIP system (Figure 14.1, left), which consists of three subsystems as follows:

S1: The cart subsystem:

$$(m_0 + m_1 + m_2)\ddot{\theta}_0(t) + \delta_0\dot{\theta}_0(t) + \left(\frac{1}{2}m_1\ell_1 + \frac{1}{2}m_2\ell_1\right)(\ddot{\theta}_1(t)\cos\theta_1(t)$$
$$-\dot{\theta}_1^2(t)\sin\theta_1(t) + \frac{1}{2}m_2\ell_1(\ddot{\theta}_2(t)\cos\theta_2(t) - \dot{\theta}_2^2(t)\sin\theta_2(t)) = F(t) \quad (14.4)$$

S2: The lower pendulum subsystem:

$$(J_1 + m_2\ell_1^2)\ddot{\theta}_1(t) + (\delta_1 + \delta_2)\dot{\theta}_1(t) - \delta_2\dot{\theta}_2(t)$$
$$+\left(\frac{1}{2}m_1\ell_1 + \frac{1}{2}m_2\ell_1\right)\ddot{\theta}_0(t)\cos\theta_1(t)$$
$$+\frac{1}{2}m_2\ell_1\ell_2(\ddot{\theta}_2(t)\cos(\theta_1(t) - \theta_2(t)) + \dot{\theta}_1^2(t)\sin(\theta_1(t) - \theta_2(t)))$$
$$-\frac{1}{2}(m_1 + m_2)g\ell_1\sin(\theta_1(t)) = 0 \quad (14.5)$$

S3: The upper pendulum subsystem:

$$J_2\ddot{\theta}_2(t) + \delta_2(\dot{\theta}_1(t) - \dot{\theta}_2(t)) + \frac{1}{2}m_2\ell_2\ddot{\theta}_0(t)\cos\theta_2(t)$$
$$+\frac{1}{2}m_2\ell_1\ell_2(\ddot{\theta}_1(t)\cos(\theta_1(t) - \theta_2(t)) - \dot{\theta}_1^2(t)\sin(\theta_1(t) - \theta_2(t)))$$
$$-\frac{1}{2}(m_1 + m_2)g\ell_2\sin(\theta_2(t)) = 0 \quad (14.6)$$

14.3.3 State-space representation

It is readily evident that, based on Lagrange's formulation, a set of nonlinear differential equations describing the IP system dynamics can be obtained. With appropriate definitions, a nonlinear state-space model of the form can be produced:

$$\dot{x}(t) = Ax(t) + Bu(t) + g(x(t), u(t), w(t)) \quad (14.7)$$
$$y(t) = Cx(t) \quad (14.8)$$

Seeking to design balancing and/or attitude controllers, a linearized model about a stable pendulum position is used. For sufficiently small angles and angular velocities, an effective quasilinearization procedure gives the following model:

$$\dot{x}(t) = \mathscr{A}x(t) + \mathscr{B}u(t) + \Gamma w(t) \quad (14.9)$$
$$y(t) = \mathscr{C}x(t) \quad (14.10)$$

where $x(t) \in \mathbb{R}^n$, $u(t) \in \mathbb{R}^m$, $y(t) \in \mathbb{R}^p$, $w(t) \in \mathbb{R}^q$ are the state, the control input, the measured output and the external disturbance, respectively.

In what follows, we examine the linearized continuous-time dynamic models of IPs without cart in terms of generalized relative coordinates.

14.4 Inverted pendulum on pivot

In this section, we examine two cases: DIP on pivot and TIP on pivot. In both cases, we present representative values for the purpose of computer simulation.

14.4.1 DIP on pivot

Reference is made to the nomenclature explained in Table 14.1 and the schematic diagram in Figure 14.2(left). Using the approach in [13,32], the following relation for the DIP is derived:

$$\begin{aligned}
&\tilde{M}\begin{bmatrix}\ddot{\theta}_1\\ \ddot{\theta}_2\end{bmatrix}+\tilde{N}\begin{bmatrix}\dot{\theta}_1\\ \dot{\theta}_2\end{bmatrix}+\tilde{P}\begin{bmatrix}\theta_1\\ \theta_2\end{bmatrix}+\tilde{M}\begin{bmatrix}\ddot{\theta}_1\\ \ddot{\theta}_2\end{bmatrix}u_1=\begin{bmatrix}O_{11}\\ O_{11}\end{bmatrix}\\
&\tilde{M}=\begin{bmatrix}I_1+m_1a_1^2+m_2\ell_1^2+I_{p1} & \ell_1 m_2 a_2-I_{p1}\\ \bullet & I_2+m_2a_2^2+I_{p1}\end{bmatrix}\\
&\tilde{N}=\begin{bmatrix}C_1+C_2+c_{p1} & -C_2-c_{p1}\\ \bullet & C_2+c_{p1}\end{bmatrix},\quad \tilde{P}=\begin{bmatrix}m_1a_1+m_2\ell_1 & O_{11}\\ \bullet & m_2a_2g\end{bmatrix},\\
&\tilde{H}=\begin{bmatrix}G_1\\ -G_1\end{bmatrix}
\end{aligned}\tag{14.11}$$

The nomenclature are explained in Figure 14.2(left) and Table 14.1. Defining the state vector $x_d=[\theta_1,\ \ \theta_2-\theta_1,\ \ \dot{\theta}_1,\ \ \dot{\theta}_2-\dot{\theta}_1]^T$ and the output vector $y_d=[\theta_1,\ \ \theta_2-\theta_1,\ \ \dot{\theta}_1]$, we cast system (14.11) into the form (14.9) and (14.10) with

$$\begin{aligned}
\mathscr{A}&=\begin{bmatrix}O_{22} & I_{22}\\ -\tilde{W}\tilde{M}^{-1}\tilde{P}\tilde{W}^{-1} & -\tilde{W}\tilde{M}^{-1}\tilde{N}\tilde{W}^{-1}\end{bmatrix},\quad \mathscr{B}=\begin{bmatrix}O_{21}\\ -\tilde{W}\tilde{M}^{-1}\tilde{H}\end{bmatrix},\\
\tilde{W}&=\begin{bmatrix}1 & 0\\ -1 & 1\end{bmatrix}
\end{aligned}\tag{14.12}$$

Table 14.1 Nomenclature of IP on pivot

u_j	Input voltage from computer to *j*th driving motor
ℓ_j	Length of *j*th arm
a_j	Center of gravity of *j*th arm
m_j	Mass of *j*th arm
I_j	Moment of inertia of *j*th arm around the center of gravity
C_j	Viscous friction coefficient of *j*th hinge
θ_j	Angle of *j*th arm from the vertical line
G_j	Static gain of *j*th motor/gearbox
c_{pj}	Viscous friction coefficient of *j*th motor/gearbox
I_{pj}	Moment of inertia of *j*th motor/gearbox reflected at the output shaft of the gearbox
g	Acceleration of gravity (9.81 m/s^2)

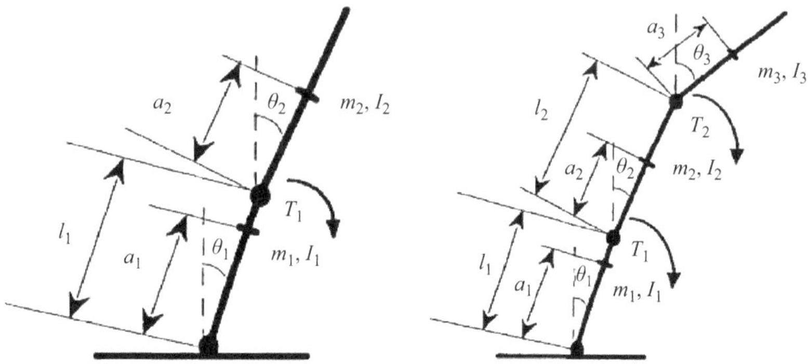

Figure 14.2 Inverted pendulum schematics: (left) double and (right) triple

Table 14.2 Parameter of double inverted pendulum

Link 1	**Link 2**
$\ell_1 = 0.174$ m $m_1 = 0.787$ kg $I_1 = 4.3 \times 10^{-3}$ kg m^2 $a_1 = 0.129$ m $C_1 = 2.69 \times 10^{-2}$ N ms	$\ell_2 = 0.174$ m $m_2 = 0.507$ kg $I_2 = 3.05 \times 10^{-3}$ kg m^2 $a_2 = 0.217$ m $C_2 = 4.04 \times 10^{-2}$ N ms

where the control input u is the input voltage to the motor, and the subscript d stands for double. In terms of the parameter values in Table 14.2, the corresponding model matrices in (14.12) are given by

$$\mathscr{A} = \begin{bmatrix} 0 & 0 & 1 & 0 \\ 0 & 0 & 0 & 1 \\ 31.475 & 3.023 & -0.411 & 96.738 \\ -13.692 & 9.495 & 0.335 & -165.042 \end{bmatrix}, \quad \mathscr{B} = \begin{bmatrix} 0 \\ 0 \\ -24.899 \\ 42.480 \end{bmatrix},$$

$$\mathscr{C} = \begin{bmatrix} 1 & 0 & 0 & 0 \\ 0 & 1 & 0 & 0 \\ 0 & 0 & 1 & 0 \end{bmatrix} \tag{14.13}$$

14.4.2 TIP on pivot

Stabilization of a TIP is developed along similar lines to the double pendulum. A schematic diagram for the triple pendulum is shown in Figure 14.2(right). Two direct current (DC) motor/gearboxes are used to provide torques to the upper and middle pendulum hinges, while the lower hinge is free to rotate. In this case, we only consider attitude control of the middle link, while the upper link is kept in alignment with the middle one. This alignment is achieved by proportional position feedback applied to

Table 14.3 Parameter of triple inverted pendulum

Link 1	**Link 2**	**Link 3**
$\ell_1 = 0.174$ m $m_1 = 0.867$ kg $I_1 = 2.824 \times 10^{-3}$ kg m^2 $a_1 = 0.124$ m $C_1 = 2.69 \times 10^{-2}$ N ms	$\ell_2 = 0.171$ m $m_2 = 0.836$ kg $I_2 = 3.078 \times 10^{-3}$ kg m^2 $a_2 = 0.217$ m $C_2 = 4.04 \times 10^{-2}$ N ms	$\ell_3 = 0.171$ m $m_3 = 0.506$ kg $I_3 = 2.91 \times 10^{-3}$ kg m^2 $a_3 = 0.222$ m $C_2 = 3.65 \times 10^{-2}$ N ms

the upper motor. The control system for the middle motor includes integral action and optimal state feedback.

The nomenclature is explained in Figure 14.2(right) and Table 14.3. The state equation in terms of the relative angles $x_t = [\theta_1,\quad \theta_2 - \theta_1,\quad \theta_3 - \theta_2,\quad \dot{\theta}_1,\quad \dot{\theta}_2 - \dot{\theta}_1,\quad \dot{\theta}_3 - \dot{\theta}_2]^T$ is given by

$$\dot{x}_t(t) = \begin{bmatrix} O_{33} & I_{33} \\ A_{21} & A_{22} \end{bmatrix} + \begin{bmatrix} O_{32} \\ B_2 \end{bmatrix}, \tag{14.14}$$

$$A_{21} = -\hat{W}\hat{M}^{-1}\hat{P}\hat{W}^{-1}, \quad A_{22} = -\hat{W}\hat{M}^{-1}\hat{N}\hat{W}^{-1}, \tag{14.15}$$

$$B_2 = -\hat{W}\hat{M}^{-1}\hat{H} \tag{14.16}$$

where

$$\hat{W} = \begin{bmatrix} 1 & 0 & 0 \\ -1 & 1 & 0 \\ 0 & -1 & 1 \end{bmatrix}, \quad \hat{P} = \begin{bmatrix} M_1 g & 0 & 0 \\ 0 & M_2 g & 0 \\ 0 & 0 & M_3 g \end{bmatrix}, \quad \hat{H} = \begin{bmatrix} G_1 & 0 \\ -G_1 & G_2 \\ 0 & -G_2 \end{bmatrix} \tag{14.17}$$

$$\hat{M} = \begin{bmatrix} J_1 + I_{p1} & \ell_1 M_2 - I_{p1} & \ell_1 M_3 \\ \bullet & J_2 + I_{p1} + I_{p2} & \ell_2 M_3 - I_{p2} \\ \bullet & \bullet & J_3 + I_{p2} \end{bmatrix},$$

$$\hat{N} = \begin{bmatrix} C_1 + C_2 + c_{p1} & -C_2 - c_{p1} & 0 \\ \bullet & C_2 + C_3 + c_{p1} + c_{p2} & -C_3 - c_{p2} \\ \bullet & \bullet & C_3 + c_{p2} \end{bmatrix} \tag{14.18}$$

$$M_1 = m_1 a_1 + (m_2 + m_3)\ell_1, \quad M_2 = m_2 a_2 + m_3 \ell_2, \quad M_3 = m_3 a_3 \tag{14.19}$$

$$J_1 = I_1 + m_1 a_1^2 + (m_2 + m_3)\ell_1^2, \quad J_2 = I_2 + m_2 a_2^2, \quad J_3 = I_3 + m_3 a_3^2 \tag{14.20}$$

Remark 14.1. *It must be observed in both double and triple pendulums that the parameters m_j, ℓ_j are measured directly, whereas a_j, I_j are determined analytically. The arms are swung freely, from the periods and the damping factor of their responses, the parameters are then determined. The remaining parameters I_{pj}, c_{pj}, G_j are calculated from values given in technical data of the motor/gearboxes.*

14.5 Rotational double inverted pendulum

The RDIP takes the classic rotational single pendulum problem to the next level of complexity [1]. The RDIP is composed of a rotary arm that attaches to a servo system which provides a torque to the base arm to control the whole system, a short bottom rod connected to the arm and a top long rod. It is an underactuated (that is, it has fewer inputs that degrees of freedom) and extremely nonlinear unstable system due to the gravitational forces and the coupling arising from the Coriolis and centripetal forces. Since the RDIP presents considerable control-design challenges, it is an attractive tool utilized for developing different control techniques and testing their performances. Related applications include stabilizing the takeoff of a multistage rocket, as well as modeling the human posture system [26,27].

A simple mathematical model has been presented in [45], which assumes that the acceleration of the base arm is able to be manipulated directly and therefore chosen as the system control input. In this dissertation, a more practical assumption is taken under which the torque of the motor to the base arm is the control signal. Moreover, the pivoting friction factors will be taken care for the goal to build a more precise model and simulate the real system we have in the lab.

We will use some additional basic assumptions of the system attributes similar with [45]:

- A1 All the link angles and the angular velocities are accessible at each time step, since we know that we can access these data with the help of the encoders on the links, and high rate of data acquisition is possible for our experiment platform.
- A2 The viscous frictions of the arm and the two pendulums are considered, while the static friction, backlash and plane slackness are ignored.
- A3 The apparatus is light weight and has low inertia resulting in a structure with low stiffness and a tendency to vibrate.
- A4 System dynamics is slow enough to be controlled.

Figure 14.3 shows the basic configurations of the RDIP. The arrows on the arcs show the positive direction for the rotary movement of the links. The straight dash lines denote the origin of the displacement of the link angles. For example, when the horizontal Link 1 is centered, and the vertical Links 2 and 3 are in the upright position, all of the position variables are zero. The state variables and related parameters of the links are listed in Table 14.4.

The Euler–Lagrange (E–L) method, introduced in details in [28], is applied in the derivation of the equations of motion for the RDIP dynamics, since the Newtonian approach of applying Newton's laws of motion is highly complicated in this case. The solutions to the E–L equation for the action of a system are capable of describing the evolution of a physical system according to the Hamilton's principle of stationary action in Lagrangian mechanics. In classical mechanics, it is equivalent to Newton's laws of motion, but it has the advantage that it takes the same form in any system of generalized coordinates, and it is better suited to generalizations.

For the purpose of modeling, the RDIP works as follows. The movement of the arm on the base, Link 1, is constrained to the x–o–z plane and rotating around

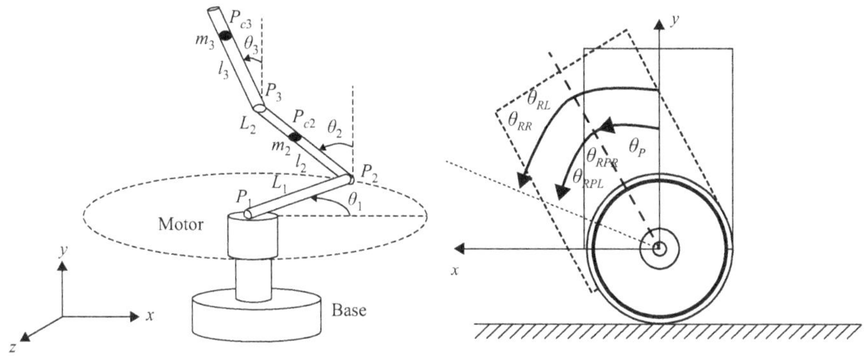

Figure 14.3 Rotational inverted pendulum: (left) schematic, (right) angle of rotation [27]

Table 14.4 Parameters of T-WIP robot

Parameter	Meaning
θ_1	Angle of Link 1 in the horizontal plane
$\dot{\theta}_1$	Velocity of Link 1 in the horizontal plane
$\ddot{\theta}_1$	Acceleration of Link 1
θ_2	Angle of Link 2 in the horizontal plane
$\dot{\theta}_2$	Velocity of Link 1 in the horizontal plane
$\ddot{\theta}_2$	Acceleration of Link 2
θ_3	Angle of Link 3 in the horizontal plane
$\dot{\theta}_3$	Velocity of Link 3 in the horizontal plane
$\ddot{\theta}_3$	Acceleration of Link 3
J_1	Moment of inertia of Link 1 about its pivot
J_2	Moment of inertia of Link 2 about its center of mass P_{c2}
J_3	Moment of inertia of Link 3 about its center of mass P_{c3}
ℓ_1	Distance from the center of rotation of Link 1 to its center of mass
ℓ_2	Distance from the center of rotation of Link 2 to its center of mass
ℓ_3	Distance from the center of rotation of Link 3 to its center of mass
m_j	Mass of link $j, j = 1, 2, 3$
L_j	Length of link $j, j = 1, 2, 3$
b_j	Viscous damping coefficient of the bearing on which Link j rotates $j, j = 1, 2, 3$
g	Gravity with the value g =9:81 m/s^2 toward the center of the earth

the y-axis. The movements of the other two links are constrained to a vertical plane perpendicular to Link 1. Link 1 is driven by a DC motor, which generates a torque to control the system and is described in [29]. Here, we will not discuss the servo system. Therefore, the control input of the RDIP is the torque applied to Link 1. The control objective is to maintain the pendulums Link 2 and 3 in the upright position with Link 1 in the origin position.

Application of (14.1) to the RDIP with $D(t) \equiv 0$, we readily obtain the following parametrized model:

$$\begin{aligned}
&h_1\ddot{\theta}_1(t) + h_2\cos\theta_2(t)\ddot{\theta}_2(t) + h_3\cos\theta_3(t)\ddot{\theta}_3(t) + b_1\dot{\theta}_1(t) - h_2\dot{\theta}_1^2(t)\sin\theta_2(t)\\
&- h_3\dot{\theta}_3^2(t)\sin\theta_3(t) = \tau(t) - h_2\cos\theta_2(t)\ddot{\theta}_1(t) - h_4\ddot{\theta}_2(t) - h_5\cos(\theta_2(t) - \theta_3(t))\ddot{\theta}_3(t)\\
&- b_2\dot{\theta}_2(t) - h_5\sin(\theta_2(t) - \theta_3(t))\dot{\theta}_3^2(t) + h_7\sin\theta_2(t) = 0 - h_3\cos\theta_3(t)\ddot{\theta}_1(t)\\
&- h_5\cos(\theta_2(t) - \theta_3(t))\ddot{\theta}_2(t) - h_6\ddot{\theta}_3(t) - b_3\dot{\theta}_3(t) + h_5\sin(\theta_2(t) - \theta_3(t))\dot{\theta}_2^2(t)\\
&+ h_8\sin\theta_3(t) = 0
\end{aligned} \tag{14.21}$$

where

$$h_1 = J_1 + L_1^2(m_2 + m_3), \quad h_2 = L_1(m_2\ell_2 + m_3L_2), \quad h_3 = L_1m_3\ell_3 \tag{14.22}$$

$$h_4 = J_2 + L_2^2m_3 + m_2\ell_2^2, \quad h_5 = L_2m_3\ell_3, \quad h_6 = J_3 + m_3\ell^2 \tag{14.23}$$

$$h_7 = (m_2\ell_2 + m_3L_2)g, \quad h_8 = m_3\ell_3g \tag{14.24}$$

In terms of the state vector $x = [\theta_1, \theta_2, \theta_3, \dot{\theta}_1, \dot{\theta}_2, \dot{\theta}_3]^T$ and the control input $u = \tau$, we quasilinearize system (14.21) about the origin $x(0) = 0$ and $u = 0$ to obtain the matrices $\mathscr{A}$ and $\mathscr{B}$ in (14.10) of the form:

$$\mathscr{A} = \frac{1}{T}\begin{bmatrix} O_{33} & I_{33} \\ \mathscr{E} & \mathscr{F} \end{bmatrix}, \quad \mathscr{B} = \frac{1}{T}\begin{bmatrix} O_{31} \\ \mathscr{J} \end{bmatrix}, \tag{14.25}$$

$$\mathscr{E} = \begin{bmatrix} 0 & (h_3h_5 - h_2h_6)h_7 & (h_2h_5 - h_3h_4)h_8 \\ 0 & (h_3^2 - h_1h_6)h_7 & (h_1h_5 - h_2h_3)h_8 \\ 0 & (h_1h_5 - h_2h_3)h_7 & (h_2^2 - h_1h_4)h_8 \end{bmatrix}, \quad \mathscr{J} = \begin{bmatrix} h_4h_6 - h_5^2 \\ h_2h_6 - h_3h_5 \\ h_3h_4 - h_2h_5 \end{bmatrix}, \tag{14.26}$$

$$\mathscr{F} = \begin{bmatrix} (h_4h_6 - h_5^2)b_1 & (h_2h_6 - h_3h_5)b_2 & (h_3h_4 - h_2h_5)b_3 \\ (h_2h_6 - h_3h_5)b_1 & (h_1h_6 - h_3^2)b_2 & (h_3h_4 - h_2h_5)b_3 \\ (h_3h_4 - h_2h_5)b_1 & (h_2h_3 - h_1h_5)b_2 & (h_1h_4 - h_2^2)b_3 \end{bmatrix},$$

$$T = h_6h_2^2 - 2h_2h_3h_5 + h_4h_3^2 + h_1h_5^2 - h_1h_4h_6 \tag{14.27}$$

14.6 Inverted pendulum-type assistant robot

Referring to Figure 14.4, the I-PENTAR is constructed to exhibit human motion. It consists of a body with a waist joint, arms designed for safety and a WIP mobile platform. It is capable to perform heavy tasks [36,37]. From a control systems standpoint, I-PENTAR is depicted as a three-dimensional robot; with controls of inclination angle, horizontal position and steering angle to achieve high mobile capability. Generally speaking, the robot dynamics is derived considering the nonholonomic constraint of the two-wheeled mobile robot. As such I-PENTAR is unstable, thus achieving the closed-loop stability is crucial for controller design.

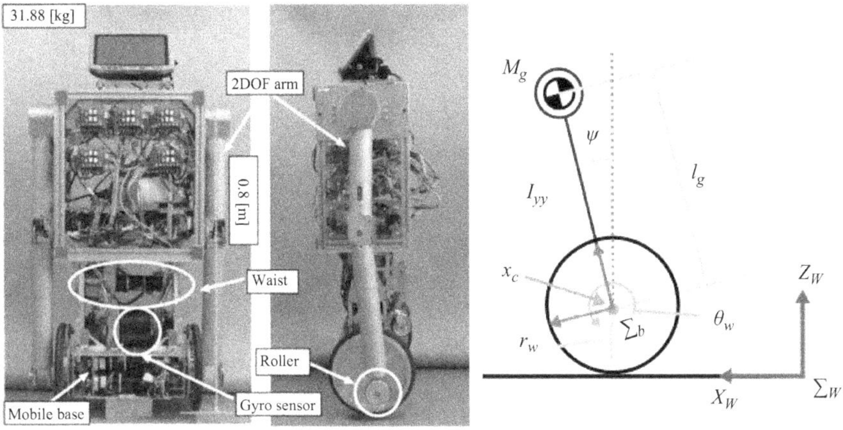

Figure 14.4 (Left) I-PENTAR robot (right) two DOF model

Balancing robot was reported in [33] using a linearized model and in [38] based on nonlinear model. Accommodation of the nonlinear disturbance based on a nonlinear observer was presented in [39]. In [40], a particular robot with movable two wheels was constructed and evaluated. A model-predictive controller was reported in [21] to balance the robot. A fuzzy control approach in [41,42] dealt with parametric uncertainties. A robust controller was proposed by [43] using a model-based sliding mode theory. For motion tracking in [44], a reference model for the yaw and tilt angle was constructed using the LQR optimization. Trajectory tracking control was treated in [47] using uncertain model. A backstepping-based adaptive control is designed in [48] to achieve output tracking for the WIP and similarly an adaptive integral backstepping controller is developed in [**?**] to stabilize the body angle. Design aspects for WIP platforms were examined in [50] to investigate the effect of design choices on balancing the performance.

Figure 14.4 shows a schematic of I-PENTAR. It has two arms with one elbow for each arm, two gyro sensors specified to define the acceleration and the inclination angle relative to the flat surface of the earth and two wheels are for movements in two degree of freedom.

Defining the state variables as

- x_v is the linear position of the vehicle (m),
- φ is the inclination angle of CoG in (rad),
- ψ is the steering angle of the vehicle in (rad),
- $\dot{x}_v$ is the linear velocity of the vehicle (m/s),
- $\dot{\varphi}$ is the inclination angular velocity of CoG in (rad/s),
- $\dot{\psi}$ is the steering angular velocity of the vehicle in (rad/s)

Reference to model (14.10), the matrices $\mathscr{A}$, $\mathscr{B}$, $\mathscr{C}$ are given by

$$\mathscr{A} = \begin{bmatrix} 0 & I_{33} \\ \mathscr{A}_1 & \mathscr{A}_2 \end{bmatrix}, \quad \mathscr{B} = \begin{bmatrix} O_{32} \\ \mathscr{B}_1 \end{bmatrix}, \quad \mathscr{C} = \begin{bmatrix} I_{22} & O_{24} \end{bmatrix} \tag{14.28}$$

$$\mathscr{A}_1 = \begin{bmatrix} 0 & 0 & a_{43} \\ 0 & 0 & 0 \\ 0 & 0 & a_{63} \end{bmatrix}, \quad \mathscr{A}_2 = \begin{bmatrix} a_{44} & a_{45} & a_{46} \\ a_{54} & a_{55} & a_{56} \\ a_{64} & a_{65} & a_{66} \end{bmatrix}, \quad \mathscr{B}_1 = \begin{bmatrix} b_1 & b_2 \\ b_3 & b_4 \\ b_5 & b_6 \end{bmatrix} \tag{14.29}$$

where

$$a_{43} = \frac{M_g^2 \ell_g^2 r_w^2 G}{V_1}, \quad a_{44} = \frac{w_1 c_1}{V_1}, \quad a_{55} = \frac{-b^2 c_1}{V_2} \tag{14.30}$$

$$a_{45} = \frac{w_1 c_1}{V_1}, \quad a_{46} = \frac{-r_w w_1 c_1}{V_1}, \quad a_{54} = \frac{-b c_2}{V_2}, \tag{14.31}$$

$$a_{56} = \frac{b r_w w_1 c_2}{V_2}, \quad a_{64} = \frac{-w_2 c_1}{r_w V_1}, \quad a_{65} = \frac{-b w_2 c_2}{r_w V_1}, \tag{14.32}$$

$$a_{63} = \frac{(M_g r_w^2 + 2m_w r_w^2 + 2(I_{wa} + \gamma I_{ra}))M_g I_g G}{V_1} \tag{14.33}$$

$$b_1 = b_2 = \frac{-r_w w_1}{V_1}, \quad b_3 = -b_4 = \frac{-r_w b}{V_2} \tag{14.34}$$

$$b_5 = b_6 = \frac{w_2}{V_1}, \quad a_{66} = \frac{-w_2 c_1}{V_1}, \tag{14.35}$$

$$V_1 = I_{zz} r_w^2 + 2(I_{wd} + I_{rd}) + 2m_w \mathbb{R}_w^2 b^2 + 2(I_{wa} + \gamma^2 I_{ra}) b^2, \tag{14.36}$$

$$w_1 = M_g \ell_g^2 + I_{yy} + M_g \ell_g r_w \tag{14.37}$$

$$w_2 = M_g \ell_g r_w + M_g r_w^2 + 2m_w r_w^2 + 2(I_{wa} + \gamma^2 I_{ra}) \tag{14.38}$$

The numerical values of the parameters are presented in Table 14.5.

14.7 LMI-based control-design methods

In this section, we develop linear control methods and formulate the design as convex optimization problems over LMIs. The common design objective is to stabilize system (14.10). All the subsequent analytical treatment are inspired by the results of [53]–[56]. The methods established hereafter are categorized into five problems:

- Linear quadratic control (LQR) design, where the objective is to determine a state-feedback gain in the nominal case (without the disturbance input). This is termed as **Problem I**.
- Linear quadratic integral control (LQRI) design, in which, we expand the state model by including the integral of the deviation of the output in the absence of disturbance. This is termed as **Problem II**.

Table 14.5 Numerical data for I-PENTAR

Parameter	Value
M_g	25.26 kg
m_g	125 kg
ℓ_g	0.4005 m
b	0.16 m
r_w	0.1 m
I_g	1.408 kg m^2
I_{wa}	0.01475 kg m^2
I_{wd}	0.0073 kg m^2
I_{ra}	0
I_{rd}	0
γ	1
c_r	0.1 N ms/rad
c_ℓ	0.1 N ms/rad
G	9.81

- Linear quadratic control with disturbance rejection (LQR+) in which we penalize the impact of disturbance by adding an extra term in the cost functional. This is termed as **Problem III**.
- Linear quadratic integral control with disturbance rejection (LQRI+), in which, we expand the state model and add extra cost term. This is termed as **Problem IV**.
- MPC is termed as **Problem V**.

In control system terminology, the problem under consideration is that of determining feedback controller that makes system (14.10) stable over a wide range of operation while achieving a prescribed performance measure. Next, we provide design techniques to achieve this goal.

14.7.1 LQR: proportional gain

With focus on the LQR design, we look at LMI-based formulation. The associated cost function is

$$J = \int_0^\infty [y^T(t)\mathbf{Q}y(t) + u^T(t)\mathbf{R}u(t)]dt \tag{14.39}$$

where $0 < \mathbf{Q}, 0 < \mathbf{R}$ are output error and control weighting matrices. In what follows, we proceed to determine a linear optimal state-feedback control $u = Lx$ that derives system (14.10) with $\Gamma \equiv 0$ while minimizing the quadratic cost (14.39). Assume that a Lyapunov functional $V(x)$ has the properties:

- $V(x) = x^T Mx, \quad M > 0,$
- There exists $\gamma_+ > 0$ such that $x_o^T Mx_o \leq \gamma_+$
- $\dot{V}(x) \leq -[x^T\mathscr{C}^T\mathbf{Q}\mathscr{C}x + u^T\mathbf{R}u]$

Then, the linear system controlled by u is asymptotically stable and $J_\infty \leq V(x_o)$. With $u = Lx$, the inequality of the derivative of the Lyapunov functional becomes

$$x^T[M(\mathscr{A} + \mathscr{B}L) + (\mathscr{A} + \mathscr{B}L)^T M^T]x \leq -x^T[\mathscr{C}^T \mathbf{Q}\mathscr{C} + L^T \mathbf{R} L]x \qquad (14.40)$$

Inequality (14.40) is satisfied for all $x \neq 0$, if there exists L and M such that

$$M(\mathscr{A} + \mathscr{B}L) + (\mathscr{A} + \mathscr{B}L)^T M^T + [C^T \mathbf{Q} C + L^T \mathbf{R} L] \leq 0 \qquad (14.41)$$

Proceeding to minimize the upper bound on the cost $x_o^T M x_o$, the under consideration for given γ_+ can be cast into

$$\min_{\gamma_+, K, L} \ \gamma_+ \ \text{ subject to (14.41)} \qquad (14.42)$$

To convexify the above problem, we first express (14.41) as

$$\Phi = M(\mathscr{A} + \mathscr{B}L) + (\mathscr{A} + \mathscr{B}L)^T M^T$$

$$\Pi = \begin{bmatrix} \Phi & \mathscr{C}^T\mathbf{Q} & L^T\mathbf{R} \\ \bullet & -\mathbf{Q} & 0 \\ \bullet & \bullet & -R \end{bmatrix} \leq 0 \qquad (14.43)$$

Pre- and postmultiply (14.43) by diag$\{Y, I, I\}$ and using $Y = M^{-1}, S = LK^{-1}$, it follows that (14.43) is equivalent to

$$\begin{bmatrix} (\mathscr{A}Y + \mathscr{B}S) + (\mathscr{A}Y + \mathscr{B}S)^T & Y\mathscr{C}^T\mathbf{Q} & YL^T\mathbf{R} \\ \bullet & -\mathbf{Q} & 0 \\ \bullet & \bullet & -\mathbf{R} \end{bmatrix} \leq 0 \qquad (14.44)$$

Additionally, inequality bound of the Lyapunov functional can be expressed as

$$\begin{bmatrix} \gamma_+ & x_o^T \\ \bullet & M^{-1} \end{bmatrix} \geq 0 \Longleftrightarrow \begin{bmatrix} \gamma_+ & x_o^T \\ \bullet & Y \end{bmatrix} \geq 0 \qquad (14.45)$$

The convex minimization problem (14.44) is cast into the form

Problem I

$$\min_{\gamma_+, Y, S} \ \gamma_+ \ \text{ subject to (14.44)} - \text{(14.45)} \qquad (14.46)$$

When a feasible solution of problem (14.46) is attained, we get $L = SY^{-1}$, $M = Y^{-1}$.

14.7.2 LQRI: proportional-integral gain

Next, a modified formulation of the LQR (14.46) is considered with an additional cost-term representing the integral of the deviation of the output from its initial state $z(t) = \int_0^t y(\tau)d\tau$. The modified cost expression becomes

$$\tilde{J} = \int_0^\infty (y(\tau)^T y(\tau) + \rho u(\tau)^T u(\tau) + \sigma z(\tau)^T z(\tau))d\tau \qquad (14.47)$$

This formulation is referred to as the LQRI. Treating $\dot{z} = \mathscr{C}x$ as additional state variable, we define $\eta = \begin{bmatrix} x & z \end{bmatrix}^T$. The augmented system dynamics becomes

$$\dot{\eta} = \begin{bmatrix} \mathscr{A} & 0 \\ \mathscr{C} & 0 \end{bmatrix} \eta + \begin{bmatrix} \mathscr{B} \\ 0 \end{bmatrix} u \quad = \quad \widetilde{\mathscr{A}}\eta + \widetilde{\mathscr{B}}u \tag{14.48}$$

and hence we can rewrite (14.47) as

$$\widetilde{J} = \int_0^\infty (\eta^T(\tau)\tilde{Q}\eta(\tau) + u(\tau)^T\tilde{R}u(\tau))d\tau, \tag{14.49}$$

$$\tilde{Q} = \begin{bmatrix} \mathscr{C}^T\mathscr{C} & 0 \\ 0 & \sigma I \end{bmatrix}, \quad \tilde{R} = \rho I \tag{14.50}$$

To ensure closed-loop system stability, we define the Lyapunov functional, $\widetilde{V}(x)$ having the properties:

- $\widetilde{V}(x) = \eta^T\bar{M}\eta, \quad \bar{M} > 0,$
- There exists $\gamma_+ > 0$ such that $\eta_o^T\bar{M}\eta_o \leq \gamma_+$
- $\dot{\widetilde{V}}(\eta) \leq -[\eta^T\tilde{Q}\eta + u^T\tilde{R}u]$

Then, the linear system controlled by u is asymptotically stable and $J_\infty \leq \tilde{V}(x_o)$. With $u = \widetilde{L}\eta$ and following parallel development to the previous section, we arrive at the convex problem

Problem II

$$\min_{\gamma_+,Y,S} \gamma_+ \quad \text{subject to}$$

$$\begin{bmatrix} (\widetilde{\mathscr{A}}\bar{Y} + \widetilde{\mathscr{B}}\bar{S}) + (\widetilde{\mathscr{A}}\bar{Y} + \widetilde{\mathscr{B}}\bar{S})^T & \bar{Y}\tilde{Q} & \bar{Y}\widetilde{L}^T\tilde{R} \\ \bullet & -\tilde{Q} & 0 \\ \bullet & \bullet & -\tilde{R} \end{bmatrix} \leq 0, \quad \begin{bmatrix} \gamma_+ & \eta_o^T \\ \bullet & \bar{Y} \end{bmatrix} \geq 0 \tag{14.51}$$

When a feasible solution of problem (14.51) is attained, we get the desired gains $\widetilde{L} = \bar{S}\,\bar{Y}^{-1}, \bar{M} = \bar{Y}^{-1}$.

14.7.3 LQR+: proportional gain with disturbance rejection

Considering system (14.10) and to penalize the effect of the disturbance, we include an extra term to the cost (14.39) to yield

$$J = \int_0^\infty [y(\tau)^T y(\tau) + \alpha u(\tau)^T u(\tau) - \lambda^2 d(\tau)^T d(\tau)]dt \tag{14.52}$$

This new formulation will be referred to as LQR+. Letting $\mu(\tau) = \begin{bmatrix} u(\tau) & d(\tau) \end{bmatrix}$, we get

$$J = \int_0^\infty [y(t)^T\mathbf{Q}y(t) + \mu(t)^T\mathbf{R}_+\mu(t)]dt \tag{14.53}$$

$\mathbf{R}_+ = \text{diag}[\alpha \; -\lambda^2]$ is a control weighting matrix. To ensure closed-loop system stability, we define the Lyapunov functional, $V_\mu(x)$ having the properties:

- $V_\mu(x) = x^T P x, \; V_\mu(x) > 0,$
- There exists $\gamma > 0$ such that $x_o^T P x_o \leq \gamma_+$
- $\dot{V}_\mu(x) \leq -[x^T \mathbf{Q} x + \mu^T \mathbf{R}_+ \mu]$

Then, the linear system controlled by u is asymptotically stable and $J_\infty \leq V_\mu(x_o)$. With $\mu = Fx$, the derivative of the Lyapunov functional satisfies

$$x^T[P(\mathscr{A} + \widehat{G}F) + (\mathscr{A} + \widehat{G}F)^T P^T]x \; \leq -x^T[\mathbf{Q} + F^T \mathbf{R}_+ F]x \tag{14.54}$$

which holds if there exist F and P such that

$$P(\mathscr{A} + \widehat{G}F) + (\mathscr{A} + \widehat{G}F)^T P^T + [\mathbf{Q} + F^T \mathbf{R}_+ F] \; \leq \; 0 \tag{14.55}$$

where $\widehat{G} = [\mathscr{B} \;\; \Gamma]$, the minimization problem is expressed as

$$\min_{\gamma, P, F} \; \gamma_+ \quad \text{subject to (14.55)} \tag{14.56}$$

Proceeding like the foregoing sections, problem (14.56) can cast into

Problem III

$$\min_{\gamma_+, Z, V} \; \gamma_+ \quad \text{subject to}$$

$$\begin{bmatrix} (\mathscr{A}Z + \mathscr{B}V) + (\mathscr{A}Z + \mathscr{B}V)^T & ZC^T Q_\mu & ZF^T R_\mu \\ \bullet & -\mathbf{Q} & 0 \\ \bullet & \bullet & -\mathbf{R}_+ \end{bmatrix} \leq 0, \quad \begin{bmatrix} \gamma_+ & \eta_o^T \\ \bullet & Z \end{bmatrix} \geq 0 \tag{14.57}$$

when a feasible solution is attained to (14.57), we get $F = VZ^{-1}, P = Z^{-1}$.

14.7.4 LQRI+: proportional-integral gain with disturbance rejection

Finally, we modify cost (14.52) by incorporating the integral of the output, $z(t) = \int_0^t y(\tau) d\tau$ to arrive at the formulation:

$$J = \int_0^\infty [y(\tau)^T y(\tau) + \rho u(\tau)^T u(\tau) + \sigma z(\tau)^T z(\tau) - \lambda^2 \Gamma(\tau)^T \Gamma(\tau)] d\tau \tag{14.58}$$

This new formulation will be referred to as LQRI+. Introducing $\mu(\tau) = \begin{bmatrix} u(\tau) & d(\tau) \end{bmatrix}$ and $\chi(\tau) = \begin{bmatrix} x(\tau) & z(\tau) \end{bmatrix}$. The augmented system dynamics becomes

$$\begin{aligned} \dot{\chi} &= \begin{bmatrix} \mathscr{A} & 0 \\ \mathscr{C} & 0 \end{bmatrix} \chi + \begin{bmatrix} \mathscr{B} \\ 0 \end{bmatrix} u + \begin{bmatrix} \Gamma \\ 0 \end{bmatrix} d \\ &= \widetilde{\mathscr{A}} \chi + \widetilde{\mathscr{B}} u + \widetilde{\Gamma} d \end{aligned} \tag{14.59}$$

and hence we can rewrite (14.58) as

$$J = \int_0^{\infty} (\chi^T(\tau)\tilde{Q}_\mu \chi(\tau) + \mu(\tau)^T \tilde{R}_\mu \mu(\tau))d\tau \tag{14.60}$$

$$\tilde{Q}_\mu = \begin{bmatrix} C^T C & 0 \\ 0 & \gamma I \end{bmatrix}, \quad \tilde{R}_\mu = \begin{bmatrix} \rho I & 0 \\ 0 & -\lambda^2 I \end{bmatrix} \tag{14.61}$$

To guarantee the closed-loop system stability, we define the Lyapunov functional, $\tilde{V}_\mu(x)$ with the properties:

- $\tilde{V}_\mu(\chi) = \chi^T \widetilde{P} \chi, \tilde{V}_\mu(\chi) > 0,$
- There exists $\gamma > 0$ such that $\chi_o^T \widetilde{P} \chi_o \leq \gamma_+$
- $\dot{\tilde{V}}_\mu(\chi) \leq -[\chi^T \tilde{Q}_\mu \chi + \mu^T \tilde{R}_\mu \mu]$

Then, the linear system controlled by μ is asymptotically stable and $J_\infty \leq \tilde{V}_\mu(x_o)$. With $\mu = \widetilde{F}\chi$, the derivative of the Lyapunov functional $\forall \chi \neq 0$ satisfies

$$\widetilde{P}(\widetilde{\mathscr{A}} + \widetilde{G}\widetilde{F}) + (\widetilde{\mathscr{A}} + \widetilde{G}\widetilde{F})^T \widetilde{P}^T + [\widetilde{Q}_\mu + \widetilde{F}^T \widetilde{R}_\mu \widetilde{F}] \leq 0 \tag{14.62}$$

where $\widetilde{G} = [\widetilde{\mathscr{B}} \quad \widetilde{\Gamma}]$. The minimization problem has the form

$$\min_{\gamma, \widetilde{P}, \widetilde{F}} \ \gamma_+ \quad \text{subject to (14.62)} \tag{14.63}$$

Convex analysis converts problem (14.63) to

Problem IV

$$\min_{\gamma_+, \widetilde{Z}, \widetilde{V}} \ \gamma_+ \quad \text{subject to}$$

$$\begin{bmatrix} (\widetilde{\mathscr{A}}\widetilde{Z} + \widetilde{\mathscr{B}}\widetilde{V}) + (\widetilde{\mathscr{A}}\widetilde{Z} + \widetilde{G}\widetilde{V})^T & \widetilde{Z}\widetilde{Q}_\mu & \widetilde{Z}\widetilde{F}^T \widetilde{R}_\mu \\ \bullet & -\widetilde{Q}_\mu & 0 \\ \bullet & \bullet & -\widetilde{R}_\mu \end{bmatrix} \leq 0$$

$$\begin{bmatrix} \gamma_+ & \eta_o^T \\ \bullet & \widetilde{Z} \end{bmatrix} \geq 0 \tag{14.64}$$

a feasible solution of which gives $\widetilde{F} = \widetilde{V}\tilde{Z}^{-1}, \widetilde{P} = \widetilde{Z}^{-1}$.

14.7.5 Model-predictive control

The model-predictive controller performs an optimization of the discrete-time system:

$$\begin{aligned} x(k+1) &= \mathscr{A}_c x(k) + \mathscr{B}_c u(k) \\ y(k) &= \mathscr{C}_c x(k) \end{aligned} \tag{14.65}$$

using the cost function

$$J = x^T(k+n_p)Sx(k+n_p) + \sum_{i=0}^{n_p-1} x(k+i)^T Qx(k+i) + u(k+i)^T Ru(k+i) \tag{14.66}$$

where $x_k \in \mathbb{R}^n$, $u_k \in \mathbb{R}^m$, $y_k \in \mathbb{R}^p$, $\mathscr{A}_c \in \mathbb{R}^{n\times n}$, $\mathscr{B}_c \in \mathbb{R}^{n\times m}$ and $\mathscr{C}_c \in \Re^{q\times n}$. Note that system (14.65) is an appropriate discretized model of (14.10) using zero-order hold (ZOH) and prescribed sampling period T_s. It is desired to synthesize the controller:

$$x_c(k+1) = A_c x(k) + B_c u(k) \tag{14.67}$$

$$y_c(k) = C_c x_c(k) \tag{14.68}$$

that stabilizes the closed-loop system. The closed-loop system can be represented as

$$\zeta(k+1) = \bar{A}\zeta(k) + \bar{B}u(k) \tag{14.69}$$

$$u(k) = K\zeta(k), \quad y(k) = \bar{C}\zeta(k) \tag{14.70}$$

where

$$\zeta = \begin{bmatrix} x(k) \\ x_c(k) \end{bmatrix}, \quad \bar{A} = \begin{bmatrix} \mathscr{A}_c & 0 \\ B_c & A_c \end{bmatrix}, \quad \bar{B} = \begin{bmatrix} \mathscr{B}_c \\ 0 \end{bmatrix} \tag{14.71}$$

$$\bar{C} = \begin{bmatrix} \mathscr{C}_c & 0 \end{bmatrix}, \quad \bar{K} = \begin{bmatrix} 0 & C_c \end{bmatrix} \tag{14.72}$$

We now consider infinite-time horizon formulation of (14.66):

$$J = \sum_{i=0}^{\infty} \zeta(k+i)^T \bar{Q}\zeta(k+i) + u(k+i)^T \bar{R}u(k+i) \tag{14.73}$$

$$\bar{Q} = \begin{bmatrix} Q & 0 \\ 0 & 0 \end{bmatrix} \; Q \geq 0, \quad R > 0 \tag{14.74}$$

Again, we define the quadratic Lyapunov function, $V(x)$ as

$$V(x) = \zeta^T P_+ \zeta, P_+ > 0 \tag{14.75}$$

$$V(x(k+i+1)) - V(x(k+i)) < -\zeta(k+i)^T \bar{Q}\zeta(k+i) - u(k+i)^T Ru(k+i) \tag{14.76}$$

The problem under consideration can be cast into the LMI form:

Problem V

$$\max \gamma_+^{-1} \quad \text{subject to} \begin{bmatrix} Y & I \\ I & X \end{bmatrix} - \gamma^{-1}\phi > 0 \tag{14.77}$$

$$\begin{bmatrix} \gamma^{-1}u_{\max}^2 I & 0 & L \\ \bullet & Y & I \\ \bullet & \bullet & X \end{bmatrix} > 0, \quad \begin{bmatrix} Y & I & \mathscr{A}_c^T \mathscr{C}_c^T \\ \bullet & X & (\mathscr{A}_c X + \mathscr{A}_c \mathscr{C}_c L)^T \\ \bullet & \bullet & \gamma^{-1} y_{\max}^2 I \end{bmatrix} > 0 \tag{14.78}$$

$$\begin{bmatrix} Y & I & \Gamma & Z & 0 & 0 \\ \bullet & X & \mathscr{A}_c & \Delta & 0 & 0 \\ \bullet & \bullet & \Pi & \Omega & 0 & 0 \\ \bullet & \bullet & \bullet & X & L^T & XQ^{1/2} \\ \bullet & \bullet & \bullet & \bullet & R^{-1} & 0 \\ \bullet & \bullet & \bullet & \bullet & \bullet & 0 \end{bmatrix} > 0 \tag{14.79}$$

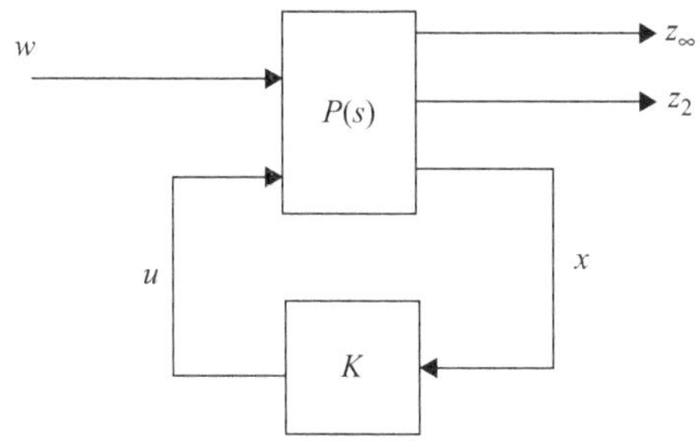

Figure 14.5 Multiobjective state feedback

where $\gamma = Y\mathscr{A}_c + F\mathscr{C}_c$, $\Delta = \mathscr{A}_c X + \mathscr{B}_c L$, $\Pi = Y - Q$, $\Omega = I - QX$. When a feasible solution is obtained, then

$$\begin{aligned} V &= (I - YX)(U^T)^{-1}, \quad C_c = L(U^T)^{-1}, \quad B_c = V^{-1}F, \\ A_c &= V^{-1}Z(U^T)^{-1} \end{aligned} \tag{14.80}$$

14.8 Multiobjective state feedback

Reference is made to Figure 14.5 where $P(s)$, K designate the linearized plant and the state-feedback gain, respectively. Denoting by $\mathbb{T}_\infty(s)$ and $\mathbb{T}_2(s)$ the closed-loop transfer functions from $w \to z_\infty$ and from $w \to z_2$, respectively, our goal is to design a state-feedback control law that meets one of the following objectives:

- $\mathscr{H}_2$ control design, in which, it is required to maintain the $\mathscr{H}_2$-norm of $\mathbb{T}_2(s)$ below some prescribed value $\nu > 0$,
- $\mathscr{H}_\infty$ control design, it is required to maintain the $\mathscr{H}_\infty$-norm of $\mathbb{T}_\infty(s)$ below some prescribed value $\gamma > 0$,
- Mixed control design, in which, it is required to minimize an $\mathscr{H}_2/\mathscr{H}_\infty$ trade-off criterion of the form $\alpha||\mathbb{T}_\infty(s)||_\infty^2 + \beta||\mathbb{T}_2(s)||_2^2$.

In the sequel, we conveniently imbed the linearized model (14.10) into the form:

$$\dot{x}(t) = \mathscr{A}x(t) + \mathscr{B}u(t) + \Gamma w(t) \tag{14.81}$$

$$z_\infty(t) = \mathscr{G}x(t) + \mathscr{S}u(t) + \Phi w(t) \tag{14.82}$$

$$z_2(t) = \mathscr{C}x(t) + \mathscr{D}u(t) + \Psi w(t) \tag{14.83}$$

with $z_\infty(t) \in \mathbb{R}^q$. $z_2(t) \in \mathbb{R}^q$ as the performance outputs. We focus on the stabilization problem under state feedback $u = \mathbb{K}x$ where the closed-loop system is given by

$$\dot{x}_s(t) = \mathbb{A}_s x_s(t) + \Gamma w(t) \tag{14.84}$$

$$z_\infty(t) = \mathbb{G}_s x(t) + \Phi w(t) \tag{14.85}$$

$$z_2(t) = \mathbb{C}sx(t) + \Psi w(t) \tag{14.86}$$

where

$$\mathbb{A}_s = \mathscr{A} + \mathscr{B}\mathbb{K}, \quad \mathbb{G}_s = \mathscr{G} + \mathscr{S}\mathbb{K}, \tag{14.87}$$

$$\mathbb{C}_s = \mathscr{C} + \mathscr{D}\mathbb{K} \tag{14.88}$$

Treated separately, our three design objectives have the following LMI formulations.

14.8.1 $\mathscr{H}_2$ performance

Here, the closed-loop $\mathscr{H}_2$-norm of $\mathbb{T}_2(s)$ does not exceed some prescribed value $\nu > 0$, if there exist two symmetric matrices $0 < \mathbb{X}_2$ and $\mathbb{Q}$ such that

$$\begin{aligned} &\text{Trace}[\mathbb{Q}] \; < \; \nu^2 \\ &\begin{bmatrix} \mathbb{A}_s\mathbb{X}_2 + \mathbb{X}_2\mathbb{A}_s^t & \Gamma \\ \bullet & -I \end{bmatrix} \; < \; 0, \\ &\begin{bmatrix} \mathbb{Q} & \mathbb{C}_s \\ \bullet & -\mathbb{X}_2 \end{bmatrix} \; < \; 0, \end{aligned} \tag{14.89}$$

Denoting the optimal solution by $(\mathbb{X}^*, \mathbb{Q}^*, \nu^*)$, the corresponding state-feedback gain is given by $\mathbb{K}^* = \mathbb{Q}^*(\mathbb{X}^*)^{-1}$.

14.8.2 $\mathscr{H}_\infty$ performance

Here, we consider minimizing the closed-loop RMS norm from $w \to z_\infty$ X_2 and $\mathbb{Q}$ does not exceed γ, if and only if there exists a symmetric matrix $0 < \mathbb{X}_\infty$ such that

$$\begin{bmatrix} \mathbb{A}_s\mathbb{X}_\infty + \mathbb{X}_\infty\mathbb{A}_s^t & \Gamma & \mathbb{X}_\infty\mathbb{G}_s^t \\ \bullet & -I & \Phi \\ \bullet & \bullet & -\gamma^2 I \end{bmatrix} \; < \; 0 \tag{14.90}$$

Denoting the optimal solution by $(\mathbb{X}^*, \mathbb{Y}^*, \gamma^*)$, the corresponding state-feedback gain is given by $\mathbb{K}^* = \mathbb{Y}^*(\mathbb{X}^*)^{-1}$.

14.8.3 Mixed $\mathscr{H}_2/\mathscr{H}_\infty$ performance

Finally, we consider minimizing a trade-off criterion of the form $\alpha||\mathbb{T}_\infty(s)||_\infty^2 + \beta||\mathbb{T}_2(s)||_2^2$. This leads to the convex optimization problem

$$\begin{aligned} &(\alpha\gamma^2 + \beta \; \text{Trace}[\mathbb{Q}])_{\mathbb{X},\mathbb{Y},\mathbb{Q},\gamma^2} \\ &\begin{bmatrix} \mathbb{A}_s\mathbb{X}_2 + \mathbb{X}_2\mathbb{A}_s^t & \Gamma \\ \bullet & -I \end{bmatrix} < 0, \\ &\begin{bmatrix} \mathbb{Q} & \mathbb{C}_s \\ \bullet & -\mathbb{X}_2 \end{bmatrix} < 0, \\ &\begin{bmatrix} \mathbb{A}_s\mathbb{X}_\infty + \mathbb{X}_\infty\mathbb{A}_s^t & \Gamma & \mathbb{X}_\infty\mathbb{G}_s^t \\ \bullet & -I & \Phi \\ \bullet & \bullet & -\gamma^2 I \end{bmatrix} < 0 \\ &\text{Trace}[\mathbb{Q}] < \nu_0^2, \quad \gamma^2 < \gamma_0^2 \end{aligned} \tag{14.91}$$

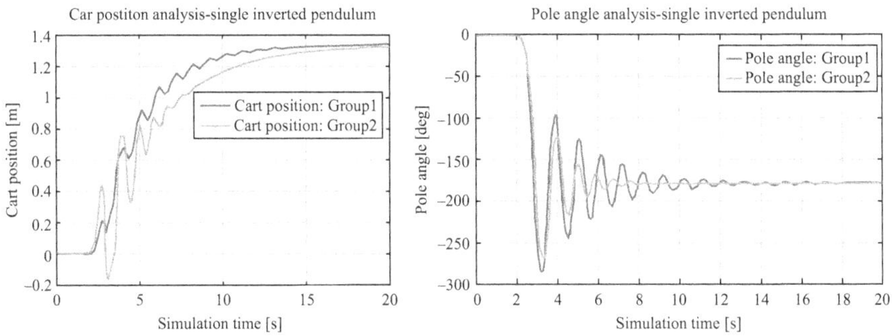

Figure 14.6 Impulse response of SIP on a cart: (left) cart position and (right) pendulum angle

Denoting the optimal solution by $(\mathbb{X}^*, \mathbb{Y}^*, \mathbb{Q}^*, \gamma^*)$, the corresponding state-feedback gain is given by $\mathbb{K}^* = \mathbb{Y}^*(\mathbb{X}^*)^{-1}$, and this gain guarantees the worst case performances:

$$||T_\infty||_\infty \le \gamma^*, \quad ||T_2||_2 \le \sqrt{\text{Trace}[\mathbb{Q}^*]} \tag{14.92}$$

Note that $\mathbb{K}^*$ does not yield the globally optimal trade-off in general due to the conservatism assuming a single Lyapunov matrix [54].

14.9 Simulation results

In this section, we aim to provide representative and meaningful simulation results of applying the different feedback control algorithms to different pendulum configurations. In computer implementation, we made use of [52].

14.9.1 SIP–DIP open-loop response

The analysis of the open-loop dynamical behavior of both the SIP and DIP systems is presented. The dynamics of SIP system is analyzed for two groups of parameters at once:

[Group I:] $m_0 = 0.3$ kg, $m_1 = 0.275$ kg, $\ell_1 = 0.5$ m, $\delta_0 = 0.3$ kg s^{-1}, $\delta_1 = 0.01148$ kg m^2 s^{-1}

[Group II:] $m_0 = 0.1$ kg, $m_1 = 1.0$ kg, $\ell_1 = 0.8$ m, $\delta_0 = 0.3$ kg s^{-1}, $\delta_1 = 0.1$ kg m^2 s^{-1}

For the DIP system, the parameters are

$m_0 = 0.3$ kg, $m_1 = 0.275$ kg, $m_2 = 0.275$ kg, $\ell_1 = 0.5$ m, $\ell_2 = 0.5$ m, $\delta_0 = 0.3$ kg s^{-1}, $\delta_1 = 0.1$ kg m^2 s^{-1}, $\delta_2 = 0.1$ kg m^2 s^{-1}

The impulse response of both SIP and DIP systems are displayed in Figures 14.6 and 14.7.

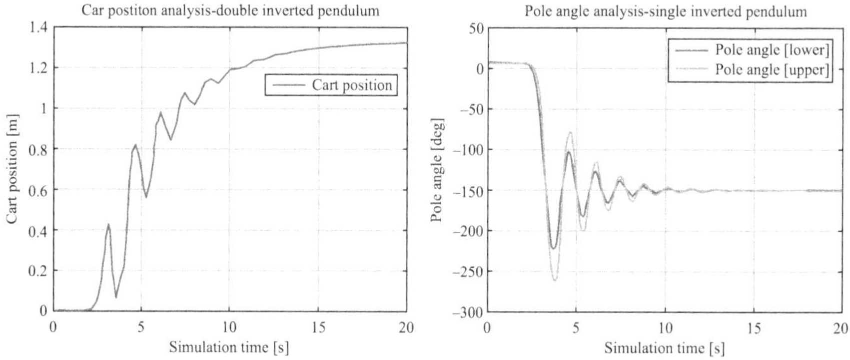

Figure 14.7 Impulse response of DIP on a cart: (left) cart position and (right) upper and lower pendulum angles

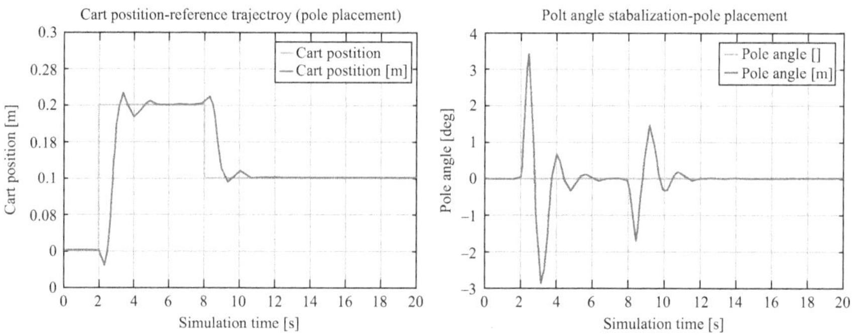

Figure 14.8 Tracking response of SIP on a cart: (left) cart position and (right) pendulum angle

14.9.2 SIP LQR response

Figures 14.8 and 14.9 document the dynamic behavior for both the cart position and the pendulum angle of the SIP system under control where the objective is to maintain the desired cart position while keeping the pendulum upright without disturbance input.

Remark 14.2. *We observe from Figures 14.6 and 14.7 that each pendulum of the system passes through oscillatory transient state until the system reaches the stable equilibrium point in which all pendulums are pointing downward. The backward impact of the pendulum/pendulums on the cart, which increases with the weight of the load, is also visible. The simulation results confirm the generally known empirical observations of pendulum behavior. The simulation results in Figures 14.8 and 14.9 reveal that both tracking control and LQR algorithms perform reasonably well. The ability to control the SIP system with respect to the all requirements was demonstrated*

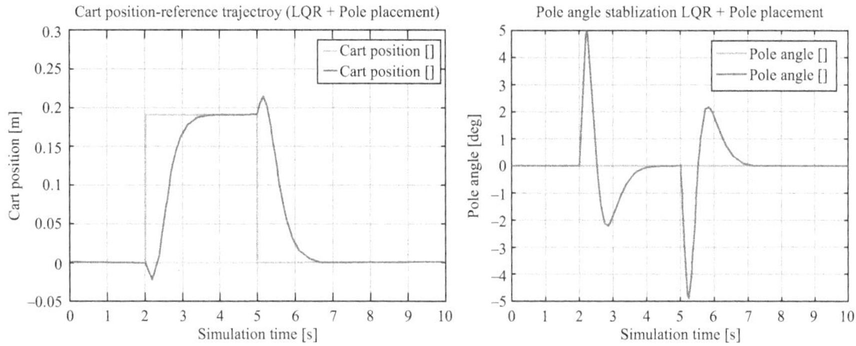

Figure 14.9 LQR response of SIP on a cart: (left) cart position and (right) pendulum angle

for both methods of feedback gain design, although LQR control produces slightly better results.

14.9.3 DIP on cart

Considering Figure 14.1 and using the following data $m_0 = 1.5\,\text{kg}$, $m_1 = 0.5\,\text{kg}$, $m_2 = 0.75\,\text{g}$, $\ell_1 = 0.5\,\text{m}$, $\ell_2 = 0.75\,\text{m}$, the state-space matrices are given by

$$\mathscr{A} = \begin{bmatrix} O_{33} & I_{33} \\ \mathscr{E} & O_{33} \end{bmatrix}, \quad \mathscr{B} = \begin{bmatrix} O_{31} \\ \mathscr{F} \end{bmatrix}, \quad \mathscr{E} = \begin{bmatrix} 0 & -7.492 & 0.799 \\ 0 & 74.927 & -33.715 \\ 0 & -59.937 & 52.121 \end{bmatrix},$$

$$\mathscr{F} = \begin{bmatrix} -0.607 \\ 1.498 \\ -0.284 \end{bmatrix} \tag{14.93}$$

We perform a MATLAB simulation based on the result of **Problem I** with $\mathbf{Q} = \text{diag}\,[5,\ 50,\ 50,\ 20,\ 700,\ 700]$, $\mathbf{R} = 1$. The ensuing results for a relatively small deflections from the equilibrium are plotted in Figure 14.10. For larger deflections toward the same direction, the state trajectories are depicted in Figure 14.11.

Turning to the application of MPC, using ZOH with sampling period $T_s = 0.002\,\text{s}$, we obtain the discrete-time model of the type (14.65). The simulation results based on the solution of **Problem V** are displayed in Figures 14.12 and 14.13 for the cases of small and large deflections, respectively.

Remark 14.3. *Looking at Figures 14.10 and 14.11, it can be readily seen that the system can be balanced, although the cart is stabilized far from the point* 0, *to which it returns to with a very low pace. From Figures 14.12 and 14.13, we observe that the lower pendulum has for a time period a large angular velocity.*

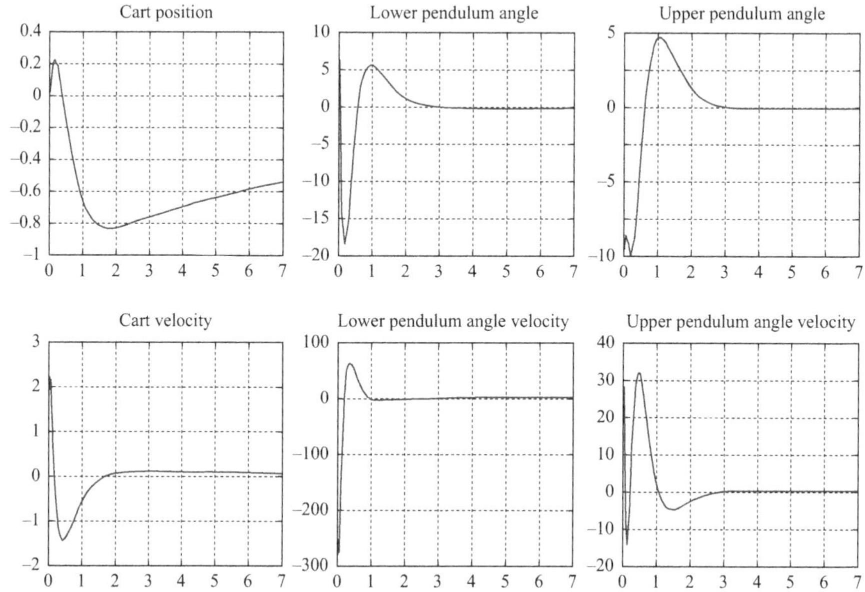

Figure 14.10 Simulation of DIP on cart based on LQR solution with small deflection

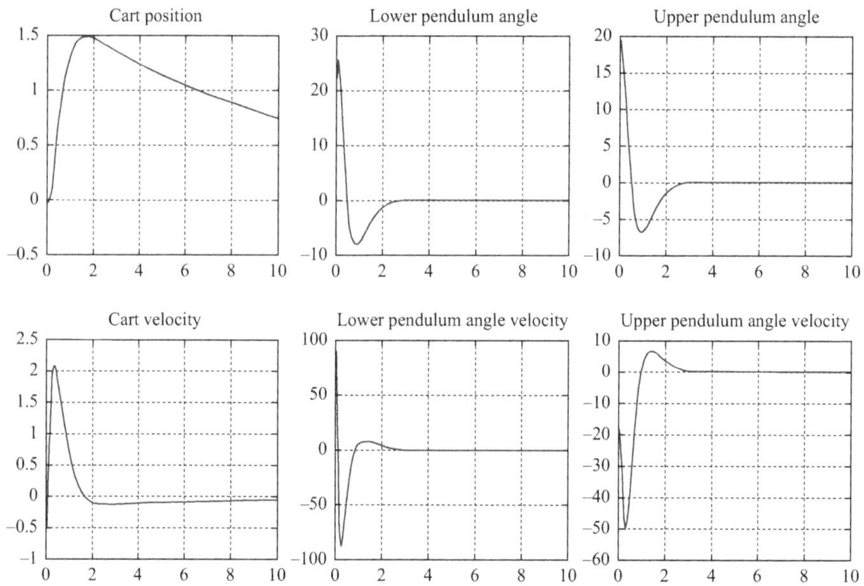

Figure 14.11 Simulation of DIP on cart based on LQR solution with large deflection

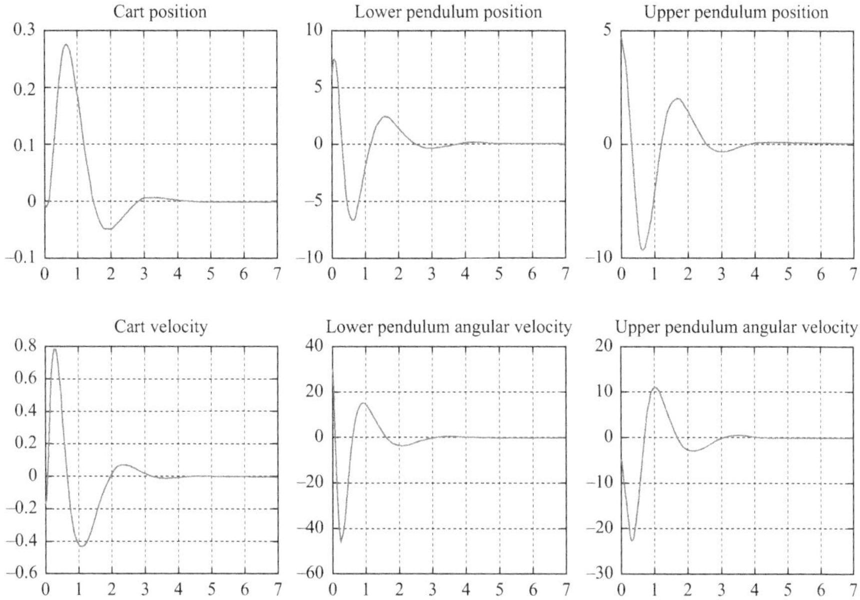

Figure 14.12 Simulation of DIP on cart based on MPC solution with small deflection

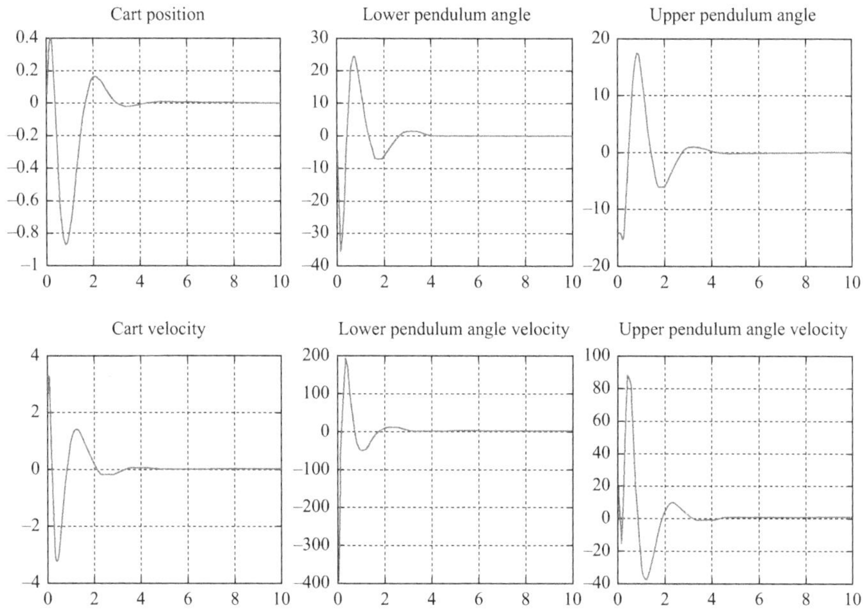

Figure 14.13 Simulation of DIP on cart based on MPC solution with large deflection

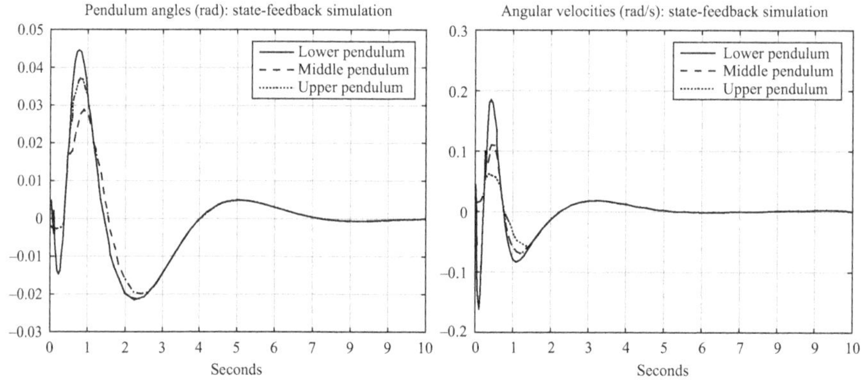

Figure 14.14 Triple IP state trajectories: (left) displacements and (right) velocities

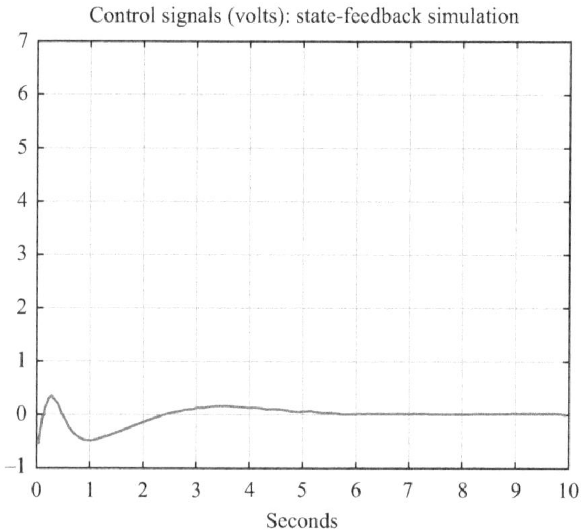

Figure 14.15 Triple IP control signal

14.9.4 Triple IP

The pendulum links are made of 50-mm-diameter plastic tube which is relatively rigid, cheap, easy to cut and weighs 0.343 kg/m. At both ends of each link, aluminum components (3 mm in thickness) are attached to provide the structures for mounting sensors and actuators. The lower hinge consists of a steel shaft mounted on ball bearings. At one end of the shaft, a potentiometer is mounted to measure the angle of the lower link, while at the other end, a small DC tacho is used to measure angular velocity. The hinges for the middle and upper links are split into two independent (suitably aligned) sections. One section is a short steel shaft mounted on ball bearings and attached to a potentiometer which measures the relative angle between adjacent links. The second section is the output shaft of a DC motor/gearbox. Planetary gearboxes with a rated continuous load capacity of 4.5 Nm were chosen.

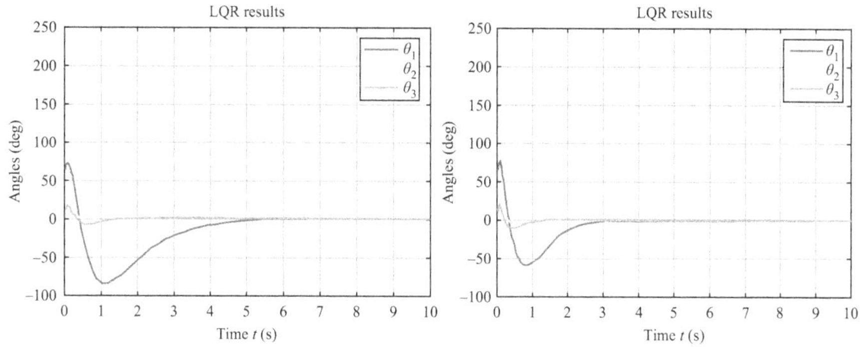

Figure 14.16 Simulation of RDIP based on LQR designs: initial (left) and tuned (right)

In Figures 14.14 and 14.15, the ensuing simulation results of TIP displacements, velocity and the corresponding feedback control based on **Problem I** are plotted.

14.9.5 RDIP simulation

The RDIP experiment platform used for simulation consists of a horizontal base arm (denoted as Link 1) driven by a servo motor and two vertical pendulums (denoted as Links 2 and 3) that move freely in the plane perpendicular to Link 1, as shown in Figure 14.3. Since we will focus on the stabilization of the pendulums, it is convenient to set the coordinate system as in Figure 14.3. In this chapter, the mathematical model of the RDIP will be developed by the use of the E–L function. A simple mathematical model was presented in [45], which assumes that the acceleration of the base arm is able to be manipulated directly and therefore chosen as the system control input. The RDIP simulation results using LQR and its tuned version are depicted in Figures 14.16 and 14.17.

14.9.6 I-PENTAR simulation

For the purpose of simulation and with reference to Figure 14.4, I-PENTAR adopts convenient set of a double motor method for mobile actuation. The robot body composed of two parts (lower and upper), a waist joint and a rate gyro sensor. In addition, a wheel-type roller adheres to the tip of the arm. In Figure 14.18, the internal signal flow of I-PENTAR is depicted. The control and sensor signals between each part and CPU board are interfaced via a field-programmable gate array (FPGA) board in 1-ms sampling time. The two angular velocity values, analog and digital, from the rate gyro sensor are merged and integrated to be used for an inclination angle of the lower body.

In order to evaluate using the stabilization methods of LQR, LQRI, $\mathscr{H}_2$ and $\mathscr{H}_\infty$, balancing experiments, which are considered one of the fundamental mobile motions of a WIP mobile robot, are conducted with I-PENTAR. In the experiments, I-PENTAR takes an upright-standing posture stretching its waist and extending its arms toward the ground. The position of CoG of the whole body, which is used for

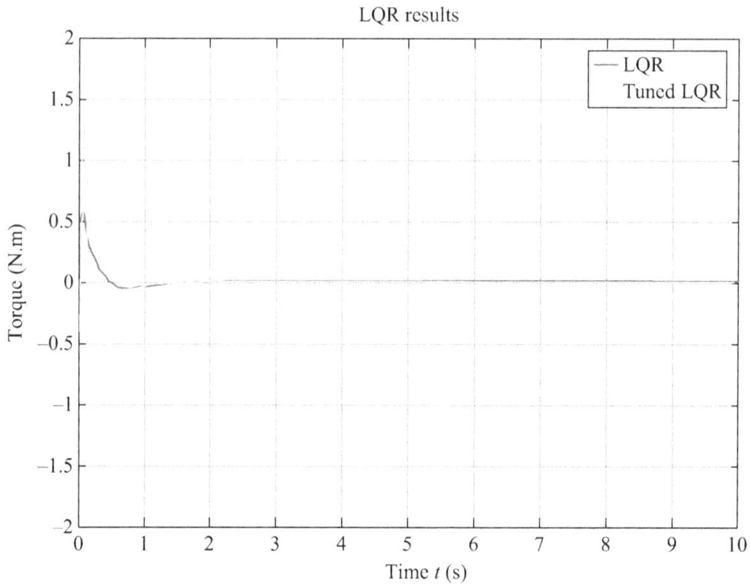

Figure 14.17 Control signals for RDIP generated by different design methods

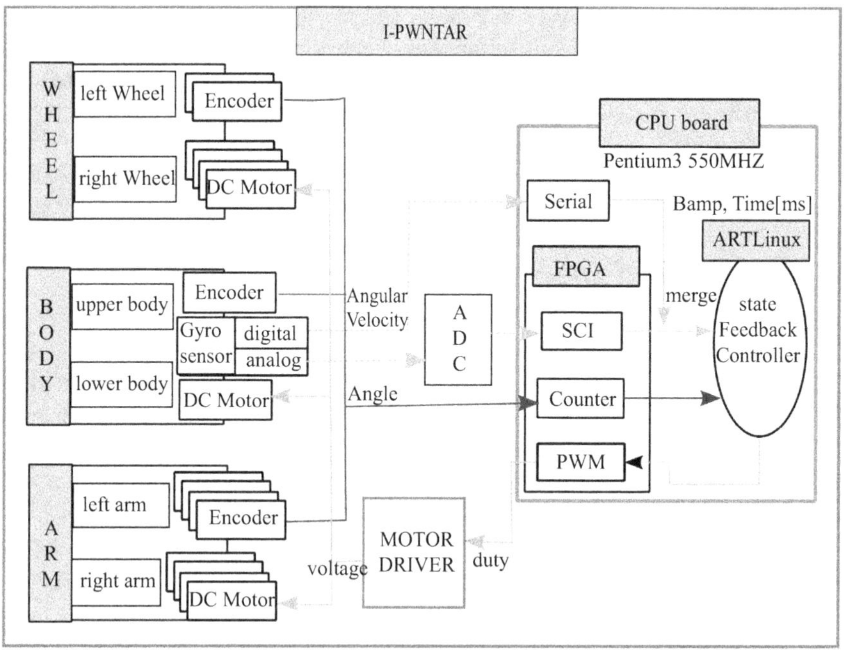

Figure 14.18 Experimental setup of I-PENTAR

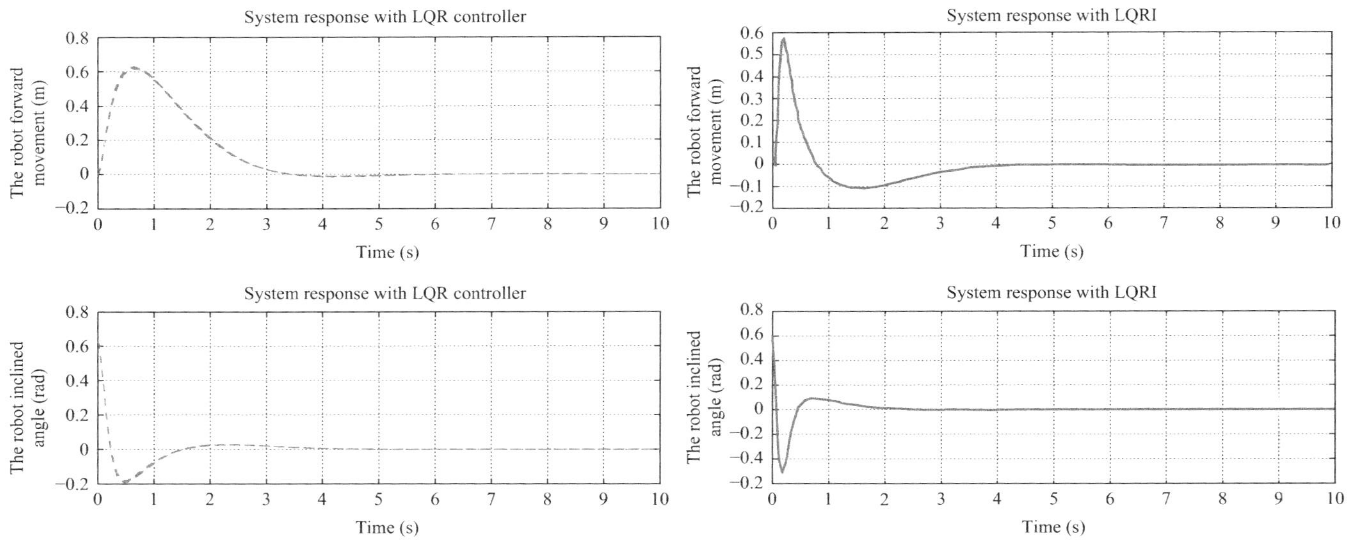

Figure 14.19 I-PENTAR closed-loop simulation: (left) LQR response and (right) LQRI response

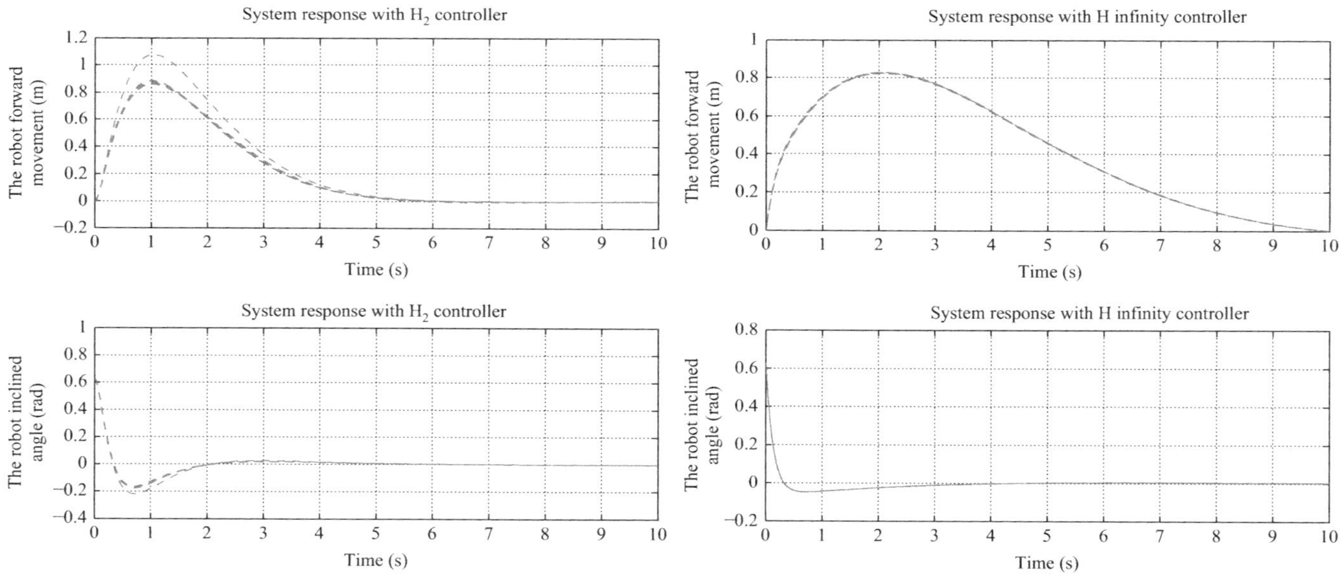

Figure 14.20 I-PENTAR closed-loop simulation: (left) H_2 response and (right) H_∞ response

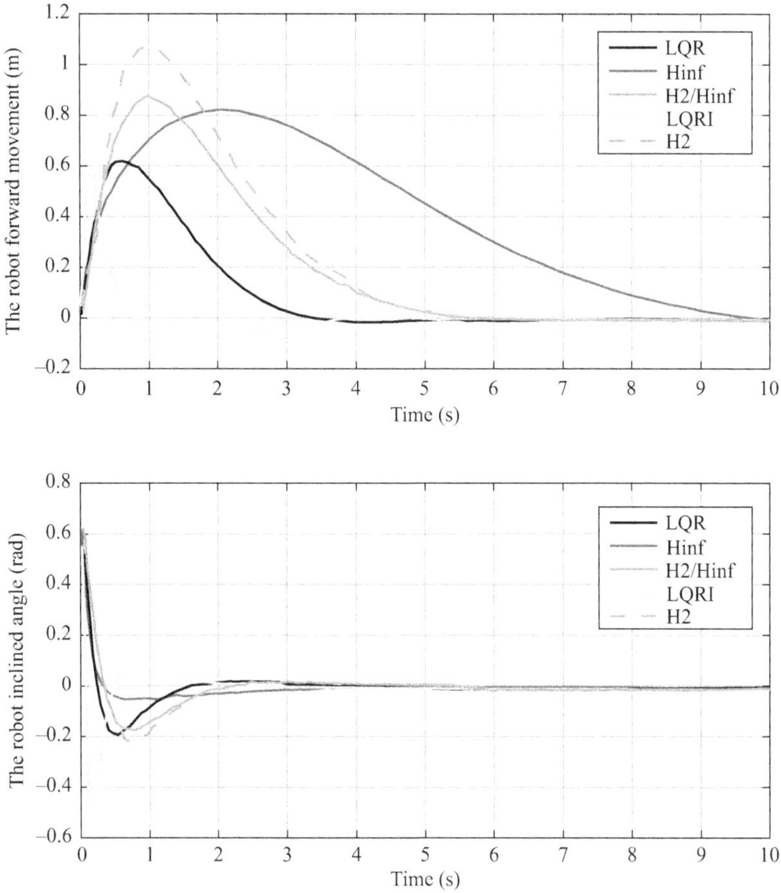

Figure 14.21 Comparison of output trajectories

defining the robot's inclination angle, is calculated by a geometrical method based on the parameter values of each part.

The ensuing results are depicted in Figures 14.19 and 14.20 for the two output variables: the robot forward movement and the robot inclined angle. In Figure 14.21, a comparison among the design methods is presented.

From the simulation results, we can arrive at the following points:

- Smooth behavior are clearly demonstrated for the patterns of the output variables,
- The LQR design determines one gain matrix for the feedback control, while the LQRI design determines two gain matrices: one for the control and the other for the integral state.

14.10 Conclusions

In this chapter, we have provided complete analytical results pertaining to dynamic modeling, control design and computer simulation of IP systems. The dynamic modeling covers different cases span from SIP to rotational and WIPs. The control design focused on the development of convex optimization over linear-LMI to yield linear controllers based on various criteria. Finally, MATLAB simulation experiments are performed to demonstrate the effectiveness of the developed methods.

Acknowledgment

This research work is supported by the deanship of scientific research (DSR) at KFUPM through research project no. IN 141048.

References

[1] V. Sukonatanakarn, M. Parnichkun, "Real-time optimal control for rotary inverted pendulum," *Am. J. Appl. Sci.*, vol. 6, no. 6, pp. 1106–1115, 2009.
[2] J. J. Craig, *Introduction to robotics: Mechanics & control*, Prentice-Hall, Upper Saddle River, NJ, 2005.
[3] J. Akesson, A. Blomdell, R. Braun, "Design and control of YAIP-an inverted pendulum on two wheels robot," *IEEE Int. Conf. Control Appl.*, vol. 2, pp. 2178–2183, 2006.
[4] M. Askari, H. A. F. Mohamed, M. Moghavvemi, S. S. Yang, "Model predictive control of an inverted pendulum," *Int. Conf. Tech. Postgraduates, TECHPOS*, pp. 1–4, 2009.
[5] R. Balan, V. Maties, O. Hancu, S. Stan, "A predictive control approach for the inverse pendulum on a cart problem," *IEEE Int. Conf. Mechatron. Autom.*, vol. 20, pp. 2026–2031, 2005.
[6] A. M. Almeshal, K. M. Goherb, M. O. Tokhi, "Dynamic modelling and stabilization of a new configuration of two-wheeled machines," *Rob. Autom. Syst.*, vol. 61, pp. 443–472, 2013.
[7] R. P. M. Chan, K. A. Stol, C. R. Halkyard, "Review of modelling and control of two-wheeled robots," *Annu. Rev. Control*, vol. 37, pp. 89–103, 2013.
[8] K. J. Astrom, K. Furuta, "Swinging up a pendulum by energy control," *Automatica*, vol. 36, no. 2, pp. 287–295, 2000.
[9] K. Yoshida, "Swing-up control of an inverted pendulum by energy-based methods," *Proc. the American Control Conf.*, San Diego, pp. 4045–4047, 1999.
[10] K. Furuta, T. Okutani, H. Sone, "Computer control of a double inverted pendulum," *Comput. Electr. Eng.*, vol. 5, pp. 67–84, 1978.
[11] S. A. Bortoff, "Robust swing control for a rotational double pendulum," *Proc. the 13th IFAC World Conf.*, San Francisco, pp. 413–419, 1996.

[12] M. Tsai, B. H. Shen, "Synchronization control of a parallel dual inverted pendulum driven by linear servomotors," *IET Control Theory Appl.*, vol. 1, no. 1, pp. 320–327, 2007.

[13] G. A. Medrano-Cedra, E. E. Eldukhri, M. Cetin, "Balancing and attitude control of double and triple inverted pendulums," *Trans. Inst. Meas. Control*, vol. 17, no. 3, pp. 143–154, 1995.

[14] H. Li, Z. Miao, J. Wang, "Variable universe adaptive fuzzy control on the quadruple inverted pendulum," *Sci. Chin., Ser. E: Technol. Sci.*, vol. 45, no. 2, pp. 213–224, 2002.

[15] J. Shen, A. K. Sanyal, N. A. Chaturvedi, D. Bernstein, H. McClamroch, "Dynamics and control of a 3D pendulum," *Proc. the 43rd IEEE Conf. Decision and Control*, Nassau, vol. 1, pp. 323–328, 2004.

[16] O. Boubaker, "The inverted pendulum benchmark in nonlinear control theory: A survey," *Int. J. Adv. Rob. Syst.*, vol. 10, pp. 233–241, 2013.

[17] S. Jeong, T. Takahashi, "Wheeled inverted pendulum type assistant robot: design concept and mobile control," *Intell. Serv. Rob.*, vol. 1, no. 4, pp. 313–320, 2008.

[18] A. Sinha, P. Prasoon, P. K. Bharadwaj, A. C. Ranasinghe, "Nonlinear autonomous control of a two-wheeled inverted pendulum mobile robot based on sliding mode," *Int. Conf. Comput. Intell. Networks (CINE)*, vol. 2015, pp. 52–57, 2015.

[19] J. Huang, S. Ri, L. Liu, Y. Wang, J. Kim, G. Pak, "Nonlinear disturbance observer-based dynamic surface control of mobile wheeled inverted pendulum," *IEEE Trans. Control Syst. Technol.*, vol. 23, no. 6, pp. 2400–2407, 2015.

[20] M. A. H. Garzn, W. O. C. Hernndez, A. P. G. Reyes, O. E. C. Quintero, "Two-wheeled inverted pendulum robot NXT Lego Mindstorms: mathematical modeling and real robot comparisons," *Rev. Politcnica*, vol. 36, no. 1, pp. 1–7, 2015.

[21] M. M. Azimi, H. R. Koofigar, "Model predictive control for a two wheeled self balancing robot," *First RSI/ISM Int. Conf. Robotics and Mechatronics (ICRoM)*, pp. 152–157, 13–15 Feb. 2013.

[22] Z. Li, C. G. Yang, L. Fan (Editors), *Advanced control of wheeled inverted pendulum systems*, Springer, London, 2013, ISBN: 978-1-4471-2962-2.

[23] Z. Li, C. G. Yang, L. Fan (Editors), *Advanced control of wheeled inverted pendulum systems*, Springer, London, 2013, ISBN: 978-1-4471-2962-2.

[24] M. G. Calkin, *Lagrangian and Hamiltonian mechanics*, World Scientific, Singapore, 1996.

[25] P. J. Larcombe, "On the control of a two-dimensional multi-link inverted pendulum: The form of the dynamic equations from choice of co-ordinate system," *Int. J. Syst. Sci.*, vol. 23, pp. 2265–2289, 1992.

[26] J. Driver, D. Thorpe, "Design, build and control of a single/ double rotational inverted pendulum," *Technical Report, School of Mechanical Engineering*, The University of Adelaide, 2004.

[27] V. Casanova, J. Salt, R. Piza, A. Cuenca, "Controlling the double rotary inverted pendulum with multiple feedback delays," *Int. J. Comput. Commun. Control*, vol. 7, pp. 20–38, 2012.

[28] C. Fox, *An introduction to the calculus of variations*, Dover Publications, New York, NY, 2010.

[29] J. R. Movellan, "DC motors," http://mplab.ucsd.edu/tutorials/DC.pdf, March 2010.

[30] Y. Wang, "Rotation double inverted pendulum," *Technical Report, School of Electrical and Computer Engineering*, The University of Dayton, April 2012.

[31] M. Demirci, "Design of feedback controllers for a linear system with applications to control of a double-inverted pendulum," *Int. J. Comput. Cogn.*, vol. 2, no. 1, pp. 65–84, 2004.

[32] K. Furuta, T. Ochiai, N. Ono, "Attitude control of a triple inverted pendulum," *Int. J. Control*, vol. 39, pp. 1351–1365, 1984.

[33] S. Jeong, T. Takahashi, "Wheeled inverted pendulum type assistant robot: design concept and mobile control," *Intell. Serv. Rob.*, vol. 1, no. 4, pp. 313–320, May 2008.

[34] A. Salerno, J. Angeles, "Nonlinear controllability of quasi-holonomic mobile robot," *Proc. IEEE ICRA*, Taiwan, 2003.

[35] F. Grasser, A. D'Arrigo, S. Comlombi, A. Rufer, "Joe: A mobile inverted pendulum," *IEEE Trans. Electron.*, vol. 49, no. 1, pp. 107–114, 2002.

[36] A. Salerno, J. Angeles, "A new family of two wheeled mobile robot: modeling and controllability," *IEEE Trans. Rob.*, vol. 23. no. 1, pp. 169–173, 2007.

[37] E. Koyanagi, S. Lida, S. Yuta, "A wheeled inverse pendulum type self-contained mobile robot and its two-dimensional trajectory control," *Proc. ISMCR92*, pp. 891–898, 1992.

[38] A. Sinha, P. Prasoon, P. K. Bharadwaj, A. C. Ranasinghe, "Nonlinear autonomous control of a two-wheeled inverted pendulum mobile robot based on sliding mode," *Int. Conf. Comput. Intell. Networks (CINE)*, vol. 1, pp. 52–57, 2015.

[39] J. Huang, S. Ri, L. Liu, Y. Wang, J. Kim, G. Pak, "Nonlinear disturbance observer-based dynamic surface control of mobile wheeled inverted pendulum," *IEEE Trans. Control Syst. Technol.*, vol. 23, no. 6, pp. 2400–2407, 2015.

[40] A. N. K. Nasir, M. A. Ahmad, R. M. T. Raja Ismail, "The control of a highly nonlinear two-wheel balancing robot: A comparative assessment between LQR and PID-PID control schemes", *World Academy of Science, Engineering and Technology*, vol. 4, pp. 176–181, 2010.

[41] Z. Li, C. Xu, "Adaptive fuzzy logic control of dynamic balance and motion for wheeled inverted pendulums," *Fuzzy Sets Syst.*, vol. 160, no. 12, pp. 1787–1803, Jun. 2009.

[42] S. Jung, S. S. Kim, "Control experiment of a wheel-driven mobile inverted pendulum using neural network," *IEEE Trans. Control Syst. Technol.*, vol. 16, no. 2, pp. 297–303, Mar. 2008.

[43] J. Huang, Z. H. Guan, T. Matsuno, T. Fukuda, K. Sekiyama, "Sliding-mode velocity control of mobile-wheeled inverted-pendulum systems," *IEEE Trans. Rob.*, vol. 26, no. 4, pp. 750–758, Aug. 2010.

[44] C. G. Yang, Z. Li, J. Li, "Trajectory planning and optimized adaptive control for a class of wheeled inverted pendulum vehicle models," *IEEE Trans. Cybern.*, vol. 43, no. 1, pp. 24–36, 2013.

[45] R. W. Brockett, H. Li, "A light weight rotary double pendulum: Maximizing the domain of attraction," *Proc. IEEE Decision and Control Conf.*, (Maui, Hawaii), pp. 3299–3304, Dec. 2003.

[46] K. Pathak, J. Franch, S. K. Agrawal, "Velocity and position control of a wheeled inverted pendulum by partial feedback linearization," *IEEE Trans. Rob.*, vol. 21, no. 3, pp. 505–512, 2005.

[47] J. M. Valenzuelaa, C. A. Avelara, S. P. Guzmnb, V. Santibezc, "Two adaptive control strategies for trajectory tracking of the inertia wheel pendulum: neural networks vis à vis model regressor," *Intell. Autom. Soft Comput.*, 2016, http://dx.doi.org/10.1080/10798587.2015.1121618.

[48] R. Cui, J. Guo, Z. Mao, "Adaptive backstepping control of wheeled inverted pendulums models," *Nonlinear Dyn.*, vol. 79, no. 1, pp. 501–511, 2015.

[49] Y. Maruki, K. Kawano, H. Suemitsu, T. Matsuo, "Adaptive backstepping control of wheeled inverted pendulum with velocity estimator," *Int. J. Control Autom. Syst.*, vol. 12, no. 5, pp 1040–1048, Oct. 2014.

[50] H. Vasudevan, A. M. Dollar, J. B. Morrell, Y. Xia, "Design for control of wheeled inverted pendulum platforms," *Trans. ASME, J. Mech. Rob.*, vol. 7, Nov. 2015, paper 041005.

[51] S. Jeong, T. Takahashi, "Wheeled inverted pendulum type assistant robot: inverted mobile, standing, and sitting motions," *IEEE/RSJ Int. Conf. Intell. Rob. Syst.*, vol. 2007, pp. 1932–1937, 2007.

[52] "Matlab Documents," MPC guide, *Mathworks*, MPC-HC vol. 1.0.8.0, 2014, USA.

[53] B. D. O. Anderson, J. B. Moore, *linear optimal control*, Prentice-Hall, Inc., Englewood Cliffs, NJ, 1971.

[54] M. S. Mahmoud, Y. Xia, *Applied control systems design: State-space methods*, Springer-Verlag, UK, 2012.

[55] K. Zhou, J. C. Doyle, *Essentials of robust control*, Prentice-Hall, New Jersey, 1998.

[56] M. S. Mahmoud, *Resilient control of uncertain dynamical systems*, Springer, Heidelberg, 2004.

Index

www.ingramcontent.com/pod-product-compliance
Lightning Source LLC
LaVergne TN
LVHW011905200926
848484LV00002B/533